T0189657

Communications in Computer and Information Science 1468

More information about this series at http://www.springer.com/series/7899

Kang Li · Tim Coombs ·
Jinghan He · Yuchu Tian ·
Qun Niu · Zhile Yang (Eds.)

Recent Advances in Sustainable Energy and Intelligent Systems

7th International Conference on Life System Modeling
and Simulation, LSMS 2021
and 7th International Conference on Intelligent Computing
for Sustainable Energy and Environment, ICSEE 2021
Hangzhou, China, October 30 – November 1, 2021
Proceedings, Part II

Editors
Kang Li
University of Leeds
Leeds, UK

Tim Coombs
University of Cambridge
Cambridge, UK

Jinghan He
Beijing Jiaotong University
Beijing, China

Yuchu Tian
Queensland University of Technology
Brisbane, QLD, Australia

Qun Niu
Shanghai University
Shanghai, China

Zhile Yang
Shenzhen Institute of Advanced Technology
Shenzhen, China

ISSN 1865-0929 ISSN 1865-0937 (electronic)
Communications in Computer and Information Science
ISBN 978-981-16-7209-5 ISBN 978-981-16-7210-1 (eBook)
https://doi.org/10.1007/978-981-16-7210-1

This Springer imprint is published by the registered company Springer Nature Singapore Pte Ltd.
The registered company address is: 152 Beach Road, #21-01/04 Gateway East, Singapore 189721, Singapore

Preface

This book series constitutes the proceedings of the 2021 International Conference on Life System Modeling and Simulation (LSMS 2021) and the 2021 International Conference on Intelligent Computing for Sustainable Energy and Environment (ICSEE 2021), which were held during October 30 – November 1, 2021, in Hangzhou, China. The LSMS and ICSEE international conference series aim to bring together international researchers and practitioners in the fields of advanced methods for life system modeling and simulation, advanced intelligent computing theory and methodologies, and engineering applications for achieving net zero across all sectors to meet the global climate change challenge. These events are built upon the success of previous LSMS conferences held in Shanghai, Wuxi, and Nanjing in 2004, 2007, 2010, 2014, and 2017, and ICSEE conferences held in Wuxi, Shanghai, Nanjing, and Chongqing in 2010, 2014, 2017, and 2018, respectively, and are based on large-scale UK-China collaboration projects on sustainable energy. Due to the COVID-19 pandemic situation, the themed workshops as part of these two conferences were organized online in 2020.

At LSMS 2021 and ICSEE 2021, technical exchanges within the research community took the form of keynote speeches and panel discussions, as well as oral and poster presentations. The LSMS 2021 and ICSEE 2021 conferences received over 430 submissions from authors in 11 countries and regions. All papers went through a rigorous peer review procedure and each paper received at least three review reports. Based on the review reports, the Program Committee finally selected 159 high-quality papers for presentation at LSMS 2021 and ICSEE 2021. These papers cover 18 topics and are included in three volumes of CCIS proceedings published by Springer. This volume of CCIS includes 51 papers covering six relevant topics.

The organizers of LSMS 2021 and ICSEE 2021 would like to acknowledge the enormous contribution of the Program Committee and the referees for their efforts in reviewing and soliciting the papers, and the Publication Committee for their editorial work. We would also like to thank the editorial team from Springer for their support and guidance. Particular thanks go to all the authors, without their high-quality submissions and presentations the conferences would not have been successful.

Finally, we would like to express our gratitude to our sponsors and organizers, listed on the following pages.

October 2021

Minrui Fei
Kang Li
Qinglong Han

Organization

Honorary Chairs

Wang, XiaoFan	Shanghai University, China
Umezu, Mitsuo	Waseda University, Japan

General Chairs

Fei, Minrui	Shanghai University, China
Li, Kang	University of Leeds, UK
Han, Qing-Long	Swinburne University of Technology, Australia

International Program Committee

Chairs

Ma, Shiwei	China Simulation Federation, China
Coombs, Tim	University of Cambridge, UK
Peng, Chen	Shanghai University, China
Chen, Luonan	University of Tokyo, Japan
Sun, Jian	China Jiliang University, China
McLoone, Sean	Queen's University Belfast, UK
Tian, Yuchu	Queensland University of Technology, Australia
He, Jinghan	Beijing Jiaotong University
Zhang, Baolin	Qingdao University of Science and Technology, China

Local Chairs

Aleksandar Rakić	University of Belgrade, Serbia
Athanasopoulos, Nikolaos	Queen's University Belfast, UK
Cheng, Long	Institute of Automation, Chinese Academy of Sciences, China
Dehghan, Shahab	Imperial College London, UK
Ding, Jingliang	Northeastern University, China
Ding, Ke	Jiangxi University of Finance and Economics, China
Duan, Lunbo	Southeast University, China
Fang, Qing	Yamagata University, Japan
Feng, Wei	Shenzhen Institute of Advanced Technology, Chinese Academy of Sciences, China
Fridman, Emilia	Tel Aviv University, Israel
Gao, Shangce	University of Toyama, Japan
Ge, Xiao-Hua	Swinburne University of Technology, Australia
Gupta M. M.	University of Saskatchewan, Canada

Gu, Xingsheng	East China University of Science and Technology, China
Han, Daojun	Henan University, China
Han, Shiyuan	University of Jinan, China
Hunger, Axel	University of Duisburg-Essen, Germany
Hong, Xia	University of Reading, UK
Hou, Weiyan	Zhengzhou University, China
Jia, Xinchun	Shanxi University, China
Jiang, Zhouting	China Jiliang University, China
Jiang, Wei	Southeast University, China
Lam, Hak-Keung	King's College London, UK
Li, Juan	Qingdao Agricultural University, China
Li, Ning	Shanghai Jiao Tong University, China
Li, Wei	Central South University, China
Li, Yong	Hunan University, China
Liu, Wanquan	Curtin University, Australia
Liu, Yanli	Tianjin University, China
Ma, Fumin	Nanjing University of Finance & Economics, China
Ma, Lei	Southwest University, China
Maione, Guido	Technical University of Bari, Italy
Na, Jing	Kunming University of Science and Technology, China
Naeem, Wasif	Queen's University Belfast, UK
Park, Jessie	Yeungnam University, South Korea
Qin, Yong	Beijing Jiaotong University, China
Su, Zhou	Shanghai University, China
Tang, Xiaopeng	Hong Kong University of Science and Technology, Hong Kong, China
Tang, Wenhu	South China University of Technology, China
Wang, Shuangxing	Beijing Jiaotong University, China
Xu, Peter	University of Auckland, New Zealand
Yan, Tianhong	China Jiliang University, China
Yang, Dongsheng	Northeast University, China
Yang, Fuwen	Griffith University, Australia
Yang, Taicheng	University of Sussex, UK
Yu, Wen	National Polytechnic Institute, Mexico
Zeng, Xiaojun	University of Manchester, UK
Zhang, Wenjun	University of Saskatchewan, Canada
Zhang, Jianhua	North China Electric Power University, China
Zhang, Kun	Nantong University, China
Zhang, Tengfei	Nanjing University of Posts and Telecommunications, China
Zhao, Wenxiao	Chinese Academy of Science, China
Zhu, Shuqian	Shandong University, China

Members

Aristidou, Petros	Ktisis Cyprus University of Technology, Cyprus
Azizi, Sadegh	University of Leeds, UK
Bu, Xiongzhu	Nanjing University of Science and Technology, China
Cai, Hui	Jiangsu Electric Power Research Institute, China
Cai, Zhihui	China Jiliang University, China
Cao, Jun	Keele University, UK
Chang, Xiaoming	Taiyuan University of Technology, China
Chang, Ru	Shanxi University, China
Chen, Xiai	China Jiliang University, China
Chen, Qigong	Anhui Polytechnic University, China
Chen, Qiyu	China Electric Power Research Institute, China
Chen, Rongbao	Hefei University of Technology, China
Chen, Zhi	Shanghai University, China
Chi, Xiaobo	Shanxi University, China
Chong, Ben	University of Leeds, UK
Cui, Xiaohong	China Jiliang University, China
Dehghan, Shahab	University of Leeds, UK
Deng, Li	Shanghai University, China
Deng, Song	Nanjing University of Posts and Telecommunications, China
Deng, Weihua	Shanghai University of Electric Power, China
Du, Dajun	Shanghai University, China
Du, Xiangyang	Shanghai University of Engineering Science, China
Du, Xin	Shanghai University, China
Fang, Dongfeng	California Polytechnic State University, USA
Feng, Dongqing	Zhengzhou University, China
Fu, Jingqi	Shanghai University, China
Gan, Shaojun	Beijing University of Technology, China
Gao, Shouwei	Shanghai University, China
Gu, Juping	Nantong University, China
Gu, Yunjie	Imperial College London, UK
Gu, Zhou	Nanjing Forestry University, China
Guan, Yanpeng	Shanxi University, China
Guo, Kai	Southwest Jiaotong University, China
Guo, Shifeng	Shenzhen Institute of Advanced Technology, Chinese Academy of Science, China
Guo, Yuanjun	Shenzhen Institute of Advanced Technology, Chinese Academy of Science, China
Han, Xuezheng	Zaozhuang University, China
Hong, Yuxiang	China Jiliang University, China
Hou, Guolian	North China Electric Power University, China
Hu, Qingxi	Shanghai University, China
Hu, Yukun	University College London, UK
Huang, Congzhi	North China Electric Power University, China

Huang, Deqing	Southwest Jiaotong University, China
Jahromi, Amir Abiri	University of Leeds, UK
Jiang, Lin	University of Liverpool, UK
Jiang, Ming	Anhui Polytechnic University, China
Kong, Jiangxu	China Jiliang University, China
Li, MingLi	China Jiliang University, China
Li, Chuanfeng	Luoyang Institute of Science and Technology, China
Li, Chuanjiang	Harbin Institute of Technology, China
Li, Donghai	Tsinghua University, China
Li, Tongtao	Henan University of Technology, China
Li, Xiang	University of Leeds, UK
Li, Xiaoou	CINVESTAV-IPN, Mexico
Li, Xin	Shanghai University, China
Li, Zukui	University of Alberta, Canada
Liu Jinfeng	University of Alberta, Canada
Liu, Kailong	University of Warwick, UK
Liu, Mandan	East China University of Science and Technology, China
Liu, Tingzhang	Shanghai University, China
Liu, Xueyi	China Jiliang University, China
Liu, Yang	Harbin Institute of Technology, China
Long, Teng	University of Cambridge, UK
Luo, Minxia	China Jiliang University, China
Ma, Hongjun	Northeastern University, China
Ma, Yue	Beijing Institute of Technology, China
Menhas, Muhammad Ilyas	Mirpur University of Science and Technology, Pakistan
Naeem, Wasif	Queen's University Belfast, UK
Nie, Shengdong	University of Shanghai for Science and Technology, China
Niu, Qun	Shanghai University, China
Pan, Hui	Shanghai University of Electric Power, China
Qian, Hong	Shanghai University of Electric Power, China
Ren, Xiaoqiang	Shanghai University, China
Rong, Qiguo	Peking University, China
Song, Shiji	Tsinghua University, China
Song, Yang	Shanghai University, China
Sun, Qin	Shanghai University, China
Sun, Xin	Shanghai University, China
Sun, Zhiqiang	East China University of Science and Technology, China
Teng, Fei	Imperial College London, UK
Teng, Huaqiang	Shanghai Instrument Research Institute, China
Tian, Zhongbei	University of Birmingham, UK
Tu, Xiaowei	Shanghai University, China
Wang, Binrui	China Jiliang University, China
Wang, Qin	China Jiliang University, China

Wang, Liangyong	Northeast University, China
Wang, Ling	Shanghai University, China
Wang, Yan	Jiangnan University, China
Wang, Yanxia	Beijing University of Technology, China
Wang, Yikang	China Jiliang University, China
Wang, Yulong	Shanghai University, China
Wei, Dong	China Jiliang University, China
Wei, Li	China Jiliang University, China
Wei, Lisheng	Anhui Polytechnic University, China
Wu, Fei	Nanjing University of Posts and Telecommunications, China
Wu, Jianguo	Nantong University, China
Wu, Jiao	China Jiliang University, China
Wu, Peng	University of Jinan, China
Xu, Peng	China Jiliang University, China
Xu Suan	China Jiliang University, China
Xu, Tao	University of Jinan, China
Xu, Xiandong	Cardiff University, UK
Yan, Huaicheng	East China University of Science and Technology, China
Yang, Aolei	Shanghai University, China
Yang, Banghua	Shanghai University, China
Yang, Wenqiang	Henan Normal University, China
Yang, Zhile	Shenzhen Institute of Advanced Technology, Chinese Academy of Sciences, China
Ye, Dan	Northeastern University, China
You, Keyou	Tsinghua University, China
Yu, Ansheng	Shanghai Shuguang Hospital, China
Zan, Peng	Shanghai University, China
Zeng, Xiaojun	University of Manchester, UK
Zhang, Chen	Coventry University, UK
Zhang, Dawei	Shandong University, China
Zhang Xiao-Yu	Beijing Forestry University
Zhang, Huifeng	Nanjing Post and Communication University, China
Zhang, Kun	Nantong University, China
Zhang, Li	University of Leeds, UK
Zhang, Lidong	Northeast Electric Power University, China
Zhang, Long	University of Manchester, UK
Zhang, Yanhui	Shenzhen Institute of Advanced Technology, Chinese Academy of Sciences, China
Zhao, Chengye	China Jiliang University, China
Zhao, Jianwei	China Jiliang University, China
Zhao, Wanqing	Manchester Metropolitan University, UK
Zhao, Xingang	Shenyang Institute of Automation Chinese Academy of Sciences, China
Zheng, Min	Shanghai University, China

Zhou, Bowen	Northeast University, China
Zhou, Huiyu	University of Leicester, UK
Zhou, Peng	Shanghai University, China
Zhou, Wenju	Ludong University, China
Zhou, Zhenghua	China Jiliang University, China
Zhu, Jianhong	Nantong University, China

Organization Committee

Chairs

Qian, Lijuan	China Jiliang University, China
Li, Ni	Beihang University, China
Li, Xin	Shanghai University, China
Sadegh, Azizi	University of Leeds, UK
Zhang, Xian-Ming	Swinburne University of Technology, Australia
Trautmann, Toralf	Centre for Applied Research and Technology, Germany

Members

Chen, Zhi	China Jiliang University, China
Du, Dajun	Shanghai University, China
Song, Yang	Shanghai University, China
Sun, Xin	Shanghai University, China
Sun, Qing	Shanghai University, China
Wang, Yulong	Shanghai University, China
Zheng, Min	Shanghai University, China
Zhou, Peng	Shanghai University, China
Zhang, Kun	Shanghai University, China

Special Session Chairs

Wang, Ling	Shanghai University, China
Meng, Fanlin	University of Essex, UK
Chen, Wanmi	Shanghai University, China
Li, Ruijiao	Fudan University, China
Yang, Zhile	SIAT, Chinese Academy of Sciences, China

Publication Chairs

Niu, Qun	Shanghai University, China
Zhou, Huiyu	University of Leicester, UK

Publicity Chair

Yang, Erfu	University of Strathclyde, UK

Registration Chairs

Song, Yang	Shanghai University, China
Liu, Kailong	University of Warwick, UK

Secretary-generals

Sun, Xin	Shanghai University, China
Gan, Shaojun	Beijing University of Technology, China

Sponsors

China Simulation Federation (CSF)
China Instrument and Control Society (CIS)
IEEE Systems, Man and Cybernetics Society Technical Committee on Systems Biology
IEEE CC Ireland Chapter

Cooperating Organizations

Shanghai University, China
University of Leeds, UK
China Jiliang University, China
Swinburne University of Technology, Australia
Life System Modeling and Simulation Technical Committee of the CSF, China
Embedded Instrument and System Technical Committee of the China Instrument and Control Society, China

Co-sponsors

Shanghai Association for System Simulation, China
Shanghai Instrument and Control Society, China
Zhejiang Association of Automation (ZJAA), China
Shanghai Association of Automation, China

Supporting Organizations

Queen's University Belfast, UK
Nanjing University of Posts and Telecommunications, China
University of Jinan, China
University of Essex, UK
Queensland University of Technology, Australia
Central South University, China
Tsinghua University, China
Peking University, China
University of Hull, UK

Beijing Jiaotong University, China
Nantong University, China
Shenzhen Institute of Advanced Technology, Chinese Academy of Sciences, China
Shanghai Key Laboratory of Power Station Automation Technology, China
Complex Networked System Intelligent Measurement and Control Base, Ministry of Education, China
UK China University Consortium on Engineering Education and Research
Anhui Key Laboratory of Electric Drive and Control, China

Contents – Part II

Power and Energy Systems

Computational Intelligence in Utilization of Clean and Renewable Energy Resources, and Intelligent Modelling, Control and Supervision for Energy Saving and Pollution Reduction

Intelligent Methods in Developing Electric Vehicles, Energy Conversion Devices and Equipment

Intelligent Control Methods in Energy Infrastructure Development and Distributed Power Generation Systems

Power and Energy Systems

Coordinative Voltage Control of Building in Supporting Electric Vehicle Integration in Distribution Networks

Yu Shen[1], Dongsheng Li[2], Wei Jiang[1], Yushu Zhu[3], Wenliang Shan[3], Xiandong Xu[3](✉), and Xiaohong Dong[4]

[1] State Grid Hubei Electric Power Research Institute, Wuhan 430074, China
{sheny20,jiangw48}@hb.sgcc.com.cn

[2] State Grid Hubei Electric Power CO., LTD., Wuhan 430074, China
lids@hb.sgcc.com.cn

[3] Key Laboratory of Smart Energy and Information Technology of Tianjin Municipality, Tianjin University, Tianjin 300072, China
{zhuyushu,wlshan,xux27}@tju.edu.cn

[4] State Key Laboratory of Reliability and Intelligence of Electrical Equipment, Hebei University of Technology, Tianjin 300041, China
dxh@hebut.edu.cn

Abstract. This paper investigates the feasibility of using buildings with heat pumps to support the integration of electric vehicles in electricity distribution networks. Behaviors of buildings with electric vehicles and heat pumps are modelled at a community scale. Based on the model, the on-off state of heat pumps along with buildings are characterized and coordinated with the charging behavior of electric vehicles from two aspects, namely coordination in a single house and coordination of multiple houses at a community scale. A community with 96 houses is used to test the performance of different coordination schemes. The results demonstrate that the new loads from electric vehicle charging would affect the electricity distribution network significantly, while coordination between houses could help mitigate this adverse impact Although the coordination requires new investment in updating IT infrastructure to support the coordination, it is worth to compare it with the cost for reinforcing distribution networks for electric vehicle integration.

Keywords: Buildings · Coordinative control · Distribution networks · Electric vehicle integration

1 Introduction

Electricity distribution network operators have been carrying out research into the impact of electric vehicle (EV) charging on the network, and it was found that the greatest impact was on their low voltage networks 1. As the uptake of EV increases, there was seen to be a significant increase in voltage violations 2. It was also noted that EV charging

K. Li et al. (Eds.): LSMS 2021/ICSEE 2021, CCIS 1468, pp. 3–12, 2021.
https://doi.org/10.1007/978-981-16-7210-1_1

can cause high harmonic current flow. This is something that will need to be taken into consideration as uptake rises, where multiple EVs can be charged on one feeder 'being manifest as distortions in the local supply network'. It was suggested that every EV charging point should be now reported to its respective distribution network, so that the penetration can be monitored and impacts can be accounted for.

Research has been carried out on the use of EVs for supporting the electricity network 3. This was carried out to help combat the increase in renewable energy. This work investigated two main control methods, the first of which was disconnection of the charging load 4 and the second was the discharging of the energy stored in the vehicle batteries 5. This work also considered three charging strategies for EV's: Dumb, off-peak and smart 6. In this case the EVs impact was seen to be positive, as it is possible to use the vehicles to decrease the network frequency variation. This research did not investigate the benefits to distribution network voltage control.

Most current EV charging in practice can be classed as uncontrolled charging. A large disadvantage to this is the lack of incentive for consumers to consider when they are charging their vehicle or at what rate 7. Time of use charging schemes could be introduced to discourage charging at peak times 8, which if implemented could have a positive impact on the EV integration of the distribution network.

Managed EV charging has been widely researched, due to its considerable potential for reducing problems that will be experience by distribution networks in the future. Managed EV charging systems generally consist of three key components; a smart charger, a control system and the service user. Work carried out by electric nation, can be used to demonstrate the acceptance by consumers of the use of domestic smart chargers 9. This is largely due to the minimal impact that smart charging can have. Customer impact assessment was carried out by electric nation to assess the impact that smart charging has on EV drivers. This demonstrated that each charge management situation would usual last for under 30 min and occur only around 4 times per year 10.

Managed charging usually aims just to move the charging away from peak times. These peaks are often at times when the vehicle is likely to return to the home when the driver has finished work, when the vehicle will likely be plugged in for charging. The managed charging can allow for this charging to be moved to avoid it clashing with the demand peak and allow for the charging to take place at a time better suited to the distribution network. Please see Fig. 1, taken from electric nation's report smart charging: a brief guide to managed EV home charging which demonstrates the movement that may be required to take charging away from peak times 10.

This paper investigates the potentials of regulating heat supply to buildings in achieving this peak shaving target caused by EV charging. Specifically, the voltage control of residential distribution networks is studied. On-off period of heat pumps (HPs) in buildings are coordinated with EV charging at a community scale in order to smooth the total power consumption. Different scenarios are constructed for EV charging in a residential distribution network. The results demonstrated that the proposed strategy could significantly reduce the voltage fluctuations of distribution networks.

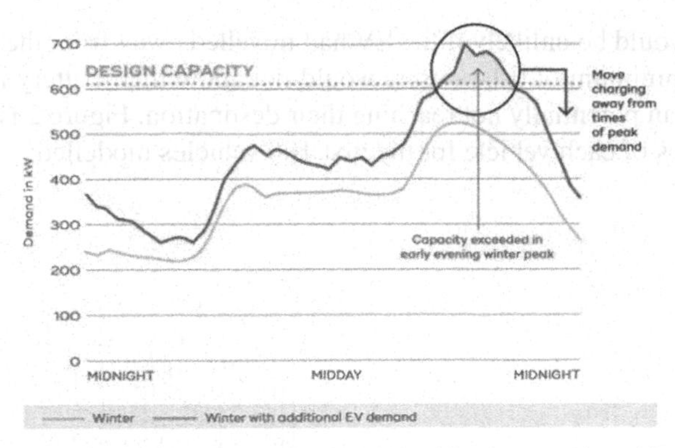

Fig. 1. Electricity demand from homes and EVs on a local feeder reproduced from smart charging: a brief guide to managed EV home charging 10.

2 Potential of HPs in Supporting EV Integration

HPs play an important role in buildings. With the increasing uptake of HPs, there will be a significant change in the domestic daily load profile 11. It is expected that by 2030, the number of HPs will have reached around 3.5 million. Within the HP, heat gain from the environment takes place in the evaporator (on the left of the figure). Even at low outside temperatures, the refrigerant within the evaporator can boil and therefore store the energy. The temperature and pressure within the HP are then increased by the gas from the boiling refrigerant passing through the compressor. The refrigerant then goes through the condenser which releases heat into the building in which the HP is installed. The gas then cools to a hot liquid which naturally has high pressure. The expansion valve then causes a drop-in temperature and pressure. Once this has taken place the HP can begin the cycle again and absorb heat from the atmosphere 12.

2.1 Load Profile

In order to show the effect of EVs on the network, an EV charger has been modelled. Research showed that there is a large variety of EVs on the market, with various battery storage capacities, so it would be difficult to model a large enough variety of EVs. The charger is the direct link to the network, which can regulate the load. MATLAB was used to model an EV charger. This model required the user to input the initial and required state of charge (SoC) of the EV, the battery storage capacity, and the maximum charging power that the charger can output. It then calculates the load until the battery is charged to the required level. This model can be implemented within other code to then coordinate the EV charging load with the other loads within the home. A start time is also input by the user, but the random number feature could also be used to vary the start time to allow for the staggered arrival home of a population of EVs. While the SoC is less than the required SoC, the charging power of the EV is 7 kW. Once the SoC has reached its required level the charging power becomes 0 kW. Initial SoC was also varied for each EV modelled, using a distribution of SoC which represented a population as to be expected. These states ranged between 0.1 and 0.9. These limits were selected as an initial SoC higher than 0.9 would be unlikely if the EV had travelled away from the home and then returned, and a minimum of 0.1 as users would not allow their battery level to reach 0 as this would mean potentially not reaching their destination. Figure 2 is an example of the charging times of each vehicle for the first 100 vehicles modelled:

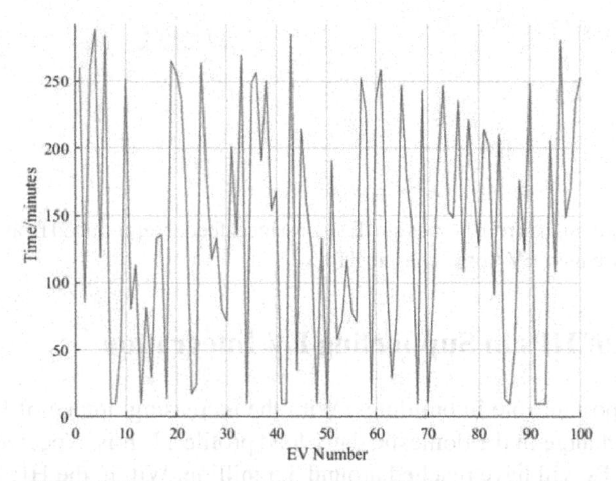

Fig. 2. Temperatures and heat pump loads over 1 week.

A simple model in 11 was used to describe load profiles for domestic HPs. This model takes into account the parameters of the home, the desired temperature, the outside temperature and the input power of the HP. For this model, 3.25 kW was selected as an appropriate power. Random number generation within MATLAB was again used to create load profiles for a number of homes. The parameters were within a range of ±5% of set values. This was then run 96 times to create profiles for all of the homes on the network to be used in later stages of the project. For this model, a value for outside temperature was required. This data was generated using the high-resolution stochastic integrated thermal-electrical domestic demand model 13. Values were input for each minute of each day of a week in January. This model and date selection was later used in the simulations for house load data (Fig. 3).

Due to the variation in temperature than can be allowed in a home, and the improvements being made in terms of heat retention using insulation, modelling of HPs has found that there are times throughout the day when the HP draws no load for space heating/cooling 14. These times will be used in coordination, as other loads could take the place of the HP load to smooth the overall loads on each feeder of the distribution network. 'The thermal storage properties of HPs allow them to be switch on and off to increase or decrease their power consumption while still maintaining the building temperature and comfortable levels'.

2.2 EV and HP Coordination Within One House

A model was created that looks at coordinated EV and HP loads within a single house. This makes use of the time when the HP is not operational. The model described above

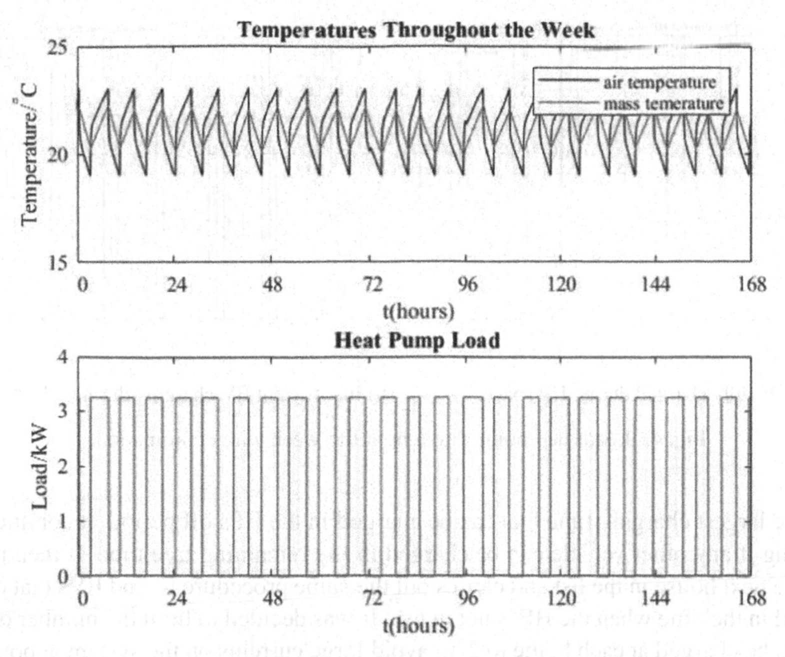

Fig. 3. Temperatures and heat pump loads over 1 week.

was first used to create a HP load profile, and this data was then imported into the model for coordination. It begins at the respective EVs start time, when it returns home, and allows for the EV to charge when the HP load is 0. It can allow for multiple vehicles to charge at the home, where the vehicles can charge simultaneously or one after the other. If the desired SoC cannot be reached, the model then allows for charging to occur when the HP is operating.

Figure 4(a) shows an example of the loads at one house. This demonstrates how the coordination works, where one load is operating at a time. In this case the vehicles could be fully charged without the need for both loads to operate simultaneously. Figure 4(b) shows an example at a house where it was not possible to charge the vehicles in just the HP space. This was due to the vehicles having a low SoC when returning home, or the parameters of the house leading to less HP space than in other simulations.

2.3 EV and HP Loads Within Multiple Houses

A model was then created for the coordination of more than one house. This model was to be used when coordination within one home did not bring the voltage levels across the modelled network within the tolerances allowed. It begins by calculating the time during a week that the HP at each home is not operational. It then orders these from the largest time to the smallest. It then calculates the time required for each of the EVs to be charged from its SoC when returning home to its required SoC. It then again ranks these into an order from largest time to smallest. The model then matches up EVs to be charged to homes. It begins by taking the house with the largest time free and matching the EV

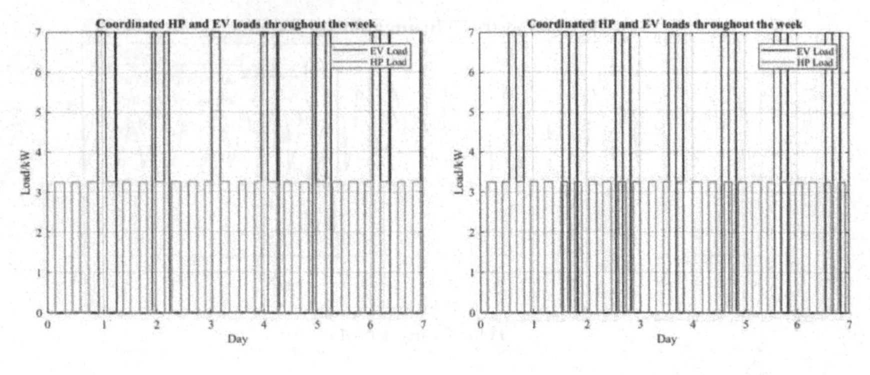

(a) EV fully charged during HP space (b) Insufficient EV charging during HP space

Fig. 4. Loads at 1 home throughout the week with coordination.

with the largest charging time that can be charged in the HPs off period. It continues by checking if any other vehicle can be charged in the remaining free time. It then moves onto the next house in the list and carries out the same procedure to find EVs that can be charged in the time when the HP is not in use. It was decided to limit the number of EVs that can be charged at each home to 2, to avoid large currents on the system at points on the network where 3 or more EVs are being charged. If future research was to be carried out on this project, the current levels on the network would also be analyzed. Please see the figures below which demonstrate a strategy created using this model. Figure 5(a) shows the time required to charge the EVs on the network.

Figure 5(b) shows the free time each HP has based on the HP model discussed above. This is the total time in one night that the HP is not operating. This figure shows an example for one night during the week. These values will change for each day of the week due to the fluctuating outside temperatures.

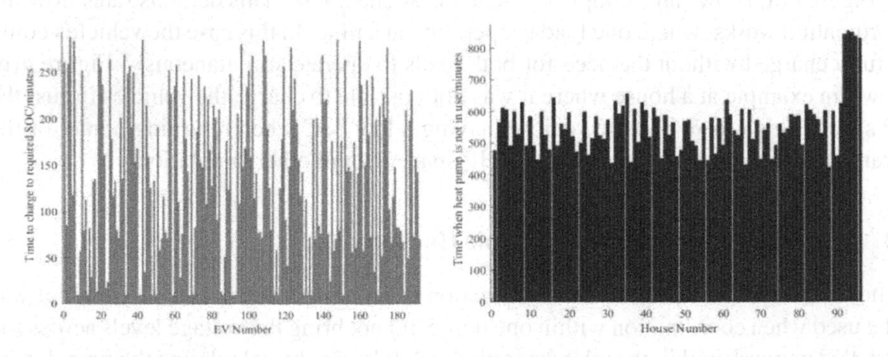

(a) Time required to charge each EV (b) HP off-state time used for EV charging

Fig. 5. Results of EV and HP coordination at one house.

Figure 6(a) shows the match between houses and EVs. This graph shows the matching for 1 day, but these matches can change on each day of the week modelled due to the

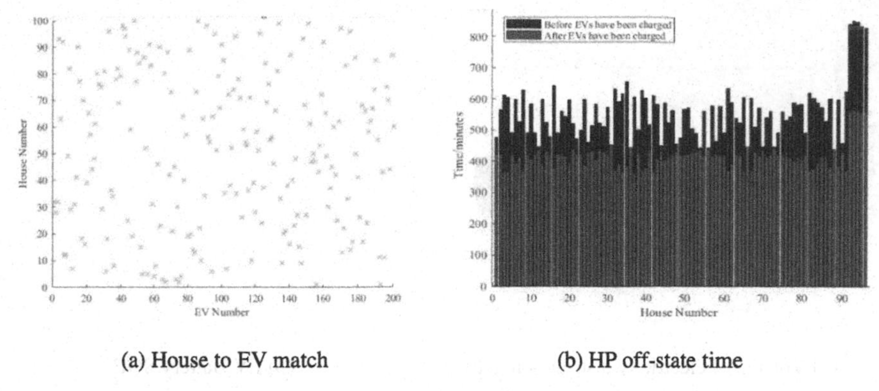

(a) House to EV match (b) HP off-state time

Fig. 6. Results of EV and HP coordination at multiple buildings.

variation in EV SoC and HP loads. In order to create this matching, the EV charging times and HP OFF state times were ordered from greatest to smallest. These were then matched from largest to smallest, with the EV with the greatest charging time being matched with the HP with the longest off-state time. A second EV was then matched to homes where enough HP off-state time remained once the first EV had been charged. Figure 6(b) shows the time in minutes when both the HP is not in use and the EVs are not charging in red. This demonstrates that not all HP free time is used up for EV charging. Shown in blue is the off-state time before any charging had taken place.

3 Impacts on the Electricity Distribution Network

The coordination methodologies described above were applied to a network model with 96 houses to analyze their effectiveness 15. By carrying out simulations of the base network, and then with both HP and EV loads within a single house and within a community scale with multiple buildings. It will be possible to directly compare the effect that the coordination of multiple buildings has.

3.1 Scenario 1 – EVs and HPs at Every House Without Coordination

The next simulation that took place included 1 eV charging at every home on the network. This charging had no form of smart or coordinated loadings, each vehicle began charging when it reached the home and charged at its full available charging power until the required SoC was reached. There were found to be voltage violations every day throughout the week. An example of an EV SoC throughout the week. It can be noted that the vehicle is charged before a user would likely need the vehicle during the next day (Fig. 7).

3.2 Scenario 2 – EVs and HPs with Coordination Within a Single House

Coordination between the HP and EV at every home was then implemented. It was noted that there was an overall rise in voltage level when compared to the previous scenario.

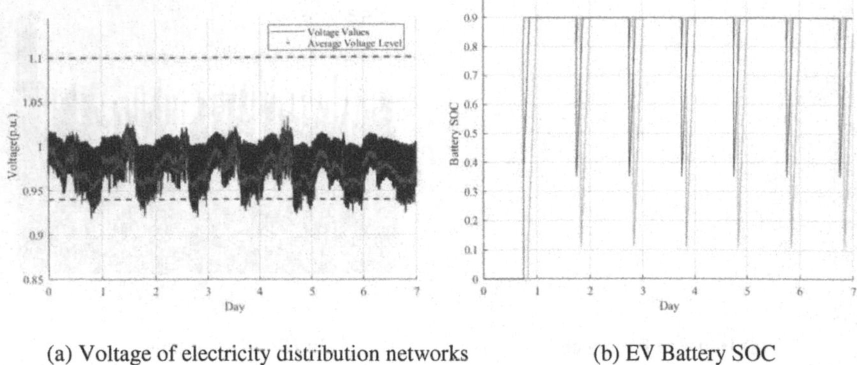

(a) Voltage of electricity distribution networks (b) EV Battery SOC

Fig. 7. Results of scenario 1.

This is due to the limit in total load which now exists as the HP and EV charger can no longer be operating simultaneously. There are still voltage violations present but less so than in the previous scenario. An example of an EV SoC throughout the week. It can be noted that the vehicle is charged before a user would likely need the vehicle during the next day (Fig. 8).

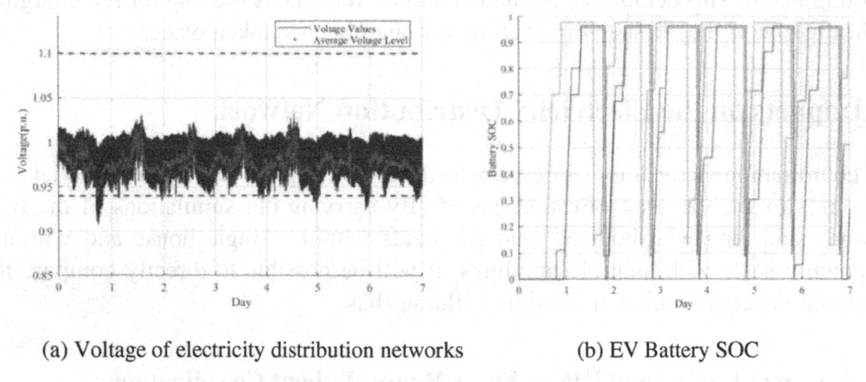

(a) Voltage of electricity distribution networks (b) EV Battery SOC

Fig. 8. Results of scenario 2.

3.3 Scenario 3 – EVs and HPs at Every House with Coordination of Multiple House on the Network

Further coordination was then carried out. This time between multiple homes on the network. This coordinated was seen to have the greatest impact on voltage levels. This demonstrates its capabilities for allowing for a larger number of EVs to be charged on the network. Please see the results for this simulation (Fig. 9).

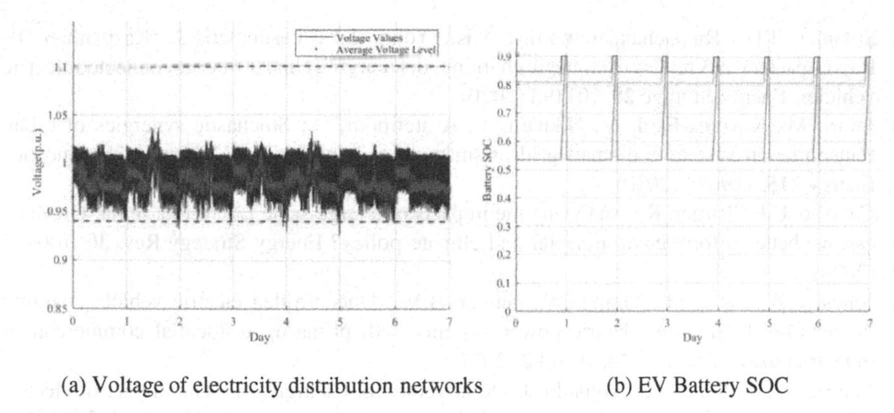

(a) Voltage of electricity distribution networks (b) EV Battery SOC

Fig. 9. Results of scenario 3.

4 Conclusion

Electricity demand of buildings accounts for a major part of the electricity network. The increasing application of EVs and HPs creates a significant burden to the power grid, especially the distribution network. The coordination of EVs and HPs enables buildings to smooth their energy demands and therefore minimize their impact on existing facilities and the required reinforcement investment. This work showed that if domestic users charge EVs whilst using HPs, without considering the constraints of distribution networks, a large rise in voltage drop would be observed. Even though current electricity demand has not led to a significant voltage regulation issue, little room was left for extra peak demand. Coordinating HPs and EVs can make use of a number of existing technologies, and the flexibility already built into these smart devices. Coordination of multiple properties was found to be the best form, as it allowed for the biggest drop in voltage violations. This system would require a greater investment to be implemented due to the requirement for communication between the properties, but this could be worthwhile when the improvement it leads to is considered. In future, more aspects on this topic, such as the current levels of distribution networks need to be considered.

Acknowledgments. This work is supported by State Grid Hubei Electric Power CO., Ltd, under Grant 521532200020.

References

1. UK Power Networks 2014: Impact of Electric Vehicle and Heat Pump Loads on Network Demand Profiles. https://innovation.ukpowernetworks.co.uk/wpcontent/uploads/2019/05/B2-Impact-of-Electric-Vehicles-and-Heat-Pump-Loads-on-Network-Demand-Profiles.pdf
2. Barbosa, T., Andrade, J., Torquato, R., Freitas, W., Trindade, F.C.L.: Use of EV hosting capacity for management of low-voltage distribution systems. IET Gener. Transm. Distrib. **14**, 2620–2629 (2020)
3. Mu, Y., Wu, J., Ekanayake, J., Jenkins, N., Jia, H.: Primary frequency response from electric vehicles in the Great Britain power system. IEEE Trans. Smart Grid **4**(2), 1142–1150 (2012)

4. Solanke, T.U., Ramachandaramurthy, V.K., Yong, J.Y., Pasupuleti, J., Kasinathan, P., Rajagopalan, A.: A review of strategic charging–discharging control of grid-connected electric vehicles. Energy Storage **28**, 101193 (2020)
5. Jafari, M., Kavousi-Fard, A., Niknam, T., Avatefipour, O.: Stochastic synergies of urban transportation system and smart grid in smart cities considering V2G and V2S concepts. Energy **215**, 119054 (2020)
6. Calvillo, C.F., Turner, K.: Analysing the impacts of a large-scale EV rollout in the UK–how can we better inform environmental and climate policy? Energy Strategy Rev. **30**, 100497 (2020)
7. Jones, C.B., Lave, M., Vining, W., Garcia, B.M.: Uncontrolled electric vehicle charging impacts on distribution electric power systems with primarily residential commercial or industrial loads. Energies **14**(6), 1688 (2021)
8. Manríquez, F., Sauma, E., Aguado, J., de la Torre, S., Contreras, J.: The impact of electric vehicle charging schemes in power system expansion planning. Appl. Energy **262**, 114527 (2020)
9. Dixon, J., Bell, K.: Electric vehicles: battery capacity, charger power, access to charging and the impacts on distribution networks. Etransportation **4**, 100059 (2020)
10. Local, P., Networks, E.: Smart charging: a brief guide to managed electric vehicle home charging. https://www.electricnation.org.uk/wp-content/uploads/2016/08/EN-Smart-Charging-Guide-Summary-SCREEN.pdf
11. Muhssin, M.T., Cipcigan, L.M., Jenkins, N., Cheng, M., Obaid, Z.A.: Modelling of a population of heat pumps as a source of load in the Great Britain power system. In: 2016 International Conference on Smart Systems and Technologies (SST), Osijek, pp. 109–113. IEEE (2016)
12. Aye, L.: Heat pumps. In: Managing Air Quality and Energy Systems, pp. 617–631. CRC Press, Los Angeles (2020)
13. Fawcett, T., Layberry, R., Eyre, N.: Electrification of heating: the role of heat pumps. In: BIEE Conference. Oxford United Kingdom (2014)
14. McKenna, E., Thomson, M.: High-resolution stochastic integrated thermal–electrical domestic demand model. Appl. Energy **165**, 445–461 (2016)
15. Taylor, Z.T., Gowri, K., Katipamula, S.: GridLAB-D technical support document: Residential end-use module version 1.0. No. PNNL-17694. Pacific Northwest National Lab. (PNNL), Richland, WA (2008)

Intellectualizing Network Distribution: A Case in the Ruyi New District in Hohhot

Zhihai Yan[2(✉)], Zhebin Sun[1], Sarengaowa[1], Chenxu Zhao[1], Hailong Fan[2], Tao Liang[1], and Pengxuan Liu[1]

[1] Inner Mongolia Electric Power Research Institute, Hohhot 0l0020, China
[2] Inner Mongolia power (Group) Co., Ltd., Hohhot 0l0020, China

Abstract. The intelligent network distribution is helpful to improve the reliability and operation economy of the power grid by converting most of the manual controlling into the active intervention of distribution automation. It covers the functions such as real-time monitoring, autonomous line fault diagnosis, and load transfer by coordinative controlling the master station and each terminal of the line. In this manuscript, taken the Ruyi New District of Huhhot as an example, the intellectualized controlling for the distribution network of the main station and related terminals has been researched. Based on the power quality requirements of the district, the specific indicators of the improved distribution automation in the main station and each terminal have been determined. The novel technology and application of the intelligent distribution network are prospected. The proposed ideas to intellectualize the distribution network will be beneficial to develop an application in the west district of Inner Mongolia regions.

Keywords: Intellectualizing distribution grid · Distribution automation · Distribution technology · Inner Mongolia regions · Mengxi area

1 Introduction

Intelligent distribution grid has always been key to attaining the enhancement of security and reliability of the city's power grid [1–3]. Over the past decade, various attempts have been made to design efficient the smart network. These studies usually merge the advanced signal processing algorithms and machine learning techniques to deal with the machine data and make diagnostic decisions intelligently, leading to impressive results in many diagnosis cases [4–6]. Conceptually, the intelligent distribution grid can intelligently integrate the actions of all users connected to it, that is, the generators and consumers, in order to efficiently deliver sustainable, economic and secure electricity supply [7–9]. Therefore, all the information and signals at the load demand will be processed directionally by the management system and they must react to any event rapidly so as to ensure efficient transmission and distribution network ongoing successfully [10, 11]. In Inner Mongolia Autonomous Region, the power grid is highly complexity due to its vast territory. Thus a higher cost of operation and maintenance have to be required for the power grid. Through the combing and transformation of the power grid in Inner

K. Li et al. (Eds.): LSMS 2021/ICSEE 2021, CCIS 1468, pp. 13–22, 2021.
https://doi.org/10.1007/978-981-16-7210-1_2

Mongolia for a long term, the intelligent distribution network technology has not be fully implemented in west district of Inner Mongolia regions (Mengxi area). In view of the requirements of the national 13th Five-Year Plan for the power industry, the overall promotion of the reliability rate, contact rate and transfer rate index need to be fully improved for the smart distribution network in Mengxi power grid. Therefore, the grid structure in the distribution network technology need to be deeply optimized and then real-time monitoring and control system of the intelligent distribution grid have to be planned to launch.

As the capital of Inner Mongolia Autonomous Region, the Huhhot city belongs to a type of power supply area, where covers a total area of about 20.6 km^2 with lots of the important administrative units, financial and commercial users. Ruyi district is a new region of Hohhot where there are many important users and high load density and thus requires high reliability of power supply. Now it involves 32 lines of 10 kV covered 6 transformers range from 220 kV Gulou, 220 kV Dongjiao, 110 kV Ruyi transformer, 110 kV Hongsheng, 110 DaTaiwan, and 110 kV clove, respectively. In this region, it layouts 228 automatic points, 87 cables with three remote points, 44 cables with two remote points, and 97 overheads with three remote points. Among them, there are equipped with power quality inspection devices and remote viewing systems for 9 important stations. The main station of distribution automation is built according to "Notice of the State Network Operation and Inspection Department on the Application of Distribution Automation Construction in the 13th Five-Year Plan" (Operation Inspection III [2017]). They have been constructed according to the types of the centralized master station and the scale of the medium master station, including 15 basic functions and 21 extended functions. The basic application functions in distribution network operational monitoring have been completed at one time. Meanwhile, the functions such as power flow calculation, unclosing loop analysis, operation ticket, distribution network emulation and training, power qualitatively monitoring, remote inspection have also been done once time. The operation states in the constructed distribution network (i.e. functional application with regional III) include data acquisition processing, distribution grounding fault analysis, power quality monitoring, data quality control, distribution terminal management, and information sharing and announce. It has been successfully connected the terminal and the main station and officially launched the distribution automation system. The information interaction scheme of distribution automation master station is adjusted from the feasibility study stage of "the distribution production system MIS, the geographic information system GIS, and the integration of distribution automation master station system" to the initial stage, including "current mode" and "long-term mode". The project has been phased implemented according these stages. The support platform of the information exchange is constructed by switching bus, which is built from the current information exchange bus in I area to the information exchange bus in III area. The main information exchange of the main station of the distribution automation with the MIS, GIS system has been completed through the information exchange bus in both the III area and the data center. The grid structures all present single ring network with a radiation type. Currently, the contact rate is 89.66%, the transfer rate is

68.97%, and the power supply reliability is 99.88%. Therefore, a degraded intellectualizing network distribution are highly demanded to match the novel standards for the distribution automation.

In the present work, we will focus on the upgradation by intellectualizing the distribution control in Mengxi area. The smart distribution network intellectualizing in Ruyi district of Hohhot as an example has been implemented. The intelligent control of the logical relationship between the main station and each terminal will be significant investigated. The promotion in essence of the distribution automation will be explored. Finally, we proposed an intelligent strategy for the application of the novel technology and construct of intelligent distribution grid in Mengxi area of Inner Mongolia Regions.

2 Results and Discuss

2.1 The Intellectualizing Upgradation of the Present Distribution Network

The reliability of power network is directly related to the perfection degree of the smart distribution network. The central district or 15 < a < 30 regions of the provincial capital city (a is the unit of load density MW/km^2 of the power supply area) are defined as the A type area of power supply according to DL/T 5729-2016 "the technical guidelines for distribution network planning and design". The power supply target of A type is that the average annual blackout time of users is not higher than 52 min (>99.990%) and the qualified rate of the comprehensive voltage >99.97%. In addition, the construction goal of 10 kV lines is to meet with the safety criteria of N-1 power supply where the cable and the ring network as the main lines and the overhead line as auxiliary. Power supply radius should not exceed 3 km and the distribution automation is constructed with the centralized or intelligent distribution mode. The optical fiber communication mode is also chosen. However, the reliability rate of the regional power supply in this investigation is only 99.8757%, even lower 0.0893% than that in B area (the average annual power outage time of users in B area is not higher than 3 h, power supply reliability >99.965), and only higher 0.0127% than that in C area (99.863%). In term of these issues, to intellectualize distribution network is necessary required and a novel strategy is presented to achieve the high reliability in this Ruyi district with a high load density of the power supply by intellectualizing upgradation of the primary grid, automation terminal and automation main station.

a. The primary grid upgradation of distribution network

To realize the reliability of power supply and the qualified rate of the comprehensive voltage to meet with the regional standards in A type area, first of all, we have accurately planned the scheme for the primary grid of the distribution network and strictly implemented the construction. Secondly, in terms of the difficulty of the construction, it is implemented in stages. Finally, it is completed and put into operation by matching with the smart distribution network. The concrete measures are as follows: (1) The connection rate and transfer rate have been raised up to 100% with the aid of the substations of Xiushui 110 kV, Fengzhou 220 kV and Horqin 220 kV (Fig. 1). The single ring network structure has also been upgraded to double ring network structure under the effective conditions of implementation. In addition, the long line radius of the power supply grid is successfully shorten to the small radius. Simultaneously, line load distribution is well optimizing and the contact point is careful combed. Further, the invalid contact and no contact points at the end are also screened and modified. (2) The power supply zoning is optimized while the cross-river, cross-rail, and cross-high bridge lines is reduced. The overhead lines are put into the ground and the cable rate is obviously increased. The communication channels are synchronously constructed. (3) The cause of the power outages in previous years has been carefully analyzed, and the equipment and cables with hidden safety issues have also been replaced. For the purpose of savings, the equipment replaced all is used in the areas where the reliability requirements of power supply are not high. (4) The capacity of load transfer belt is well improved and the planned blackout time is largely reduced. At the same time, the warning signs, inspection work are conventionally carried out to prevent external forces from destroying the line.

b. The upgrading of the distribution automation terminal

The distribution automation terminal is built according to the three remote standards where the optical cable can be laid. The configuration is added in the three remote stations and the blackout range is reduced. Moreover, the number and time of power outages are decreased while the reliability is well increased. The intelligent state sensing system and local discharge device have been installed in the important stations, and an intelligent inspection robot has also been installed on distribution room to be responsible for the maintenance at the important civil construction stations.

The intelligent sensing system is an assembly included the temperature sensor, humidity sensor, water level sensor, door magnetic sensor, and video monitoring, which have been installed in the distribution station. The data there could be transmitted to the data platform in time. The schematic diagram of the intelligent sensing system is shown in Fig. 1.

c. The upgradation of the automation distribution main station

Here, the existing functions of the regional distribution automation main station in the present work have compared with the required functions of the main station in the file of “the notice about the construction and application of distribution automation in the 13th Five-Year Plan issued by the operation and inspection department of the state grid company” No. 6 [2017]. The extra upgrading has been implemented on the original basis as follows. (1) The monitoring and controlling of the distribution network with multi-power and two-way power flow have been achieved when 10 kV distributed power/energy storage device/micro-grid was accessed. This function could be added by utilizing the allowanced internet access conditions in the target area of the photovoltaic power generation, in case that 10 kV distributed power access scale is significantly increased and supported by the vigorously national photovoltaic poverty alleviation project. (2) The function of thematic chart generation is based on the whole network model and the local extraction is simplified by using topology analysis technology. By generating related electrical graphics, it is convenient to combine and improve grid structure efficiency.

(3) Using the redundancy of real-time measurement, the estimation algorithm is used to detect and eliminate the bad data, then to improve the accuracy of the data, and to maintain the consistency of the data. Also it can realize the identification of the bad measurement data of distribution network and further repair the data by loading estimation and the compatibility analysis methods. (4) The network reconfiguration function has been settled in the present distribution network system. The goal of distribution network reconstruction is to change the operation mode of distribution line, eliminate branch overload and voltage overrun, balance feeder load, and reduce line loss by the methods such as switching operation under the premise of satisfying safety constraints. (5) Combining with the fault processing, safe operation analysis, state estimation, power flow calculation, and cycle diagnosis for the current operating state of distribution network, the control strategy is proposed. The automatic controls of the primary and secondary equipment for distribution network have also been achieved. Thus, with the timely removal of the fault and the elimination of the running hidden trouble in the distribution network, the impaired running state could be quickly self-healed. (6) The economic operation function in the present distribution network mode is mainly through the economic and security aspects of the operation. (7) Based on the data of distribution automation operation and the existed distribution network model and parameters, the power supply capacity of this distribution network is evaluated and analyzed.

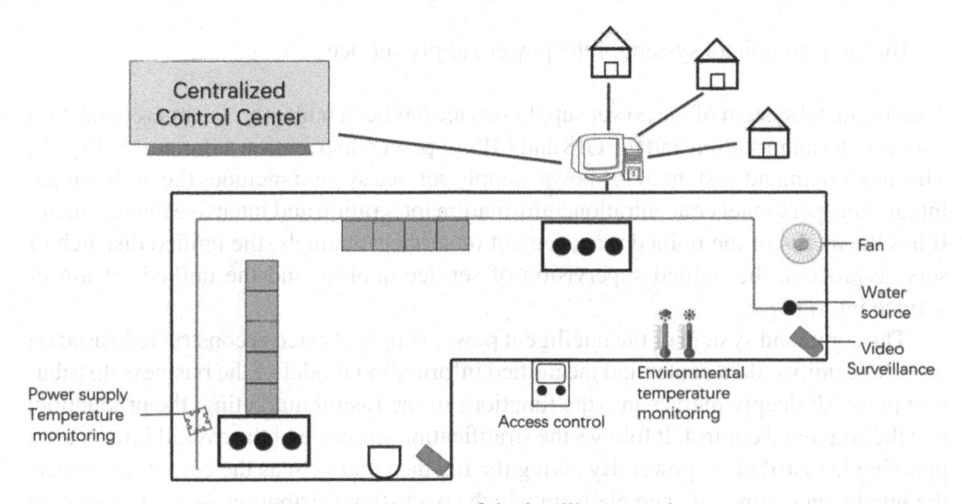

Fig. 1. Schematic diagram of the intelligent sensing system

2.2 Construction of the Intelligent Distribution Network

a. Intellectualizing the distribution main station

According to the construction of the data center of Mengxi power company, the information exchange scheme of distribution automation master station is adjusted from the research stages of "the integrating of the distribution automation master station system with the production system (MIS), the geographic information system (GIS), the energy management system (EMS)" to the initial stages of "the current mode" and "the long-term mode". It could match with the construction of the related projects in stages.

Here, the construction of distribution automation in the current mode follows the IEC 61968/61970 standard. The support platform of the information exchange was built by using the exchange bus in the three different levels. At present stage, only the information exchange bus in I area has been built according the Fig. 1, and the construction of the III area will be considered uniformly in the later stage by Inner Mongolia power (Group) Co., Ltd. The long-term mode adopts the following scheme: after the completion of GIS, the graphic and function module of MIS is firstly installed and then the information interaction of the main station of distribution automation with MIS and GIS is completed through the information exchange bus III area and the data center. All these could avoid the inconsistency of information update among the systems and therefore it could realize the direct import of the graphics and modules from MIS. Consequently, the workload of the graphic and function drawing by master station is significantly reduced. The functions such as the distribution line, grid structure, equipment, line equipment geographical location, line and even superior substation transient steady state information, switch status in the main station system could be easily reviewed. It is very convenient for operators to read data and make decisions quickly.

b. Building command system of the power supply service

The command system of the power supply service has been built after the interconnection between the main station and the GIS and MIS of power distribution automation (Fig. 2). This new command system of the power supply service system includes the professional integration, personnel concentration, information integration and intensive management. It has the merits of the unified management of service channels, the unified dispatch of service process, the unified supervision of service quality, and the unified release of service information.

The command system of the intelligent power supply service is constructed based on the whole unified data center and the unified information model of the business distribution power. It deeply merges into the functions of the fusion marketing, the production, and the command control. It follows the stratification structure of the overall information planning in distribution power. By taking the big data platform as the core, it constructs the intelligent command open platform which covers the distribution network operation management and the customer service. The following sub-systems are built, as showed in Fig. 2: (1) The control system of the distribution power status: the functions of the defect prediction model and the prediction warning are constructed by comprehensive combing the information and data, including the equipment account, defect processing and filling, meteorological information, distribution live detection, real-time abnormal data, multi-dimensional statistical analysis of distribution network equipment defects, cause analysis of the important equipment defects. The defect causes is auxiliary analyzed and the equipment inspection plan is reasonably arranged. The elimination work of the hidden trouble defect is carried out and the equipment quality is well improved. The related analysis such as the defects, hidden trouble, patrol operation and maintenance, can be achieved in the intelligent command system of distribution network. Finally, the operation and maintenance job of thc distribution network of each sub-company are objectively evaluated and an improved suggestion is put forward. (2) The operation monitoring system of the distribution network: based on the geographical map, thematic map, and single line map, all kinds of graphical controls are illustrated with various forms. These maps also include the abnormal, alarm and other control scenarios, power conservation tasks, production planning, climate environment, mountain fire, defect hidden dangers. Therefore, each sub-company could directly grasp the abnormal equipment state of the distribution network, operation and maintenance management, abnormal equipment due to weather etc. The risk state of the network equipment operation is well controlled to auxiliary support the policy analysis from a macro point of view. (3) Repair system for the fault command: using the concept of "customer-centered", and the comprehensive utilizing of the camp distribution and adjustment depth fusion, the repair work is efficiently commanded. It is also driving from the combination with power grid topology, two-way check, rapid acquisition fault, accurate judgment, accurate location and resource allocation. Moreover, the passive repair is changed into active service. The fault repair efficiency and service quality are significantly improved. (4) Control system for the customer service: the customer service mainly include the functions of several work orders, such as the reminder, marketing service, complaint, praise, supervision of the purchase electricity not issued during 24 h, fault processing for zero no jump of the fee control intelligent meter. It also contains the distribution of the work order that failure collection with the intelligent meter continuous 72 h. In addition, the functions is covered with the processing situation tracking assessment, and the process monitoring. The integration of the present system with 95598 marketing customer service work order, and combining with the customer service closed-loop control, support customer service command function.

c. Intellectualizing the active distribution network

The active distribution network is a distribution system that manages power flow by using flexible network topology [12]. It could realize the active control and active management for the local distributed energy (DER). On the basis of the appropriate regulatory environment and access criteria, DER can assume a certain degree of support to the system. In this research, the designed pattern of an active distribution network is shown in Figure 2, including communication network active control combining power transmission.

d. Intellectualizing the distribution internet of things system

The core equipment of the logistics distribution internet is the intelligent distribution transformer terminal. According to the State Grid technical standard [13], the typical design scheme of the distribution internet of things with the intelligent distribution transformer terminal as showed in Fig. 3.

The strategic pilot period is set at the three states. Firstly, the project plan of the distribution internet of things is thoroughly completed, and the construction scheme is

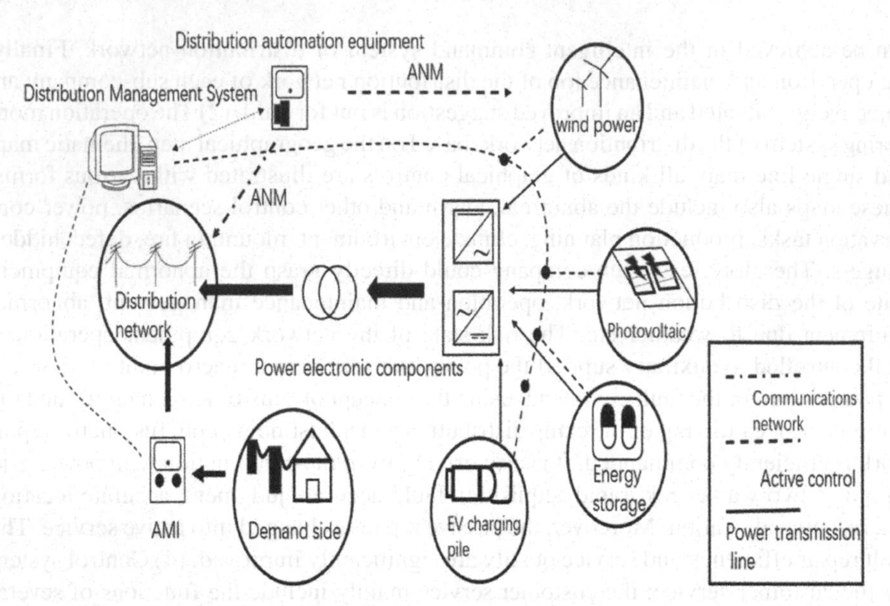

Fig. 2. Schematic diagram of the intelligent sensing system

proposed. In this state, the operation flows include: integrated resource services, virtual power plant, energy finance, resource commercialization management, big data network management, integration of three certificate, customer service, implementation of electric wing, enterprise transportation management. And then, by data sharing and capacity opening, the construction of the logistics distribution internet will be unremitting driven on advance. It connected various ends, tubes, and edges with big data management with wired or wireless modes. The construction and transformation of about 20% demonstration area for the logistics distribution internet in the public transformer region is planned to complete, in which the key technologies will be tackled, such as terminal plug, terminal cloud cooperation, cloud distribution automation master station. These key technologies have been practiced in engineering and they support the applications including the lean operation and maintenance of the distribution network, the intelligent control, the customized application, the network cooperation, the precision investment, etc. Finally, the logistics distribution internet was built to form the completely ecosystem. The technical standard and typical design scheme of logistics distribution internet is established with the intelligent distribution and transformer terminal as a center. The construction of the strong smart internet based an energy flow from generate electricity, transmit electricity, power transformation, power distribution, and finally to users. Also the new energies have been integrated into the strong ring smart grid. The intelligent distribution network and logistics distribution internet have been integrated to develop in the present strategy

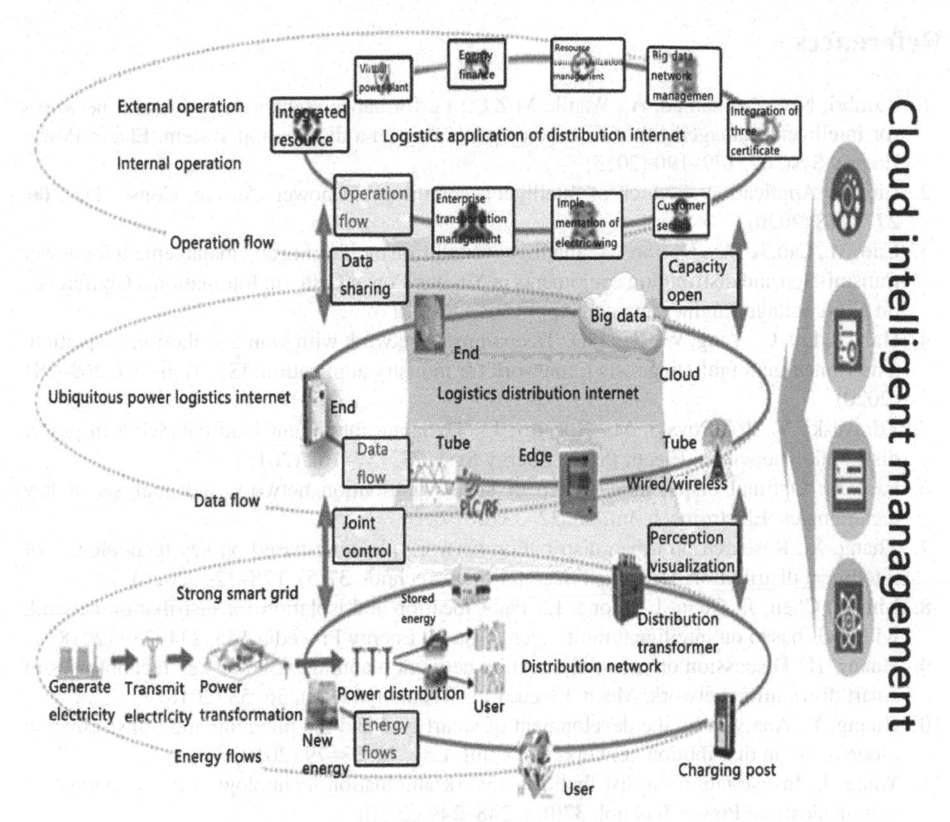

Fig. 3. The typical scheme of the distribution internet of things

3 Conclusion

The development and application of smart distribution network has changed the operation idea of the traditional power grid with completely different approaches of the regulation and control. Taking the main control station of distribution network and its related terminals in the Ruyi district of Hohhot as an example, this wok makes clear the specific index to intellectualize upgrading the main station and terminal of distribution automation, and proposes the application of the new technology of smart distribution network in combination with the development of power supply, distribution technology and industrial policy. It have maximized the value and efficiency of the power grid. By optimizing the self-healing ability of power grid, providing higher security and power quality, visual management of distribution network and its equipment, and the management of the information and power distribution, the utilization ratio of assets of power grid enterprises has been larger improved. It also provides technical guarantee for the high reliability requirements of distribution network in several important areas of Mengxi.

References

1. Zambri, N.A., Mohamed, A., Wanik, M.Z.C.: Performance comparison of neural networks for intelligent management of distributed generators in a distribution system. Electr. Power Energy Syst. **67**, 179–190 (2015)
2. Liu, S.: Application research of intelligent technology in power. Autom. Constr. Des. **06**, 277–278 (2020)
3. Cao, M., Cao, K. Wu, B., Tan, M.: Intelligent condition monitoring and management for power transmission and distribution equipments in Yunnan Power Grid. In: International Conference on High Voltage Engineering and Application (2012)
4. Han, T., Liu, C., Yang, W., Jiang, D.: Deep transfer network with joint distribution adaptation: a new intelligent fault diagnosis framework for industry application. ISA Trans. **97**, 269–281 (2020)
5. Zdraveski, V., Todorovski, M., Kocarev, L.: Dynamic intelligent load balancing in power distribution networks. Electr. Power Energy Syst. **73**, 157–162 (2015)
6. Pan, X.: Optimal dispatching design of smart distribution network and analysis of key technologies. Electromech. Inf. **35**, 32–33 (2019)
7. Zhang, X.: Research on urban distribution network planning based on key technologies of intelligent distribution network. Telecom Power Technol. **37**(5), 128–129 (2020)
8. Lin, L., Chen, J., Guan, L., Dong, L.: Fault location and isolation for distribution network with DGs based on intelligent multi-agent system. Energy Procedia **145**, 234–239 (2018)
9. Huang, H.: Discussion on urban distribution network planning based on key technologies of smart distribution network. Mech. Electr. Eng. Technol. **48**(12), 56–58 (2018)
10. Huang, Y.: Analysis on the development of smart grid and the planning and transformation mode of urban distribution network. Netw. Inf. Eng. **12**, 78–79 (2019)
11. Wang, J.: Investigation on distribution network automation technology and its application communication. Power Technol. **37**(02), 248–249 (2020)
12. McDonald, J.: Adaptive intelligent power systems: active distribution networks. Energy Policy **36**, 4346–4351 (2008)
13. Lv, J., Luan, W., Liu, R.: Distribution internet of things architecture based on comprehensive perception and software definition. Power Grid Technol. **42**(10), 3108–3115 (2018)

Classifications of Lithium-Ion Battery Electrode Property Based on Support Vector Machine with Various Kernels

Kailong Liu[1], Zhile Yang[2(✉)], Haikuan Wang[3], and Kang Li[4]

[1] WMG, University of Warwick, Coventry CV4 7AL, UK
[2] Shenzhen Institute of Advanced Technology, Chinese Academy of Sciences, Shenzhen 518055, People's Republic of China
zl.yang@siat.ac.cn
[3] Shanghai Key Laboratory of Power Station Automation Technology, School of Mechatronic Engineering and Automation, Shanghai University, Shanghai 200072, People's Republic of China
[4] School of Electronic and Electrical Engineering, University of Leeds, Leeds LS2 9JT, UK

Abstract. Manufacturing chain of lithium-ion batteries belongs to a significantly complex process with many coupled product parameters and intermediate products. To well monitor and optimize battery manufacturing process, it is vital to design a data-driven approach for effectively modelling and classifying the product properties within this complicated production chain. In this paper, a support vector machine (SVM)-based framework, through using four various and powerful kernels including linear kernel, quadratic kernel, cubic kernel and Gaussian kernel, is proposed to well classify the electrode mass loading property of battery. The effects of four crucial variables including three product features from mixing step and one product parameter from coating step on the electrode property classification are also investigated. Comparative results illustrate that electrode mass loading can be effectively classified by the designed SVM framework while Gaussian kernel-based SVM achieves the best classification for all labelled classes. This is the first time to systematically evaluate and compare the performance of different kernel-based SVMs on the battery electrode property classification. Due to data-driven nature, the proposed SVM-based framework can be easily extended to classify other product properties and analyze other variables in battery production domain.

Keywords: Lithium-ion battery · Battery production chain · Support vector machine · Kernel functions · Electrode property classifications

1 Introduction

Lithium-ion (Li-ion) batteries have been widely utilized as main power source for sustainable energy applications such as smart grids, electrical vehicles and electrical trains, owing to their competitive properties such as high energy density and low self-discharge rate [1]. Due to the battery performance would be highly and directly affected by the

K. Li et al. (Eds.): LSMS 2021/ICSEE 2021, CCIS 1468, pp. 23–34, 2021.
https://doi.org/10.1007/978-981-16-7210-1_3

related manufacturing processes, an effective production chain that can monitor and analyze battery intermediate product properties is thus vital for boosting the development of Li-ion batteries [2].

However, due to the complexity of containing lots of intermediate processes, numerous variables would be generated in battery production chain. These variables would highly affect the properties of intermediate products, further playing a key role in determining the final battery performance. Therefore, it is crucial to design suitable solutions for better investigating the effects of intermediate product variables (IPVs) on the classification performance of battery product properties.

With the rapid developments of machine learning (ML) algorithms, data-driven approaches are becoming the powerful tools for handling lots of issues within battery managements [3]. To date, numerous data-driven methods have been effectively adopted to estimate dynamics states [4, 5], forecast service lifetime [6, 7], achieve charging managements [8, 9] and energy managements [10, 11] of batteries. Overall, reliable analyses can be done by well designing data-driven solutions in battery management domain. However, the analyses of battery manufacturing are still mainly obtained by expert knowledge as well as trial and error methods. It should be known that battery production will also generate a large number of related data, deriving suitable data-driven models to analyze these data should be also considered as a promising way to achieve battery smarter manufacturing.

In comparison with battery managements, fewer attempts have been done so far to derive advanced machine-learning strategies in battery production chain [12]. Among many corresponding themes (process monitoring and adjustments) of battery manufacturing, deriving suitable data-driven model to predict or classify the properties of battery intermediate products is a hot research topic. For examples, a data-driven approach was proposed in [13] to determine the internal parameters for quality controls in battery production. Based upon the cross-industry standard process, Schnell et al. [14] designed linear and neural network (NN) models to forest the product properties of battery manufacturing. Turetskyy et al. [15] proposed the decision tree based models to conduct feature selections and predict the maximal capacity of battery production. In [16], the dependencies of three parameters from mixing step within battery production chain are mainly analyzed by the 2D graphs from a SVM and experimental data. For the aforementioned researches, some common ML algorithms such as SVM has presented powerful potentials to derive suitable models for classifying the properties of battery production. However, many works mainly focus on simply using a common ML algorithm without in-depth investigating its performance in battery production domain. It should be known that the kernel function plays an important role in SVM and also needs to be carefully selected for solving battery manufacturing issues. Therefore, how to design a suitable kernel within SVM to not only achieve acceptable classification accuracy but also present high generalization ability of battery production is still a key but challenging issue.

According to the above discussions, driven by the purpose to effectively classify the product properties of battery production chain, a kernel-based SVM framework is designed in this study. Specifically, after well labelling electrode mass loading into four classes, four crucial variables including three product features from mixing step and one product parameter from coating step are selected as the inputs of SVM to investigate their effects on the classifications of battery electrode mass loading. Instead of simply using a linear kernel, other three powerful kernels including quadratic kernel, cubic kernel and Gaussian kernel are also coupled within SVM for achieving better classification performance. This is the first time to systematically evaluate and compare the classifications of various kernel-based SVMs on the battery electrode property. Experimental results from confusion matrix and ROC curve illustrate the effectiveness of this SVM-based framework, paving a promising way to better analyze and classify other product properties of battery production domain.

2 Key Steps in Battery Electrode Manufacturing

The electrode manufacturing of Li-ion batteries belongs to a highly complicated chain which involves many disciplines such as electrical, chemical, and mechanical engineering. Figure 1 systematically illustrates several key steps in electrode manufacturing.

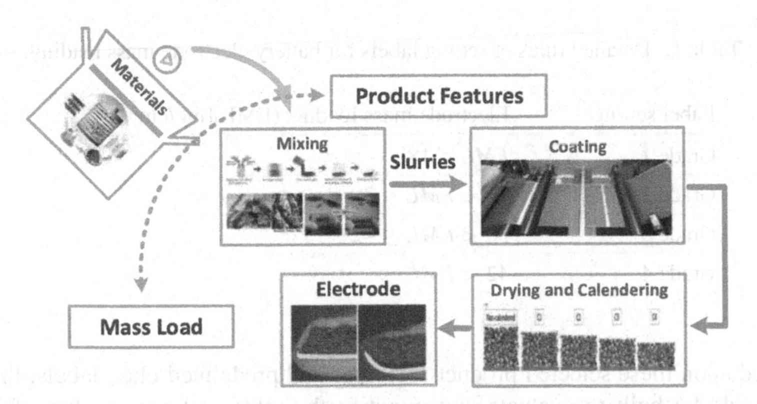

Fig. 1. Key steps in battery electrode manufacturing.

According to Fig. 1, the battery electrode manufacturing chain mainly contains mixing, coating, drying and calendaring steps. In the mixing step, the slurries of both anode and cathode would be produced through mixing the active materials (graphite and Li-NCM-Oxide), the conductive additives (carbon black), the solvent (NMP) and the binder (PVDF) within a soft blender. Then the well-mixed slurries will be coated into a mental foil (generally copper for anode and aluminium for cathode) by the coating machine, followed by a drying process to dry the coating product through using built-in ovens. Finally, the electrodes can be obtained after the calendaring process. It should be known that a large number of variables and parameters can be generated during such complex electrode production chain. The product features and parameters from key steps such as mixing and coating are significantly important for the electrode properties, which would further affect the final performance of battery products and must be carefully analyzed.

In light of this, some representative product features from mixing as well as coating steps are selected to develop suitable kernel-based SVM for classifying the electrode properties in this study. To be specific, four product features or parameters including the active material mass content (AMMS), solid-to-liquid ratio (STOLR), viscosity and comma gap (CG) are utilized to build the SVM models for investigating their effects on the classification results of one battery electrode property named electrode mass loading (electrode mass per unit area). In theory, STOLR is the mass ratio between slurry solid (active material, conductive carbon as well as binder) and mass (solid component as well as solvent). Viscosity impacts the shear rate within coating step. CG is the gap between comma roll and coating roll within a coater. According to these selected product variables and electrode property, the original manufacturing data from Laboratoire de Reactivite et Chimie des Solides (LRCS) are explored. The effectiveness of these data has been proven in [16], which would not be repeated here due to space limitation. For these data, because eight same samples of product features are utilized to obtain one mass loading of battery electrode, the original data would be first compressed to 82 samples through averaging them with the same observations. Then the electrode mass loading would be classified into four grades with the labels as Grade 1, Grade 2, Grade 3 and Grade 4, respectively. The detailed rules of setting these labels are described in Table 1.

Table 1. Detailed rules of setting labels for battery electrode mass loading.

Label setting	Electrode mass loading (EML)[mg/cm^2]
Grade 1	$EML \leq 18$
Grade 2	$18 < EML \leq 30$
Grade 3	$30 < EML \leq 42$
Grade 4	$42 < EML$

Based upon these selected product variables and predefined class labels, the SVM model could be built to evaluate and quantify the effects of various kernels on the classification performance of battery electrode mass loading.

3 Technology

In this section, the fundamental of SVM classification is first presented, followed by the descriptions of four various kernel functions. Then the indicators to evaluate the classification performance are also described.

3.1 Support Vector Machine Classification

SVM belongs to a powerful ML tool for both classification and regression [17]. To achieve reasonable classification, the best classification hyper-plane would be searched in training process of SVM. The hyper-plane is determined by an orthogonal weight vector ω that could give the wider margin of separations. Supposing the training dataset is noted as $TD = (X_i, Y_i), i = 1, 2, ..., l, X \in R^m$, while hyper-plane is $(\omega \cdot X_i + b) = 0$.

To ensure all observations can be classified correctly by the hyper-plane, following constraints should be satisfied as:

$$Y_i(\omega \cdot X_i + b) \geq 1, i = 1, 2, ..., l. \tag{1}$$

Then the process to maximize the classification margin is defined by:

$$\begin{cases} \min\limits_{\omega,b} \|\omega\|_2^2 / 2 \\ s.t. \quad Y_i(\omega \cdot X_i + b) \geq 1, i = 1, 2, ..., l \end{cases} \tag{2}$$

After constructing Lagrange function, this process can be expressed by the Lagrange multiplier α_i as:

$$\begin{cases} \min\limits_{\alpha} \ \frac{1}{2}\sum_{i=1}^{N}\sum_{j=1}^{N}\alpha_i\alpha_j Y_i Y_j (X_i \cdot X_j) - \sum_{i=1}^{N}\alpha_i \\ \sum_{i=1}^{N}\alpha_i Y_i = 0 \\ \alpha_i \geq 0, \quad i = 1, 2, ..., N \end{cases} \tag{3}$$

Based upon Eq. (3), SVM is capable of not only guaranteeing the accuracy of classification, but also maximizing the blank ranges on all sides of hyper-plane [18].

In order to improve the nonlinear classification performance of SVM, kernel functions should be coupled within SVM. Specifically, through using proper kernel functions, raw data from the original space could be effectively transferred to a high-dimensional space, then the SVM-based classification model could be trained through using the data from this high-dimensional space with the linear classification approach. Supposing $\phi(e)$ is a function to map the input space to a new feature space, the kernel function can be expressed by:

$$K(e, g) = \phi(e) \cdot \phi(g). \tag{4}$$

According to Eq. (3), the cost function to maximize the classification margin through involving the kernel functions becomes:

$$W(\alpha) = \frac{1}{2}\sum_{i=1}^{N}\sum_{j=1}^{N}\alpha_i\alpha_j Y_i Y_j K(X_i \cdot X_j) - \sum_{i=1}^{N}\alpha_i. \tag{5}$$

Based upon the above discussions, kernel functions play the key important roles in determining the classification performance of SVM. It should be known that for different applications, various kernel functions would present different performance, which should be carefully selected. In our study, to well classify the battery electrode property of mass loading, the SVM classification model with four typical and powerful kernel functions are designed and compared.

The first one is the linear kernel with the following form as:

$$K_{linear}(x, z) = a_1 \cdot x^T \cdot z + b_1. \tag{6}$$

Next, two polynomial kernels contain the quadratic kernel and cubic kernel are also utilized in this study as:

$$\begin{cases} K_{quadratic}(x, z) = \left(a_2 \cdot x^T \cdot z + b_2\right)^2 \\ K_{cubic}(x, z) = \left(a_3 \cdot x^T \cdot z + b_3\right)^3 \end{cases} \tag{7}$$

The last one is the Gaussian kernel with the following form as:

$$K_{Gaussian}(x, z) = \exp\left[-\frac{\|x - 2\|^2}{2\sigma^2}\right]. \tag{8}$$

3.2 Performance Indicators

To quantify and investigate the classification performance of proposed SVM with various kernels, some typical performance indicators including the positive predictive value (PPV), false discovery rate (FDR), confusion matrix, true positive rate (TPR), false positive rate (FPR), area under curve (AUC) and receiver operating characteristic (ROC) curve [19] are adopted in this study.

For a multiple classification application, let positive corresponds to an interested class while negative corresponds to other classes, four elements contain true positive (TP), false positive (FP), true negative (TN) and false negative (FN) could be derived for each class. Then for the class C_k (here $k = 1 : 4$), its PPV to quantify the correct rate of class can be calculated by:

$$PPV = \frac{TP}{TP + FP}. \tag{9}$$

FDR to quantify the rate of all false discovery of this class can be obtained by:

$$FDR = \frac{FN}{TP + FN}. \tag{10}$$

Based upon these two performance indicators, a $(N + 1) \times (N + 1)$ confusion matrix (CM) to reflect the accuracy of each classification within multi-class problem can be generated. All terms on the primary diagonal of CM represent the correctly-classified results while other terms stand for the incorrect cases of SVM classification.

Next, to further investigate the classification performance of battery multi-class electrode mass loading, the ROC curves of all classes through using different kernel-based SVM is also utilized in this study. It should be known that the ROC curve is generated by plotting TPR against FPR for different threshold settings. Specifically, TPR is used to reflect the number of correct positive classifications happen among all positive observations. FPR provides the amount of incorrect positive classifications happen among all negative observations. The equations of calculating TPR and FPR are expressed as follows:

$$TPR = \frac{TP}{TP + FN}. \tag{11}$$

$$FPR = \frac{FP}{FP + TN}. \quad (12)$$

After adopting the normalized unit, the AUC of ROC curve can be utilized to reflect the probability that a SVM would rank a randomly-selected positive case higher than a randomly-selected negative case. Better classification approach would generate a classifier point closer to the left-upper corner with a larger AUC value.

4 Result and Discussion

To investigate and quantify the performance of SVMs and the effects of kernel functions on the classification of battery electrode property, the SVM-based framework with various kernels are built and compared for the classification of electrode mass loading in this section.

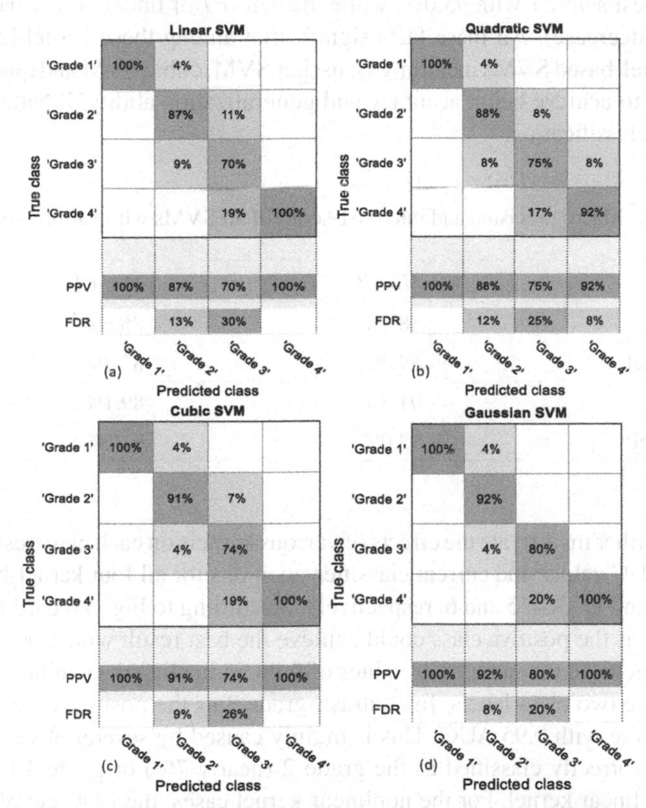

Fig. 2. Confusion matrix for SVM-based classifications with various kernels. a) Linear SVM, b) Quadratic SVM, c) Cubic SVM and d) Gaussian SVM.

Specifically, four product variables consist of the AMMS, STOLR, viscosity and CG are selected as the inputs while the electrode mass loading of battery is utilized as output for all SVM models. After preparing the suitable inputs-output pairs, five folds cross-validation is conducted to train and validate the accuracy and generalization ability of derived SVM models. Figure 2 illustrates the confusion matrices of all kernel-based SVM classifications. It can be seen that all SVMs present the reliable classification results of all four classes (the PPVs of four SVMs are all larger than 70%), indicating that SVM is capable of effectively classifying the electrode mass loading. Quantitatively, Gaussian kernel-based SVM provides a best classification result with lowest PPV of 80%, indicating the strong nonlinear capture ability of Gaussian kernel. Both cubic kernel-based SVM and quadratic kernel-based SVM present the slight worse classification performance with 74% lowest PPV (7.5% decrease) and 75% lowest PPV (6.3% decrease), respectively. Table 2 illustrates the macro-precision (*Mac-P*) and micro-F1 score (*Mic-F1*) of all SVM classification results. It can be seen that the *Mac-P* of Gaussian kernel-based SVM reaches 93.0%, which is 1.9%, 4.1% and 4.8% more than that of cubic, quadratic and linear kernel-based SVMs, respectively. The similar trend can be also observed for the *Mic-F1* values of all SVMs. Here the Gaussian kernel-based SVM achieves the best *Mic-F1* with 93.0%, while the *Mic-F1* of linear kernel-based SVM is 86.6% (7.4% decrease). All these facts signify that among these kernel-based SVMs, nonlinear kernel-based SVMs including Gaussian SVM, cubic SVM and quadratic SVM are preferable to achieve better accuracy and generalization ability of battery electrode mass loading classification.

Table 2. Macro precision and micro F1-score of all SVMs with various kernels.

Kernel types	*Mac-P*	*Mic-F1*
Linear kernel	88.8%	86.6%
Quadratic kernel	89.3%	87.8%
Cubic kernel	91.3%	89.1%
Gaussian kernel	93.0%	91.5%

Next, to further investigate the effects of various kernels on each class result, the ROC curves with AUC values and current classifier positions for all four kernel-based SVMs are illustrated in Fig. 3, 4, 5 and 6, respectively. According to Fig. 3 regarding the linear SVM, grade 1 as the positive class could achieve the best result with 1.00 AUC. Grade 2 and grade 4 achieve the same AUC values of 0.98, indicating the similar classification results for these two class labels. In contrast, grade 3 as the positive class leads to the worst ROC curve with 0.95 AUC. This is mainly caused by several observations from grade 3 are incorrectly classified as the grade 2 (nearly 7%) or grade 4 (nearly 19%) through using linear kernel. For the nonlinear kernel cases, the ROC curves have been effectively improved with larger AUCs. Quantitatively, from Fig. 4 to Fig. 6, the AUCs of grade 4 become 0.99, implying the nonlinear kernels can benefit the classification of grade 4. For the cubic and Gaussian based SVMs, the AUC of grade 3 both become 0.97, which is 2.2% and 3.2% more than that of linear and quadratic cases, respectively. Interestingly, all AUCs of grade 1 are 1.00, which means that grade 1 can be exactly

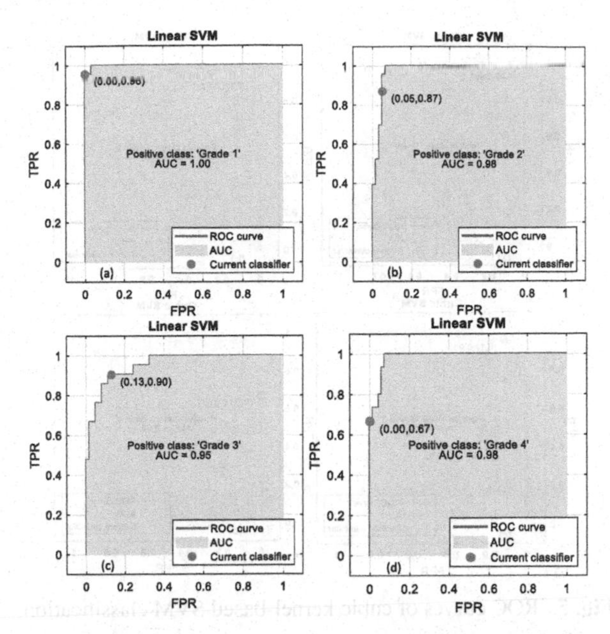

Fig. 3. ROC curves of linear kernel-based SVM classification.

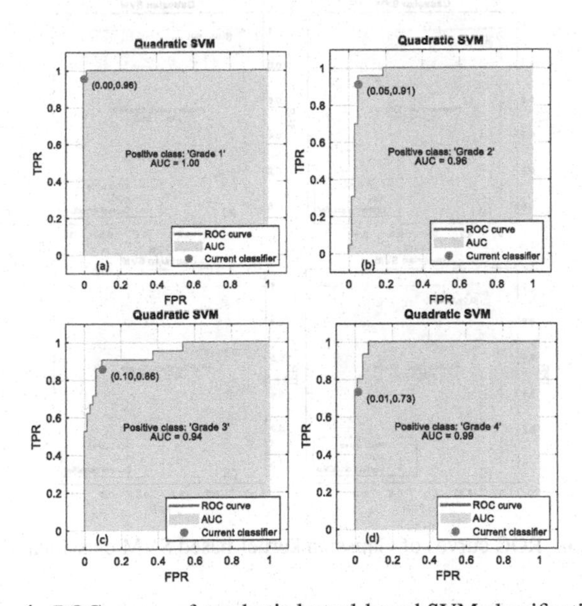

Fig. 4. ROC curves of quadratic kernel-based SVM classification.

classified through using SVMs. In summary, all kernel-based SVMs can well classify grade 1, while all other grades prefer the nonlinear kernels. Gaussian-based kernel can achieve best classification results for all grades, which is recommended for designing the corresponding SVM framework for electrode mass loading classification.

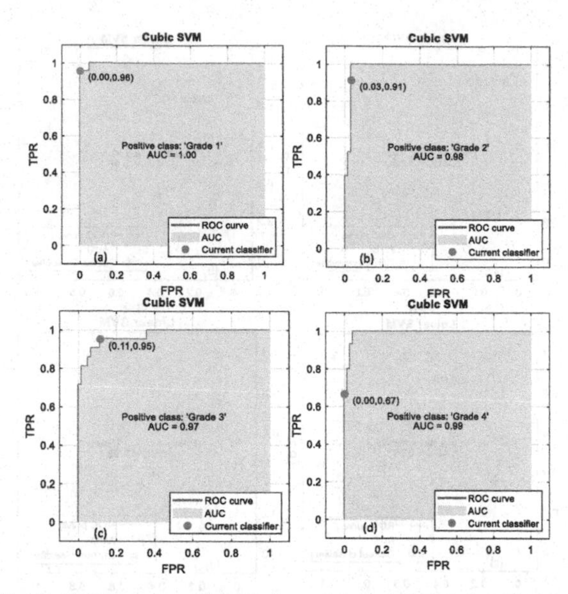

Fig. 5. ROC curves of cubic kernel-based SVM classification.

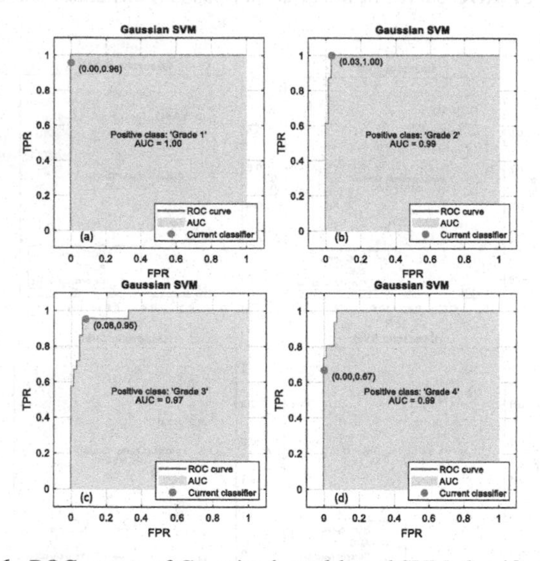

Fig. 6. ROC curves of Gaussian kernel-based SVM classification.

5 Conclusions

Electrode property plays an important role in determining the final battery performance, which should be carefully classified and analyzed. In this study, an effective data-driven classification method, based on the SVM with various kernels, is proposed to well classify the battery electrode mass loading and analyze the effects of four product features

(AMMC, STOLR, viscosity, CG) from mixing and coating steps. The classification performance of linear, quadratic, cubic and Gaussian kernel-based SVMs are all systematically investigated and compared through using different performance indicators and ROC curves. Illustrative results demonstrate that Gaussian kernel based SVM is capable of achieving the best classification results among four kernels (here is 93.0% *Mac-P* and 91.5% *Mic-F1*). Besides, the labelled grade 1 of electrode mass loading can be exactly classified by SVM with all kernels. The labelled grade 3 presents the worst classification results (here the worst AUC is 0.94) of all kernel cases, which should be reset to further improve the classification performance. This proposed kernel-based SVM framework actually belongs to a data-driven approach, which can be conveniently extended to classify other product properties and analyze other variables in battery production domain.

Acknowledgments. This work was supported by the Natural Science Foundation of China (No. 52077213 and 62003332) and Youth Innovation Promotion Association CAS 2021358.

References

1. Liu, K., Li, K., Peng, Q., Zhang, C.: A brief review on key technologies in the battery management system of electric vehicles. Front. Mech. Eng. **14**(1), 47–64 (2018). https://doi.org/10.1007/s11465-018-0516-8
2. Kwade, A., Haselrieder, W., Leithoff, R., Modlinger, A., Dietrich, F., Droeder, K.: Current status and challenges for automotive battery production technologies. Nat. Energy **3**(4), 290–300 (2018)
3. Li, Y., et al.: Data-driven health estimation and lifetime prediction of lithium-ion batteries: a review. Renewable Sustain. Energy Rev. **113**, 109254 (2019)
4. Guo, Y., Yang, Z., Liu, K., Zhang, Y., Feng, W.: A compact and optimized neural network approach for battery state-of-charge estimation of energy storage system. Energy **219**, 119529 (2021)
5. Hu, X., Feng, F., Liu, K., Zhang, L., Xie, J., Liu, B.: State estimation for advanced battery management: key challenges and future trends. Renewable Sustain. Energy Rev. **114**, 109334 (2019)
6. Liu, K., Li, Y., Hu, X., Lucu, M., Widanage, W.D.: Gaussian process regression with automatic relevance determination kernel for calendar aging prediction of lithium-ion batteries. IEEE Trans. Ind. Inform. **16**(6), 3767–3777 (2019)
7. Tang, X., Liu, K., Wang, X., Gao, F., Macro, J., Widanage, W.D.: Model migration neural network for predicting battery aging trajectories. IEEE Trans. Transp. Electrification **6**(2), 363–374 (2020)
8. Liu, K., Li, K., Zhang, C.: Constrained generalized predictive control of battery charging process based on a coupled thermoelectric model. J. Power Sources **347**, 145–158 (2017)
9. Ouyang, Q., Wang, Z., Liu, K., Xu, G., Li, Y.: Optimal charging control for lithium-ion battery packs: a distributed average tracking approach. IEEE Trans. Ind. Inform. **16**(5), 3430–3438 (2019)
10. Shang, Y., Liu, K., Cui, N., Wang, N., Li, K., Zhang, C.: A compact resonant switched-capacitor heater for lithium-ion battery self-heating at low temperatures. IEEE Trans. Power Electron. **35**(7), 7134–7144 (2019)

11. Wu, J., Wei, Z., Liu, K., Quan, Z., Li, Y.: Battery-involved energy management for hybrid electric bus based on expert-assistance deep deterministic policy gradient algorithm. IEEE Trans. Veh. Technol. **69**(11), 12786–12796 (2020)
12. Liu, K., Wei, Z., Yang, Z., Li, K.: Mass load prediction for lithium-ion battery electrode clean production: a machine learning approach. J. Clean. Prod. **289**, 125159 (2021)
13. Schnell, J., Reinhart, G.: Quality management for battery production: a quality gate concept. Procedia CIRP **57**, 568–573 (2016)
14. Schnell, J., et al.: Data mining in lithium-ion battery cell production. J. Power Sources **413**, 360–366 (2019)
15. Turetskyy, A., Thiede, S., Thomitzek, M., Drachenfels, N.V., Pape, T., Herrmann, C.: Toward data-driven applications in lithium-ion battery cell manufacturing. Energy Technol. **8**(2), 1900136 (2019)
16. Cunha, R.P., Lombardo, T., Primo, E.N., Franco, A.A.: Artificial intelligence investigation of NMC cathode manufacturing parameters interdependencies. Batteries Supercaps **3**(1), 60–67 (2020)
17. Noble, W.S.: What is a support vector machine? Nat. Biotechnol. **24**(12), 1565–1567 (2006)
18. Rebentrost, P., Mohseni, M., Lloyd, S.: Quantum support vector machine for big data classification. Phys. Rev. Lett. **113**(13), 130503 (2014)
19. Liu, K., Hu, X., Zhou, H., Tong, L., Widanalage, D., Macro, J.: Feature analyses and modelling of lithium-ion batteries manufacturing based on random forest classification. IEEE/ASME Trans. Mechatron. **99** (2021)

Robot Platform for Live Working in Distribution Network

Xiaoqiang Yuan[1], Dahong Wang[1], Ning Ding[1](✉), Chengxiang Zhang[1], and Qing Zhou[2]

[1] Institute of Robotics and Intelligent Manufacturing, The Chinese University of Hong Kong (Shenzhen), Shenzhen 518172, Guangdong, China
dingning@cuhk.edu.cn

[2] State Grid Chongqing Electric Power Research Institute, Chongqing 401123, China

Abstract. Live work in a distribution network is limited due to factors such as the large size of the insulated bucket-arm truck and the necessity of flat ground. This paper proposes a hybrid arm-based live-line auxiliary working robot for live working in a distribution network. One-button automatic leveling can be achieved by applying remote control and local autonomy strategies. The mechanical and electrical properties of the composite insulating arm are analyzed, and the electrical performance, maneuverability, and operation function of the entire vehicle are tested. This study realizes a flexible and intelligent platform for full-scale live working in a distribution network.

Keywords: Robot · Distribution network · Climbing platform · Insulating arm · Live-line work

1 Introduction

Owing to the technology boom, power consumption is continuously increasing, while power grid development is facing growing demands for high reliability, load density, and power quality. Users have less tolerance for outage time, and increasingly fewer opportunities are available for maintenance outages of key lines. The distribution network is an essential part of a power system and the key link to ensure continuous electricity supply. The reliability of the distribution network is vital to the entire power supply system. Live working in a distribution network has thus become a vital aspect to better serve the current economic construction and people's pursuit for a better life [1, 2]. However, the design, specifically the height, of insulation platforms is constrained by factors such as space and ground leveling, which make full-scale live work on a distribution network difficult [3]. There is therefore an urgent need for a work robot platform that occupies a small area and can move, raise and lower the insulating bucket, and rotate in all directions to achieve live working in a distribution network.

Resrach Supported by the National Natural Science Foundation of China (U1813216) and National Grid Science and Technology Project (52200018000G).

K. Li et al. (Eds.): LSMS 2021/ICSEE 2021, CCIS 1468, pp. 35–46, 2021.
https://doi.org/10.1007/978-981-16-7210-1_4

Aerial work vehicle manufacturers are mainly concentrated in developed countries (e.g., some European countries, the United States, and Japan). The boom structure of these vehicles can be categorized into folding arms, telescopic arms, and hybrid arms [4]. Well-known manufacturers of hybrid arm products include Bronto (Finland), Palfinger (Austria), and Oil & Steel (Italy) [5]. North American aerial work vehicles are mainly insulated. The three largest manufacturers of insulated aerial work vehicles in the world are located in the United States, and their products are the market leaders worldwide [6]. With the development of the Chinese national economy and urban construction, a variety of work platforms have been successfully developed in China. For example, Zhejiang Dingli Machinery Co. Ltd. [7] has developed six series of aerial work platforms, including the straight-arm type, crank-arm type, and scissor type. Xuzhou Heavy Machinery Co., Ltd. and Hunan Xingbang Heavy Industry Co., Ltd. have successively developed more than 10 series of aerial work platform products in only a few years, which provide more choices for the installation and maintenance of domestic ships, construction, airports, and mining enterprises.

Jiatian Guo et al. [8] of Tianjin Vocational and Technical Normal University investigated a fully electric aerial work vehicle modified with a hydraulic lifting platform and showed that its handling stability can be improved by optimizing the center of mass, suspension stiffness, and cornering stiffness of the tires. Chongxin Xu et al. [9] of the State Grid Shandong Yantai Power Supply Company researched a live-working robot for a master–slave teleoperation distribution network; they applied a neural network teleoperation force feedback signal delay prediction algorithm to solve the significant delay problem to complete live-working tasks in various complex circuit environments. Shi Li et al. [10] of the Institute of Automation of the Chinese Academy of Sciences analyzed and compared the technical advantages and existing problems of current-powered robots and highlighted gaps between current robot systems and the desired capabilities. Hangyu Sun [11] of the Shanwei Power Supply Bureau of Guangdong Power Grid experimentally studied insulation platform–based live operation technology for a distribution network and summarized that although insulation platform–based technology is highly effective, it suffers from poor mobility and inconvenient transfer.

In summary, there are many types of aerial work vehicles worldwide, and new Chinese products are rapidly emerging. Non-Chinese insulated aerial work vehicles are leading the market. Nevertheless, domestic production has been achieved for the core insulating parts of insulated mobile equipment used for live work. Aerial work vehicles are powered by bulky engines that produce noise and pollute the environment, while the vehicles themselves are constrained by their requirements for flat terrain and large space. Therefore, it is difficult to adapt existing insulated aerial vehicles to full-scale live distribution network operations.

This remainder of the paper is organized as follows. Section 2 outlines the mechanical structure and control system of the auxiliary robot platform. Section 3 presents the mechanical analysis of the telescopic insulating arm. The experiments and testing are presented in Sect. 4. Finally, the conclusions are drawn in Sect. 5.

2 Robot Platform Design

In light of the rapid development of robotics technology, current live-working requirements, and high-altitude working equipment technology, this paper investigates a distribution network live auxiliary work robot with a folding and telescopic arm structure. A high-capacity battery is used as the power source to increase the working radius and reduce the space occupied by the robot. The inner arm and working bucket insulation system are composed of glass fiber–reinforced plastic, which ensures the insulation safety of live workers. A schematic of the machine structure of this robot is shown in Fig. 1.

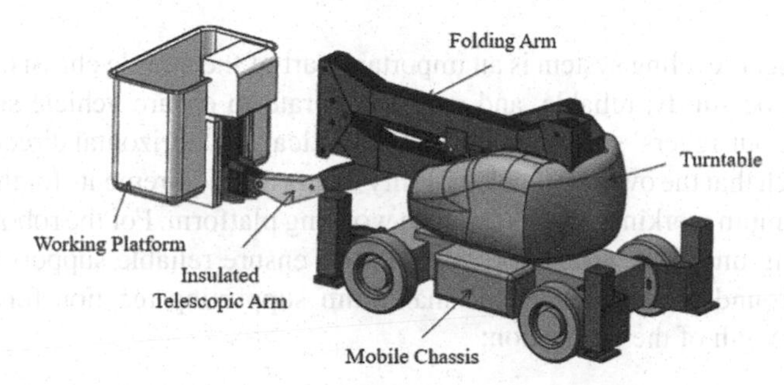

Fig. 1. Live working robot for distribution network

The proposed robot includes a mobile chassis, an upper installation system, and a control system and meets the following functional requirements.

1) The robot weighs less than 7 tons, and its footprint is 30% smaller than that of similar models.
2) The robot can work around obstacles and adapt to different terrain operations.
3) The maximum working height is 13.7 m, and the insulation voltage rating is 10 kV.
4) The lifting stroke of the insulating platform is 500 mm.
5) The battery life of the robot is ≥ 10 h.
6) The maximum horizontal working radius is 3.5 m.
7) The robot has a human–computer interaction safety monitoring function.
8) The robot has a battery capacity management function.
9) The working platform can be manually shut off in case of failure.
10) The walking speed of the robot is ≤ 5 km/h.

2.1 Mobile Chassis

The mobile chassis is mainly responsible for driving and steering the vehicle and for supporting the turntable, boom, and hydraulic and electrical systems. Its main components are four solid tires, a walking motor, a reducer, an outrigger leveling system, steering cylinder, and an underframe. The mobile chassis is driven by a motor that has stable walking characteristics, timely response, low noise, and limited obstacle-crossing ability. The overall chassis design is shown in Fig. 2.

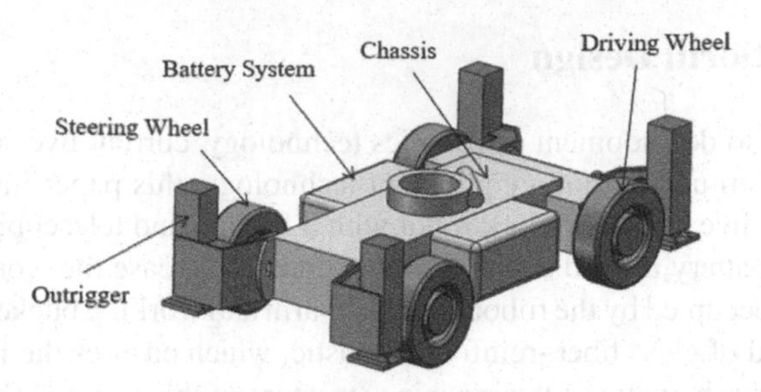

Fig. 2. Moving chassis

The outrigger leveling system is an important part of the mobile chassis. The system is required to be sturdy, reliable, and easy to operate to ensure vehicle stability. The position of the outriggers' support points in the vertical and horizontal directions should be selected such that the overall vehicle stability meets the requirements for the calibrated load and maximum working amplitude of the working platform. For the robot to function without sinking under the ground pressure and to ensure reliable support on different terrain, the ground pressure under the maximum supporting reaction force must not exceed the strength of the foundation:

$$S_j \geq \frac{F_{\max}}{[\rho_d]}. \tag{1}$$

Where S_j is the grounding area, $F_{\max}$ is the maximum support reaction force, and $[\rho_d]$ is the foundation strength (typically, 1.6 MPa). The grounding area is thus calculated to be 6000 mm^2.

2.2 Upper Installation System

The primary function of the upper installation system is to support the work platform to reach the predetermined position and to offer insulation protection. This system adopts a three-joint arm and a telescopic arm, and the structure size is evenly arranged. Small folding arms and large telescopic arms are common in the modern market. Compared with such arms, the proposed upper installation system has the following advantages. (1) The folding arm moves faster and is more efficient. (2) Smaller size requirements under the same load. (3) The horizontal operating range is larger than that of the traditional bucket truck. (4) The folding arm and telescopic arm can be connected to achieve one-key lifting. The structure of the upper installation system is shown in Fig. 3.

The working bucket must be maintained level during lifting and operation. Four automatic leveling mechanisms are commonly used: (1) self-weight leveling; (2) parallel four-bar leveling; (3) equal-volume hydraulic cylinder leveling; (4) and electro-hydraulic servo leveling. Considering safety, the self-weight leveling mechanism is not suitable for application in aerial work vehicles. The equal-volume hydraulic cylinder leveling mechanism is generally applied only to the telescopic lifting mechanism, whereas the electro-hydraulic servo leveling mechanism is challenging to realize. Therefore, the parallel four-bar leveling mechanism is adopted for the proposed working bucket.

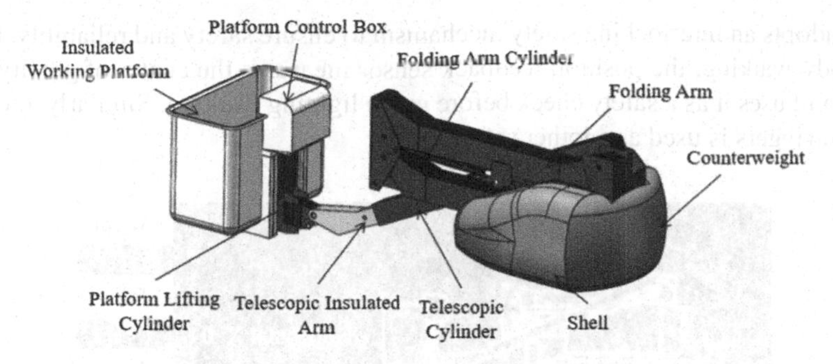

Fig. 3. Moving chassis

The insulating arm and insulated working bucket are made from an epoxy resin–glass fiber composite through a double-sided steel-mold compression molding process. These components thus are lightweight and have smooth inner and outer surfaces, no burrs, no bubbles, high density, high mechanical strength, and good electrical insulation.

2.3 Control System

This paper adopts remote control and local autonomous control strategies to execute the functions that robots require to efficiently and reliably complete the live-line auxiliary operation of a distribution network. These strategies also reduce the operational difficulty and improve the system intelligence. To implement these strategies, the control system adopts a three-layer structure (Fig. 4) that can process a variety of front-end sensor information.

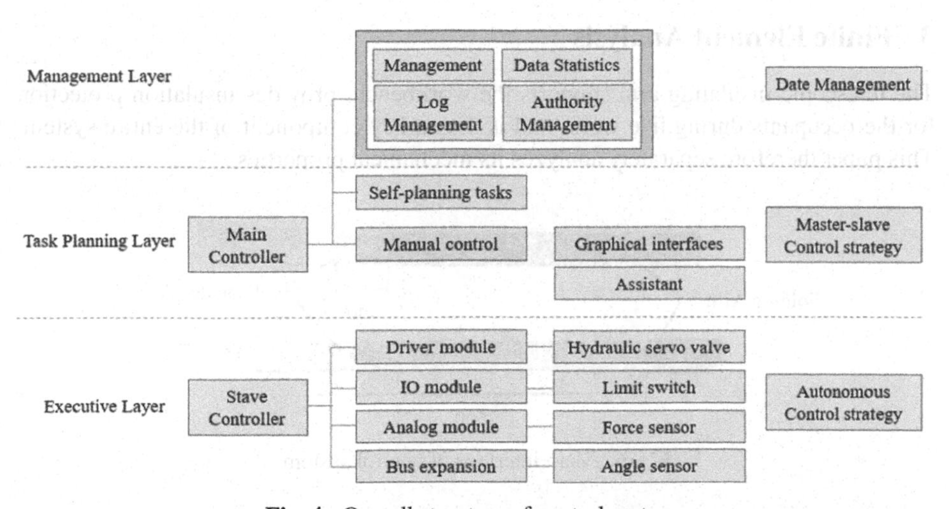

Fig. 4. Overall structure of control system

The live-line auxiliary operation robot of a distribution network comprises two parts: (1) walking of the main body and (2) lifting and lowering of the working arm. The control system adopts an interlocking safety mechanism to ensure safety and reliability. During main body walking, the position feedback sensor measures the center of gravity of the vehicle and uses it as a safety check before green lighting walking. Similarly, the return of the outriggers is used as another safety check.

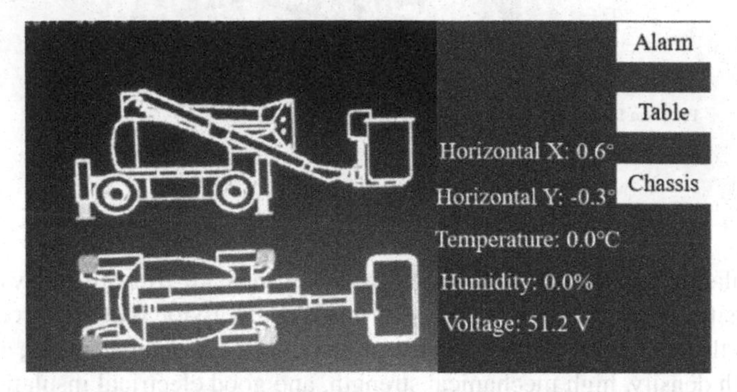

Fig. 5. Main interface of control system

The main control program of the proposed robot is divided into the following modules: task planning, kinematic calculation, compliance control, human–computer interaction, master–slave communication, peripheral drive, thread management, exception handling, and log management. The master and slave control systems use TCP/IP protocol to communicate. The master control system issues control commands calculated by the kinematics module to the slave control system and reads back the hybrid arm's movement status in real-time. The main interface is depicted in Fig. 5.

3 Finite Element Analysis

The telescopic insulating arm supports the workbench, provides insulation protection for the occupants during live work, and is thus a key component of the entire system. This paper therefore separately analyzes its mechanical properties.

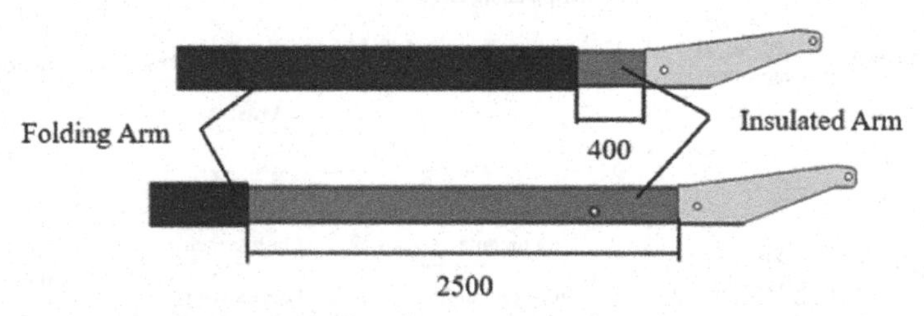

Fig. 6. Main interface of control system

The physical position of the insulating arm is shown in Fig. 6. An effective insulation length of 400 mm is retained when contracted to ensure effective insulation level for low-height operations. The maximum length of the cantilever at full extension is 3530 mm, 640 mm of which is inserted into the folding arm.

Table 1. Force statistics of insulating arm.

No	Force type	Force arm (mm)	Force (N)	Moment (Nm)
1	Telescopic arm weight	2372	1225	2905.7
2	Telescopic cylinder thrust	1767	617.4	1090.95
3	Leveling cylinder thrust	3115	245	763.17
4	Seat Self-weight	3554	176.4	626.93
5	Swing motor thrust	3700	274.4	1015.28
6	Work bar weight	4060	1715	6962.9
7	Load	4505	2254	10154.27

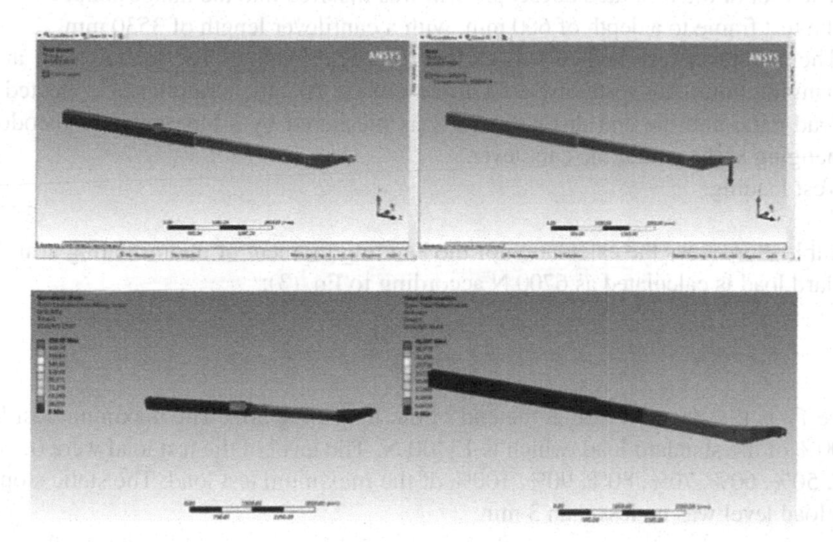

Fig. 7. Finite element analysis of insulating arm

The force statistics of the insulating arm are listed in Table 1. A safety factor of 2.0 is adopted for the first and seventh items, and a safety factor of 1.1 is adopted for the other (second to sixth) items. To improve the analysis accuracy, the mechanical analysis of the insulating arm is conducted on Ansys Workbench software. In addition, the limit working condition (i.e., the insulating arm extends the farthest and receives the greatest force when placed horizontally with a maximum force arm of 3530 mm) is adopted. The end load is calculated as 10,700 N according to Eq. (2):

$$F = \frac{\sum n_i M_j}{l}. \tag{2}$$

Where F is the end load of the insulating arm, n_i is the safety factor, M_j is the moment, and l is the moment arm.

The resulting stress cloud and displacement cloud are shown in Fig. 7. The maximum stress of the insulating arm is 216.83 MPa, the maximum displacement is 41.6 mm, and the tensile strength of the FRP is 290.77 MPa. These results indicate that the strength of the insulating arm meets the required standards.

4 Experiments

4.1 Mechanical Performance Test of Insulating Arm

The insulating arm must not only have good mechanical properties but also be insulated from the ground. Its mechanical properties affect the success and failure of the entire equipment. We therefore conducted inspection tests at Huazhong University of Science and Technology using the following test protocol.

1) The root of the insulated telescopic arm was inserted into the flange connected with the test frame to a depth of 600 mm, with a cantilever length of 3530 mm.
2) The cantilever test method was used for vertical loading. The load applied at the hanging point was tested by a 5T force sensor. A static strain tester collected the load data, and the end displacement was measured by a tape measure theodolite hanging at the end of the cantilever.
3) Test loading.

Table 1 provides the calculation of the bending moment of the insulating arm. The standard load is calculated as 6700 N according to Eq. (3):

$$F_0 = \frac{\sum M_j}{l}. \tag{3}$$

where F_0 is the standard load at the end of the insulating arm. The maximum test load is 200% of the standard load, which is 13400 N. The level of the test load were 0, 20%, 40%, 50%, 60%, 70%, 80%, 90%, 100% of the maximum test load. The static stop for each load level was no less than 3 min.

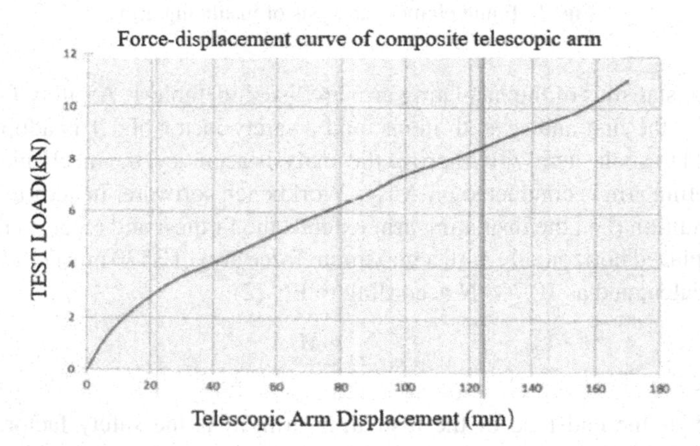

Fig. 8. Test results of mechanical properties of insulating arm

4) Test results and conclusions.

The test results of the mechanical properties of the insulating arm are shown in Fig. 8. The actual applied load of the composite telescopic arm was found to be 11.11 kN, which exceeds the required test load, and the maximum end displacement was 128.7 mm. There was no abnormality in the loading process of the test piece, and no overall damage features were observed.

4.2 Electrical Performance Test

It is essential to evaluate whether the electrical insulation performance of the vehicle and its live-working equipment meet the standard requirements. For the insulated bucket arms, we conducted electrical performance tests on the hydraulic oil, hydraulic tubing, insulated arms, insulated working buckets, and the entire vehicle. The test results meet the requirements of the relevant standards. The test process is as follows.

1) Electrical performance test of hydraulic oil and hydraulic tubing

Table 2. Test results of electrical properties of hydraulic oil.

Average breakdown voltage (kV)	Standard value	Result
22.6	≥20 kV	Qualified

Table 3. Electrical performance test results of pipe with oil after standing still for 20 min.

	Test value					Standard value
Applied voltage (kV)	10	22	30	40	45	20
Leakage current (μA)	72	157.2	214.12	287	312	≤200

Table 4. Electrical performance test results of pipe with oil after standing still for 3 weeks.

	Test value					Standard Value
Applied voltage (kV)	10	20	30	40	20	10
Leakage current (μA)	72.27	142.82	215.66	289.1	≤200	72.27

The insulation test of the hydraulic oil and tubing was conducted following the regulations in GB/T 9465-2008 "Vehicle-Mounted Mobile Elevating Work Platform" and

Table 5. Electrical performance test results of insulating arm and insulating bucket

1 min insulation withstand voltage test					
Part	Item	Test distance	Test voltage		Result
			Standard value	Test value	
Insulated arm	Power frequency withstand voltage	0.4 m	45.0	45	Qualified
	Surface power frequency withstand voltage	0.4 m	45.0	45	Qualified
Working bucked	Laminar power frequency Withstand Voltage	/	45.0	45	Qualified

GBT 507-2002 "Insulating liquids–Determination of the breakdown voltage at power frequency." The test results are listed in Tables 2, 3 and 4.

2) Electrical performance test of the insulating bucket and insulating arm.

Electrical performance tests were performed on the insulated telescopic boom, insulated bucket, and entire vehicle according to the requirements of the GB/T 9465-2008 "Aerial Operating Vehicles" standard. The test site was the high-voltage test hall of Chongqing Electric Power Research Institute. The insulated arms and insulated buckets met the withstand voltage test. The results are shown in Table 5, and the field test is depicted in Fig. 9.

Table 6. Electrical performance test results of whole vehicle

1 min insulation withstand voltage test					
Part	Item	Test distance	Test voltage		Result
			Standard value	Test value	
Engineering prototype	Power frequency withstand voltage	1 m	45.0	45	Qualified

3) Vehicle electrical performance test.

The whole vehicle passed the withstand voltage test at a test distance of 1 m and voltage of 45 kV. The complete machine withstand voltage test is shown in Table 6. The on-site test is depicted in Fig. 9.

Fig. 9. Field test of electrical performance of insulating arm, insulating bucket and whole vehicle

4.3 Application Test

After the robot is assembled, we test the motion performance of the power distribution network assisted working robot. According to the design index, we test the climbing ability of robot in the 30° road. The results show that the robot can pass the slope smoothly. Next, the robot can cross a height of 10 cm obstacles. Finally, we test the turning ability of the robot in ordinary parking spaces, and the performance is good.

In addition, we tested the prototype on the spot with professional testers, and the performance is remarkable (see Fig. 10).

Fig. 10. Test results of mechanical properties of insulating arm

5 Conclusion

The live-line auxiliary work robot proposed in this paper adopts a hybrid arms and a battery power supply. The control strategy is mainly based on remote control, local autonomy, and one-key outrigger leveling. The proposed robot has the advantages of omnidirectional rotation and self-moving operation, a small footprint, and considerable lifting height of the insulated bucket. Although the lifting operation is manual, the proposed system improves efficiency, reduces labor intensity, and is a flexible and intelligent platform for full-scale operations on a live distribution network.

In the new future, we will integrate a cooperative manipulator into this robot platform to replace the manual completion of the task of distribution network electrification. We believe that this goal will be achieved soon.

References

1. Xiao, B., Guo, B.: Review and prospect of distribution network planning. Electr. Power Autom. Equip. **38**(12), 200–211, 217 (2018)
2. Liu, Y., Ji, K., Fu, X.: Review on development of live working robot technology in distribution network. Power Energy **40**(4), 446–451 (2019)
3. Li, H., Yang, Z., Li, X., Shao, Y., Deng, S.: Research and application of integrated live working new technology based on insulation platform. Modern Electron. Tech. **39**(09), 153–159 (2016)
4. Li, W., Gang, Y.: Advantages and disadvantages of insulated bucket arm trucks with different structural forms. Sci. Technol. Innov. **28**, 55 (2016)
5. Li, R.: Structural analysis and improvement of working apparatus for a hybrid arm aerial work vehicle. Electric Power Automation Equipment (2017)
6. Huang, G., Chen, M.: Review of research progress of aerial work vehicles in China. Mech. Electr. Technol. **35**(1), 2–5 (2012)
7. Li, T.: Research on active vibration control for boom of telescopic boom aerial working vehicle. Huazhong University of Science and Technology (2018)
8. Guo, J., Jin, Y.: Optimization design for the electric vehicle high altitude working vehicle based on ADAMS/Car. Chin. J. Constr. Mach. **18**(3), 253–263 (2020)
9. Xu, C., Zhao, Y., Zhao, S.: Research on distribution network live working robot with master-slave tele-operation. Mach. Tool Hydraul. **47**(23), 102–105 (2019)
10. Li, S., Liu, B., Han, G.: The hot-line work robot for power distribution network. Autom. Panorama **3**, 68–70 (2018)
11. Sun, H.: Research on live working technology of distribution network based on insulation plat form. Electr. Eng. **20**, 79–82 (2019)

A Novel Optimization Algorithm for Solving Parameters Identification of Battery Model

Junfeng Zhou[1,2], Yubo Zhang[1], Zhile Yang[2], Wei Feng[2], and Yanhui Zhang[2](✉)

[1] School of Electrical Engineering, Zhengzhou University, Zhengzhou 450001, China
[2] Shenzhen Institutes of Advanced Technology, Chinese Academy of Sciences, Shenzhen, China
zhangyh@siat.ac.cn

Abstract. The battery energy storage system (BESS) has received broad concern with the development of photovoltaic technology. The unknown internal parameters of the battery model in BESS will affect its output, so extract the parameters accurately is essential. In order to extract the parameters of the battery model, an improved JAYA (IJAYA) optimization algorithm is presented. In IJAYA, a self-adaptive weight, an experience-based learning strategy, and a chaotic elite learning method are adopted. The three strategies can increase the probability of obtaining the best solution, enhance the population diversity, and improve the quality of the solution. The 16 A discharging experiment is performed in BESS. The parameters are identified according to the second-order RC equivalent circuit model. The results indicate that the IJAYA algorithm can accurately and effectively extract parameters of the battery model.

Keywords: Parameter identification · Battery energy storage system · IJAYA algorithm

1 Introduction

Nowadays, energy has become a vital driving force in social development and progress, the application of renewable energy is becoming more and more widespread [1]. The battery energy storage system is an application of store energy, dramatically developed in recent years [2]. To improve the conversion efficiency and better utilize its characteristics, modeling the operation of battery energy storage systems has become an important research topic in recent years [3]. Some mathematical models with unknown parameters have been widely used in practice. These parameters that can reflect the actual performance of the battery are often difficult to extract due to their uncertainty and variability, which inevitably affects the role of the built model in optimal operation and actual control. It is indispensable to study the parameters estimation problem of the battery energy storage model. Therefore, there is an urgent need for a feasible optimization way to ascertain the battery energy storage model's parameter. Based on the introduced battery energy storage model, the parameters extraction method is studied in this paper.

In this paper, IJAYA algorithm is adopted to solve the problem of parameters identification better. In the algorithm, three strategies are proposed: a self-adaptive weight

K. Li et al. (Eds.): LSMS 2021/ICSEE 2021, CCIS 1468, pp. 47–54, 2021.
https://doi.org/10.1007/978-981-16-7210-1_5

is introduced to improve the probability of closing the global optimal solution; A learning strategy that makes the individual learn from the other individuals is developed to enhance the population diversity; A chaotic learning way is explained to increase the quality of the global optimal at each iteration.

2 Model and Problem Approach

2.1 Battery Energy Store Model

A suitable model can accurately simulate the functional characteristics of the battery [4]. The primary purpose of building the battery model is to obtain the mathematical relationship between the internal parameters of the battery and the external features, then establish a suitable equation to estimate the output voltage of the energy storage battery in the solar storage system. There are mainly three types of battery models: the electrochemical model, the mathematical model, the equivalent circuit model. Among them, the electrochemical model has high accuracy, but the model is complex. The mathematical model has nothing to do with the external volt-ampere characteristics of the battery, and the model is uncertain due to significant errors. The equivalent circuit model can better simulate the exterior features of the battery so that it is widely used in practice. The second-order RC equivalent circuit model (SECM) as one of the equivalent circuit models is researched in the parameters identification issue [5]. The mathematical model is shown in Fig. 1, and the Second-order RC equivalent circuit model is used here. For the terminal voltage hysteresis phenomenon exhibited by the battery, the equivalent circuit model can be used to simulate the external characteristics of the battery better. Considering the electrochemical polarization characteristics and concentration polarization characteristics of the battery, two RC networks are used to simulate the two dynamic processes of the situation. While considering the influence of internal resistance, the internal resistance is connected in series in the circuit.

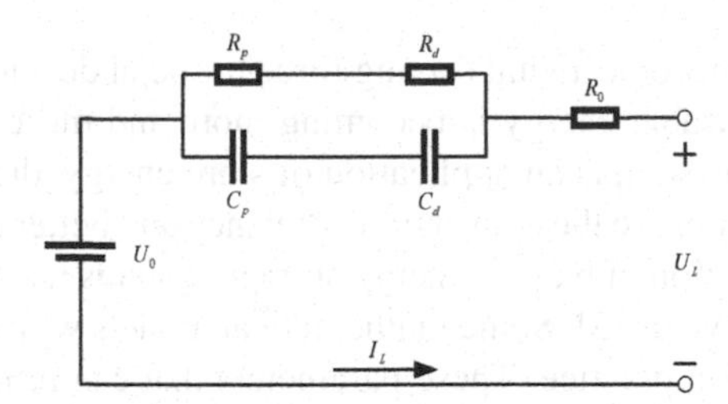

Fig. 1. Schematic of battery energy storage model: SECM

For the SECM model, according to Kirchhoff's law of voltage and current, the equation could be known as follows (Eq. 1):

$$\begin{cases} U_L = U_{OCV} - U_0 - U_p - U_e \\ I_L = \frac{U_p}{R_p} + C_p \frac{dU_p}{dt} \\ I_L = \frac{U_e}{R_e} + C_e \frac{dU_e}{dt} \\ I_L = \frac{U_0}{R_0} \end{cases} \tag{1}$$

Where U_L is the terminal voltage of the battery energy storage output, U_{OCV} is the open-circuit voltage, U_0 is voltage, U_p and U_e respectively expresses the terminal voltages of the two RC networks. R_p represents the resistance of electrochemical polarization, C_p denotes the capacitance of electrochemical polarization. R_e and C_e are the resistance and the capacitance of concentration polarization, respectively. R_p, C_p, R_e, C_e constitute two RC parallel loops are used to simulate polarization effects. R_0 is the internal resistance and I_L is the load current. There are five unknown parameters (R_p, C_p, R_e, C_e, R_0) that should be calculated to observe the actual behavior of energy storage battery.

2.2 Relation Curve Between OCV and SOC

The parameters identification of resistance and capacitance in the battery model requires the value of U_{OCV} measurement first. The electromotive force and the SOC maintain a relatively stable monotonic relationship in a specific environment. The electromotive force voltage source adopts a controlled voltage source whose behavior is controlled by SOC. The experimental data can be used to fit U_{OCV} and SOC, then the functional relationship between them can be obtained [6]. According to the practical principle, the experimental steps of the static experiment are as follows: first, use the standard charging method to charge the battery fully, let it stand for 1 h, and measure the open-circuit voltage at SOC = 1; Discharge the battery with a current of 6 A for 20 min each time; Let it stand for 1 h, and approximate the sampled voltage obtained at this time as the open-circuit voltage. The discharge data of SOC and U_{OCV} measured by experiment are shown in table 1. Data fitting is performed on the above experimental data to obtain the functional relationship between U_{OCV} and SOC. The fourth-order polynomial function is selected to fit the experimental data when considering the accuracy and computational complexity. The fitting function is shown as follows (Eq. 2):

$$\begin{array}{c} U_{OC} = -0.925263 * SOC^4 + 2.671602 * SOC^3 \\ -2.614026 * SOC^2 + 1.118892 * SOC + 3.118363 \end{array} \quad (2)$$

Table 1. SOC and U_{OCV} experimental data.

SOC	U_{OCV}	SOC	U_{OCV}
1	3.375	0.5	3.296
0.9	3.334	0.4	3.293
0.8	3.332	0.3	3.283
0.7	3.315	0.2	3.257
0.6	3.303	0.1	3.212

2.3 Objective Function

The experimental data will be compared with simulated parameter data. Therefore, choosing an appropriate objective function could minimize the error between the two data. A continuous discharge experiment is used for the battery, and the sampling time: T = 1 s. According to Eq. 1, the state space equation of the energy storage battery can be known and then discretize the state space equation formula. Equation 1 could be written as Eq. 3. Equation 4 is the objective function used to express the error of the experimental and estimated data for the SECM model.

$$\begin{cases} U_p(k) = e^{-\frac{T}{\tau_1}} U_p(k-1) - (1 - e^{-\frac{T}{\tau_1}}) R_p I_L(k-1) \\ U_e(k) = e^{-\frac{T}{\tau_2}} U_e(k-1) - (1 - e^{-\frac{T}{\tau_2}}) R_e I_L(k-1) \\ U_L(k) = U_{OC}(SOC(k)) - U_p(k) - U_e(k) - R_0 I_L(k-1) \\ \tau_1 = R_p C_p \\ \tau_2 = R_e C_e \end{cases} \cdot \tag{3}$$

$$\begin{cases} f_k(U_L, I_L, SOC, x) = U_{OC}(SOC(k)) - \\ e^{-\frac{T}{R_p C_p}} U_p(k-1) - (1 - e^{-\frac{T}{R_p C_p}}) R_p I_L(k-1) - \\ e^{-\frac{T}{R_e C_e}} U_e(k-1) - (1 - e^{-\frac{T}{R_e C_e}}) R_e I_L(k-1) \\ -R_0 I_L(k-1) - U_L \\ x = \{R_p, C_p, R_e, C_e, R_0\} \end{cases} \cdot \tag{4}$$

In this study, the root mean square error (RMSE) represents the overall difference between the experimental and estimated current data. RMSE defined by Eq. 5.

$$RMSE(x) = \sqrt{\frac{1}{N} \sum_{k=1}^{N} f_k(V_L, I_L, x)^2}. \tag{5}$$

3 Improved JAYA Algorithm

JAYA (a Sanskrit word meaning victory) algorithm is a new population-based optimization algorithm for solving optimization problems [7]. Unlike other mainstream algorithms, this algorithm is popular because it has few initial parameters and requires no additional adjustments at run time [8]. There are many variations on it, and the improved IJAYA (IJAYA) algorithm is one of them. In IJAYA, the main improvement is the use of three different strategies. These strategies are improved in the particle solving speed, the individual exploration ability, and the population quality, respectively. First, a self-adaptive weight is introduced to increase the rate of convergence. Second, a learning strategy is employed to change the population diversity. Third, a chaotic learning method is proposed to improve the quality of solutions [9]. By introducing these three strategies, the IJAYA algorithm maintains the same and simple structure as the original JAYA algorithm and does not need to set any additional parameters. The improvement method of the algorithm is as follows.

3.1 Self-adaptive Weight

In the searching area of the algorithm, the individual should reach an idea area of the search space at the early stage. At the later stage, the local search in ideal areas should be implemented to refine the quality of the population. To this end, a self-adaptive weight is presented in Eq. 6 to achieve the goal. Then, $x_{i,j}$ can be updated by Eq. 7 by adding the weight.

$$w = \begin{cases} (\frac{f(x_{best})}{f(x_{worst})})^2, \text{ iff}(x_{worst}) \neq 0 \\ 1, \text{ otherwise} \end{cases}. \quad (6)$$

$$x'_{i,j} = x_{i,j} + rand_1 * (x_{best,j} - |x_{i,j}|) - w * rand_2 * (x_{worst,j} - |x_{i,j}|). \quad (7)$$

where $x_{best,j}$ and $x_{worst,j}$ are the values of the jth individual for the best and worst performance, respectively. $x'_{i,j}$ is the updated value of $x_{i,j}$, and $|x_{i,j}|$ represents the absolute value of $x_{i,j}$. $rand_1$ and $rand_2$ are two random numbers within [0, 1].

3.2 Experience-Based Learning Strategy

When individuals don't behave well enough, they could learn from the individuals that act well. These experiences can help individuals acquire a stronger searching ability to overcome the influence of local convergence and increase the diversity of the population. In other words, the other two individuals x_k and x_l are randomly selected from the population. The potential search direction determined by them is used to update the current individual x_i as shown in Eq. 8.

$$x'_{i,j} = \begin{cases} x_{i,j} + rand * (x_{k,j} - x_{l,j}), \text{ iff}(x_k) < f(x_l) \\ x_{i,j} + rand * (x_{l,j} - x_{k,j}), \text{otherwise} \end{cases}. \quad (8)$$

Where $x_{k,j}$ and $x_{l,j}$ are the values of the jth variable for the k and l individuals ($k \neq l \neq i$), respectively. $rand \in (0, 1)$ is a random number.

3.3 Chaotic Elite Learning Method

The chaotic elite learning method can improve the quality of the solution on the basis of the existing solution. The mathematical expression for this strategy can be expressed by Eq. 9. Then, the global optimal value is calculated by using Eq. 10.

$$z_{m+1} = 4 * z_m * (1 - z_m). \quad (9)$$

$$x'_{best,j} = x_{best,j} + rand * (2 * z_m - 1). \quad (10)$$

Where m is the generation number, z_m is the value of mth chaotic iteration, and the initial value $z_0 \in (0, 1)$ is randomly selected. $x'_{best,j}$ is the updated value of jth variable for the global optimal value. The original solution will be replaced if the updated best solution has a better fitness value.

4 Results and Analysis

In this section, IJAYA algorithm is tested on parameters identification of the battery energy storage model. A continuous current discharge experiment at 16 A current is implemented, the internal parameters of the battery are identified and verified by using the measured data. The unknown model parameters are shown in Table 2, which lists the lower and upper values.

Table 2. Parameters range for SECM models.

Parameter	Lower bound	Upper bound
$R_p(\Omega)$	0.00213	0.28017
$C_p(F)$	6.32838	70.87500
$R_d(\Omega)$	0.00541	0.00865
$C_d(F)$	27.58706	225.25246
$R_0(\Omega)$	0.01268	0.01414

Table 3 lists the values of $R_p(\Omega)$, $C_p(F)$, $R_d(\Omega)$, $C_d(F)$, $R_0(\Omega)$ extracted by IJAYA algorithm under 16 A current. The RMSE value is also listed in the table. Besides, the parameter values extracted by IJAYA algorithm are substituted into the formula to calculate the output voltage value. The experimentally measured voltage value and the calculated voltage value curve as shown in Fig. 2. The experimental voltage data and simulated voltage data are in a consistent relationship in the figure. The error of voltage is within an acceptable range. The relationship between SOC and voltage is depicted in Fig. 3. It can be known that the simulated voltage could replace the experimental voltage to calculate SOC. The output voltage calculated by using the identified parameters is the same as the actual voltage, which can well correspond to the SOC. In other words, the voltage value obtained from the parameters identified by the IJAYA algorithm could be used to calculate the value of the SOC.

Table 3. Parameters range for SECM models.

Current	$R_p(\Omega)$	$C_p(F)$	$R_d(\Omega)$	$C_d(F)$	$R_0(\Omega)$	RMSE
16 A	0.00235	11.0927	0.00722	225.252	0.01298	0.01271

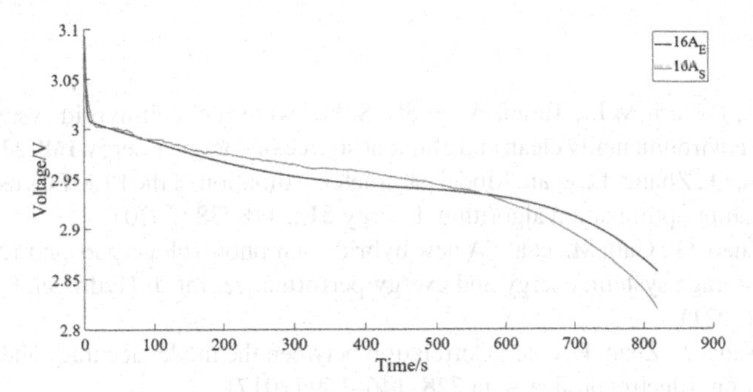

Fig. 2. Experimental and simulated voltage data comparison under 16 A current

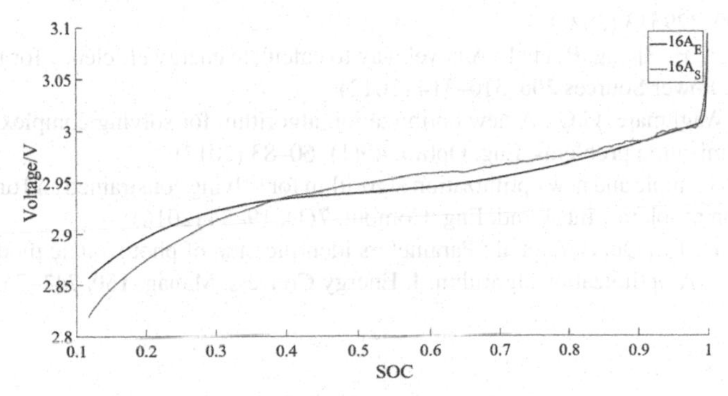

Fig. 3. $SOC - U_{OCV}$ curve diagram.

5 Conclusions

In this paper, an effective method is used to solve the battery energy storage parameters identification problem. Three strategies are adopted by the IJAYA algorithm to increase the probability that the individual obtains the global optimal solution. The second-order RC equivalent circuit model is tested and verified in the battery energy storage system. The battery parameters under 16 A discharge current are identified, and the output voltage of the battery is estimated according to the identified parameters. The estimated voltage value is compared with the experimental voltage value, the practical voltage-estimated voltage curve and SOC-voltage curve are plotted. In general, the IJAYA algorithm is an excellent method to identify unknown parameters of the battery energy storage system.

Acknowledgements. The authors are grateful for the financial support from Guangdong-Hong Kong-Macao Joint Laboratory of Human-Machine Intelligence-Synergy Systems (\#2019B121205007), the Science and Technology Innovation Commission of Shenzhen (No. ZDSYS20190902093209795), National Natural Science Foundation of China (No. U1813222), Shenzhen Science and technology project (JSGG20191118115801739), Guangdong Frontier and Key Technological Innovation (No. 2017B090910013), the Science and Technology Innovation Commission of Shenzhen (No. JCYJ20170818153048647 and No. JCYJ20180507182239617, JCYJ2018050718 223961).

References

1. Damo, U.M., Ferrari, M.L., Turan, A., et al.: Solid oxide fuel cell hybrid system: a detailed review of an environmentally clean and efficient source of energy. Energy **168**, 235–246 (2019)
2. Yang, Z., Liu, Q., Zhang, L., et al.: Model parameter estimation of the PEMFCs using improved barnacles mating optimization algorithm. Energy **212**, 118738 (2020)
3. Cheng, S., Zhao, G., Gao, M., et al.: A new hybrid solar photovoltaic/phosphoric acid fuel cell and energy storage system; energy and exergy performance. Int. J. Hydrogen Energy **46**(11), 8048–8066 (2021)
4. Wang, Q., Wang, J., Zhao, P., et al.: Correlation between the model accuracy and model-based SOC estimation. Electrochimica Acta **228**, 146–159 (2017)
5. Bruch, M., Millet, L., Kowal, J., et al.: Novel method for the parameterization of a reliable equivalent circuit model for the precise simulation of a battery cell's electric behavior. J. Power Sources **490**, 229513 (2021)
6. Kang, J., Yan, F., Zhang, P., et al.: A novel way to calculate energy efficiency for rechargeable batteries. J. Power Sources **206**, 310–314 (2012)
7. Rao, R.V., Waghmare, G.G.: A new optimization algorithm for solving complex constrained design optimization problems. Eng. Optim. **49**(1), 60–83 (2017)
8. Jaya, R.R.: A simple and new optimization algorithm for solving constrained and unconstrained optimization problems. Int. J. Ind. Eng. Comput. **7**(1), 19–34 (2016)
9. Yu, K., Liang, J.J., Qu, B.Y., et al.: Parameters identification of photovoltaic models using an improved JAYA optimization algorithm. J. Energy Convers. Manag. **150**, 742–753 (2017)

Steady-State Modeling and Power Flow Calculation of Integrated Electric-Gas Coupling Energy System Based on Energy Circuit Theory

Shuai Liu[1(✉)], Xiaoju Yin[1], Yan Zhao[1,2], and Donglai Wang[1,2]

[1] Shenyang Institute of Engineering, Shenyang 110136, Liaoning, China
[2] Key Laboratory of Regional Multi-Energy System Integration and Control of Liaoning Province, Shenyang Institute of Engineering, Shenyang 110136, Liaoning, China

Abstract. With the rapid development of P2G technology and gas turbine technology, the connection between the power system and natural gas system has become closer. Power flow calculation, as a basic application of energy network analysis, has formed mature but not uniform calculation models and methods in various energy networks. In order to promote the interdisciplinary integration of different energy networks, a new unified energy system model based on the unified energy loop theory is proposed in this paper. Particle Swarm Optimization algorithm (PSO) is used to numerically optimize the initial pressure, and the Newton method is used to solve the power flow of the electric-gas coupling integrated energy system. Finally, the feasibility and accuracy of the proposed method are verified in an electric-gas coupling system composed of the IEEE 14-node distribution system and 14-node natural gas system.

Keywords: Unified energy circuit theory · PSO-Newton algorithm · Integrated energy system modeling · Power flow calculation

1 Introduction

In recent years, a variety of energy networks represented by electricity, heat and gas are closely coupled, forming a comprehensive energy system with multi-energy optimization and complementarity, which improves the comprehensive energy utilization efficiency, and has become a hot spot and front of research [1, 2].

The planning and operation of an integrated energy system are based on the modeling and analysis of individual energy networks. Among them, the analysis of power network is based on the simplification from "field" to "circuit", and has formed a mature circuit theory, while the analysis of the heating network and the natural gas network has not formed a unified mature theory. Taking the comprehensive energy system with electric-gas coupling as an example, the steady-state natural gas flow model based on the Weymouth equation was adopted in [3, 4] for analysis, but it could not describe the dynamic process of gas network and was not suitable for the general model of the gas network. Reference [5, 6] used the "pipeline storage" model to approximate the

K. Li et al. (Eds.): LSMS 2021/ICSEE 2021, CCIS 1468, pp. 55–63, 2021.
https://doi.org/10.1007/978-981-16-7210-1_6

dynamic process of natural gas, but could not accurately describe the "partial differential equations composed of the momentum conservation equation of pipeline storage to accurately describe the dynamic process of natural gas.

Based on the above research, this paper applies the deductive methodology of power networks from "field" to "circuit" to guide the analysis of other energy networks and puts forward the intuitive unified energy road model, which has clear physical significance. A unified theory is established for the coupling analysis of heterogeneous energy networks, which helps to break the knowledge barrier between different disciplines. Based on the unified integrated energy system model, particle swarm optimization and Newton algorithm are used to calculate the power flow of the system, and two coupling modes of IEEE 14-node power distribution system and 14-node natural gas system are used to verify the feasibility of the proposed method.

2 Integrated Energy System Modeling Based on the Energy Circuit Theory of Electric-Gas Coupling

2.1 Unified Energy Approach Methodology

The methodology of unified energy circuit is derived from the deductive law of power network from "electromagnetic field" to "distributed parameter time domain circuit", then to "distributed parameter frequency domain circuit", and finally to "lumped parameter frequency domain circuit" [7, 8]. Based on the above deduction law, the gas network model can also complete the above deduction according to the unified energy path theory, and build the same model as the power grid.

2.2 Mathematical Model of Electrical Interrotation Based on Unified Energy Circuit Theory

2.2.1 P2G Device

The power to gas (P2G) refers to the conversion of electricity to natural gas. The main equations of the P2G model are described as:

$$E_{TR} = \frac{m_{LHV}}{\eta_{TR}} V_{CH_4}. \tag{1}$$

$$M_{TR} = \frac{V_{CH_4}}{V_{CH_4}^M} M_{CO_2}^M. \tag{2}$$

where V_{CH_4} and E_{TR} are respectively the volume and energy consumption of natural gas synthesized by P2G; M_{TR} is the amount of CO_2 absorption; η_{TR} is the conversion efficiency of P2G equipment; m_{LHV} is the calorific value of natural gas; $V_{CH_4}^M$ and $M_{CO_2}^M$ are the molar volume of CH_4 and the molar mass of CO_2, respectively.

2.2.2 Separable Thermoelectric Gas Turbine Model

In this paper, the separable thermoelectric unit is adopted as the key coupling element of the integrated energy system, and the combination of convex poles in the polyhedral feasible operating region is used to describe the power generation and heat generation of the separable thermoelectric unit, which is expressed as:

$$P_{i,t}^{chp} = \sum_{k=1}^{N} a_{i,t}^{k} P_{i}^{k}. \tag{3}$$

$$h_{i,t} = \sum_{k=1}^{N} a_{i,t}^{k} H_{i}^{k}. \tag{4}$$

$$\sum_{k=1}^{N} a_{i,t}^{k} = 1, 0 \leq a_{i,t}^{k} \leq 1. \tag{5}$$

where $P_{i,t}^{chp}$ is the power generation of power of combined heat and power (CHP) generation unit; $h_{i,t}$ is the heating CHP unit; P_{i}^{k} is the peak power generation within the operating area of CHP unit; H_{i}^{k} is the peak calorific value in the operating area of CHP unit; $a_{i,t}^{k}$ represents the working state of CHP unit, also known as the working point within the operating area.

2.2.3 Energy Hub Model Based on Unified Energy Circuit Theory

At the mathematical level, the multi-input multi-output function of the energy hub can be expressed as:

$$\begin{bmatrix} L_1 \\ L_2 \\ \vdots \\ L_n \end{bmatrix} = \begin{bmatrix} \eta_{11} & \eta_{12} & \cdots & \eta_{1m} \\ \eta_{21} & \eta_{22} & \cdots & \eta_{2m} \\ \vdots & \vdots & \ddots & \vdots \\ \eta_{n1} & \eta_{n2} & \cdots & \eta_{nm} \end{bmatrix} \begin{bmatrix} \upsilon_{11} & \upsilon_{12} & \cdots & \upsilon_{1m} \\ \upsilon_{21} & \upsilon_{22} & \cdots & \upsilon_{2m} \\ \vdots & \vdots & \ddots & \vdots \\ \upsilon_{n1} & \upsilon_{n2} & \cdots & \upsilon_{nm} \end{bmatrix} \begin{bmatrix} S_1 \\ S_2 \\ \vdots \\ S_n \end{bmatrix}. \tag{6}$$

where η_{ij} is the efficiency factor; υ_{ij} is the distribution factor; the Corresponding η and υ are the efficiency matrix and distribution matrix respectively.

The energy in the system is normalized as energy, energy pressure and energy flow respectively, which correspond to the gas energy, pressure and airflow in the gas path and the heat energy, temperature and heat flow in the heat path, as shown in Table 1.

Table 1. Circuit comparison of the integrated energy system.

Type	Circuit		Gas path	
Energy	Electric energy	W	Gas energy	WG
Potential	Voltage	U	Air pressure	P
Stream	Electric current	I	Gas current	G

The energy conversion relationship of the energy coupling part is described in the form of a circuit as shown in Fig. 1.

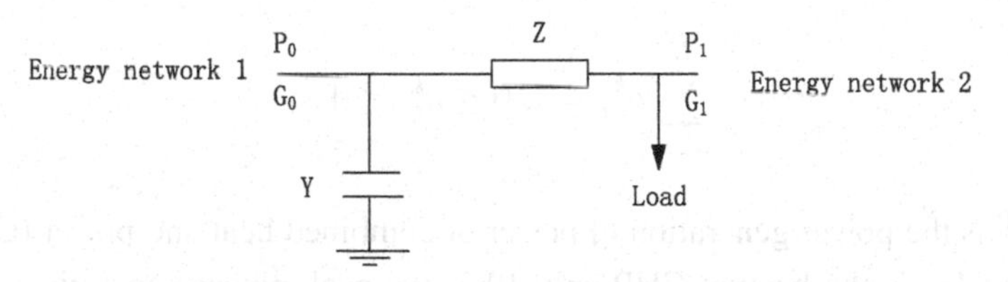

Fig. 1. Circuit comparison of the energy coupling part.

The above coupling part model has a clear physical meaning. *Z* represents the conversion effect of the "potential" transmitted in the energy network from 1 to 2. *Y* represents the conversion effect of transferring "flow" in the energy network from 1 to 2, and LOAD represents the loss generated in the two energy conversions.

2.3 Steady-State Model of Gas Path

The gas-path model based on a unified energy path can be represented by a two-port network as shown in Fig. 2.

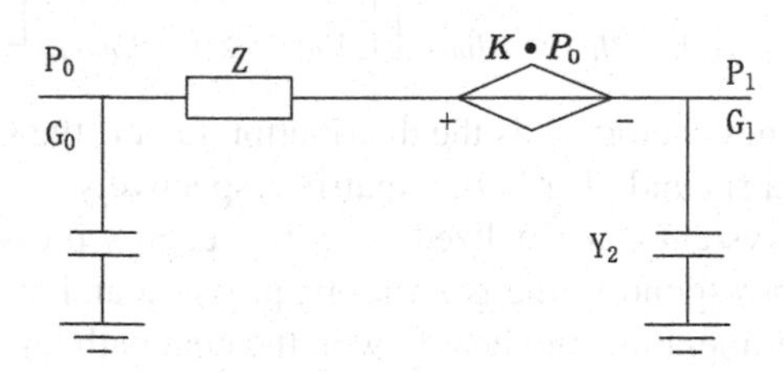

Fig. 2. Gas path model.

where *Z* and *Y* represent the equivalent gas impedance and gas admittance in the natural gas pipeline; *K* is the equivalent controlled pressure source; *P* is the air pressure, and *G* is the airflow.

2.4 Power Grid Model

The nodal voltage in the power system can be expressed in two ways: Cartesian coordinates and polar coordinates. The polar coordinate method is adopted in the paper, and its expression is as follows:

$$\dot{U}_i = U_i(\cos\delta_i + j\sin\delta_i). \quad (7)$$

The power expression of the node is:

$$P_i = U_i \sum_{j=1}^{n} U_i(G_{ij}\cos\delta_{ij} + B_{ij}\sin\delta_{ij}). \quad (8)$$

$$Q_i = U_i \sum_{j=1}^{n} U_i(G_{ij}\cos\delta_{ij} - B_{ij}\sin\delta_{ij}). \quad (9)$$

where $\dot{U}_i$ is node voltage; U_i is the voltage amplitude of nodes; P_i and Q_i are respectively the active and reactive power of the node; δ_{ij} is the phase angle difference between two nodes; G_{ij} and B_{ij} are the real and imaginary parts of the nodal admittance matrix respectively. n is the number of nodes in the power system.

3 PSO-Newton Power Flow Calculation Method

PSO algorithm is an intelligent optimization algorithm developed after being inspired by the foraging of birds. Each particle in the particle swarm updates itself by tracking and comparing the optimal value found by itself and the optimal value found by the whole population.The update equation of position and velocity of each particle is:

$$\begin{cases} v_{id} = wv_{id} + c_1r_1(\xi_{id} - x_{id}) + c_2r_2(\xi_{gd} - x_{id}) \\ x_{id,k+1} + v_{id,k} \end{cases}. \quad (10)$$

where v_{id} and x_{id} are the velocity and position of the particle respectively; w is the inertia factor, 0.6; c_1 and c_2 are learning factors, 2; ξ_{id} is the best position found for the i-th particle itself; r_1, r_2 is a random number between [0,1]; ξ_{gd} is the best location for the entire population.

In order to improve the speed of power flow calculation by Newton method, the PSO algorithm is applied to optimize the initial power flow calculation value, which is named as PSO-Newton power flow calculation method. Based on gas-to-electricity and power-to-gas technology, two coupling modes of "gas-to-electricity" and "power-to-gas" are proposed to verify the proposed method. The calculation flow of the two modes is shown in Fig. 3.

Table 2. Natural gas system load.

Nodal gas load $\times 10^6/(m^3 \cdot d^{-1})$		Nodal gas load $\times 10^6/(m^3 \cdot d^{-1})$	
2	0.28	12	2.26
3	1.13	13	0.85
5	0.57	14	0.85

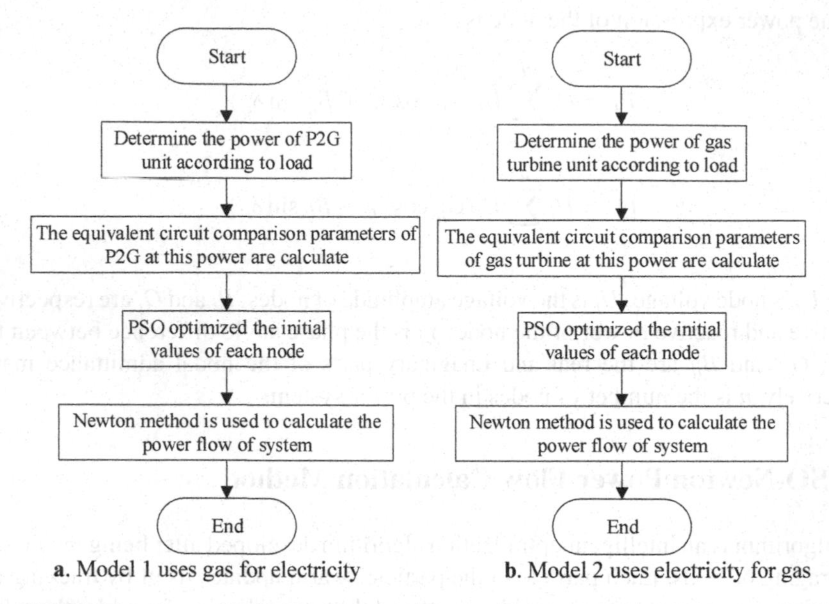

a. Model 1 uses gas for electricity **b.** Model 2 uses electricity for gas

Fig. 3. Flowchart of PSO-Newton algorithm applied to two computational modes.

4 Analysis of Calculation Examples

The method proposed in this paper has been verified on IEEE14-node distribution system and 14-node natural gas system. The gas load of the natural gas system used is shown in Table 2. The structure of the distribution system and the natural gas system is shown in Fig. 4.

In the "power-by-gas" mode, P2G equipment is used to make up for the missing natural gas load in the natural gas system, while in the "gas-by-power" mode, the gas turbine is used to make up for the missing power load in the distribution system.

In the IEEE 14-node distribution system, Node 1 is the balance node, Nodes 2, 3, 6, and 8 are PV nodes, and the rest nodes are PQ nodes. Except for coupling nodes, loads of other nodes of the two systems remain unchanged. The node voltage of the power system, node pressure of the natural gas system and pipeline flow in the two modes are shown in Fig. 5, Fig. 6 and Table 3 respectively.

In this paper, the convergence accuracy of the Newton method is 10^{-4}. In the mode of "determined by gas and electricity", the pressure initial value obtained in the PSO

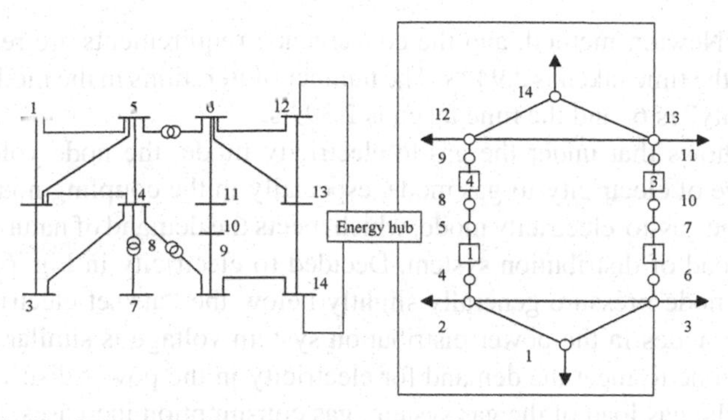

Fig. 4. Schematic diagram of power system and natural gas system.

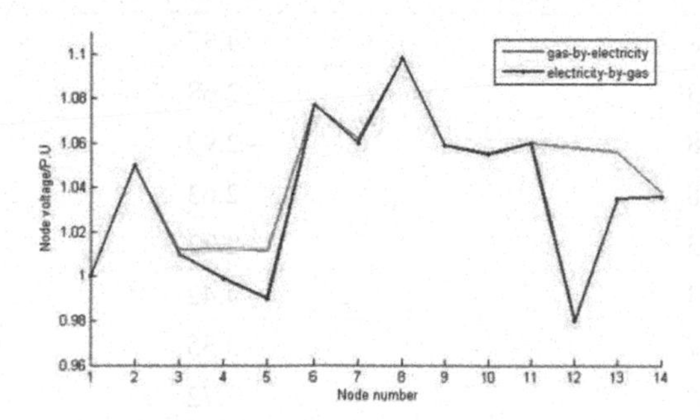

Fig. 5. Node voltage of power system.

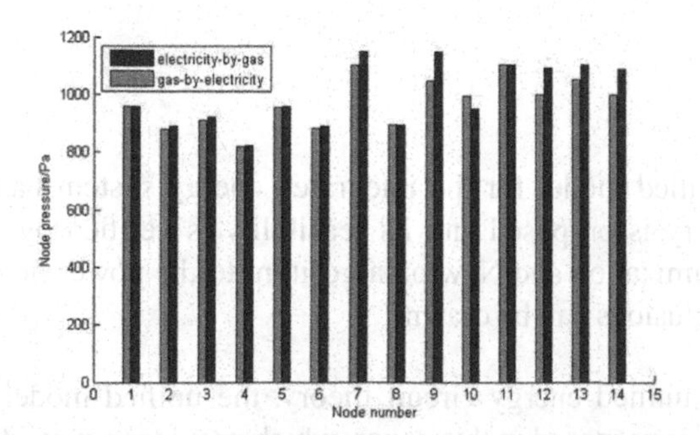

Fig. 6. Node pressure of natural gas system.

is put into the Newton method, and the convergence requirements are reached after 8 iterations, and the time taken is 2.942 s. The number of iterations in the mode of "holding gas by electricity" is 6, and the time taken is 2.399 s.

Figure 5 shows that under the gas-to-electricity mode, the node voltage is lower than the voltage of electricity-to-gas mode, especially in the coupling nodes. The main reason is that the gas-to-electricity mode, which meets the demand of natural gas system, increases the load of distribution system. Decided to electricity in Fig. 6, "gas" mode of gas system node pressure generally slightly below the "air set electric" mode, the pressure of the nodes in the power distribution system voltage is similar, "surely with electric gas" mode to meet the demand for electricity in the power distribution system that increases the gas load of the gas system, gas consumption increases, leading to the node pressure drop.

Table 3. Flow of natural gas pipeline.

Pipe	Power-to-gas $\times 10^6/(m^3 \cdot d^{-1})$	Gas-to-power $\times 10^6/(m^3 \cdot d^{-1})$
1–2	4.18	4.37
1–3	2.43	2.68
2–3	−2.98	−2.99
2–4	2.68	2.63
3–6	2.83	2.77
5–8	3.03	−4.42
7–10	4.77	3.35
9–12	4.15	3.72
11–13	1.68	3.67
12–13	−1.66	2.45
12–14	0.74	0.71
13–14	1.70	2.34

5 Conclusion

In this paper, a unified model for the integrated energy system based on the unified energy circuit theory is proposed and its feasibility is verified by the application of particle swarm optimization and Newton algorithm to the power flow calculation. The following two conclusions can be drawn:

(1) Based on the unified energy circuit theory, the unified model of the integrated energy system is proposed in this paper, which provides a new idea for the analysis of the integrated energy system. The energy hub model is constructed in the form of a generalized circuit model. The experimental results show that this method is feasible.

(2) PSO-Newton algorithm is used to analyze the example. An example is given to verify the effectiveness of the proposed integrated energy system model in a small electricity-gas coupled integrated energy system. Compared with the traditional method, the proposed model has a faster calculation speed and better convergence.

Acknowledgments. This work was supported by Liaoning PhD Initial Scientific Research Fund (2020-BS-179), Liaoning Revitalization Talents Program (XLYC1907138), the Key R&D Program of Liaoning Province (2018220017, 2020JH2/10300101), the Natural Science Foundation of Liaoning Province (2019-MS-239), the Scientific Research Fund of Liaoning Provincial Education Department (JL-1901, JL-2021), the Technology Innovation Talent Fund of Shenyang (RC190360, RC200252) and Liaoning BaiQianWan Talents Program.

References

1. Li, Q., Zhao, N.: Probabilistic power flow calculation based on importance-hammersley sampling with eigen-decomposition. Int. J. Electric. Power Energy Syst. **130**, 106947 (2021)
2. Mongkoldee, K., Kulworawanichpong, T.: Optimal sizing of ac railway power conditioner in autotransformer-fed railway power supply system. Int. J. Electric. Power Energy Syst. **127**, 106628 (2021)
3. Wu, T., Han, J., Cai, C., Fan, A.J.: Research on load transfer based on power supply unit of distribution network. In: E3S Web of Conferences, vol. 233, p. 01001 (2021)
4. Wei, Q.L., Zheng, J.H., Tang, W.H., Wu, Q.H.: Equivalent model of active distribution network considering uncertainties of wind turbines, photovoltaics and loads. IOP Conf. Ser. Earth Environ. Sci. **645**, 012086 (2021)
5. Sun, L.J., Su, Z.X., Tian, M.X., Xing, D.F., Pan, C.L.: Power flow calculation method of traction network considering geographical and climatic conditions. Energy Rep. **6**, 228–235 (2020)
6. Li, H., Li, H.J., Lu, W.H., Wang, Z.H., Bian, J.: Optimal power flow calculation considering large-scale photovoltaic generation correlation. Front. Energy Res. **8**, 590418 (2020)
7. Wang, Y.A.: Research on modeling of new energy power plant interface characteristics based on computer power flow calculation mode. J. Phys. Conf. Ser. **1650**, 032046 (2020)
8. Wang, L., Zhang, C.: Load stability analysis of kazakhstan photovoltaic power system based on data visualization. J. Phys. Conf. Ser. **1634**, 012094 (2020)

Study on the Prediction of Indoor Temperature and Thermal Load on Different Floors in a Community

Lidong Zhang[1], Jiao Li[1], Heng Zhang[1], Tianyu Hu[1], Yuanjun Guo[2(✉)], Zhile Yang[2], and Xiandong Xu[3]

[1] Engineering Research Centre of Oil Shale Comprehensive Utilization, Ministry of Education, Northeast Electric Power University, Jilin 132012, China

[2] Shenzhen Institute of Advanced Technology, CAS, Shenzhen 518055, China
yj.guo@siat.ac.cn

[3] Key Laboratory of Smart Energy and Information Technology of Tianjin Municipality, Tianjin 300072, China

Abstract. In order to meet the required heat supply of each household, save energy, as well as avoid wasting excessive heat, it is necessary to accurately predict the thermal load (*TL*) in a community. In this paper, we focus on cross-floor prediction over a week for three thermal users located on different floors. Three machine learning methods including Back Propagation Neural Network (BPNN), Extreme Learning Machine (ELM), and Support Vector Machines (SVM) are applied for indoor temperature (T_{in}) and thermal load prediction. The experimental results show that Extreme Learning Machine outperforms the other two methods in terms of thermal load prediction.

Keywords: Machine learning · Prediction · Indoor temperature · Thermal load

1 Introduction

In recent years, centralized heating is widely used in residential buildings with a fast expanded scale [1]. In a centralized thermal system, Thermal load is a major basis for the design of the thermal management system and the planning of urban thermal supply. For thermal users, thermal demands management is necessary to meet the thermal comfort of residents. For the whole thermal system, it is an important task to improve the efficiency of energy use, and to reduce pollutant emissions [2]. Considering household energy consumption problem, it is necessary for the thermal system to forecast thermal load based on outdoor environmental parameters and historical operating data.

The thermal supply is unbalanced due to the different flow rates at the proximal end and the distal end. The heating imbalance will be worse in the operation of the secondary heating networks [3]. If all thermal users reach the appropriate indoor temperature simply by changing the flow rate of the secondary network pipeline, the overall thermal load will be too high and operating costs will be significantly increased. thermal metering

K. Li et al. (Eds.): LSMS 2021/ICSEE 2021, CCIS 1468, pp. 64–75, 2021.
https://doi.org/10.1007/978-981-16-7210-1_7

devices can be installed at thermal exchange stations and thermal consumers' side to monitor thermal supply data [4]. Prediction of thermal load based on monitoring data and outdoor meteorological parameters can be realized. Therefore, unbalanced heating temperature can be avoided in order to save more energy and reduce cost.

To improve thermal efficiency, different technologies have been successfully developed and applied to predict indoor temperature, including artificial neural networks [5], online learning algorithm [6], long short-term memory [7], and so forth. Thermal load prediction has been mentioned in many literatures. Wang [8] used seven shallow learning, two deep learning and three heuristics to predict the thermal load of residential buildings, comparing mainly Shallow machine learning and deep learning. In addition, machine learning method is widely used in thermal load prediction, for example, Multiple Linear Regression [9], support vector machines prediction models with discrete wavelet transform algorithm [10], grey-box and black-box models [11]. Justus [12] has developed a new topology analysis tool that plays an important role in the application of the fifth-generation central thermal networks to optimize the use of low temperature wasted thermal and minimize pollutant gas emissions Song [13] proposed a hybrid spatiotemporal prediction algorithm based on convolution neural net-work long–short term memory for the thermal load forecasting problem. And a time series model introduced by Oluwaseyi [14] based on the physical resistance-capacitance model was used to predict the thermal load of residential buildings.

In the previous studies, indoor temperature and thermal load were predicted only according to the outdoor meteorological parameters the following improvements are made in this study:

1) The indoor temperature of thermal users in the district and the overall thermal demand of the district are collected to calculate the thermal consumption of a single household.
2) A correlation analysis on the meteorological factors affecting the thermal load is made based on thermal system operation parameters and outdoor meteorological parameters.
3) Indoor temperature of typical users on different floors are chosen to predict thermal load with higher precision.

The rest of the paper is organized as follows: Part 2 provides an analysis of typical community users. Part 3 mainly introduces the prediction method used in this paper. Part 4 describes the prediction results of the model, and Part 5 is the conclusion.

Nomenclature

BPNN	Back Propagation Neural Network	T_{sr}	second return water temperature[°C]
ELM	Extreme Learning Machine	T_{avin}	average indoor temperature[°C]
SVM	Support Vector Machines	T_{in}	indoor temperature[°C]
RMSE	Root-mean-square error	T_{out}	outdoor temperature[°C]
MAE	Mean absolute error	$T_{A\text{-}1}$	A-1 indoor temperature[°C]

(continued)

(*continued*)

Nomenclature			
MAPE	Mean absolute percentage error	$T_{A\text{-}2}$	A-2 indoor temperature[°C]
R^2	R-Square	$T_{A\text{-}6}$	A-6 indoor temperature[°C]
T_{pr}	Primary return water temperature[°C]	$TL_{A\text{-}1}$	A-1 thermal load[MW]
F	Primary back water flow[t/h]	$TL_{A\text{-}2}$	A-2 thermal load[MW]
T_{ps}	Primary supply water temperature[°C]	$TL_{A\text{-}6}$	A-6 thermal load[MW]
T_{ss}	Second supply water temperature[°C]	TL	Thermal load[MW]

2 Typical User Analysis of the Community

For a community, it is difficult to ensure that the T_{in} of every household in the area is in a comfortable state. In this paper, the room temperature of all users with T_{in} monitoring instruments is collected, and Fig. 1 shows the distribution map of residents in the district and the location of typical users in this prediction is marked.

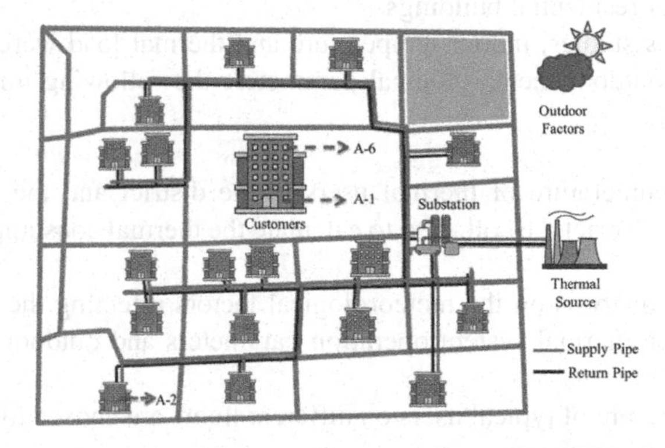

Fig. 1. Graph model of community distribution.

Figure 2 shows the T_{avin} that can be collected from the neighborhood occupants, including those on the first, second, third and fourth floors as well as the sixth floor. On one hand, for residential blocks with thermal exchange stations, it is considered that the blocks closer to the thermal exchange stations get the lower water temperature losses and lower resistance along the pipeline, so the thermal effect is better. The farther away from the thermal exchange station, the more thermal will be dissipated in the transmission process of the pipe network. In the thermal system, considerations from all aspects should be done in order to make the whole district thermal users satisfied. Especially for the distant users, the thermal company needs to pay more attention and make a success of advance to meet the thermal demand.

On the other hand, the room temperature of thermal users located in the same building varies greatly from lower floors to higher floors. For example, the room temperature of

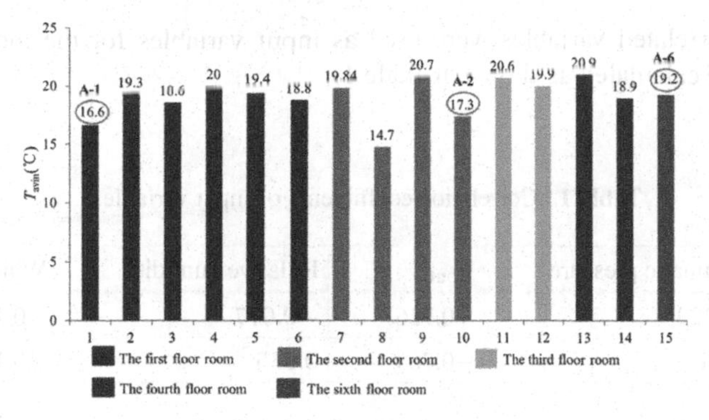

Fig. 2. T_{avin} in different rooms.

thermal users in the middle floors will be influenced by the thermal dissipated from the surrounding neighbors to reach the appropriate T_{in}, but thermal users in the top or bottom floors will lose thermal and cause the T_{in} to be substandard. Considering the above two aspects, A-1, A-2 and A-6, which are the farthest from the thermal exchange station, are selected as typical users for predicting the *TL*, to provide a basis for the regulation data of the thermal exchange station.

3 Methods

3.1 Correlation Analysis of Input Variables

Based on the pre-processed data set, the Pearson correlation coefficient analysis method was introduced to analyze each influencing factor with respect to *TL* before determining the input variables for the model. The Pearson correlation coefficient is defined as follows.

$$\gamma_{X,Y} = \frac{\sum_{i=1}^{n}(X_i - \overline{X})(Y_i - \overline{Y})}{\sqrt{\sum_{i=1}^{n}(X_i - \overline{X})^2 \sum_{i=1}^{n}(Y_i - \overline{Y})^2}}. \quad (1)$$

Where, n is the sample length, X, Y are random variables $\overline{X}, \overline{Y}$ are the average of the sequences X_i, Y_i.

- Firstly, the possible input influencing factors are selected based on the analyses of *TL* influence by thermal system characteristics. Correlation analysis of environmental factors including atmospheric pressure, relative humidity T_{out}, relative humidity and wind.
- For these factors Pearson's formula is used to calculate the correlation of possible input variables with T_{in} and *TL*. The strength of correlation of the variables is judged by the range of values taken, when $|\gamma_{X,Y}| \in [0, 0.09)$, no correlation between possible input variables and predicted values, when $|\gamma_{X,Y}| \in [0.09, 0.5)$, Consider a weak correlation between possible input variables and predicted values, when $|\gamma_{X,Y}| \in [0.5, 1]$, it is possible that there is a strong correlation between the input variables and the predicted values.

- Strongly correlated variables were used as input variables for the model, and the results were calculated as shown in Table 1.

Table 1. Correlation coefficients of input variables.

	Pneumatic pressure	T_{out}	Relative humidity	Wind Power
T_{in}	−0.229	0.726	−0.017	0.227
TL	0.151	−0.768	0.035	−0.149

According to the analysis, it is found that T_{out} is strongly correlated with T_{in} and TL. Therefore, T_{out} and system data are used as input factors to the model. The T_{in} and TL will be predicted seven days in advance.

3.2 Methods

In this section, three neural networks are used for T_{in} and TL prediction, mainly BPNN, ELM and SVM, and three algorithms are compared.

BPNN: BPNN is a machine learning algorithm commonly used for processing and analyzing data, it learns how to map the input data to the appropriate output with strong generalization ability [15]. It is universally applicable to all data, and given an input there is a corresponding output. The intermediate process from the input layer to the result of the output layer are implemented by an activation function, and the number of neurons in the hidden layer is determined by an iterative method.

ELM: ELM is a new single hidden layer Feed Forward Neural Network learning algorithm[16]. ELM consists of three parts: the input layer, the implicit layer, and the output layer. There are n input layers, there is an n input variables $(x_1,\ldots,x_n)$, m output layers, there is an m output variables $(y_1,\ldots,y_m)$, w_{ij} is the connection weight between the i-th neuron in the input layer and the j-th neuron in the hidden layer. Generally, ELM is inseparable from the number of nodes in the hidden layer. Through the research tests on the number of nodes, it is concluded that 11 nodes is the optimal number of nodes in the hidden layer, and then the activation function g(x) of the hidden neuron is set [17]. In the whole training process, the model does not need to iterate and connect the input weights and hidden thresholds, and the output layer weights are solved without iterative computation. The training speed of such a model is significantly higher.

SVM: SVM is one of the soft computing learning algorithms, which is primarily used in pattern recognition, forecasting, classification, and regression analysis [10]. When SVM is applied to regression fitting analysis, the basic idea is no longer to find an optimal classification surface that separates the two classes of samples, but to find an optimal classification surface that minimizes the error of all training samples from that optimal classification surface. Universally, suppose the training set sample pair containing l training samples is $\{(x_i, y_i), i = 1, 2, \ldots, l\}$, in which, $x_i(x_i \epsilon R^d)$ is the input column vector of the i-th training sample.

In this paper, the first 1032 sets of data are selected as training samples, and the last 168 groups of data are used as test sets, and the prediction model with satisfactory accuracy is finally obtained after repeated training.

In order to better and more intuitively reflect the fitting level of different methods, RMSE, MAE, MAPE and R^2 are evaluated for the accuracy of the prediction model.

1) RMSE

$$RMSE = \sqrt{\frac{1}{n}\sum_{i=1}^{n}\left(y_i - y_i^{'}\right)^2}. \quad (2)$$

2) MAE

$$MAE = \frac{1}{n}\sum_{i=1}^{n}\left|y_i - y_i^{'}\right|. \quad (3)$$

3) MAPE

$$MAPE = \frac{\sum_{i=1}^{n}\left|\frac{y_i - y_i^{'}}{y^{'}}\right|}{n}. \quad (4)$$

4) R^2

$$R^2 = \frac{\left[\sum_{i=1}^{n}\left(y_i^{'} - \overline{y_i^{'}}\right)\cdot\left(y_i - \overline{y_i}\right)\right]^2}{\sum_{i=1}^{n}\left(y_i^{'} - \overline{y_i^{'}}\right)\cdot\left(y_i - \overline{y_i}\right)}. \quad (5)$$

Where y_i is the predicted output variable, $y_i^{'}$ is the actual output variable corresponding to the i-th variable in the database, and n is the number of samples in the training subset. Figure 3 gives the detailed process steps of this article.

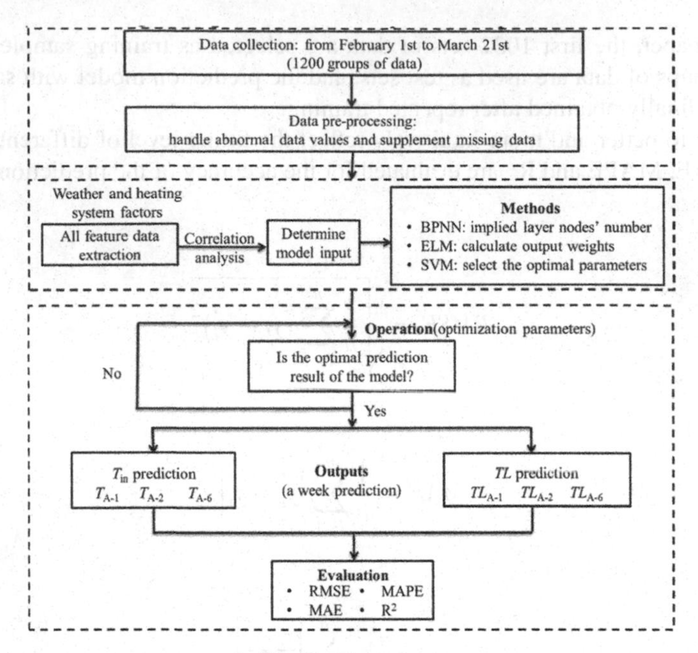

Fig. 3. Flow chart.

4 Analysis of Results and Discussion

4.1 Thermal Demand

Jilin Province is located in the northeastern part of China, the thermal period lasts from mid-October to the end of March next year. At the end of the thermal season, in early spring, the temperature starts to rise but is often unstable. In order to ensure that customers remain comfortable, thermal companies need to monitor thermal parameters to provide heat required by thermal users. Therefore, we select 50 days of heating data from February to the end of March, these data are in one-hour steps, with a total of 1200 data sets. Figure 4 shows the final selected raw data of the T_{avin} and the *TL*.

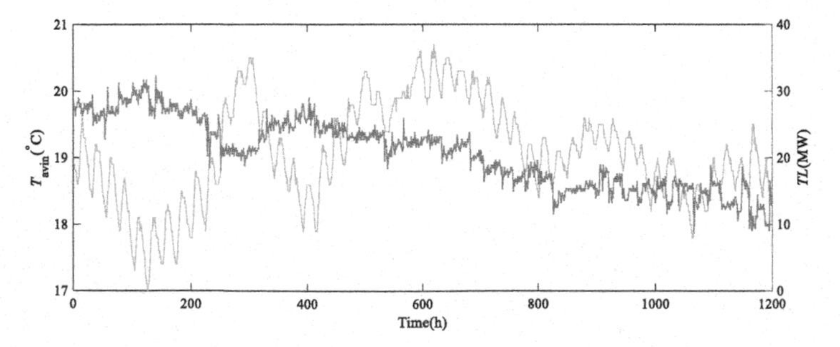

Fig. 4. T_{avin} and *TL* data.

4.2 $T_{A\text{-}1}$,$T_{A\text{-}2}$,$T_{A\text{-}6}$ Predictive Analysis

The T_{in} of a single-family home is a measure of the degree of heating merit and comfort for consumers, which inevitably leads to the prediction of specific single-family T_{in}. The characteristic parameters are given in Table 2. And the scatter plot of A-1, A-2 and A-6 is shown in Fig. 5.

Table 2. Description of characteristic parameters of T_{in} prediction.

Name	Object	Variable
T_{in} prediction	$T_{A\text{-}1}$	$T_{A\text{-}1}$, T_{out}, T_{ss}, T_{sr}
	$T_{A\text{-}2}$	$T_{A\text{-}2}$, T_{out}, T_{ss}, T_{sr}
	$T_{A\text{-}6}$	$T_{A\text{-}6}$, T_{out}, T_{ss}, T_{sr}

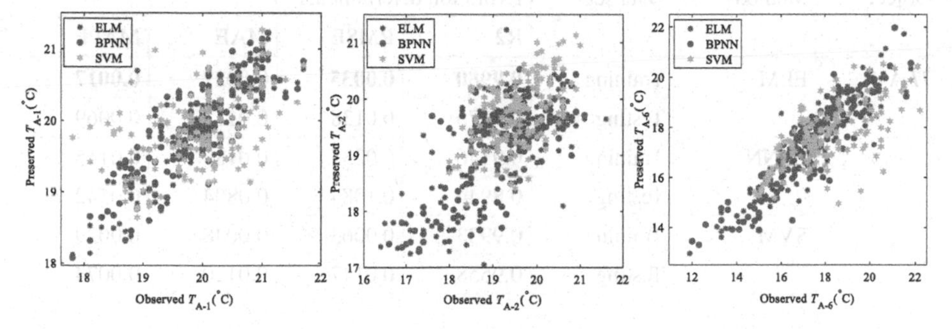

Fig. 5. T_{in} prediction of typical users.

The scatter represents the fitting effect of Tin, and when the scatter distribution is on the diagonal indicates the stronger correlation between the data of the two axes, as shown in Fig. 5. ELM is the best prediction method among the three models, and the scatter distribution is closest to the diagonal shape. Some points of BPNN and SVM are more scattered.

4.3 *TL* Prediction

This section mainly describes the *TL* prediction. The characteristic parameters are given in Table 3.

For the *TL* prediction, it is necessary to adapt to different types of buildings, residential buildings in different locations and different floors. Thermal consumption between the households and the floors of each residential building are also different. *TL* predict model should be established for single-family thermal users. The location of T_{in} prediction is the same as that of a single household. The historical data of *TL* of typical users combined with the influencing factors for one week *TL* prediction have a practical significance of saving energy and reducing consumption.

Table 3. Description of characteristic parameters of *TL* prediction.

Name	Object	Variable
TL prediction	$TL_{A\text{-}1}$	T_{in}, T_{out}, T_{ss}, T_{sr}, F, T_{pr}, T_{ss}, T_{ps}, $TL_{A\text{-}1}$
	$TL_{A\text{-}2}$	T_{in}, T_{out}, T_{ss}, T_{sr}, F, T_{pr}, T_{ss}, T_{ps}, $TL_{A\text{-}2}$
	$TL_{A\text{-}6}$	T_{in}, T_{out}, T_{ss}, T_{sr}, F, T_{pr}, T_{ss}, T_{ps}, $TL_{A\text{-}6}$

According to the indoor and outdoor data collected by each household and the total *TL* data, the least square method is utilized to calculate the *TL* of typical users. The results are displayed in the table below.

Table 4. $TL_{A\text{-}1}$ prediction results.

Object	Methods	Data set	Evolution determinants			
			R2	RMSE	MAE	MAPE
*TL*A-1	ELM	Training	**0.9980**	**0.0035**	**0.0027**	**0.0017**
		Testing	0.9875	0.0138	0.0112	0.0069
	BPNN	Training	0.9146	0.0235	0.0187	0.0115
		Testing	0.8944	0.1084	0.0884	0.0542
	SVM	Training	0.9933	0.0063	0.0048	0.0029
		Testing	0.9558	0.0147	0.0120	0.0074

Table 5. $TL_{A\text{-}2}$ prediction results.

Object	Methods	Data set	Evolution determinants			
			R2	RMSE	MAE	MAPE
*TL*A-2	ELM	Training	**0.9989**	**0.0088**	**0.0057**	**0.0033**
		Testing	0.9978	0.0223	0.0167	0.0159
	BPNN	Training	0.9896	0.0173	0.0143	0.0083
		Testing	0.9639	0.0977	0.0699	0.0674
	SVM	Training	0.9840	0.0588	0.0380	0.0247
		Testing	0.9818	0.0655	0.0455	0.0429

Table 6. $TL_{A\text{-}6}$ prediction results.

Object	Methods	Data Set	Evolution determinants			
			R2	RMSE	MAE	MAPE
$TL_{A\text{-}6}$	ELM	Training	**0.9993**	**0.0059**	**0.0039**	**0.0021**
		Testing	0.9991	0.0238	0.0168	0.0151
	BPNN	Training	0.9964	0.0305	0.0247	0.0124
		Testing	0.9947	0.0491	0.0395	0.0338
	SVM	Training	0.9985	0.0111	0.0088	0.0047
		Testing	0.9834	0.0702	0.0508	0.0414

Table 4, Table 5 and Table 6 give the *TL* prediction results of three rooms respectively. As far as prediction accuracy is concerned, ELM is slightly better than the other two methods. It can be seen from the above table that the RMSE vary from 0.0035–0059. The smaller the RMSE, the higher the prediction accuracy. It shows that the prediction accuracy is within a tolerable range. This prediction accuracy shows that the heat supply can be adjusted in advance by predicting the thermal load of the room, further, achieving the purpose of heating on demand.

It can be seen from Fig. 6 that among the three methods selected in this paper, ELM performs the best, The actual load value almost coincides with the predict load value, and BPNN and SVM show certain fluctuations. So, the heating parameters can be adjusted in advance based on ELM load prediction.

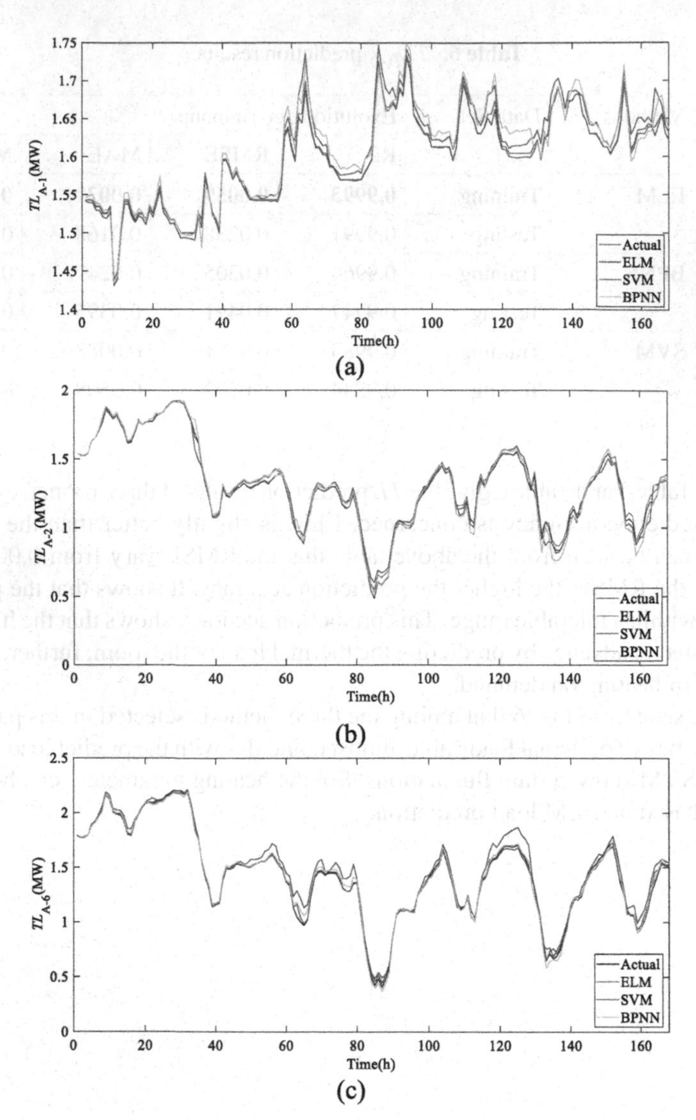

Fig. 6. (a), (b) and (c) compare $TL_{A\text{-}1}$, $TL_{A\text{-}2}$ and $TL_{A\text{-}6}$ prediction results by different models.

5 Conclusion

In this paper, from the overall analysis of the residential area, multiple floors are selected to predict the T_{in} and *TL*. With the aim to make the T_{in} of households in the district reach the standard and improve the comfort level. In addition, the thermal company can supply thermal on demand and avoid wasting. A-1, A-2 and A-6 are selected as typical thermal users for prediction. Three common machine learning algorithms are chosen for prediction. The prediction errors of the three methods are within acceptable limits. Further, ELM showed the best performance.

References

1. Zhang, L., Li, Y., Zhang, H., Xu, X., Yang, Z., Xu, W.: A review of the potential of district heating system in Northern China. Appl. Thermal Eng. **188**, 116605 (2021)
2. Maljkovic, D., Basic, B.D.: Determination of influential parameters for heat consumption in district heating systems using machine learning. Energy **201**, 117585 (2020)
3. Huang, K., Feng, G., Li, H., Yu, S.: Opening window issue of residential buildings in winter in north China: a case study in Shenyang. Energy Build. **84**, 567–574 (2014)
4. Bai, W., Liang, Y., Sun, Y., Zhu, Y., Sun, S., Zhao, Y., Yang, H.: Application of heat metering data in balancing regulation of secondary network. District Heat. **2**, 9–13 (2020). (In Chinese)
5. Li, X., Han, Z., Zhao, T., Zhang, J., Xue, D.: Modeling for indoor temperature prediction based on time-delay and Elman neural network in air conditioning system. J. Build. Eng. **33**, 101854 (2021)
6. Zamora-Martínez, F., Romeu, P., Botella-Rocamora, P., Pardo, J.: On-line learning of indoor temperature forecasting models towards energy efficiency. Energy Build. **83**, 162–172 (2014)
7. Fang, Z., Crimier, N., Scanu, L., Midelet, A., Alyafi, A., Delinchant, B.: Multi-zone indoor temperature prediction with LSTM-based sequence to sequence model. Energy Build. **245**, 111053 (2021)
8. Wang, Z., Hong, T., Piette, M.A.: Building thermal load prediction through shallow machine learning and deep learning. Appl. Energy **263**, 114683 (2020)
9. Geysen, D., De Somer, O., Johansson, C., Brage, J., Vanhoudt, D.: Operational thermal load forecasting in district heating networks using machine learning and expert advice. Energy Build. **162**, 144–153 (2018)
10. Al-Shammari, E.T., et al.: Prediction of heat load in district heating systems by Support Vector Machine with Firefly searching algorithm. Energy **95**, 266–273 (2016)
11. Afram, A., Janabi-Sharifi, F.: Black-box modeling of residential HVAC system and comparison of gray-box and black-box modeling methods. Energy Build. **94**, 121–149 (2015)
12. Von, R.J., Henze, G.P., Long, N., Fu, Y.: Development of a topology analysis tool for fifth-generation district heating and cooling networks. Energy Convers. Manag. **196**, 705–716 (2019)
13. Song, J., Zhang, L., Xue, G., Ma, Y., Gao, S., Jiang, Q.: Predicting hourly heating load in a district heating system based on a hybrid CNN-LSTM model. Energy Build. **243**, 110998 (2021)
14. Ogunsola, O.T., Song, L., Wang, G.: Development and validation of a time-series model for real-time thermal load estimation. Energy Build. **76**, 440–449 (2014)
15. Vikas, V.K., Ashu, J.: Knowledge extraction from trained ANN drought classification model. J. Hydrol. **585**, 124804 (2020)
16. Naji, S., et al.: Estimating building energy consumption using extreme learning machine method. Energy **97**, 506–516 (2016)
17. Kumar, S., Pal, S.K., Pal, R.: SinghIntelligent energy conservation: indoor temperature forecasting with extreme learning machine. Intell. Syst. Technol. Appl. **530**, 977–988 (2016)

Multi-objective Optimization of Electric Vehicle and Unit Commitment Considering Users Satisfaction: An Improved MOEA/D Algorithm

Ping Shao[1], Zhile Yang[2(✉)], Xiaodong Zhu[1], and Shihao Zhao[1]

[1] School of Electrical Engineering, Zhengzhou University, Zhengzhou, China
[2] Shenzhen Institute of Advanced Technology, Chinese Academy of Sciences, Shenzhen, China
zl.yang@siat.ac.cn

Abstract. In this paper, unit commitment and the scheduling problem of plug-in electric vehicles (PEV) under the condition of power grid feedback are jointly considered, formulating a many objective problem. The objectives involve economic and environmental impact, carbon emission as well as the user satisfaction. To solve this multi-objective problem, an improved MOEA/D algorithm is proposed where levy flight is integrated to improve the algorithm performance. The improved MOEA/D algorithm, NSGA-II algorithm and the traditional MOEA/D algorithm are compared. The experimental results show that the improved MOEA/D algorithm is comprehensive optimal in the multi-objective function, which shows the effectiveness of the improved algorithm. Moreover, the four objectives are comprehensively considered to balance the multiple factors in the application.

Keywords: PEV · MOEA/D · Unit commitment · Multi-objective optimization

1 Introduction

With the popularization of electric vehicles and the rapid development of V2G technology [1], the research of intelligent coordinated power generation has become a hot issue. However, the PEV charge is random and intermittent [2], the economy and safety of the power grid will be affected to a certain extent after electric vehicles are incorporated into the power grid [3]. Therefore, it is of profound significance to combine electric vehicles with traditional units to participate in scheduling, which can achieve reasonable operation of power grid [4]. Unit commitment problems refers to the unit switch state and the output power for scheduling object, in a series of constraint conditions, to achieve the economic target, the optimal environmental objectives [5]. UC problem is considered as a high-dimensional, non-convex, discrete, nonlinear mixed integer optimization problem due to its typical complexity and constraint [6]. When the charging deviation function target is added, the super multi - target problem is formed.

Traditional methods to solve this kind of super-multi-objective problems include vector evaluation genetic algorithm (VEGA), non-dominated sorting genetic algorithm

K. Li et al. (Eds.): LSMS 2021/ICSEE 2021, CCIS 1468, pp. 76–85, 2021.
https://doi.org/10.1007/978-981-16-7210-1_8

(NSGA-II) [7], decomposition-based multi-objective optimization method (MOEA/D), etc. In recent years, many novel algorithms have been proposed. For example, in reference [8], a multi-objective adaptive differential evolution (MOSADE) algorithm is proposed. The external elitist archive method and asymptotic comparison algorithm are used to optimize the fuel consumption and emissions. The Benders Decomposition Technology optimization model is adopted in literature [9], and a multi-objective algorithm framework is proposed to minimizes the total operating cost and emissions of the system. A multi-objective dynamic economic dispatch (MODEED) model with demand side management for electric vehicles is proposed in literature [10], The multi-objective particle swarm optimization (MOPSO) algorithm is used to minimize the dynamic economic cost and emissions of the system. In reference [11], a real-time energy management system is proposed, which uses binary particle swarm optimization (BPSO) algorithm to minimize energy cost function and CO2 emission function. In reference [12], a dynamic non-dominated sorting multi-objective biogeography optimization algorithm (Dy-NSBBO) is proposed. Non-dominated sorting technology, change detection and memory selection strategy is adopted to significantly improve the economic and environmental benefits of the system. The ε- constraint method is adopted in reference [13] to generate the Pareto optimal solution, which reduces the emission and cost by about 10% and 6% respectively. In reference [14], the water cycle algorithm and gravity search algorithm are used to minimize the generation cost and emission cost. Although these algorithms play a certain role in economy and environment, whether the users are satisfied with the scheduling results is not taken into account.

Aiming at the multi-objective optimization problem of unit commitment under grid feedback [9], an improved MOEA/D algorithm is proposed in this paper. Compared with the traditional MOEA/D algorithm, improvements have been made in the decomposition method and solution update method, and a method for selecting the current optimal solution is also proposed. Comparing the improved MOEA/D algorithm with the traditional MOEA/D algorithm and the NSGA-II algorithm under the conditions of 10, 50, and 90 units respectively, the performance of IMOEA/D is comprehensively optimal, which also proves the validity of the proposed method.

The rest of the thesis is arranged as follows: The second section introduces the problem model under the grid feedback strategy. The third section proposes an improved MOEA/D algorithm. The experimental results and analysis are given in the fourth section. Finally, the fifth part is a summary of this paper.

2 Problem Model

The model takes the charging charge of PEV, switching state and output power of the unit as scheduling objects [10]. Considering that economic goals alone cannot effectively meet the needs of society, targets such as carbon dioxide, sulfur compounds and charging deviation function are also taken into account [11]. The specific mathematical model is as follows.

2.1 The Objective Function

Fobject 1 (F1) is the power generation cost function:

$$TPC_{Tn} = \min \sum_{t=1}^{T} \sum_{j=1}^{n} [F_j(P_{j,t})u_{j,t} + SU_{j,t}(1 - u_{j,t-1})u_{j,t}]. \quad (1)$$

The economic cost includes the fuel cost and the unit start-up cost, $u_{j,t}$ is the binary decision variable of the state of the jth unit at the t hour. $F_j(P_{j,,t})$ is the coal cost of the jth unit, $P_{j,t}$ is the generating amount of the unit, And $SU_{j,t}$ is the start-up and cost of the jth unit within the period of t.

Fobject 2 (F2) is the CO2 emission function:

$$C_{\text{emi}} = \min \sum_{t=1}^{T} \sum_{j=1}^{n} \left[F_{j,t}^{emi}(P_{j,t})u_{j,t} \right]. \quad (2)$$

$F_{j,t}^{emi}(P_{j,t})$ represents 24 h total CO2 emissions, $u_{j,t}$ is the binary decision variable of the state of the jth unit at the t hour and $P_{j,t}$ is the generating amount of the unit.

Fobject 3 (F3) is the sulfur emission function:

$$C_{\text{ce}} = \min \sum_{t=1}^{T} \sum_{j=1}^{n} \left[F_{j,t}^{ce}(P_{j,t})u_{j,t} \right]. \quad (3)$$

C_{ce} represents the total sulfur emission of all units within 24 h, $u_{j,t}$ is the binary decision variable of the state of the jth unit at the t hour and $P_{j,t}$ is the generating amount of the unit.

Fobject 1 (F4) is the charging deviation function:

$$DC_{\text{PEV}} = \sum_{t=1}^{T} \left[P_{PEVdispatch,t} - P_{PEVpredict,t} \right]. \quad (4)$$

The charging load under the completely unconstrained condition is represented by $P_{PEVpredict,t}$, and $P_{PEVdispatch,t}$ represents the PEV charging load after optimized scheduling. The difference between the two has the same effect as charging satisfaction.

2.2 Limitations of the Model

Power balance constraints:

$$\sum_{j=1}^{n} P_{j,t}u_{j,t} = P_{D,t} + P_{PEVdispatch,t}. \quad (5)$$

$P_{PEVdispatch,t}$ represents the sum of PEV charges that the system needs to schedule under the condition of charge and discharge, $P_{D,t}$ represents the load demand of the system. This constraint means that the output power of the system is the same as the demand load of the system to achieve the purpose of supply and demand balance.

Rotation standby constraint:

$$\sum_{j=1}^{n} P_{j,\max} u_{j,t} \geq P_{D,t} + SR_t + P_{PEVdispatch,t}. \tag{6}$$

SR_t is the rotating reserve in time t, and $P_{j,max}$ is the maximum output of the j unit. There is a proportional relationship between the rotating reserve and the load of the system. In this paper, the proportional value is set as 0.1.

Power constraints of electric vehicles:

$$\sum_{t=1}^{T} P_{PEVdispatch,t} = P_{PEV,total}. \tag{7}$$

$P_{PEV,total}$ is the total charging load of electric vehicle.This constraint requires that the load involved in PEV dispatching be equal to the load demand of electric vehicles.

Charging and discharging constraints of electric vehicles:

$$P_{PEVC,t,\max} \geq P_{PEVdispatch,t} \geq P_{PEVD,t,\max}. \tag{8}$$

$P_{PEVC,t,max}$ refers to the maximum amount of load that electric vehicle can obtain from the power grid every hour, while $P_{PEVD,t,max}$ refers to the maximum amount of discharge of PEV.

3 Improvement of MOEA/D Algorithm

MOEA/D is a decompose-based multi-objective evolutionary algorithm. Its main idea is to transform the multi-objective optimization problem into a series of single-objective optimization sub-problems. And then these subproblems are optimized simultaneously by using the information provided by the neighborhood. The MOEA/D algorithm proposed in this paper will be improved in the following three aspects to achieve the optimal value of multiple objectives such as economic function, carbon dioxide, sulfide, and charging variance.

3.1 Limitations of the Model

Chebyshev weighted sum method: the decomposition is as follows:

$$\min g^{wst}(x|\lambda, z^*) = \frac{(m-1)^2}{m^2} \max_{1\leq i\leq m} \lambda_i |f_i(x) - z_i^*| + \left[1 - \frac{(m-1)^2}{m^2}\right] \sum_{i=1}^{m} \lambda_i f_i(x). \tag{9}$$

λ is a set of weight vectors, $f_i(x)$ is the ith objective function. Z^* is the ideal value, which also represents the point corresponding to the minimum value of the objective function. The weighted sum decomposition method focuses on the global information in the evolutionary process, while the chebyshev decomposition method focuses on the local information in the evolutionary process. By unifying the two methods through linear combination, the respective advantages of the two methods will be combined, and the weights of the two methods are set as $1-(m-1)^2/m^2$ and $(m-1)^2/m^2$ respectively.

3.2 Improvement of Solution Update Mode

Traditional MOEA/D is prone to premature convergence due to lack of diversity in the later stage of the algorithm, the optimal solution may be missed. In response to this phenomenon, this paper temporarily leave the evolved solutions out of the next generation, and carry them on Levy flight to find better solutions. Levy flight can effectively jump out of the local optimum and increase the diversity of solutions and the ability of global optimization. The location update method is as follows:

$$x_i^{t+1} = x_i^t + \alpha \oplus s(\lambda). \tag{10}$$

α is the step size scaling factor and it is equal to 1, ⊕ is the dot product. x_i^t stands for the position of time t and $S(\lambda)$ is a random search path. In order to make the direction of updating closer to the direction of optimal solution, the optimal solution is added into the formula of solution updating.

$$\begin{cases} stepsize = \alpha \oplus levy(\lambda) \oplus \left(x_i^t - best\right) \\ x_i^{t+1} = x_i^t + stepsize \oplus randon \end{cases} \tag{11}$$

Best is the current optimal solution, levy (λ) is Levy random path, which follows the traditional Levy flight formula, will not be described too much here. *randon* is a random sequence which obeys the standard normal distribution. the connection between the update of the solution and the optimal solution will effectively enhance the ability of the algorithm.

3.3 Selection of the Optimal Value

In the single objective optimization algorithm, the optimal value can be selected directly by using the objective function. However, for multi-objective optimization, the search for the optimal solution is often complicated. In this paper, the optimal solution selection method is: The solutions generated after each iteration are sorted in a fast non-dominated manner, and then the individuals with Pareto rank 1 are extracted for Euclidean distance sorting, and the individuals with the smallest Euclidean distance are found and recorded as the current optimal solution. The Euclidean distance formula is as follows:

$$d_i = \sqrt{(f_{1i} - f_{1\min})^2 + (f_{2i} - f_{2\min})^2 + (f_{3i} - f_{3\min})^2 + (f_{4i} - f_{4\min})^2}. \tag{12}$$

f_{mi} represents the value of the ith individual on the objective function m, and f_{mmin} represents the minimum value of the objective m in this iteration process.

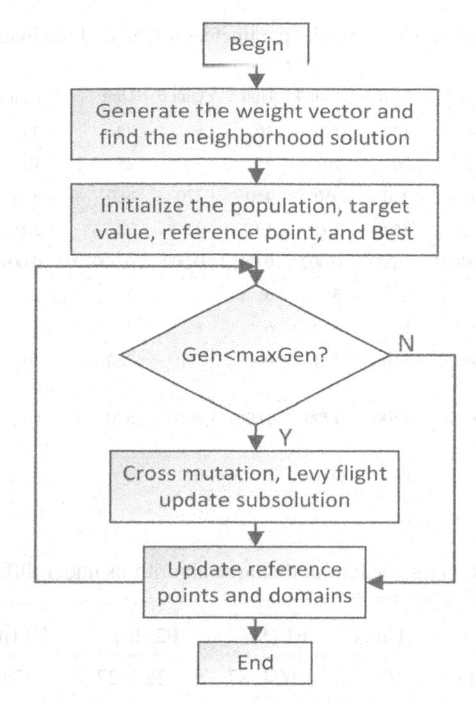

Fig. 1. IMOEA/D algorithm flow chart.

3.4 Algorithm Steps

See Fig. 1.

4 Experimental Results and Analysis

The unit commitment model under the grid feedback strategy is optimized under the conditions of 10, 50, and 90 units respectively, and the experimental results of the three algorithms are compared to prove the effectiveness of the improvement of the MOEA/D algorithm.

4.1 Parameter Description

The parameters of the unit in the experiment are shown in Table 1. The population is 50, the number of iterations is 800, the total load of the PEV is 1002.80 MW, and the hourly charge and discharge boundaries are 122.1494 MW and −122.1494 MW respectively.

4.2 The Experimental Results Show

The optimal and worst values of the four objective functions optimized by MOEA/D, NSGA-II and IMOEA/D are shown in Table 2. In order to explore the improvement effect

Table 1. Unit combination model parameters under grid feedback condition.

Parameter	Unit 1	Unit 2	Unit 3	Unit 4	Unit 5	Unit 6	Unit 7	Unit 8	Unit 9	Unit 10
Pmax (MW)	455	455	130	130	162	80	85	55	55	55
Pmin (MW)	150	150	20	20	25	20	25	10	10	10
a ($/MWh)	1000	970	700	680	450	370	480	660	665	670
b ($/MWh)	16.19	17.26	16.6	16.5	19.7	22.26	22.74	25.92	27.27	27.79
c ($/MWh)	0.00048	0.00031	0.002	0.002	0.003	0.007	0.000793	0.00413	0.002221	0.00173
MU (h)	8	8	5	5	6	3	3	1	1	1
MD (h)	8	8	5	5	6	3	3	1	1	1
Hot start cost (/$)	4500	5000	550	560	900	170	260	30	30	30
Cold start cost (/$)	9000	10000	1100	1200	1800	340	520	60	60	60
Initial state (h)	1	1	0	0	0	0	0	0	0	0

Table 2. The optimization target values of the three algorithms under different number of units.

Most value	Method	Units	F1 ($)	F2 (lb)	F3 (lb)	F4 (MW)
Best	MOEA/D	10	604387	226127	474028	7411
	NSGA-II		579233	218654	474028	1484
	IMOEA/D		600617	221125	469528	397
Worst	MOEA/D		635234	262352	550090	41361
	NSGA-II		635234	276951	584139	109860
	IMOEA/D		620482	231569	496420	6738
Best	MOEA/D	50	3028922	1141273	2354684	9825
	NSGA-II		2865306	1139329	2351573	1135
	IMOEA/D		2980712	1139329	2361702	8164
Worst	MOEA/D		3104503	1185687	2443180	46396
	NSGA-II		3118728	1317606	2770183	137059
	IMOEA/D		3094600	120008	2523589	62119
Best	MOEA/D	90	5500908	2050586	4211344	8204
	NSGA-II		5148387	2032939	4173547	2681
	IMOEA/D		5414027	2032939	4173547	16510
Worst	MOEA/D		5590750	2119667	4352343	51105
	NSGA-II		5610876	2365905	4962755	140846
	IMOEA/D		5620101	2175604	4470780	78522

under different dimensions, the optimization results of 10 and 90 units were specially selected for comparison. And the Fig. 2 and Fig. 3 are the comparison pictures of the optimization results of the three algorithms in three-dimensional space.

According to the data in Table 2, when the number of units is 10, the range of fitness value of charging variance obtained by IMOEA/D algorithm is 397 MW–6738 MW. However, the fitness value of charging variance obtained by MOEA/D algorithm ranges from 7411 MW to 41361 MW. It can be clearly observed that the worst value of charging variance of IMOEA/D algorithm is also superior to that of MOEA/D algorithm, while the worst value of charging variance of NSGA-II algorithm reaches 109860 MW, which is much higher than the corresponding values of the other two algorithms. When the number of units increased to 50 and 90, the charging variance of IMOEA/D was higher than that of MOEA/D, but the range of increase was not large.

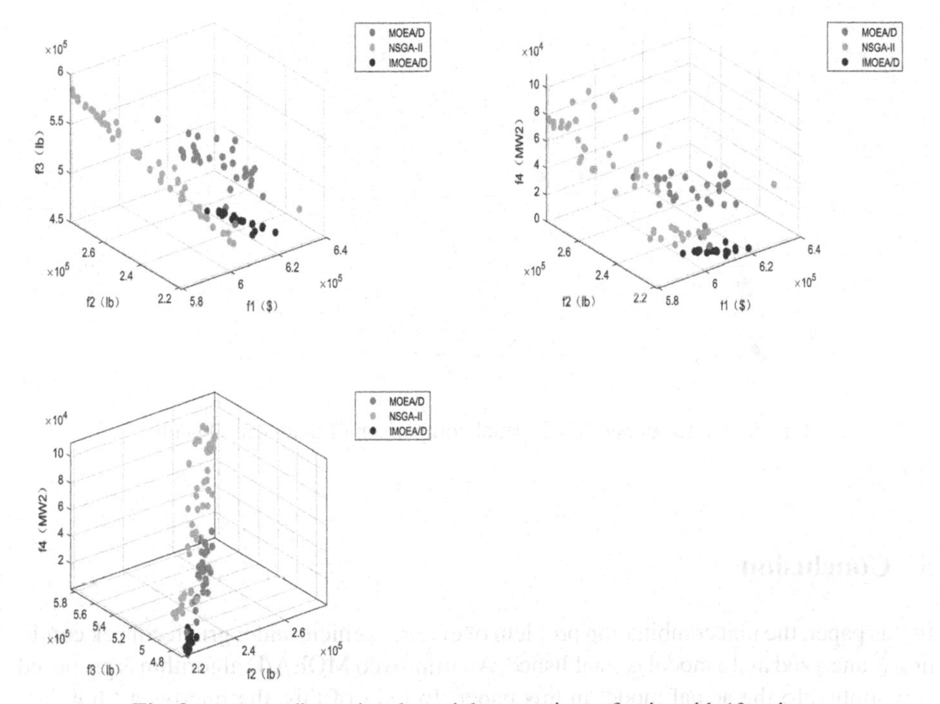

Fig. 2. A three-dimensional spatial comparison of units with 10 units.

When the number of units is 90, the optimal economic cost of NSGA-II algorithm after optimization is 5148387$, while the optimal value of MOEA/D algorithm and IMOEA/D algorithm are 604387 $ and 5414027 $ respectively, under the condition that the worst value of economic cost of the three algorithms is not significantly different. This indicates that the NSGA-II algorithm is relatively superior when only economic objectives are considered.

Combined with the data in Fig. 2 and Fig. 3, when the number of units is 10, the optimized carbon dioxide and carbon dioxide emissions of IMOEA/D algorithm are better than those of the other two algorithms. and when the number of units increased from 50 to 90, This phenomenon has been maintained. In one word, no matter if the number of units is 10 or 90, the optimization results of IMOEA/D in all four target directions are better than that of MOEA/D, that is, when the optimal value of the charging deviation function is satisfied, the IMOEA/D algorithm can ensure that other objective function values are relatively good.

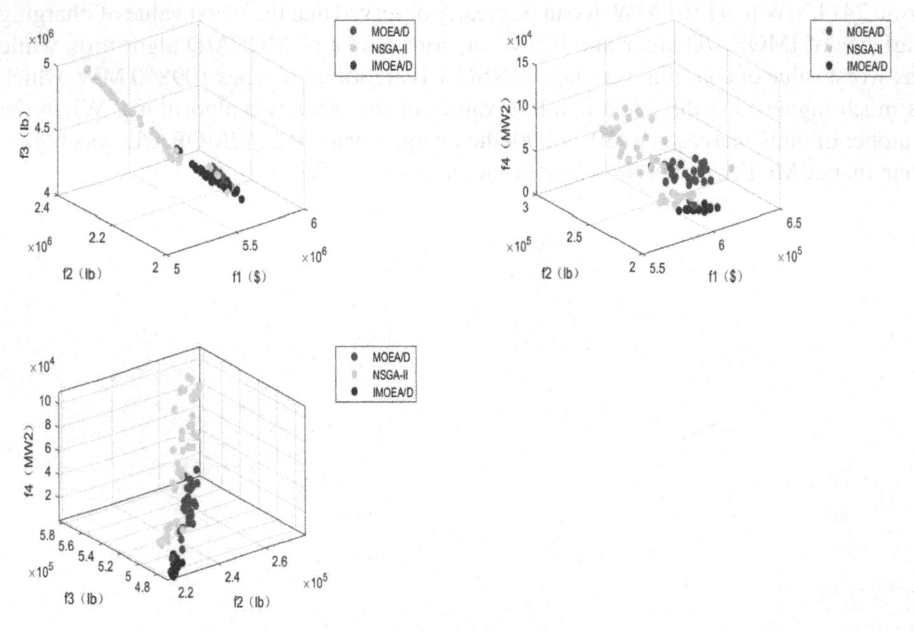

Fig. 3. A three-dimensional spatial comparison of units with 90 units.

5 Conclusion

In this paper, the unit combination problem of electric vehicle under grid feedback condition is analyzed and a model is established. An improved MOEA/D algorithm is proposed and applied to the actual model in this paper. In light of this, the improved MOEA/D algorithm is compared with traditional MOEA/D algorithm and NSGA-II algorithm. In the analysis of economic goal, carbon dioxide emissions, sulfur emissions and charging variance. By comparing the optimization results of the three algorithms in this model when the number of units is 10, 50 and 90, it is concluded that the IMOEA/D algorithm is more suitable to solve this problem and performs better in the low-dimensional optimization. This also fully proves the success of the improvement of the MOEA/D algorithm.

Acknowledgments. This work was supported by the Natural Science Foundation of China (No. 52077213 and 62003332) and Youth Innovation Promotion Association CAS 2021358.

References

1. Yang, Z., Li, K., Niu, Q., et al.: A novel parallel-series hybrid meta-heuristic method for solving a hybrid unit commitment problem. Knowl.-Based Syst. **134**, 13–30 (2017)
2. Jian, L., Zheng, Y., Shao, Z.: High efficient valley-filling strategy for centralized coordinated charging of large-scale electric vehicles. Appl. Energy **186**, 46–55 (2017)
3. Wang, Y., Yang, Z., Mourshed, M., et al.: Demand side management of plug-in electric vehicles and coordinated unit commitment: a novel parallel competitive swarm optimization method. Energy Convers. Manag. **196**, 935–949 (2019)
4. Li, Y.F., Pedroni, N., Zio, E.: A memetic evolutionary multi-objective optimization method for environmental power unit commitment. IEEE Trans. Power Syst. **28**(3), 2660–2669 (2013)
5. Trivedi, A., Srinivasan, D., Pal, K., et al.: Enhanced multiobjective evolutionary algorithm based on decomposition for solving the unit commitment problem. IEEE Trans. Ind. Inform. **11**(6), 1346–1357 (2017)
6. Shukla, A., Singh, S.N.: Multi-objective unit commitment with renewable energy using hybrid approach. IET Renewable Power Gener. **10**(3), 327–338 (2016)
7. Jian, L.N., Yan, C., et al.: High efficient valley-filling strategy for centralized coordinated charging of large-scale electric vehicles. Appl. Energy **186**, 46–55 (2017)
8. Wu, L., Wang, Y., Yuan, X., et al.: Multiobjective optimization of HEV fuel economy and emissions using the self-adaptive differential evolution algorithm. IEEE Trans. Veh. Technol. **60**(6), 2458–2470 (2011)
9. Zakariazadeh, A., Jadid, S., Siano, P.: Multi-objective scheduling of electric vehicles in smart distribution system. Energy Convers. Manag. **79**, 43–53 (2014)
10. Lokeshgupta, B., Sivasubramani, S.: Multi-objective dynamic economic and emission dispatch with demand side management. Int. J. Electr. Power Energy Syst. **97**, 334–343 (2018)
11. Elsied, M., Oukaour, A., Gualous, H., et al.: Optimal economic and environment operation of micro-grid power systems. Energy Convers. Manag. **122**, 182–194 (2016)
12. Ma, H., Yang, Z., You, P., et al.: Multi-objective biogeography-based optimization for dynamic economic emission load dispatch considering plug-in electric vehicles charging. Energy **135**, 101–111 (2017)
13. Hadayeghparast, S., Shayanfar, H., Farsangi, A.S., et al.: Stochastic multi-objective economic/emission energy management of a microgrid in presence of combined heat and power systems. In: 2019 IEEE/IAS 55th Industrial and Commercial Power Systems Technical Conference. IEEE (2019)
14. Swief, R.A., Amary, N.H., Kamh, M.Z.: Optimal energy management integrating plug in hybrid vehicle under load and renewable uncertainties. IEEE Access **8**, 176895–176904 (2020)

Modeling of Supercritical Once-Through Boiler-Turbine Units Based on Improved Sine and Cosine Algorithm

Guolian Hou[1], Huilin Su[1(✉)], Hongqun Gu[2], Linjuan Gong[1], Ting Huang[1], and Xuming Lv[2]

[1] School of Control and Computer Engineering, North China Electric Power University, Beijing 102206, China

[2] State Grid Liaoning Electric Power Supply Co. Ltd., Shenyang 110004, China

Abstract. Under the background of flourishing development of renewable energy, higher requirements are put forward for the operation of supercritical once-through boiler-turbine (OTB) units. It is necessary but challenging to establish a precise dynamic model of supercritical OTB units to ensure safe and efficient operation of units. In this paper, a dynamic mathematical model of supercritical OTB units is derived from mass and energy conservation. Then, an improved sine and cosine algorithm (ISCA) fusing chaotic initialization and cloud theory is proposed to identify the parameters combined with operating data of a 600 MW supercritical unit. The experimental results illustrate that the model outputs can approach the actual data over a wide operating range with satisfactory accuracy.

Keywords: Supercritical once-through boiler-turbine (OTB) units · Improved sine and cosine algorithm (ISCA) · Mathematical model · Parameter identification

1 Introduction

In order to absorb more renewable energy generation into power grid, supercritical once-through boiler-turbine (OTB) units are required to maintain operating with high safety and efficiency [1, 2]. Thus, it is necessary to develop a dynamic mathematical model of OTB units. The accurate model can not only reflect the dynamic characteristics of the units correctly, but also facilitate the design of the controller. More precisely, the effectiveness and accuracy of the model determine whether the satisfactory control performance can be obtained.

At present, domestic and foreign scholars have carried out extensive research on modeling. It can be summarized into three kinds of modeling methods: white box modeling, black box modeling and gray box modeling. White box modeling, also known as mechanism modeling, is to establish a mathematical model based on the conservation relationship of the system and related thermodynamic processes on the premise of a profound understanding of the described process. Inspired by the nonlinear dynamic model

K. Li et al. (Eds.): LSMS 2021/ICSEE 2021, CCIS 1468, pp. 86–96, 2021.
https://doi.org/10.1007/978-981-16-7210-1_9

of drum boiler established by Åström [3, 4], a simplified nonlinear model suitable for the controller design of the coordinated control system of a OTB unit was established [5, 6]. However, the model lacks closed-loop verification [7, 8]. However, the contradiction between modeling complexity and accuracy is still worth further discussion. Black box modeling, also known as data modeling, refers to modeling by using process data. The initial clustering center was modified through quantum artificial bee colony algorithm, and the accuracy of the model was further improved [9]. For all that, the reliability of the black box model needs further study.

Grey box modeling is a combination of white box modeling and black box modeling. A performance analysis and evaluation model was established based on the interaction of operating parameters of thermal power units, and the accuracy of model was verified [10]. However, the accuracy of grey box modeling is still unsatisfactory.

In the past ten years, researches on swarm intelligence algorithms have emerged in an endless stream, and they have been successfully used in parameter identification in order to improve the accuracy of grey box modeling, such as particle swarm optimization (PSO), immune genetic algorithm (IGA), and grey wolf optimizer (GWO) etc. However, the solution is likely to converge to the local optimum due to the weak global search ability. Therefore, this paper improves the sine and cosine algorithm proposed by Mirjalili in 2016, and applies it to parameter identification to improve the optimization effect.

The overall outline of this paper is presented as follows. A 600MW supercritical OTB unit is taken as the research object, the mechanism model is established with three-input-three-output in the second section through reasonable assumptions and simplification on the basis of previous studies. An improved Sine and Cosine (ISCA) algorithm fusing chaotic initialization and cloud theory is proposed in Sect. 3. After that, Sect. 4 combines the model and the improved algorithm in the previous two sections to carry out parameter identification in combination with the operation data, and verifies the accuracy of the model through simulation. At last, the paper is summarized in Sect. 5.

2 Model Description

A 600 MW supercritical OTB unit is taken as the research object. Direct-fired pulverizing system is adopted. In the boiler-turbine system, pulverized coal is mixed with the primary air to combustion, and the working medium is heated to produce high temperature and pressure steam. Afterwards, the steam does work while expanding in a turbine and then generate electricity through a generator.

It is necessary to make reasonable simplifications and assumptions in order to establish a suitable model. Thus, the economizer, the water wall and the superheaters are regarded as a whole heating tube, and only the state changes of inlet point, outlet point, and the representative point of the boiler are taken into account. Therefore, the simplification can be given as:

1. Flue gas:

(1) Axial heat conduction between flue, pipeline and working fluid is ignored.
(2) The dynamic changes of the flue gas are negligible.

(3) Heat transferred to the wall by the gas is proportional to the heat released from fuel combustion.

2. Boiler-turbine section:

(1) The fluid characteristics on the cross section of any tube are uniform.
(2) Integrate all levels of spray attemperators into one part and relocated at the downstream of the superheaters.
(3) Regard the heating section (economizer, water wall and the separator) as a whole.
(4) A set of high/intermediate/low pressure turbines is represented as a unit.

Therefore, the simplified OTB unit composed of three sections that can be modeled independently for the pulverizing system, the boiler and attemperator section, and the turbine section. The simplified diagram of the OTB units is shown in Fig. 1.

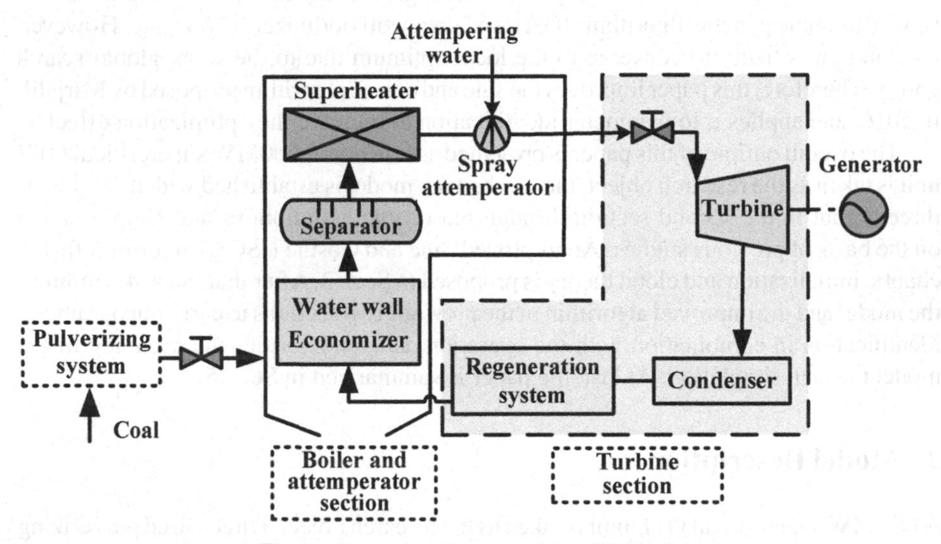

Fig. 1. Simplified diagram of the OTB units.

2.1 The Pulverizing System

The process includes three dynamic links: the coal feeder, the coal mill and the primary air pipe. Among them, the dynamic characteristics of coal feeder and primary air pipe are mainly shown as pure delay, while the dynamic characteristics of coal mill are mainly shown as inertia. So, there is the following expression:

$$C_0 \frac{dr_B}{dt} = -r_B + u_B(t - \tau). \tag{1}$$

where C_0 is the inertia time of the pulverizing system, *s*; r_B is fuel quantity, *kg*/*s*; u_B is the fuel command, *kg*/*s*; τ is the delay time, *s*.

2.2 The Boiler and Attemperator Section

The division of this section according to the properties of working medium will inevitably involve the position of phase transition points. The state space expression of the model established according to this scheme is too complicated. Therefore, this section is divided according to their physical locations.

The enthalpy or temperature of the working medium in the separator is the earliest signal that can reflect the imbalance of coal water ratio, so it is the key measurement and control parameter of the unit coordination control system. Based on the global conservation of mass and energy, following equations are displayed:

$$s_1 \frac{d\rho_m}{dt} = D_{fw} - D_s. \tag{2}$$

$$s_2 \frac{d(\rho_m u_m)}{dt} = D_{fw} h_{fw} - D_s h_s + Q. \tag{3}$$

where s_1 and s_2 are dynamic parameters; ρ_m is the average density of separator, kg/m^3; u_m is steam internal energy of steam separator, kJ/kg; D_{fw} is feedwater flow at economizer inlet, kg/s; h_{fw} is the specific enthalpy of feedwater, kJ/kg; seeing the constant value here, $h_{fw} = 1025$ kJ/kg [8]. D_s is the steam flow rate superheater outlet, kg/s; h_s is steam enthalpy at separator outlet, kJ/kg; Q is the energy absorbed by the working medium, kJ, $Q = k_Q r_B$ and k_Q is the endothermic gain, kJ/kg.

Assuming that the superheater resistance and pressure drop are concentrated at its outlet, the superheater pipe pressure loss model can be obtained according to the flow formula as follows:

$$p_m - p_s = \sigma D_s^2. \tag{4}$$

where p_m is the steam pressure at the outlet of the separator, *MPa*; p_s is superheater outlet steam pressure, *MPa*; σ is superheater pipe resistance coefficient.

The internal dynamics of the part of attemperating water is ignored, and its mass and energy conservation equation is:

$$D_s + D_{sw} = D_{st}. \tag{5}$$

$$D_s h_s + D_{sw} h_{fw} = D_{st} h_{st}. \tag{6}$$

where, D_{sw} is total equivalent attemperating water flow, kg/s; D_{st} is the main steam flow, kg/s; h_{st} is the enthalpy of main steam, kJ/kg.

The partial pressure variation of attemperating water is ignored: the main steam pressure p_{st} is equal to the superheater outlet steam pressure p_s.

Since the pressure, enthalpy and temperature of the working medium are very important controlled parameters, the mass and energy conservation Eqs. (2) and (3) in the heated section are rewritten as:

$$b_{11} \frac{dp_m}{dt} + b_{12} \frac{dh_m}{dt} = D_{fw} - D_s. \tag{7}$$

$$b_{21}\frac{dp_m}{dt}+b_{22}\frac{dh_m}{dt}=D_{fw}h_{fw}-D_slh_m+Q. \tag{8}$$

where $b_{11}=s_1\frac{\partial\rho_m}{\partial p_m}$, $b_{12}=s_1\frac{\partial\rho_m}{\partial h_m}$, $b_{21}=s_2(h_m\frac{\partial\rho_m}{\partial p_m}-1)$, $b_{22}=s_2(\rho_m+h_m\frac{\partial\rho_m}{\partial h_m})$, $h_s=lh_m$.

The superheater differential pressure can be regarded as a function of steam pressure p_m at the outlet of the steam separator as an independent variable.

$$\Delta p=g(p_m). \tag{9}$$

2.3 The Turbine Section

The main equipment on the steam turbine side is the steam turbine, reheater, extraction steam and regenerative heating system. According to the previous studies, there are the following relations:

$$D_{st}=\mu_T f(p_{st},h_{st}). \tag{10}$$

where μ_T is the valve opening of turbine, %;

When the control object model is established, the main steam temperature can be regarded as stable, so the following equation can be obtained:

$$h_{st}=S(p_{st}). \tag{11}$$

$$D_{st}=\mu_T f(p_{st}). \tag{12}$$

Ignoring inertia, energy loss of working medium and change of feedwater of steam turbine part, active power is modeled as:

$$N_e=\eta\left[D_{st}h_{st}-(D_{st}h_{fw}+H_n)+D_r(h_{r_2}-h_{r_1})\right]. \tag{13}$$

where N_e is active power, MW; η is the turbine energy conversion efficiency; D_r is reheated steam flow, *kg/s*; h_{r1} is the steam enthalpy of reheater inlet, *kJ/kg*; h_{r2} is the steam enthalpy of reheater outlet, kJ/kg; H_n is the transport enthalpy of circulating water in condenser, kg/s; $D_{st}(h_{st}-h_{fw})$ represents heat absorption of superheated steam; $D_r(h_{r_2}-h_{r_1})$ represents heat absorption of reheated steam.

It is assumed that the heat release of flue gas to the pipe wall is a forced heat flow uniformly distributed along the pipe length, and the ratio of $D_{st}(h_{st}-h_{fw})$ and $D_r(h_{r_2}-h_{r_1})$ can be regarded as a constant. Therefore, it can be written as:

$$N_e=\eta\left[\delta(D_{st}h_{st}-D_{st}h_{fw})-H_n\right]. \tag{14}$$

where δ is the coefficient. When the turbine is operating at such an efficiency, it can be assumed that the carrying enthalpy of the circulating water in the condenser is proportional to the unit load. It can be obtained:

$$N_e=k_2(D_{st}h_{st}-D_{st}h_{fw}). \tag{15}$$

where k_2 is the gain of turbine, kJ/kg.

Substituting the attemperating water model into Eq. (7, 8), the following equation can be obtained:

$$c_1\frac{dp_m}{dt} = (h_{fw} - d_1)D_{fw} + \frac{(d_1 - lh_m)(h_{st} - h_{fw})}{(lh_m - h_{fw})}D_{st} + Q. \tag{16}$$

$$c_2\frac{dh_m}{dt} = (h_{fw} - d_2)D_{fw} + \frac{(d_2 - lh_m)(h_{st} - h_{fw})}{(lh_m - h_{fw})}D_{st} + Q. \tag{17}$$

where $c_1 = b_{21} - (b_{11}/b_{12}) * b_{22}, c_2 = b_{22} - (b_{12}/b_{11}) * b_{21}, d_1 = b_{22}/b_{12}, d_2 = b_{21}/b_{11}$.

In summary, the state variables, input variables and output variables of the supercritical unit model are as follows:

$$X = \begin{bmatrix} x_1 \\ x_2 \\ x_3 \end{bmatrix} = \begin{bmatrix} r_B \\ p_m \\ h_m \end{bmatrix}, U = \begin{bmatrix} u_1 \\ u_2 \\ u_3 \end{bmatrix} = \begin{bmatrix} u_B \\ D_{fw} \\ \mu_T \end{bmatrix}, Y = \begin{bmatrix} y_1 \\ y_2 \\ y_3 \end{bmatrix} . = \begin{bmatrix} p_{st} \\ h_m \\ N_e \end{bmatrix}. \tag{18}$$

$$\begin{aligned} \dot{x}_1 &= -\tfrac{1}{C_0}[x_1 - u_1(t - \tau)] \\ \dot{x}_2 &= \tfrac{1}{c_1}\left[(h_{fw} - d_1)u_2 + \tfrac{f[x_2 - g(x_2)](d_1 - lx_3)\{S[x_2 - g(x_2)] - h_{fw}\}}{(lx_3 - h_{fw})}u_3 + k_0x_1\right] \\ \dot{x}_3 &= \tfrac{1}{c_2}\left[(h_{fw} - d_2)u_2 + \tfrac{f[x_2 - g(x_2)](d_2 - lx_3)\{S[x_2 - g(x_2)] - h_{fw}\}}{(lx_3 - h_{fw})}u_3 + k_0x_1\right]. \\ y_1 &= x_2 - g(x_2) \\ y_2 &= x_3 \\ y_3 &= k_2u_3f(y_1)\{S[x_2 - g(x_2)] - h_{fw}\} \end{aligned} \tag{19}$$

3 Improved Sine and Cosine Algorithm

Sine and cosine algorithm was proposed by Australian scholar Mirjalili in 2016 with the advantages of fewer parameters, easy implementation, simple structure and fast convergence rate [11]. Then several improved algorithms are brought up [12, 13]. However, the convergence of the algorithm needs to be further improved and its application field is worthy of being further broadened.

Some improvements have been made on the basis of previous research. There are four parameters r_1, r_2, r_3 and r_4 in the sine and cosine algorithm. Among them, the most critical is the adaptive parameter r_1. Therefore, parabolic function is selected as the update strategy of r_1's nonlinear decline. Secondly, Logistic, a classical chaotic map, is used for initialization operation. Last but not least, it is combined with cloud theory to swarm particles through different weight evolution strategies.

The steps of the ISCA are as follows:

1) Initialization parameters: particle swarm size N, dimension D, maximum iteration number T;
2) Start iteration and let $t = 1$;

3) Calculate the value of r_1 according to the formula (20) and random generation r_2, r_3 and r_4, where $r_2 \in [0, 2\pi]$, $r_3 \in [0, 2]$, $r_4 \in [0, 1][0, 1]$;

$$r_1 = 2 \times (1 - t/T)^2. \tag{20}$$

4) Initialization location;
5) Calculate the fitness value, and divide the particles into three layers according to the fitness value;
6) Assign values to each layer ω, in which the ω value of the layer with a smaller fitness value is 0.2, which is larger than 0.9. The middle layer is calculated by using cloud theory;

The formula of cloud adaptive inertia weight ω is as follows:

$$\omega = 0.9 \sim 0.5 \times \exp(-(f_i - Ex)^2/(2(En')^2)). \tag{21}$$

where f_i is the fitness value corresponding to *i-th* particle; *Ex* is the expected value; *En'* is a normal random number.

(7) Iterate according to the sine and cosine algorithm formula (22–23) update the position of the particle, and get the current optimal solution of the particle; Let $t = t+1$, if the maximum number of iterations is reached, the optimal solution and the optimal fitness value will be output; otherwise, turn to Step 5.

if $r_4 < 0.5$, the following update Eq. (22) is used, otherwise the Eq. (23) is used.

$$X_i^j(t+1) = X_i^j(t) + r_1 \sin(r_2)\left|r_3 P^j(t) - X_i^j(t)\right|. \tag{22}$$

$$X_i^j(t+1) = X_i^j(t) + r_1 \cos(r_2)\left|r_3 P^j(t) - X_i^j(t)\right|. \tag{23}$$

where $X(t)$ is individual position; $P(t)$ is the optimal individual position.

4 Parameter Identification and Simulation

Parameters to be identified can be divided into two categories, eight parameters to be determined are: C_0, c_1, c_2, d_1, d_2, k_0, k_2, l; the three functions to be determined are: $f(x)$, $g(x)$, $S(x)$.

4.1 Unknown Function

Through regression analysis, unknown equations can be obtained according to the historical data of different load ranges [5]. In order to facilitate the design of the controller, the equation form as simple as possible is selected in this paper, and the following unknown equation form can be obtained.

$$\begin{aligned} f(x) &= k_3 x - k_4 \\ g(x) &= k_1 x^b \\ S(x) &= k_5 x + k_6 \end{aligned} \tag{24}$$

To sum up, the 14 parameters to be determined are: $C_0, c_1, c_2, d_1, d_2, k_0, k_1, k_2, k_3, k_4, k_5, k_6, l, b$.

4.2 Illustrative Example

By using the operation data of 600 MW supercritical unit, the parameters of the unit are identified by the improved sine and cosine algorithm. The actual input data is added into the model to construct the optimization function:

$$E = \sum_{i=1}^{N} \left(\left| \frac{\Delta P_{st}}{P_{st0}} \right| + \left| \frac{\Delta h_m}{h_{m0}} \right| + \left| \frac{\Delta N_e}{N_{e0}} \right| \right). \tag{25}$$

where $\Delta P_{st}, \Delta N_e, \Delta h_m$ represent the deviation between the calculated value of the model and the actual measured value; P_{st0}, N_{e0}, h_{m0} are the outputs of the unit when it is in normal operation.

The steady-state values are shown in Table 1.

Table 1. Steady-state values at different steady conditions.

P_{st}(MPa)	h_m(kJ/kg)	N_e(MW)
13.5	3587.8	185
14.1	3581.5	235
15.1	3572.9	345
20.9	3519.1	455
23.5	3483.3	565
24.5	3490.7	600

The on-site data used in this section were collected from a 600 MW supercritical unit in a power plant in North China. Parameters were identified based on the improved sine and cosine algorithm, and the corresponding fitting curves of actual data and model output were obtained, as shown in Fig. 2. After the closed-loop identification, a set of dynamic parameters are as follows:

$$\begin{array}{l} C_0 = 0.0052,\ c_1 = 0.0084, c_2 = 0.0423, d_1 = 0.0284,\ \ d_2 = 0.0231 \\ k_0 = 0.0002,\ k_1 = 0.0369, k_2 = 0.0219, k_3 = 0.0132, k_4 = 0.0025 \\ k_5 = 0.0327, k_6 = 0.0121,\quad l = 0.0327,\quad b = 0.0175 \end{array}. \tag{26}$$

As can be seen from Fig. 2, when the data of 185–410 MW load period and 450–600 MW load period are respectively verified, the error between the identified model output curve and the actual output curve is small. Therefore, the feasibility of the proposed scheme based on the mechanism analysis modeling combined with the data-driven ISCA parameter identification is verified.

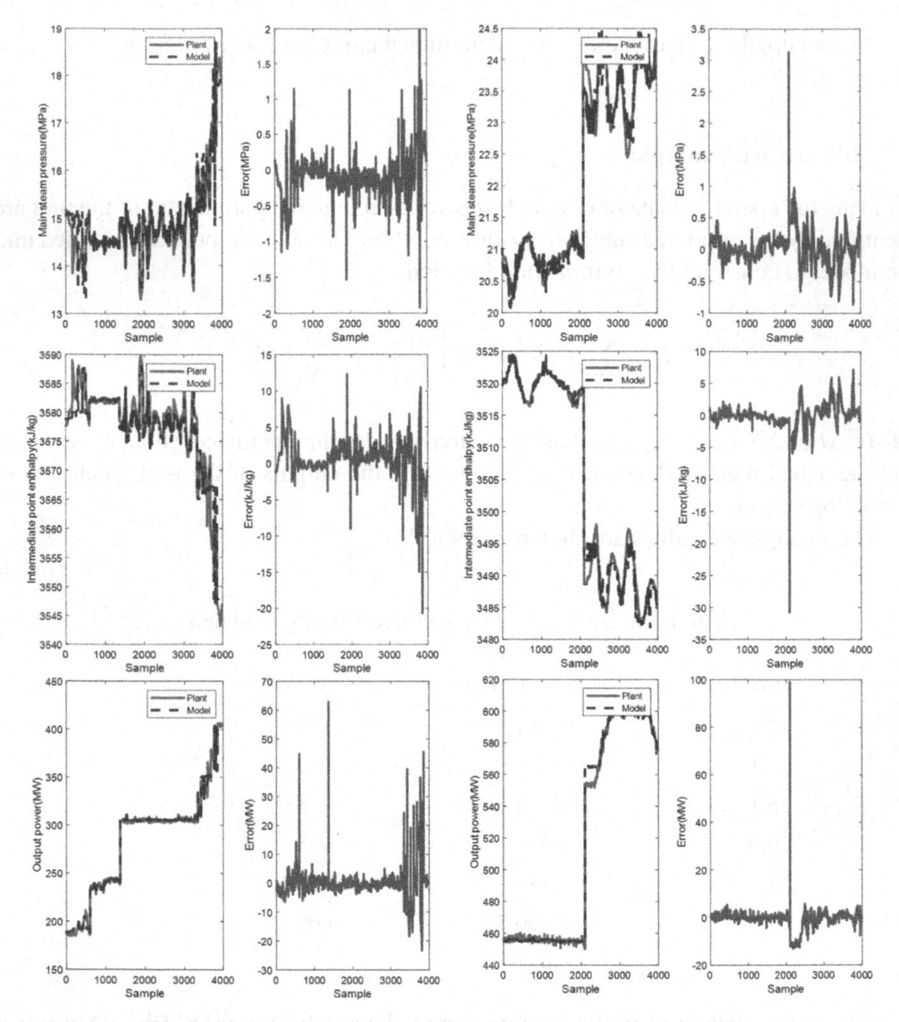

Fig.2. Fitting curves between the actual data and model output.

Obviously, the correctness and accuracy of the established model are verified in Fig. 2. Furthermore, the modeling performance of coordinated PSO, GWO, and conventional SCA is compared in order to more intuitively show the performance of the proposed approach.

To perform the analytical evaluation of the models, the root mean square error (RMSE) is used as follows:

$$RMSE = \sqrt{\frac{1}{n}\sum_{i=1}^{n}\left(y_i^* - y_i\right)^2}. \tag{27}$$

In addition, some quantitative results for CPU elapsed time and some root mean square error (RMSE) are given in Table 2.

The values in Table 2 demonstrated the phenomenal precision of the model which combining the mechanism model and the ISCA has better accuracy and efficiency.

Table 2. CPU elapsed time and root mean square error (RMSE)

Algorithm		PSO	GWO	SCA	Proposed
TIME(s)		188.1	143.4	59.2	67.3
RMSE	P_{st}	0.884	0.379	0.305	0.105
	h_m	2.583	3.061	2.769	0.720
	N_e	2.694	1.852	2.306	0.807

5 Conclusion

In this paper, a comprehensive modeling scheme based on the improved sine and cosine algorithm is proposed. Firstly, the nonlinear model of supercritical boiler-turbine unit is established based on the simple mechanism analysis and reasonable assumptions. Secondly, the ISCA algorithm is used to identify the unknown parameters of the unit according to the operation data of the unit. Finally, the simulation results show that the model can reflect the dynamic characteristics of the unit correctly. However, the application and precision of the model need to be further improved.

Acknowledgement. This work is supported by the National Key Research and Development Project (Grant No. 2019YFB1505403). The reviewers' insightful comments and valuable suggestions are also greatly appreciated.

References

1. Han, S.W., Shen, J., Pan, L., et al.: A L1-LEMPC hierarchical control structure for economic load-tracking of super-critical power plants. ISA Trans. **96**, 415–428 (2020)
2. Zeng, D.L., Gao, Y.K., Yong, H., et al.: Optimization control for the coordinated system of an ultra-supercritical unit based on stair-like predictive control algorithm. Control Eng. Pract. **82**, 185–200 (2019)
3. Åström, K.J., Eklund, K.: A simplified non-linear model of a drum boiler-turbine unit. Int. J. Control **16**, 145–169 (1972)
4. Åström, K.J., Bell, R.D.: Drum-boiler dynamics. Automatica **36**, 363–378 (2000)
5. Liu, J.Z., Yan, S., Zeng, D.L., et al.: A dynamic model used for controller design of a coal fired boiler-turbine unit. Energy **93**, 2069–2078 (2015)
6. He, F.A., Zgs, A., Phw, A., et al.: A dynamic mathematical model for once-through boiler-turbine units with superheated steam temperature. Appl. Thermal Eng. **170**, 114–912 (2020)
7. Fan, H., Zhang, Y.F., Su, Z.G., et al.: A dynamic mathematical model of an ultra-supercritical coal fired once-through boiler-turbine unit. Appl. Energy. **189**, 654–666 (2017)

8. Niu, Y.G., Du, M., Ge, W., et al.: A dynamic nonlinear model for a once-through boiler-turbine unit in low load. Appl. Thermal Eng. **161**, 113–880 (2019)
9. Hou, G.L., Gong, L.G., Huang, C.Z., et al.: Novel fuzzy modeling and energy-saving predictive control of coordinated control system in 1000 MW ultra-supercritical unit. ISA Trans. **86**, 48–61 (2019)
10. Wang, Y.H., Cao, L.H., Hu, P.F., et al.: Model establishment and performance evaluation of a modified regenerative system for a 660 MW supercritical unit running at the IPT-setting mode. Energy **179**, 890–915 (2019)
11. Mirjalili, S.: SCA: a sine cosine Algorithm for solving optimization problems. Knowl. Based Syst. **96**, 120–133 (2016)
12. Nenavath, H., Jatoth, R.K.: Hybridizing sine cosine algorithm with differential evolution for global optimization and object tracking. Appl. Soft Comput. **62**, 1019–1043 (2018)
13. Long, W., Wu, T.B., Liang, X.M., et al.: Solving high-dimensional global optimization problems using an improved sine cosine algorithm. Exp. Syst. Appl. **123**, 108–126 (2019)

Energy Storage Capacity Optimization for Deviation Compensation in Dispatching Grid-Connected Wind Power

Jian-hong Zhu[1(✉)], Kang Li[2(✉)], Renji Xu[1], Juping Gu[1], Lei Zhang[1(✉)], and Chenxiang Sun[3]

[1] School of Electrical Engineering, NanTong University, Nantong 226019, Jiangsu, China
{jh.zhu,jp.gu,nttzzl}@ntu.edu.cn

[2] School of Electronic and Electrical Engineering, University of Leeds, Leeds LS2 9JT, UK
K.Li1@leeds.ac.uk

[3] College of Energy and Electrical Engineering, HoHai University, NanJing 210098, Jiangsu, China

Abstract. Many uncertain factors in wind power forecasting lead to large prediction errors. Various prediction technologies have been developed to reduce errors and improve the dispatch-ability of grid-connected wind power. To install energy storage systems is an effective approach to reduce the scheduling deviation in dispatching the grid-connected wind power. This paper considers the optimal capacity allocation, a key issue in smoothing the grid wind power generation and integration. Based on the analysis of wind power prediction technologies and the resultant prediction deviations, the relationship between the distribution characteristics of wind power prediction errors and energy storage capacity demand is first investigated. Then, an optimization method is proposed, considering the stability of grid operation and the relationship between compensation necessity and load changes. Further, load fitness factor is introduced in processing the deviation data samples, together with an economic dispatch model for the deviation compensation, considering the operation costs. Finally, based on the analysis of various factors, the technical route to achieve energy storage capacity allocation for scheduling deviation compensation is proposed. Case studies are also presented to demonstrate the effectiveness of the proposed approach.

Keywords: Wind power dispatching · Deviation compensation · Prediction error distribution characteristics · Storage capacity

1 Introduction

Wind power generation and integration has been intensively researched and rolled out worldwide in the last decade. The power generated from a fixed wind turbine mainly depends on the local wind speed, direction, wind pressure and other weather conditions, as well as geographical environmental factors. The performance of wind turbine models are often limited by the uncertainties imposed on both the model inputs and outputs

K. Li et al. (Eds.): LSMS 2021/ICSEE 2021, CCIS 1468, pp. 97–109, 2021.
https://doi.org/10.1007/978-981-16-7210-1_10

while the conventional numerical weather forecasting methods also have low accuracy on localized regions. It is therefore difficult to accurately predict the actual wind power generation.

However, the power system dispatch always relies on the predicted data, it is therefore vital to improve the wind power prediction techniques. The early wind power prediction models often use wind speed information and experimental curves, which sometimes are also referred to as the physical methods. The second category is statistical methods, most of which rely on historical data [1, 2]. The third category is intelligent algorithms, where machine learning rules are used to establish the relationship between input and output, such as SVM and ELM, etc. [3–5]. Now, kinds of intelligent combination algorithms are emerged [6, 7].

All the existing literatures demonstrated that the prediction accuracy has improved with the development of technology, but the demand for data processing speed and memory has also been increased simultaneously. Although the prediction error has been decreased to a certain extent, the absolute error is still on the rise inevitably with the significantly increased installed capacity of wind power. Consequently, the increased un-schedulable wind power deviation greatly affects the balance of supply and demand of the power grid, bringing unstable factors to the grid. The increased reserve capacity must be installed to suppress the unpredictable deviations originated from the wind power integrated in the grid, which increases the cost of grid operation. On the other hand, the inefficient use of standby units is either a waste, especially, the thermal power generation units that consume coal resources has exacerbated the depletion of limited nonrenewable energy resources and their environment impact. As a countermeasure, national system operators such as the National Energy Administration in China issue functional specifications on the wind power prediction systems [8]. As an effective measure to tackle the problem of wind power scheduling deviations, energy storage technologies have drawn much attention in recent years due to their great potential of flexible throughput characteristics. However, due to the sheer demand for a significantly huge amount of storage capacity for deviation compensation, it is practically challenging to adopt energy storage to reduce the scheduling deviation of wind power and turn un-schedulable into schedulable. The concept was proposed for about 20 years, and a number of results have been presented for scheduling deviation compensation. As shown in Table 1, under the consideration of different factors for scheduling deviation compensation, the capacity allocation of energy storage accounts from 19.8% to 60% of the wind power capacity.

The dynamic throughput of energy storage depends on the positive and negative offset of the scheduling deviation, not only the polarity of the bias, but also the magnitude, so the cumulative amount of deviation directly impacts on the overcharge and over-discharge characteristics of the energy storage system. In the case that the predicted wind power is in the power dispatch, the input and output throughput of energy storage not only depends on the prediction technology, but also the uncertainties of the key factors such as the wind speed, the superimposed effect arisen from these important factors directly determines the energy storage capacity allocation. It is therefore necessary to investigate the deviation distribution characteristics.

Table 1. Energy storage configuration for deviation compensation

Scheduling compensation technology	Wind power capacity/MW	Statistical span/h	Storage configuration/MAh	Compensation target	Actual results
Kernel density estimation followed by GA-ANFIS prediction error statistics [9]	148.5	48	89, confidence level 80%	Small prediction error,	MAE and RMSE 6.70%, 9.15%
Based on feature extraction and LSSVM MPC [10]	49.5	1/4, 1/3, 1/2	9.8	Grid-connected operation cost-effectively	Full compensation
Exceeding error frequency weight (EEFW) combined with three-dimensional optimization method [11]	148.5	24	35.3	Degree of wind power compensation, energy storage life, economic ratio of hybrid storage capacity	Prediction error ≤ 15%, wind power fluctuation ≤ 20%
Two-stage stochastic optimization framework [12]	3.6, 2 × 170 kW diesel generators at minimum load	24	1.81	wind speed and the load growth rate, load balancing and frequency control	Life cycle cost varies very slightly with the number of scenarios (less than 1%)
Based on comprehensive cost-benefit model [13]	200	24	60.5	Ease the peaking burden, smooth load curve, and reduce the thermal power coal consumption	Adapting to presupposed dispatching schedule

2 The Impact of Scheduling Deviation Distribution Characteristics on Energy Storage Capacity Allocation

Energy storage capacity is used to deliver the throughput of the entire charging and discharging cycle. Due to the irregular and continuous alternation of positive and negative deviations, it is quite challenging to determine the desired energy storage capacity. Grid scheduling performance is heavily affected by the prediction technique. Positive and negative polarity changes of scheduling deviations imply the dynamic requirements on energy storage charging and discharging. Therefore, it is necessary to investigate the relationship between deviation distribution characteristics and energy storage capacity demand.

2.1 Error Distribution Analysis

The capacity analysis of the energy storage system is obviously a key prerequisite for the realization of schedulable wind power. Positive or negative deviation fluctuation within a sampling period renders the fluctuation of the capacity amplitude, while continuous positive or negative deviations implies that the energy storage system needs to

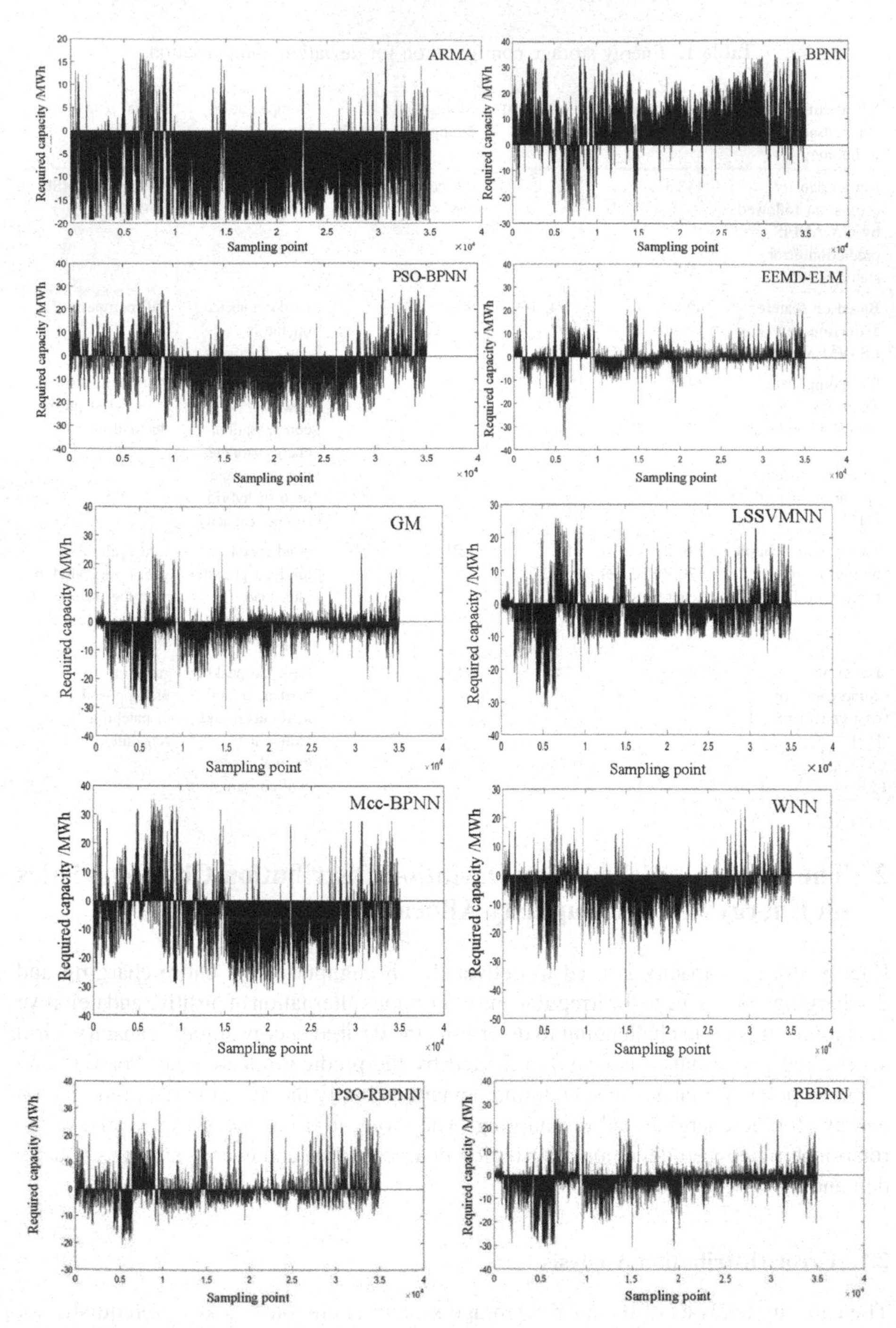

Fig. 1. Sampling time prediction deviation compensation demand distribution

absorb or release the accumulated energy continuously. Obviously, if the wind power prediction is directly used in the grid power dispatch, the distribution characteristics of the prediction errors determine the demand for energy storage capacity. Continuous positive or negative errors will inevitably increase the energy storage capacity for deviation compensation. Several error compensation distribution characteristics are shown in Fig. 1, such as ARIM (Autoregressive Integrated Moving Average model) [14], BPNN (Back-propagation Neural Network) [15], PSO (Particle Swarm Optimization) [16], SVM (Support Vector Machines) [17], EEMD (Ensemble Empirical Mode Decomposition) [18], ELM (Extreme Learning Machine) [19], WNN (Wavelet-based Neural Network) [20], RBPNN (Radial Basis Function Neural Network) [21], RBF (Radial Basis Function) [22], LSSVM (Least Squares Support Vector Machines) [23], GM (Gray Forecast Model) [24], MCC (Matthews Correlation Coefficient) [25]. As shown in Fig. 1, the error distribution characteristics varies in different time periods, and some continuous intervals have the same polarity, which implies that the energy storage system needs to be charged or discharged continuously, leading to a large cumulative value.

Thus the energy storage system needs to have a larger rechargeable or redischargeable capacity requirement. While for some intervals, the polarity of the deviations alternate frequently, which implies that the energy storage system is subject to alternating charge and discharge. The energy storage capacity is determined by the maximal cumulative value of the positive and negative deviations. If the positive and negative deviations are symmetric, the capacity of the energy storage system only needs to accommodate the maximal cumulative energy of a single charge or discharge phase, and thus the capacity requirement is relative small. However, in practice it is difficult to develop a wind power prediction technology which exhibits a symmetric error distribution characteristics, and in most cases the prediction error distribution is asymmetrical. If the overall error amplitude is small, the prediction technology still has lower requirements for the energy storage capacity for deviation compensation. On the other hand, frequent

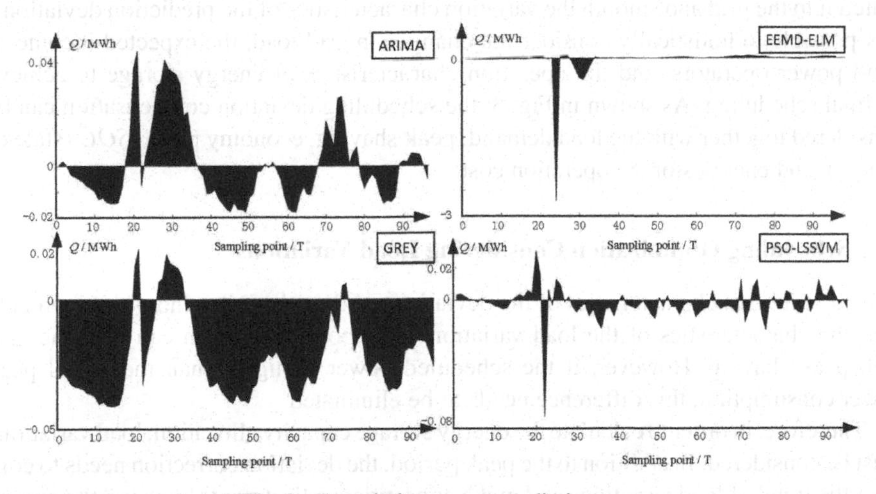

Fig. 2. Cumulative power prediction error

positive and negative changing in prediction errors implies that the energy storage system has to be charged and discharged frequently, which will affect the service life of energy storage.

Overall, a general conclusion can be drawn that storage capacity demand for error compensation largely depends on the error distribution characteristics, while large errors requires large storage capacity, while small errors do not necessarily imply small capacity as their cumulative values can still be very large. Symmetrical error distribution but with frequent polarity alternations is not necessarily a good prediction technique from the energy storage configuration prospective. Take 24 h statistics of 4 algorithms as an example, the cumulative capacities are shown in Fig. 2.

A desired error distribution for energy storage configuration could be the case where polarity of the prediction errors alternate at certain interval, symmetrically, and the amplitude should not be too large as well.

2.2 The Energy Storage Capacity Demand for Deviation Compensation

The statistics of the actual cumulative power prediction errors of typical day, typical month, and a year have been analyzed for a 1.5 MW wind turbine using different wind power prediction methods. As shown in Table 2, it is evident that the prediction error variations are uncertain for different time intervals, but the cumulative errors increase with the time scales. It is also evident that different prediction algorithms and different time scales require different energy storage configurations.

3 Optimal Energy Storage Capacity Allocation Considering Load Variation and Operating Costs in Power System Dispatch

Since the integration of energy storage can support the scheduling of wind power integrated into the grid and smooth the variation characteristics of the prediction deviations, it is possible to holistically consider the changes in grid load, the expected income of wind power operators, and the operation characteristics of energy storage to achieve optimal scheduling. As shown in Fig. 3, the scheduling deviation compensation can be considered together with the load demand, peak shaving, economy profit, *SOC* (state of charge), and energy storage operation cost.

3.1 Scheduling Optimization Considering Load Variations

If the distribution characteristics of the deviations of the schedule wind power coincide with the characteristics of the load variations, this power deviation can help the grid with peak shaving. However, if the scheduled power is higher than the actual peak power consumption, this difference needs to be eliminated.

Therefore, in order to calculate the energy storage capacity allocation, load variations must be considered. In addition to the peak period, the deviation correction needs to consider the trend of load variations and make appropriate adjustments by using the energy storage, while reducing the number of charging and discharging switching times, thus increase the service life of the energy storage system. Further, the dispatching results and

Table 2. Capacity demand for power deviation compensation of all periods (kWh/kW)

Statistical span	Predictive technology										
	BPNN	MCC-BPNN	PSO-BPNN	RBFNN	PSO-RBFNN	W-NN	LS-SVM	ARIMA	EEMD-ELM	Grey	PSO-LSSVM
Typical day	191/74	219/67	429/78	1393/111	1233/78	1052/62	1097/99	2130/125	948/65	2357/134	1242/112
48 h	1558/77	1822/95	1087/72	1752/80	1499/70	1751/65	1139/96	3047/124	1247/64	3558/135	891/99
Typical month	6381/136	27314/ 141	12060/126	5057/121	9508/112	11463/134	5315/113	11440/95	6778/142	14646/100	7618/123
Year	436918/142	260867/141	232675/138	41789/121	12756/123	206406/145	117619/126	171510/97	22799/142	245293/100	99510/123

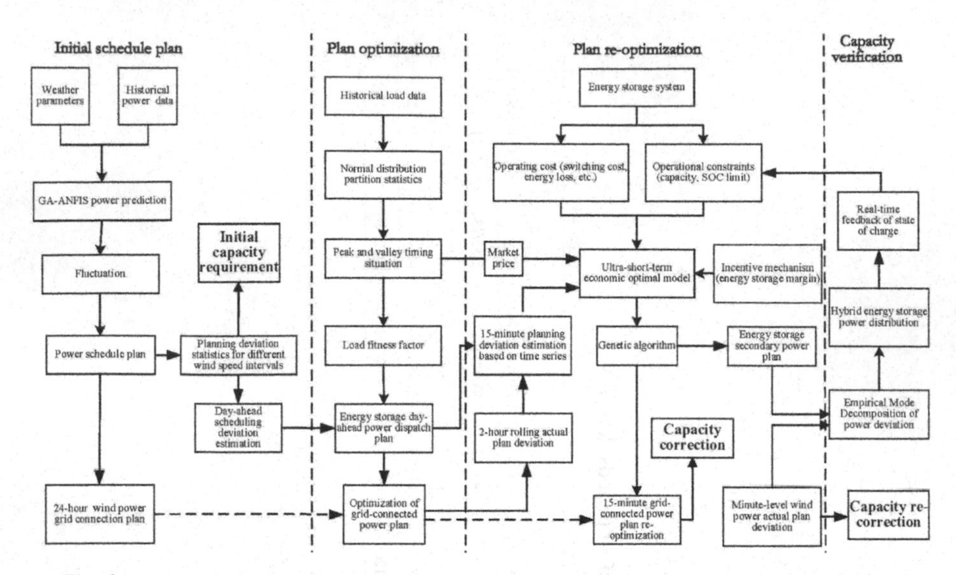

Fig. 3. Energy storage capacity configuration considering different scheduling methods

revenue costs are also affected by load variations to some certain extent. Wind power prediction deviations and dispatching behavior therefore need to can be considered holistically to determine the energy storage capacity. The deviation compensation method considering the load variations is given as follows.

$$\begin{cases} P^s_{s,\mathrm{t}} = P^s_{c,\mathrm{t}} + (1 - r_n) \cdot \Delta P \\ \Delta P = P^s_{c,\mathrm{t}} - P^s_{\mathrm{a,t}} \end{cases} . \tag{1}$$

where $P^s_{s,\mathrm{t}}$ is the compensated scheduling power after considering the load variations, $P^s_{c,\mathrm{t}}$ is the initial scheduled power, $P^s_{\mathrm{a,t}}$ is the actual wind power output during the period (refer to historical data), r_n is the load fitness factor considering the deviations. The value in the table not only considers the load change trend in different periods, but also considers the polarity of the deviation. The value of the load fitness factor is given shown in Table 3.

3.2 Scheduling Optimizing Considering Operating Costs

From the electricity market operation perspective, in addition to the grid-side load variations, the energy storage configuration must also consider the economic benefits of wind power operators. Therefore, it is necessary to study the energy storage operating costs and grid-connected power generation benefits of the deviation compensation scheme, and optimize the energy storage configuration to achieve high-accuracy schedule implementation. Aimed at maximizing the profit Z of the wind power system, and the following formula (2) is arrived.

$$\max Z\left[\sum_{t\in T} f(s, P_{d,t})\right] = \max \sum_{t=1}^{T}\left[Z^s_{\mathrm{a,t}} - Z^s_{\mathrm{bq,t}} + Z^s_{\mathrm{se,t}} - Z^s_{\mathrm{loss,t}} - Z^s_{qs,t} - Z^s_{\mathrm{tz,t}} + Z^s_{st,t}\right]. \tag{2}$$

Table 3. Load fitness factor value r_n

Load period distribution	Scheduling deviation (ΔP) and grid load change trend (K)								
	$\Delta P > 0$ and $K \geq 1$	$\Delta P > 0 and$ $0 < K < 1$	$\Delta P \neq 0 and$ $K = 0$	$\Delta P > 0 and$ $-1 < K < 0$	$\Delta P > 0$ and $K \leq -1$	$\Delta P < 0$ and $K \geq 1$	$\Delta P < 0$ and $0 < K < 1$	$\Delta P < 0$ and $-1 < K < 0$	$\Delta P < 0$ and $K \leq -1$
Valley periods	1/3	1/4	1/8	0	0	1/4	1/2	3/4	1
Transition periods	1	2/3	¾	1/4	1/8	1/2	3/4	1/2	1
Peak periods	1	1	1	1/2	1/4	0	0	1/4	1/2

where, s represents the uncertain scenario of the scheduling scheme, T is the number of time segments in a day, $Z^s_{\mathrm{a,t}}$ is the daily benefit of a scheduling output, $Z^s_{\mathrm{bq,t}}$ is the penalty cost of daily scheduling of wind energy, $Z^s_{\mathrm{se,t}}$ is the exchange cost of energy storage power, $Z^s_{\mathrm{loss,t}}$ is the loss cost of energy storage exchange power, $Z^s_{qs,t}$ is the frequent switching cost of energy storage, $Z^s_{\mathrm{tz,t}}$ is the upfront investment cost of energy storage, and $Z^s_{st,t}$ is the incentive premium.

4 A Case Study

The historical operation data of a wind farm are used in the case study, and the initial storage capacity is 148.5 kW. The rated voltage of the battery and the super capacitor are 180 V and 160 V respectively. One year data are selected to train and test energy storage capacity required to compensate for the deviations of different grid-accessed schedule, including the direct scheduling for stabilizing the grid after prediction, primary optimization combined with grid load sequence distribution, and the re- optimization under joint consideration of economic dispatch costs. Several scheduled power curves and actual power curves, the original predicted power curve, and the wind/storage power curve are illustrated in Fig. 4. The change curve of energy storage *SOC* used for scheduled power deviation compensation of one day is as shown in Fig. 5, where the number of 10–11 h sampling points is around 1.25–1.375 (10^5), while the number of 14–15 h sampling points is around 1.75–1.875 (10^5), for 21–22 h it is about 2.75–2875 (10^5).

As shown in Fig. 4, the power prediction curve exhibits dramatic fluctuations. If the load variation is considered, the compensation is increased during peak hours (10–11 h, 14–15 h, 21–22 h), while the volatility of the scheduled curve is slightly higher than that of the initial scheduling, the maximum volatility, which has increased by about 3%. After increasing the compensation, the overall tendency is the discharge behavior, as shown in Fig. 5(b). This reveals that the degree of compensation increases significantly after the load fitness factor is added into the cost function. This implies that it is insufficient to just consider the load factor and it is also necessary to consider various economic factors in the objective function. After the model is optimized, charge and discharge behavior can maintain the *SOC* of the energy storage system at about 50%, as shown in Fig. 5(c).

Among three different compensation schemes, the variations in the *SOC* curves due to the scheduling amendment are the smoothest, thus significantly reduced the energy

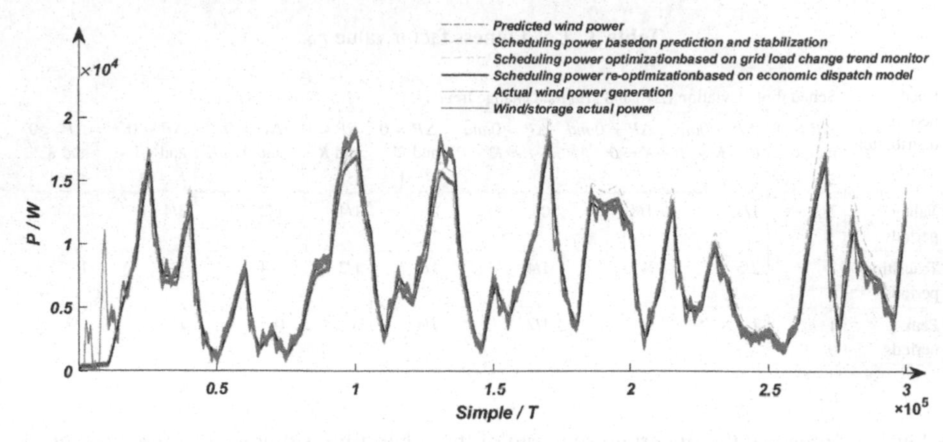

Fig. 4. Power curves of scheduling and actual output

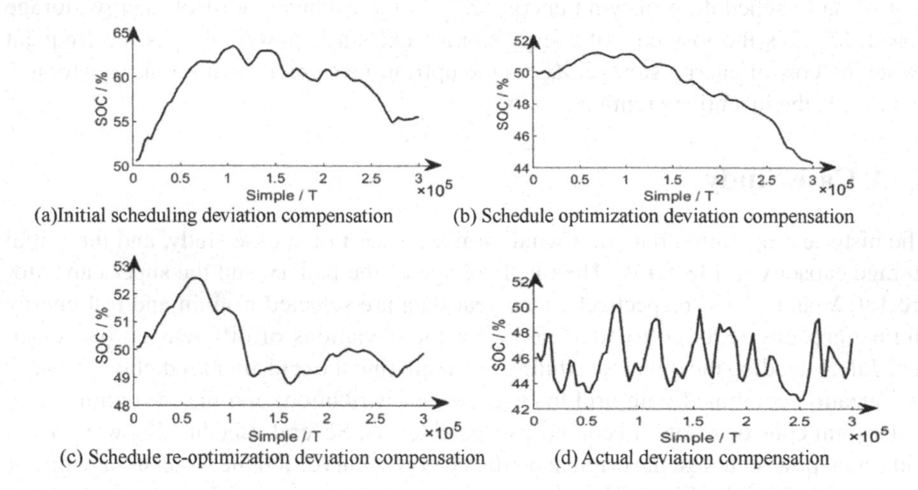

(a)Initial scheduling deviation compensation (b) Schedule optimization deviation compensation

(c) Schedule re-optimization deviation compensation (d) Actual deviation compensation

Fig. 5. Hybrid energy storage SOC changes corresponding to several scheduling strategies

storage capacity requirement. The deviation index is also decreased, and MAE and RMSE under the re-optimized scheduling is reduced by around 1.6% and 0.22%, leading to the reduction in energy storage capacity allocation, as shown in Table 4.

Table 4 reveals that the energy storage capacity requirement of optimized scheduling deviation compensation is lower than the capacity requirement before optimization, total actual capacity be reduced by about 15% and 36% respectively. Meanwhile, the proportion of super-capacitors in the total capacity has also increased. This can effectively utilize the characteristics of super-capacitors and have certain economic benefits. At the same time, the MAE and the RMSE after the re-optimization are reduced by about 1% and 0.12% than that of the initial schedule, respectively, and more than 0.7% and 0.04% of the primary optimized scheduling. The re-optimization is better from the aspect of accuracy. However, from the point of view of volatility, the initial schedule is the largest, while the maximum volatility and average volatility are reduced by 0.7%

Table 4. Power scheduling indexes

Technology	Capacity		Accuracy		Volatility		Reported pass rate	
	Battery/kWh	Super capacitor/kWh	MAE/%	RMSE/%	Maximum volatility/%	Average volatility/%	r1/%	r2/%
Stabilization followed by prediction	3.78	2.24	9.4624	1.1689	2.0349	0.1031	81.36	97.68
Consider load fitness	3.52	1.6	8.7692	1.1085	4.9724	0.0634	95.48	100
Consider economic dispatch	1.26	2.0	6.9959	0.8488	3.6459	0.0622	98.97	100

and 0.01% respectively after model re-optimized. In regards to the accuracy and pass rate, the accuracy of the schedule has been steadily increased after consideration of the load adaptability and model re-optimization.

5 Discussions and Conclusions

Due to the cumulative effect of the throughput characteristics of energy storage operation and the wind power scheduling deviation distribution, the capacity allocation technology used for scheduling deviation compensation mainly depends on the scheduling method. Different scheduling methods will lead to different distribution characteristics of the power deviations. Therefore, the demand for energy storage capacity is also different due to the following specific influencing factors.

1) The distribution characteristics of scheduling deviations corresponding to different prediction technologies, including changes in amplitude and polarity.
2) The scheduling technology, which leads to different distribution characteristics of the scheduled deviations.
3) The intervals of the wind power dispatch.
4) The initial value setting of the *SOC* of energy storage according to different scheduling strategies.

It should be noted that other factors also need to be considered, such as the feasibility of capacity configuration which directly depends on the operation mode and energy management method of the subsequent energy storage. Inappropriate operation control may directly deteriorate the efficiency of existing energy storage system. To achieve economic operations, effective energy management system combined with on-site wind power operation must be put in place.

6 Patents

Chinese invention patents are resulting from the work as follows:

1. Wind/storage integrated power dispatching scheduling optimization realization method. Authorization number: ZL 202010782421.x. Authorization date: 2021/03/26.
2. A hybrid energy storage power distribution method for improving wind power dispatch reliability. Authorization number: ZL 201911165452.4. Authorization date: 2020/12/08.
3. A method for determining hybrid energy storage capacity of Microgrid system load reliable power supply. Authorization number: ZL 201911397312.X. Authorization date: 2020/12/08.

Acknowledgements. This work is a part of the National Natural Science Research Programs of China (No. 61673226), considerable Natural Science Research Projects of Colleges and Universities in Jiangsu Province of China (No. 18KJA470003), Natural Science Foundation of Jiangsu Province (BK20200969), Nantong Science and Technology Bureau Project (JC2018116), and Jiangsu Province's fifth '333 high-level talent training objects' Project. The authors would like to thank for the supports from both the Ministry of Science and Technology and National Natural Science Foundation of China.

References

1. Yan, J., Li, K., Bai, E., et al.: Hybrid probabilistic wind power forecasting using Temporally Local Gaussian Process. IEEE Trans. Sustain. Energy. **7**(1), 87–95 (2016)
2. Yan, J., Li, K., Bai, E., et al.: Analytical iterative multistep interval forecasts of wind generation based on TLGP. IEEE Trans. Sustain. Energy. **10**(2), 625–636 (2019)
3. Lahouar, A., Slama, J.B.H.: Hour-ahead wind power forecast based on random forests. Renew. Energy **109**, 529–541 (2017)
4. Wang, H.Z., Li, G.Q., Wang, G.B., et al.: Deep learning based ensemble approach for probabilistic wind power forecasting. Appl. Energy **188**, 56–70 (2017)
5. Yi, J., Lin, W.F., Hu, J.X., et al.: An integrated model-driven and data-driven method for on-line prediction of transient stability of power system with wind power generation. IEEE Access. **8**, 83472–83482 (2020)
6. Peng, H.W., Liu, F.G., Yang, X.F.: A hybrid strategy of short term wind power prediction. Renew. Energy. **50**, 590–595 (2013)
7. Azimi, R., Ghofrani, M., Ghayekhloo, M.: A hybrid wind power forecasting model based on data mining and wavelets analysis. Energy Conv. Manag. **127**, 208–225 (2016)
8. QGDW10588–2015 Specification of wind power forecasting function. State Grid Corporation (2016)
9. Li, H.B., Zou, H.R., Zhu, J.H.: Optimization of energy storage configuration for error compensation of wind power prediction. In: 2019 6th IEEE International Conference on Systems and Informatics, Shanghai, China, pp. 02–04 (2019)
10. Shi, J., Zhang, G., Liu, X.: Generation scheduling optimization of wind-energy storage generation system based on feature extraction and MPC. Energy Procedia **158**, 6672–6678 (2019)
11. Tao, S., Zhang, B., Liu, J., et al.: Economic storage ratio and optimal control of hybrid energy capacity combining stabilized wind power fluctuations with compensated predictive errors. Power Syst. Technol. **40**(02), 477–483 (2016). (in Chinese)

12. Nguyen-Hong, N.H., Nguyen-Duc, H., Nakanishi, Y.: Optimal sizing of energy storage devices in isolated wind-diesel systems considering load growth uncertainty. IEEE Trans. Ind. Appl. **50**(3), 983–1991 (2018)
13. Wu, Y., Chen, H., Xu, K.: Determination of optimal capacity of battery energy storage system for large-scale wind farm integration considering peaking demand. In: 2018 IEEE Innovative Smart Grid Technologies - Asia (ISGT Asia), Singapore, 22—25 May 2018 (2018)
14. Liu, Y.X., Zhang, Y.Y.: A rolling ARMA method for ultra short term wind power prediction. In: 2017 13th IEEE Conference on Automation Science and Engineering. Xi'an, China, 20--23 August 2017 (2017)
15. Suykens, J.A.K., Vandewalle, J.: Least squares support vector machine classifiers. Neural Process. Lett. **9**(3), 293–300 (1999)
16. Zhang, C., Ding, M., Wang, W.S., et al.: An improved ELM model based on CEEMD-LZC and manifold learning for short-term wind power prediction. IEEE Access. **7**, 121472–121481 (2019)
17. Zhu, Q.Y., Qin, A.K., Suganthan, P.N., et al.: Evolutionary extreme learning machine. Pattern Recogn. **38**(10), 1759–1763 (2005)
18. Catalo, J.P.S., Pousinho, H.M.I., Mendes, V.M.F.: Short-term wind power forecasting in Portugal by neural networks and wavelet transform. Renew. Energy **36**(4), 1245–1251 (2011)
19. Xu, P.C., Li, Y., Zhao, Y.L.: Short-term wind power forecasting based on adaptive mutant bat optimized BP neural network. Electric. Meas. Instrument. **58**(4), 97–104 (2021). (in Chinese)
20. Cui, S., Peng, D.G., Qian, Y.L.: Short-term wind power prediction based on the optimization of radial basis function by adaptive chaos fruit fly optimization algorithm. Renew. Energy. **35**(1), 80–85 (2017). (in Chinese)
21. Jiang, Y.C., Yang, X.Q., He, F., et al.: Super-short-time wind power forecasting based on EEMD-IGSA-LSSVM. J. Hunan Univ. Nat. Sci. **43**(10), 70–78 (2016)
22. Hao, Y., Dong, L., Liao, X.Z., et al.: A novel clustering algorithm based on mathematical morphology for wind power generation prediction. Renew. Energy. **136**, 572–585 (2019)
23. Zeng, G.M., Ye, S.Z.: A grey model for river water qualification and its grey parameters' optimal estimation. J. Grey Syst. **1**(1), 53–64 (1989)
24. Wang, Y.R., Wang, D.C., Tang, Y.: Clustered hybrid wind power prediction model based on ARMA, PSO-SVM and Clustering methods. IEEE Access **8**, 7071–17079 (2020)
25. Junho, L., Wu, W., Fouzi, H., et al.: Wind power prediction using ensemble learning-based models. IEEE Access **8**, 61517–61527 (2020)

An H-ELM Based Equivalent Droop Control for Hybrid Power Systems

Jianhua Zhang[1(✉)], Bin Zhang[2], Hongqun Gu[3], Bin Li[3], and Lei Wang[3]

[1] State Key Laboratory of Alternate Electrical Power System With Renewable Energy Sources, North China Electric Power University, Beijing 102206, China

[2] School of Control and Computer Engineering, North China Electric Power University, Beijing 102206, China

[3] State Grid Liaoning Electric Power Supply Co. Ltd., Shenyang 110006, China

Abstract. Primary frequency regulation plays an important role in a hybrid power system which is composed of battery energy storage systems (BESSs), photovoltaic (PV) systems and thermal power plants. Droop control is a basic primary frequency control method. In this paper, an equivalent droop control method based on hierarchical extreme learning machine (H-ELM) is proposed for hybrid power plants. Therefore the $P-f$ droop characteristics can be fitted by an H-ELM rather than traditional fixed droop control curve. Finally, the effectiveness of the proposed method is justified by the simulation results.

Keywords: Frequency regulation · Hybrid power plant · Battery energy storage system · Photovoltaic · Hierarchical extreme learning machine · Droop control

1 Introduction

In recent years, due to rapid development of renewable energy, the penetration rate of new energy resources in the power system has continued to increase. However, the power system with new energy resources is usually complex in terms of intermittent, volatility, uncertainty and low or non-existent inertia. It is necessary to manage inherent flexibilities in demand to follow variable renewable generation. Hybrid power plants (HPPs) might be one of promising way to optimally balance variable generation and flexible demand.

Frequency regulation plays an important role in maintaining the balance between power generation and load [1]. Besides photovoltaic (PV) power plants [2], battery energy storage systems (BESSs) [3] can be utilized to improve frequency response as well. In this work, an HPP which consists of thermal power plants, photovoltaic power plants and battery energy storage systems is investigated to improve droop control performance.

Primary frequency control is an automatic governor response which is proportional to the frequency deviation. For a thermal power plant, it is typically provided by governor droop control to regulate its power output. To imitate the traditional governor response for primary frequency regulation, droop control has been implemented in PV power plants

K. Li et al. (Eds.): LSMS 2021/ICSEE 2021, CCIS 1468, pp. 110–119, 2021.
https://doi.org/10.1007/978-981-16-7210-1_11

and BESSs respectively [2–4]. However, researchers have not dealt with the equivalent droop control characteristics of HPPs.

Static droop control can provide additional active power based on the grid frequency deviation. It usually features a droop curve with a fixed slope and a pre-defined dead band. Both simulation tests and engineering practice have illustrated that droop control can prevent system frequency decline and reduce the steady-state frequency deviation.

In this work, the equivalent droop control characteristics of HPPs is investigated. In particular, dynamic droop control law is implemented by a hierarchical extreme learning machine (H-ELM), hence the slope and dead band of droop curve can be adjustable in real-time.

The rest of the paper is organized as follows: Sect. 2 revisits droop control of PV power plants and BESS respectively. Section 3 proposes an equivalent droop control strategy for an HPP which is comprised of thermal power plants, PV power plants and BESSs using H-ELM. Section 4 presents a simulation example to testify the effectiveness of the proposed approach. The last section concludes this paper.

2 Droop Control of PV Power Plant and BESS

Generally, the power system should continuously meet the instantaneous balance between power generation and demand. Any disturbances to the balance may cause frequency fluctuations. Droop control is a commonly used primary frequency regulation strategy. In this section, the droop control method of both PV power plants and BESS are revisited respectively.

2.1 Droop Control of BESS

The power frequency static characteristic means that the corresponding output power increases when the grid frequency drops, and vice versa. Although BESS does not have the droop control like traditional power units, it can emulate the traditional governor response to adjust the active power output of the BESS according to the frequency deviation. The power reference signal related to the frequency deviation ΔP_B is as follows:

$$\Delta P_B = -\frac{1}{R_B}\Delta f \tag{1}$$

where the droop gain $1/R_B$ is the inverse of droop coefficient R_B, Δf stands for the system frequency deviation.

The droop control strategy of BESSs is shown in Fig. 1. The input signals of this controller are the deviation of the measured frequency f and a reference frequency f_{ref}. A dead zone and a low-pass filter (LPF) blocks are included into the control strategy. The droop controller is a proportional controller whose gain is $\frac{1}{R_B}$. Finally, the output can be obtained through a saturation module.

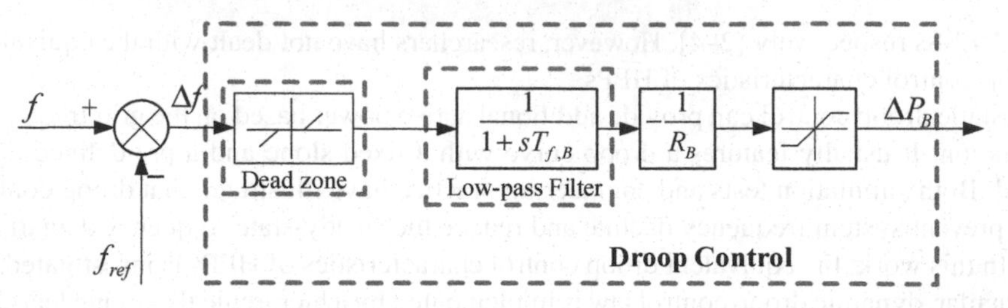

Fig. 1. The frequency control strategy of BESS

2.2 Droop Control of PV

The droop controller of PV power plants is similar to that of BESSs. Nevertheless, photovoltaic power plants cannot provide additional power support to cope with the change of frequency at the maximum power point, therefore, the de-loading control method is adopted, so that PV power plants can provide reserve power to participate in the frequency regulation. PV power plants involved in frequency regulation operate in the mode of de-loading control, furthermore, the de-loading rate of PV is defined as follows:

$$\sigma\% = \frac{P_d}{P_m} \tag{2}$$

where $\sigma\%$ is the de-loading rate and set to 20%. P_m and P_d are the maximum output power of PV power plants and the de-loading power respectively.

Under de-loading control mode, the photovoltaic output power is:

$$P_{pv} = P_m(1 - \sigma\%) \tag{3}$$

PV power plants also use droop control to adjust the active power output of the PV according to the frequency deviation. The power reference signal related to the frequency deviation ΔP_{pv} is as follows:

$$\Delta P_{pv} = -\frac{1}{R_{pv}}\Delta f \tag{4}$$

where the droop gain $1/R_{pv}$ is the inverse of droop coefficient R_{pv}.

In summary, the droop control strategy of PV power plants is shown in Fig. 2. The frequency control is a series combination of droop gain, a LPF and a dead zone. The output signal of the frequency controller is then added to the active power reference of the output power of de-loading control.

3 Equivalent Droop Control of Hybrid Power System

Hybrid power plants constituted by combining PV and conventional energy sources with BESSs can provide economic, environment friendly and reliable supply of electricity. The thermal power plants in the hybrid power system shown in Fig. 3 do not participate in frequency regulation. In this HPPs, m PV power plants and n BESSs provide frequency responsive reserve.

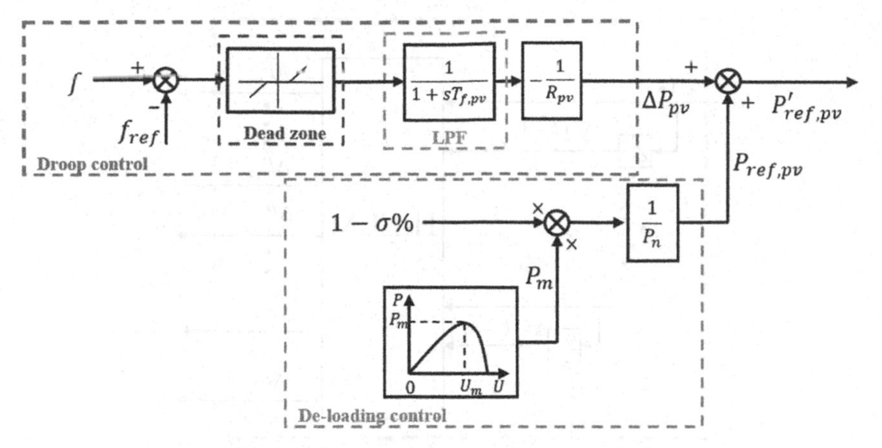

Fig. 2. The frequency control strategy of PV

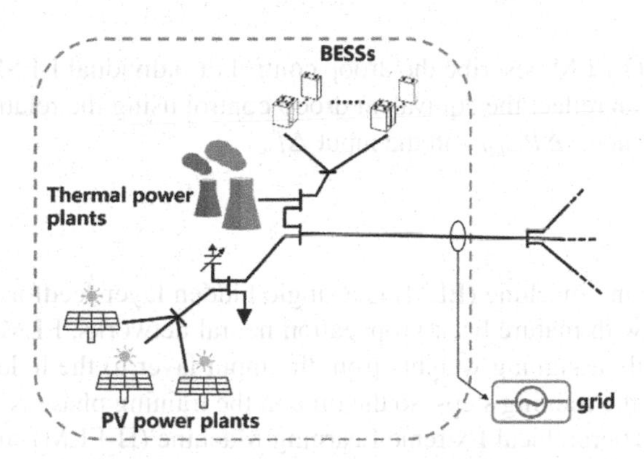

Fig. 3. Hybrid power system with BESS

A hybrid power plant is expected to optimize the generation portfolio. In this work, PV power plants and BESSs provide ancillary service for frequency regulation. Droop control plays an important role in mitigating the steady-state frequency deviation before secondary frequency response. Therefore, it is necessary to study the equivalent droop control strategy for the HPP. Moreover, H-ELM shown in Fig. 4 is employed to fit the $P - f$ droop relationship curve instead of the traditional droop curve with a fixed slope and a pre-defined dead band. At k instant, the input of H-ELM neural network is $\left[\Delta f(k), \Delta f(k-1), \cdots \Delta f(k-M)\right]^T$, the output of H-ELM neural network is $\left[\Delta P_{B,1}, \cdots \Delta P_{B,n}, \Delta P_{pv1}, \cdots \Delta P_{pv,m}, \Delta P_{total}\right]^T$, where $\Delta P_{B,i}(i = 1, \cdots, n)$, $\Delta P_{pv,j}(j = 1, \cdots, m)$ and ΔP_{total} satisfies

$$\Delta P_{total} = \sum_{i=1}^{n} \Delta P_{B,j} + \sum_{j=1}^{m} \Delta P_{pv,j} \tag{5}$$

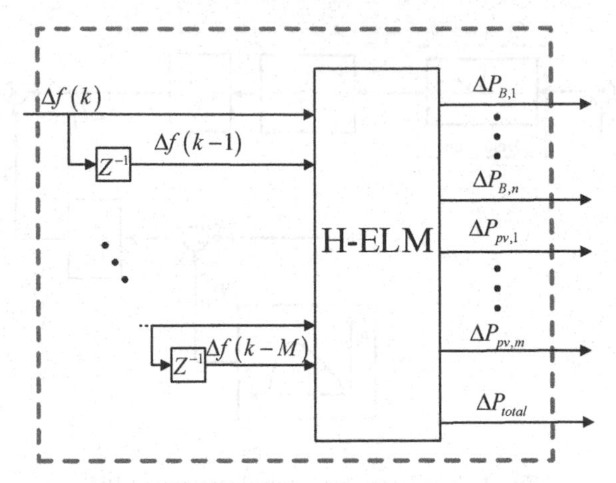

Fig. 4. An equivalent droop control strategy

Not only can H-ELM describe the droop control of individual BESS and PV power plant, but also it can reflect the equivalent droop control using the relationship between the bottom output node ΔP_{total} and the input Δf.

3.1 H-ELM

The extreme learning machine (ELM) is a single hidden layer feedforward neural network. Compared with mature back-propagation neural networks, ELM has the characteristic of randomly assigning weights from the input layer to the hidden layer, which greatly simplifies the training steps, so the time in the training phase is greatly reduced. In this paper, the Hierarchical Extreme Learning Machine (H-ELM) shown in Fig. 5 is applied to fit $P-f$ droop relationship curve.

H-ELM neural network is a simple encoder-decoder framework [5–7]. The encoder-decoder framework is one of the most extensive network frameworks currently used by neural networks in the fields of classification, regression and prediction. Because the input of the neural network has the irregularity of the training data, the uncertainty of the data size, the inconspicuousness of the data dimension and the insignificance of the data characteristics, in this case, the input data can be converted into another standard easy-to-use data, and then train the neural network, the transformation process is called the encoding process, and the process of using the encoded data for training is called the decoding process. The decoder network constructs a mapping between the output of the encoder network and the training output of the neural network, so that the mapping relationship between non-standard input data and output can be obtained. This ELM-based hierarchical learning framework is more efficient than deep learning due to its excellent feature representation ability.

It can be observed from Fig. 5 that H-ELM is divided into two parts: the forward encoder layer and the ELM training layer. When the estimated value of the state is obtained, it is then normalized as a vector and input to the neural network, subsequently, the input is passed through several hidden layers for feature extraction in the forward

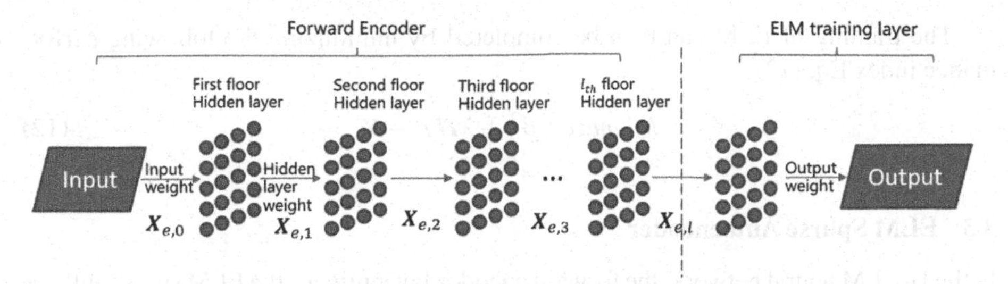

Fig. 5. H-ELM neural network framework diagram

coder layer to form a standard, neat and characteristic vector. Finally, the encoded vector is passed to the ELM training layer for training and testing to get the desired output.

3.2 ELM Neural Network

For the training data set $\{x_i, y_i | x_i \in \mathbb{R}^{M+1}, y_i \in \mathbb{R}^{m+n+1}\}_{i=1}^{N}$, where N is the size of the data set, $M+1$ is the number of features of the input data, and $m+n+1$ is the number of features of the output data. The ELM Lets β be the output weight matrix. $\boldsymbol{H}$ is the hidden layer matrix that is a mapping from input $\boldsymbol{X}$ to the ELM solution space, its goal is to form the relationship between input $\boldsymbol{X}$ and output $\boldsymbol{Y}$:

$$Y = H\beta \tag{6}$$

where

$$H = \begin{bmatrix} \boldsymbol{h}(x_1) \\ \vdots \\ \boldsymbol{h}(x_N) \end{bmatrix} = \begin{bmatrix} h_1(x_1) & \dots & h_L(x_1) \\ \vdots & \ddots & \vdots \\ h_1(x_N) & \dots & h_L(x_N) \end{bmatrix} \tag{7}$$

where L is the number of hidden layer nodes, and let

$$h_j(x_i) = g\left(a_j^T x_i + b_j\right), \ (1 \le i \le N, 1 \le j \le L) \tag{8}$$

Equation (8) indicates that the input weight a and bias b of the j_{th} hidden layer node are randomly generated for the input x_i, and then mapped to the high-dimensional space through the activation function $\boldsymbol{g}$.

The output weight matrix β can be solved as follows

$$\beta = \boldsymbol{H}^{\dagger} Y \tag{9}$$

If $\boldsymbol{H}^{\dagger}\boldsymbol{H}$ is non-singular, the output weight matrix β can be calculated by adding a positive value $\frac{1}{\lambda}$, it leads to

$$\beta = \left(\tfrac{1}{\lambda} + \boldsymbol{H}^{\dagger}\boldsymbol{H}\right)^{-1} \boldsymbol{H}^T Y \tag{10}$$

Consequently, the output equation of ELM is as follows

$$f(x_i) = h(x_i)\left(\tfrac{1}{\lambda} + \boldsymbol{H}^{\dagger}\boldsymbol{H}\right)^{-1} \boldsymbol{H}^T Y \tag{11}$$

The training of ELM can then be completed by minimizing the following performance index Eq. (12).

$$Minmize : \beta^2 + \lambda \boldsymbol{H}\beta - \boldsymbol{Y} \tag{12}$$

3.3 ELM Sparse Autoencoder

In the H-ELM neural network, the forward encoder layer utilizes the ELM sparse autoencoder. It means that the ELM autoencoder with sparse constraints is used for forwarding encoder. The ELM neural network has the characteristics of fast training and high degree of fitting, therefore this autoencoder can obtain better results.

As shown in Fig. 5, let the input of the i_{th} hidden layer be $\boldsymbol{X}_{e,i-1}$, then it's corresponding reconstruction matrix in the forward encoder layer is $\beta_{e,i}$, hence,

$$\boldsymbol{X}_{e,i} = \boldsymbol{H}_{e,i}\beta_{e,i} \tag{13}$$

where $\boldsymbol{H}_{e,i} = \begin{bmatrix} \boldsymbol{h}(x_{e,i-1_1}) \\ \vdots \\ \boldsymbol{h}(x_{e,i-1_N}) \end{bmatrix}$ and $\boldsymbol{h}_j(x_{e,0_k}) = \boldsymbol{g}\left(a_j^T x_{e,0_k} + b_j\right)$, $(1 \le i \le N,\ 1 \le j \le L)$. N is the scale of the input data set. L represents the number of hidden layer nodes in the k_{th} layer.

From Eq. (13), we have

$$\beta_{e,i} = \boldsymbol{H}_{e,i}^{\dagger}\boldsymbol{X}_{e,i} \tag{14}$$

In the forward encoder layer, the input is encoded to form a certain mapping relationship, and β can obtain the key features of the input. In order to minimize the reconstruction error, the constraint index of the ELM sparse autoencoder can be designed as follows:

$$O_{\beta_{e,i}} = \underset{\beta_{e,i}}{arg\ min}\left\{ \left\|\boldsymbol{H}_{e,i}\beta_{e,i} - \boldsymbol{X}_{e,i}\right\|^2 + \left\|\beta_{e,i}\right\|_{l1} \right\} \tag{15}$$

where $\boldsymbol{X}_{e,i}$ is the output of the i_{th} forward encoder layer. In this autoencoder, random mapping is used to extract features from the hidden layer. Equation (15) can be reformulated as follows:

$$O_{\beta_{e,i}} = p(\beta_{e,i}) + q(\beta_{e,i}) \tag{16}$$

where $p(\beta_{e,i}) = \boldsymbol{H}_{e,i}\beta_{e,i} - \boldsymbol{X}_{e,i}^2$,and $q(\beta_{e,i}) = \left\|\beta_{e,i}\right\|_{l1}$.

3.4 Training Data of H-ELM

The $P - f$ droop relationship curve can be formulated through a series of rules, such as the frequency regulation time and response speed of individual BESS or PV power plant. The rules of battery energy storage and photovoltaic systems have been investigated and reported in [8] and [9] respectively. The training data of the H-ELM in this work is collected based on the nonlinear $P - f$ droop control curve [8, 9].

4 Simulation Results

In this section, some simulations are conducted to verify the feasibility of the proposed equivalent droop control method. The investigated hybrid power system is shown in Fig. 6. The HPP consists of a thermal power plant, a photovoltaic power station and a battery energy storage system, where the thermal power plant does not participate in frequency regulation. The parameters of the HPP and H-ELM are listed in Table.1.

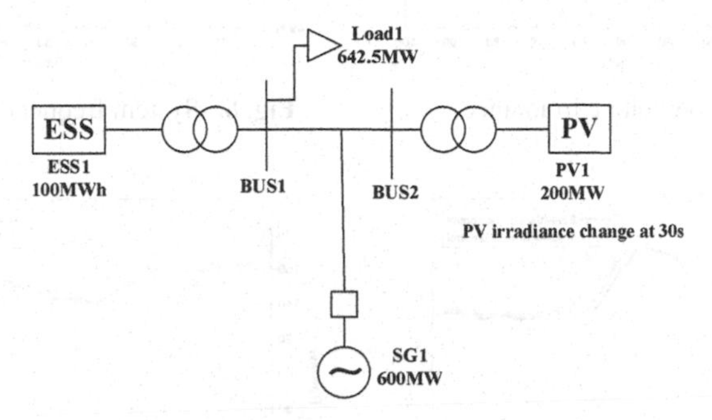

Fig. 6. Simulation system

Table.1. Parameters of the HPP and H-ELM

Parameter	Value
Power of thermal power plant	600 MW
Power of photovoltaic power station	200 MW
Rated capacity of battery energy storage system	100 MWh
Value of initial load	642.5 MW
Number M of delays of frequency deviation	9
Number l of hidden layers of H-ELM	4
Hidden layer iterations	300

In the simulation, the solar irradiance with random disturbance is shown in Fig. 7. To compare conventional droop controller [8, 9] with the proposed equivalent droop controller, the conventional droop control coefficients of BESS and PV are set to 0.02 and 0.05 respectively. Figure 8 demonstrates the system frequency responses under the traditional droop control and the presented equivalent droop control method respectively. When the droop control coefficient is fixed, the minimum value of the frequency is 49.8296 Hz. While the equivalent droop control method is utilized, the minimum value of the frequency is 49.8628 Hz. In addition, Fig. 9 illustrates the active power of battery energy storage system and photovoltaic power station under the traditional droop control and the equivalent droop control respectively. Therefore, it can be observed from Fig. 8 and Fig. 9 that the equivalent droop control method has better control performance.

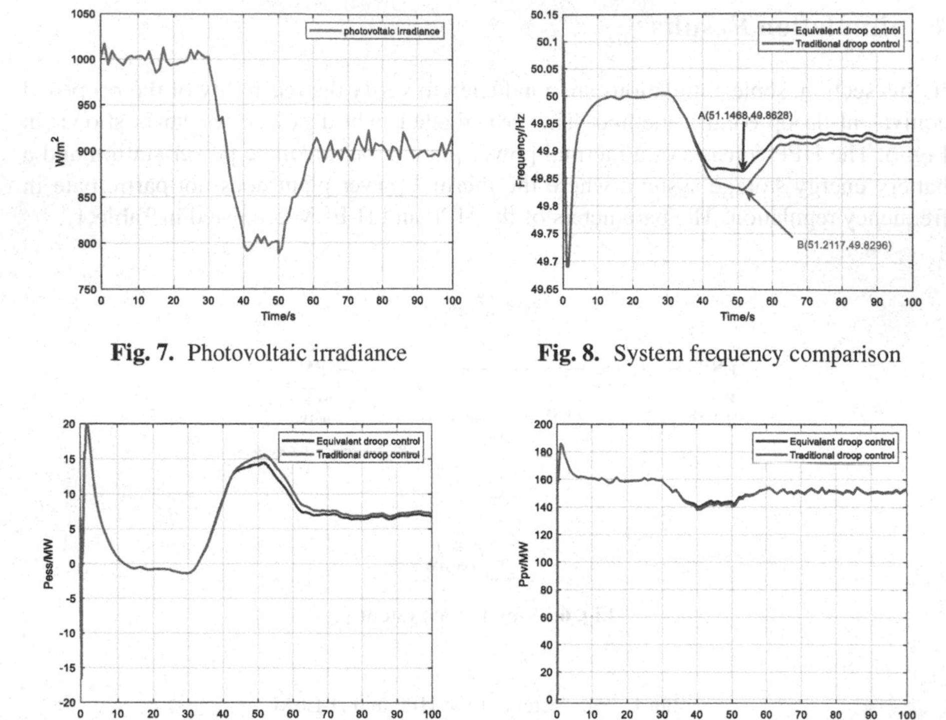

Fig. 7. Photovoltaic irradiance

Fig. 8. System frequency comparison

(a) Active power of BESS comparison

(b) Active power of PV comparison

Fig. 9. Active power of BESS and PV comparison

5 Conclusion

This paper presented an equivalent droop control method for HPPs with PV power plants, thermal power plants and BESSs. An H-ELM is utilized to fit the $P - f$ droop relationship curve. Simulation results show that the proposed equivalent droop controller outperforms the traditional droop controller.

Acknowledgments. This work was supported by National Key R&D Program of China No.2019YFB1505400. These are gratefully acknowledged.

References

1. Akram, U., Nadarajah, M., Shah, R.: Review on rapid responsive energy storage technologies for frequency regulation in modern power systems. Renew. Sustain. Energy Rev. **120**, 109626 (2020)
2. Mohapatra, A., Nayak, B., Das, P., et al.: A review on MPPT techniques of PV system under partial shading condition. Renew. Sustain. Energy Rev. **80**, 854–867 (2017)

3. Datta, U., Kalam, A., Shi, J.: Battery energy storage system control for mitigating PV penetration impact on primary frequency control and state-of-charge recovery. IEEE Trans Sustain. Energy. **11**, 746–757 (2019)
4. Li, J., Xiong, R., Yang, Q., et al.: Design/test of a hybrid energy storage system for primary frequency control using a dynamic droop method in an isolated microgrid power system. Appl. Energy **201**, 257–269 (2017)
5. Lebanoff, L., Song, K., Liu, F.: Adapting the neural encoder-decoder framework from single to multi-document summarization. In: Proceedings of the 2018 Conference on Empirical Methods in Natural Language Processing (2018)
6. Cambria, E., Huang, G.B., Kasun, L., et al.: Extreme Learning Machines [Trends & Controversies]. IEEE Intell. Syst. **28**(6), 53–59 (2013)
7. Tang, J., Deng, C., Huang, G.B.: Extreme Learning Machine for Multilayer Perceptron. IEEE Trans. Neural Netw. Learn. Syst. **27**, 5809–5821 (2017)
8. Northeast Regulatory Bureau of National Energy Administration. Grid code for auxiliary service management of grid connected power plants in Northeast China, Shenyang (2019)
9. Zhao, D.W., Ma, J., Qian, M.H., et al.: Research on control gain for photovoltaic power plants participating in primary frequency regulation of large power grid. Power Syst. Technol. **43**(02), 5425–5435 (2019). (in Chinese)

Energy-Efficient Operation Curve Optimization for High-Speed Train Based on GSO Algorithm

Wei Li[1], Sizhe Zhao[1(✉)], Kang Li[1,2], Yi Xing[1,3], Gaofeng Liu[1], and Jilong Liu[1,4]

[1] School of Traffic and Transportation Engineering, Central South University, Changsha, China
[2] School of Electronic and Electrical Engineering, University of Leeds, Leeds, W Yorkshire, England
[3] CRSC Research and Design Institute Group Co., Ltd., Beijing, China
[4] CRRC QINGDAO SIFANG Co., Ltd., Shandong, China

Abstract. The demand for high-speed automatic train operation (ATO) system brings new opportunities and challenges for the high-speed railway field. The requirements of energy consumption, punctuality, safety and smoothness are also increasing, especially the research on energy saving of high-speed trains ushers in a broader development space. This paper proposes quantitative functions of four energy-saving performance indexes and the multi-objective optimization model of train operation curve considering the characteristics of high-speed train ATO system. An energy-saving optimization method of train operation curve based on Glowworm Swarm Optimization (GSO) algorithm is proposed. The simulation results show that the train operation using the high-speed train operation curve optimization method based on the GSO algorithm can save about 16.9% of electrical energy consumption per kilometer compared with that by other optimization algorithms, which verifies the effectiveness of the method. The result from this paper provides theoretical support for practical application.

Keywords: High-speed train · Energy-efficient operation · Train operation curve · Multi-objective optimization · GSO algorithm

1 Introduction

The development of high-speed trains has made the research on energy-saving operation more and more urgent and demanding. With regard to energy consumption optimization, operation mode optimization of the train is one of the easiest and feasible ways to achieve energy consumption reduction in high-speed train operation, and a large number of studies have been conducted by domestic and foreign scholars.

The continuous development of computer control technology and intelligent control theory has provided new ideas and in-depth development for the research of train energy saving. T Xie [1] carried out an energy-efficient method of the problem by using simulated annealing method. Y Bai [2] developed a fuzzy predictive control method based on previous research to reduce the energy consumption of train operation. L Zhou *et al.* [3] searched the speed/acceleration curve using dynamic programming algorithm and

K. Li et al. (Eds.): LSMS 2021/ICSEE 2021, CCIS 1468, pp. 120–132, 2021.
https://doi.org/10.1007/978-981-16-7210-1_12

developed a modeling framework, and the method was tested practically on the Beijing-Shanghai line. Scholars Shuqi Liu and *et al.* [4] conducted a related study based on the ATO control strategy, and based on this, they proposed a single-vehicle energy-saving optimization model to optimize the recommended operating curve and ATO system tracking control measurements. Zhang Miao *et al.* [5] designed a train operation energy consumption optimization model based on the Q-learning algorithm. Tang Tao *et al.* [6] used an ant colony algorithm to optimize the train ATO control strategy, which ultimately reduced the energy consumption of single-train operation by 5.67%. Many scholars have provided new ideas and in-depth development for the research of train energy saving.

Most of the existing studies have confronted with problems such as complicated modeling, influenced by parameters, slow solving process and easy to fall into the optimal solution, while the group intelligent algorithm can adapt well to the complex and nonlinear train operation process. In this paper, based on the research of other scholars, we introduce the GSO algorithm to solve the energy-saving train operation optimization problem, and verify the effectiveness of the method through simulation.

2 Analysis of Energy-Saving Operation and Its Control Impact Based on ATO

Table 1. Symbols.

Symbol	Description
$v(s)$	Speed of the train at position s (m/s)
F_f, F_b	Traction force and braking force on the train (N)
F_r, F_g, F_c	Basic resistance, additional resistance on ramps and additional resistance (N)
M	Mass of the train (t)
a, b, c	Davis index based on the structure of the train
q	The thousandth of slope
A_c	Experience coefficient about the resistance of the curve
R	Radius of curve (m)

The architecture of the onboard ATO system for high-speed trains is composed of two main layers, as shown in Fig. 1. The upper layer is the optimization layer, which is mainly responsible for generating the reference operation curve under the given constraints; the lower layer is the control layer, which is responsible for comparing the current speed of the train with the target speed in the reference operation curve and outputting control commands to realize the traction operation of the train.

In the actual operation of high-speed trains, the operation status is changed according to the control of the train operation curve, to carry out energy-saving train operation optimization, we need to analyze and model the train operation process in the system optimization layer first.

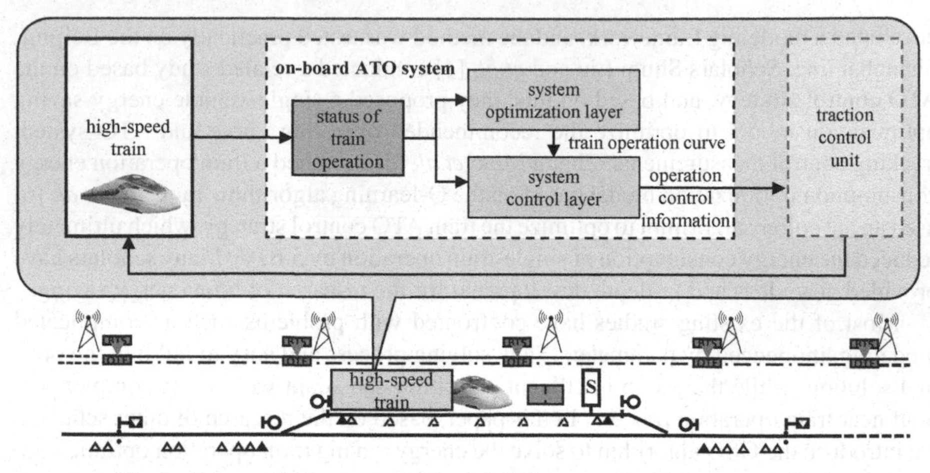

Fig. 1. Schematic structure of onboard ATO system for high-speed trains.

2.1 Analysis of High-Speed Train Operation Process

The train in the process of operation by the locomotive traction and a variety of resistance, which plays an impact on the train traction energy consumption, the train in the driving process of the kinetic equation is shown in Eq. (1). Among them, $F_r = a + b \cdot v + c \cdot v^2$, $F_g = \pm M \cdot g \cdot q$ and $F_c = M \cdot A_c/R$.

$$\frac{dv(s)}{ds} = \frac{F_f - F_b - F_r - F_g - F_c}{v(s) \cdot M \cdot (1 + \rho)} \tag{1}$$

According to the magnitude and direction of the combined force on the train, the train operation condition is composed of six driving conditions: constant-torque traction mode, constant-power traction mode, cruising mode, costing mode, braking mode, and stopping mode. Figure 2 shows an ideal train running speed curve.

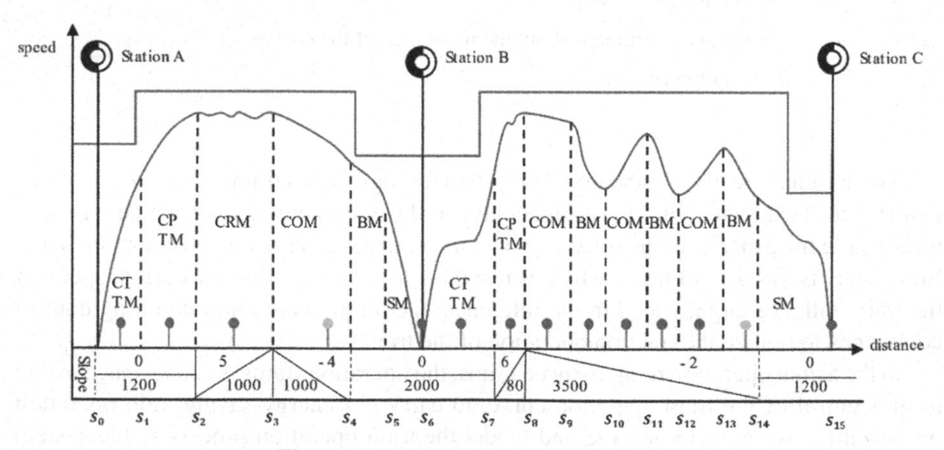

Fig. 2. The running process of the train.

In the actual operation process, different operating curves correspond to different train maneuvering sequences, and different sets of maneuvering sequences have different performances in terms of energy consumption, passenger comfort, and safety. Even under the same line conditions, the energy consumption corresponding to different train maneuvering strategies varies up to 30% [7].

By analyzing the process of the high-speed train operation, we can get the conclusion that the object of operation curve optimization is the operation manipulation strategy.

2.2 Analysis of High-Speed Train Operation Manipulation Strategy

According to the operational needs of the driving strategy classification, common train maneuvering strategy are the minimum time strategy, the minimum energy strategy, and the integrated optimization strategy, three different strategies corresponding to the operating curve schematic as shown in Fig. 3.

Among them, the minimum time strategy highlights the importance of punctuality of the train, but damages the train's condition in the long term, and also brings a lot of energy consumption. The minimum energy strategy takes longer time compared with the shortest time strategy, but it does not consider the smoothness of train operation and the accuracy of the stopping position.

In response to the shortcomings of the above two driving strategies, the integrated optimization strategy takes several indicators such as time consumption, energy consumption, and smoothness into consideration. The integrated optimization strategy can save train energy while ensuring punctuality and smooth train operation, is a more complex but more realistic real-time variable parameter strategy.

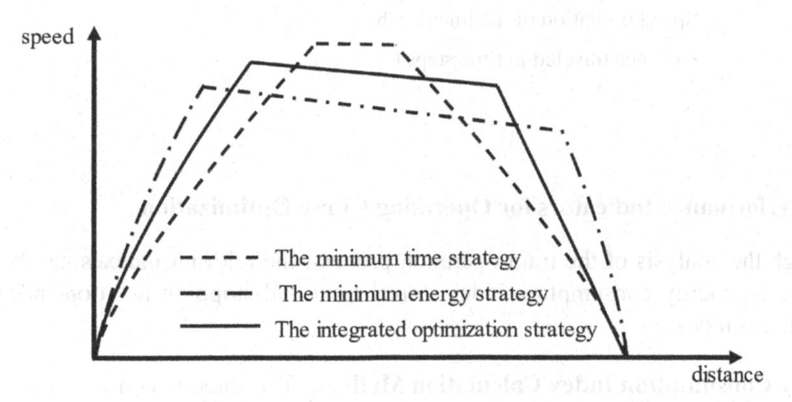

Fig. 3. Three typical driving strategies.

3 Energy-Efficient Operation System Method and Basic Control Strategy

The goal of train running curve optimization is to generate a train running speed curve that meets the energy-saving target. The train is affected by factors such as the operating line, signal control, train characteristics, and train smoothness requirements when traveling on the line, so it is necessary to first calculate the train operating curve optimization index and derive the corresponding energy consumption optimization model when researching the curve optimization problem.

Table 2. Symbols.

Symbol	Description
μ_f	Electrical energy conversion efficiency in motoring mode
s_0	The end of the interval of motoring mode (km)
A	Auxiliary energy efficiency (kWh/h)
$\Delta T, \Delta S$	Deviation value of actual time/distance from planned time/distance (s)/(m)
P_t, P_s	Travel time and stopping position index penalty function
τ	Maximum acceptable travel time index, taken as 120s
σ	Maximum acceptable stopping position index, taken as 0.3m
A_{pt}, A_{ps}	A positive integer greater than or equal to 1, taken as 1000
G	Train smoothness index value (maximum value is 1)
$\overline{a}$	The average acceleration in a certain time interval (m/s^2)
f_e	Objective function of the optimization model
$f(v), w(s),$	Unit traction/ resistance of the train (N/kN)
$b(v)$	Unit braking force on the train (N/kN)
v_{max}	Speed limitation of the line (km/h)
s_i	Distance traveled in time steps (s)

3.1 Performance Indicators for Operating Curve Optimization

Through the analysis of the train operation process, the relevant indexes can be summarized as energy consumption index, travel time and stopping position index, and smoothness index.

Energy Consumption Index Calculation Method. The energy consumed by high-speed train operation is composed of traction energy consumption and non-traction energy consumption. Traction energy consumption is expressed as the integral of the tractive force against the distance under each working condition, and the non-traction train energy consumption is calculated in terms of time. The operation energy consumption E during the process is calculated by the following formula.

$$E = \frac{1}{\mu_f} \int_0^{s_0} F(v)ds + A \times \frac{T}{3600} \tag{2}$$

Travel Time and Stopping Position Index Calculation Method. In actual running condition, the train runs on the line according to the fixed running plan, and the interval running time and station stopping position are pre-set fixed values. In the process of optimizing the train running curve, the running time error and stopping position error of the train is critical, and when the error value exceeds the set range, the additional penalty value should avoid unnecessary safety hazards.

$$\Delta T = |T - T_0| \tag{3}$$

$$P_t = \begin{cases} 1, \Delta T \le \tau \\ A_{pt}, \Delta T > \tau \end{cases}, P_s = \begin{cases} 1, \Delta S \le \sigma \\ A_{ps}, \Delta S > \sigma \end{cases} \tag{4}$$

Smoothness Index Calculation Method. The intense speed changes during the train operation will react with the inertia of the passengers and bring uncomfortable riding experience for them. In this paper, the smoothness of the train driving is taken into consideration as one of the performance indicators, and the average acceleration G value of the high-speed train is calculated as follows.

$$G = \frac{\overline{a}}{\Delta t} \tag{5}$$

According to the train dynamics theory introduced in this section and the indexes of energy consumption, the mathematical model of train operation curve optimization is established as follows.

$$\min J = k_1 \sum\nolimits_{i=1}^{n} E_k + k_2|T_0 - \sum\nolimits_{i=1}^{n} T_k| + k_3|S_0 - \sum\nolimits_{i=1}^{n} S_k| + k_4 max\{G\}$$

$$s.t. \begin{cases} v = ds/dt, \\ m \cdot dv/dt = f(v) - w(s) - b(v), \\ 0 \le s_i \le S, i = 1, 2, \cdots, n s_1 = 0, s_n = S \\ v(0) = v(S) = 0, v(s_i) \le v_{\max} \\ t(0) = 0, t(S) = T_0, 0 \le t(s_i) \le T_0. \end{cases} \tag{6}$$

3.2 Analysis of Train Operation Curve Optimization Process

The train operation optimization process analysis is the basis for combining the train operation optimization problem with the optimization algorithm. The train operation curve optimization problem can be translated into solving the optimal maneuvering scheme from the departure point s_0 to the destination s_n.As shown in Fig. 4, the state of the train at working condition changeover point s_2 is determined by the working condition taken at working condition changeover point s_1 and the duration of this working condition. If the train takes working condition u_1 at working condition changeover point s_1 and continues to travel, then the expected benefit achieved when the train travels to working condition changeover point s_2 is Q_1, the lower the expected benefit the better the optimization effect.

According to the principles of energy-saving train operation [8–10], the train adopts the maximum tractive force in the starting phase $[S_0, S_1]$ and the maximum braking force in the braking phase $[S_{n-1}, S_n]$ to achieve the effect of energy-saving and consumption reduction. When the train is in the intermediate operation stage, the following strategies can effectively improve the performance indexes to obtain the desired benefits of the optimization function.

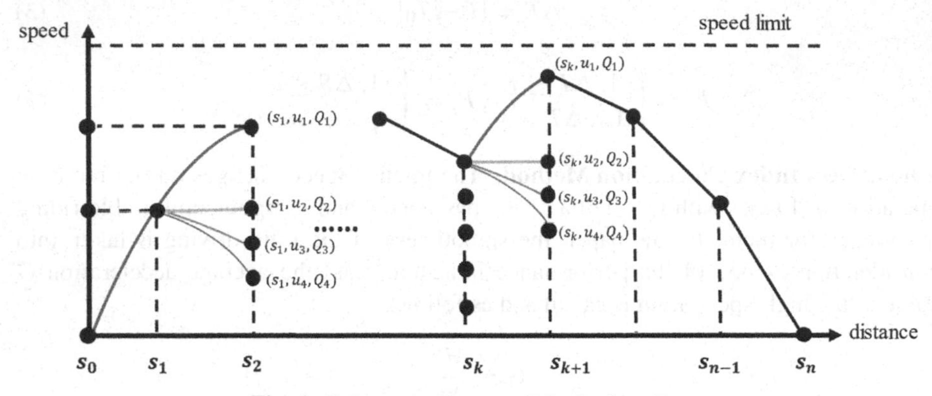

Fig. 4. Schematic diagram of the basic policy

(1) Operation at constant acceleration on short-distance level tracks.
(2) Switching to cruising mode on long-distance level tracks if the train reaches the upper operating speed limit ahead of schedule.
(3) On uphill grades, applying partial traction to the train to maintain a state of uniform motion in preparation for downhill coasting.
(4) On the downhill slope, taking full advantage of the slope of the ramp and adopts the coasting mode to achieve the purpose of energy-saving maneuvering, and if the distance of the ramp is long, it needs to switch to the cruising mode in the middle to run to the end of the ramp.
(5) On undulating ramps, by alternately adopting three working conditions of traction, cruising and coasting. And also avoids braking to achieve the effect of energy-saving and smooth operation.

4 Solution Methodology Based on GSO Algorithm

The optimization algorithm has an important impact on the optimization effect of the problem. At present, traditional optimization algorithms such as genetic algorithm encounter bottlenecks in the optimization of train operation, while the GSO algorithm, as a new population intelligence algorithm, has the characteristics of fewer parameters and not easy to fall into the optimal solution [0], in this way, it can better construct and find the optimal solution according to the train operation optimization problem, the following section shows the way to combine the train optimization problem with the GSO algorithm.

Table 3. Basic parameters of the GSO optimization algorithm.

Parameter	Description	Parameter value
ρ	Fluorescein volatility coefficient	0.40
γ	Fluorescein growth coefficient	0.60
B	Perceptual radius coefficient of variation	0.08
n_i	Number of outstanding individuals in the perceptual range	5
l_0	Initial fluorescein value	5
N	Population size	30

Setting of Algorithm Parameter. When using the GSO algorithm to optimize the problem, we should first determine the relevant parameters, which can be directly referred to and the initial population-related parameters that play a key role in the optimization of the algorithm. After multiple comparisons and considerations [0, 0, 0], the basic parameters of the GSO optimization algorithm in the train running curve optimization problem are set as follows.

Determination of Fitness Value. Individual fireflies rely on brightness to attract others to achieve position shifting in order to search for the optimal solution, the Q can be calculated as follows. k is the corresponding weight coefficient of each index, which can be adjusted according to the actual need.

$$Q = \frac{1}{k_1 \sum_{i=1}^{n} E_k + k_2 \left| T_0 - \sum_{i=1}^{n} T_k \right| + k_3 \left| S_0 - \sum_{i=1}^{n} S_k \right| + k_4 max\{G\}} \tag{7}$$

Update of Manipulation Sequence. Each individual firefly represents a manipulation scheme, which consists of a sequence of conditions and the location of the transition point, i.c., $x_i = \{U, S\}$. When attraction is generated between fireflies for position transfer, individual fireflies need to consider whether to update the condition sequence, i.e., $x_i(n+1) = \{U_i(n), S_i(n+1)\}$ or $\left\{U_j(n), S_i(n+1)\right\}$.

The steps of the GSO algorithm for getting an efficient operation strategy are shown in Fig. 5.

5 Simulation and Verification

The simulation train adopts CR400BF type locomotive with 4 moving and 4 dragging groups, and the train mass is 913t. The simulation data refers to the actual line between Heishan North Station and Fuxin Station of Beijing-Shenzhen Passenger Dedicated Line. The train parameters and line parameters involved in the algorithm are the same as those in literature [5], showed in Table 4.

Combining the GSO algorithm with the line situation, we set the population number to 30, the maximum number of iterations to 200, the energy consumption performance

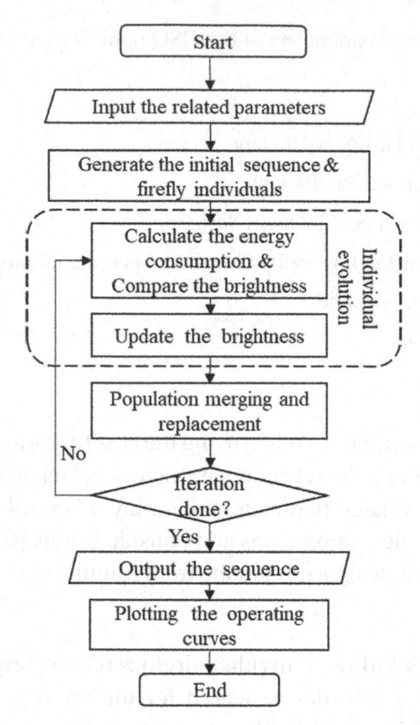

Fig. 5. Flowchart of Solution methodology based on GSO algorithm.

Table 4. Line parameters and train-related parameters.

Parameter	Description	Parameter value
S_0	Length of the line	58 km
T_0	Planned running time of the train	23 min
v_{max}	Speed limit for the whole road	288 km/h
$w_0(v)$	Basic resistance of the train	$0.399 + 0.0013 \cdot v + 0.000109 \cdot v^2$ N/kN
$F(v)$	Train motoring characteristics	$\begin{cases} 267 - 0.243 \cdot v\,\text{kN} & (0 \le v < 160\,\text{km/h}) \\ 0.0021v^2 - 1.7308 \cdot v + 446.75\,\text{kN} & (v \ge 160\,\text{km/h}) \end{cases}$
$B(v)$	Train braking characteristics	$\begin{cases} 280 \cdot v/10\ \text{kN} & (0 \le v < 10\,\text{km/h}) \\ -0.2241v + 281.74\ \text{kN} & (10 \le v < 200\,\text{km/h}) \\ 0.0017v^2 - 1.5602v + 475.38\,\text{kN} & (v \ge 200\,\text{km/h}) \end{cases}$

index coefficient $k_1 = 0.35$, the time performance index coefficient $k_2 = 0.35$, the stopping position performance index coefficient $k_3 = 0.20$, and the smoothness performance index coefficient $k_4 = 0.10$.

At the beginning of the simulation, the program randomly generates 30 train running curves according to the principle of train operation, 错误!未找到引用源。depicts the distribution of energy and time consumption of each individual firefly in the population at different iterations.

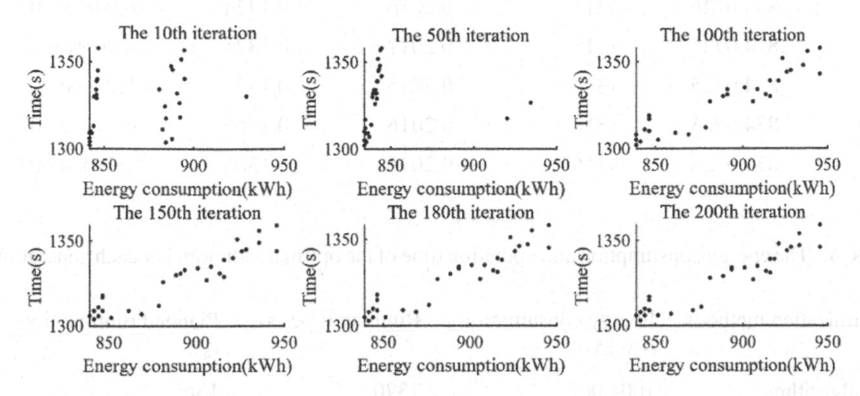

Fig. 6. Distribution of energy consumption-run time during population iteration.

When the iteration is just started, the fireflies closer to each other are more likely to attract each other, and the individual fireflies have a blocky distribution in 错误!未找到引用源。. The energy consumption of the firefly population ranges from 800 to 3000 kWh. As the population continues to iterate, the whole population is concentrated in the reasonable range of operating energy and time consumption values.

As can be seen from Table 5, the GSO algorithm takes advantage of the population intelligence algorithm to perform several optimization solutions at the same time, and by about the 50th generation, the fitness value of the population tends to stabilize and has searched for an operating curve that reduces the operating energy consumption by approximately 12 kWh compared with the optimal solution of the initial population, and when the iteration proceeds to about the 100th generation, the fitness value of the firefly population experiences another increase, and the time consumption and fitness values both The time consumption and fitness values meet the constraint requirements and finally show a stable state with only small adjustments to the performance indicators. By the 200th generation, the population converges to the final solution.

The operation curve of the last generation population is shown in Fig. 7 shows the strategies proposed by literature [5] using the Q-learning method and the DP algorithm.

As can be seen from Table 6 and Fig. 7, compared with the DP algorithm and Q-learning algorithm, the train operation curve optimized using the GSO algorithm consumes less energy, which is more efficient and can provide research ideas for more intelligent automatic train driving in the future.

Table 5. The energy consumption and operation time of the optimal individual of each generation.

Iteration number	Energy consumption (kWh)	Running time (s)	Stop position index	Smoothness index	Fitness value
1	846.4612	1302	0.1865	0.1567	6.6852e−07
10	834.0726	1312	0.2016	0.1424	6.8931e−07
50	834.0726	1314	0.2016	0.1424	6.9086e−07
100	834.0725	1316	0.2017	0.1371	6.9096e−07
150	834.0725	1316	0.2015	0.1371	6.9096e−07
180	834.0725	1316	0.2016	0.1366	6.9107e−07
200	834.0725	1316	0.2016	0.1366	6.9107e−07

Table 6. The energy consumption and operation time of the optimal individual of each generation.

Optimization methods	Energy consumption (kWh)	Running time (s)	Planned running time (s)
DP algorithm	1004.062	1370	1365
Q-learning algorithm (1 * 10^7)	922.367	1370	1369
Q-learning algorithm (6 * 10^6)	886.261	1370	1377
GSO algorithm	834.072	1370	1316

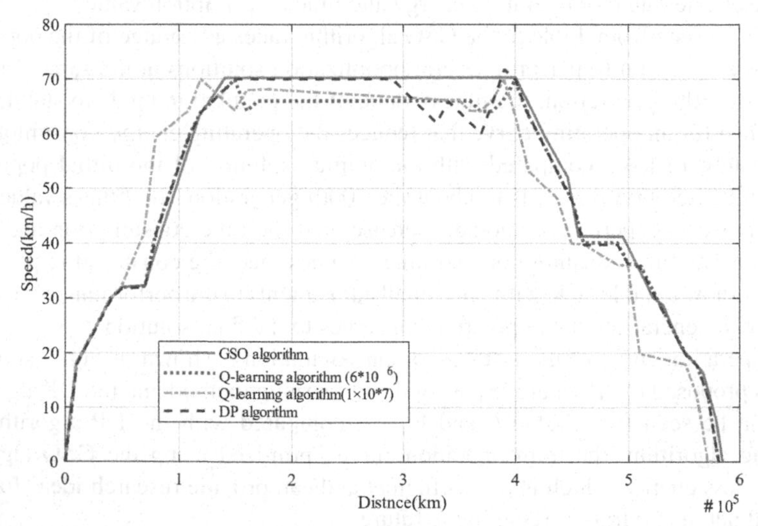

Fig. 7. The running curve obtained from the solution of GSO algorithm, DP algorithm, and Q-learning algorithm.

6 Conclusion and Summary

The main work of this paper is to apply the GSO algorithm, a new intelligent group optimization algorithm, to the problem of energy-saving optimized operation of high-speed trains, analyze the operation process and maneuvering strategy of high-speed trains, construct four quantitative energy-saving performance index functions, establish a multi-objective optimization model of the train operation curve, and design a train operation curve optimization method based on the GSO algorithm. Finally, the paper shows the effectiveness of the proposed method by simulating the actual line data of the Beijing-Shenzhen Passenger Dedicated Line. In the future, the optimization method can be improved by combining more complex operation environments, and considering the condition of tracking operation.

Acknowledgment. The authors wish to convey their sincere sense of gratitude to the support of the National Key Research and Development Program of China under Grant 2016YFB1200602-34.

References

1. Xie, T., Wang, S., Zhao, X., Zhang, Q.: Optimization of train energy-efficient operation using simulated annealing algorithm. In: Li, K., Li, S., Li, D., Niu, Q. (eds.) ICSEE 2012. CCIS, vol. 355, pp. 351–359. Springer, Heidelberg (2013). https://doi.org/10.1007/978-3-642-37105-9_39
2. Bai, Y., Ho, T.K., Mao, B.: Energy-efficient locomotive operation for Chinese mainline railways by fuzzy predictive control. IEEE Trans. Intell. Transp. Syst. **15**(3), 938–948 (2014)
3. Zhou, L.S., Tong, L. (Carol), Chen, J.H., et al.: Joint optimization of high-speed train timetables and speed profiles: a unified modeling approach using space-time-speed grid networks. Transp. Res. Part B **97**, 157–181 (2017)
4. Liu, S., Cao, F., Xun, J., Wang, Y.: Energy-efficient operation of single train based on the control strategy of ATO. In: IEEE International Conference on Intelligent Transportation Systems, pp. 2580–2586 (2015)
5. Miao, Z., Qi, Z., Zi, X.Z.: A study on energy-saving optimization for high-speed railways train based on Q-learning algorithm. Rail Transp. Econ. **41**(12), 111–117 (2019). (in Chinese)
6. Tang, T., Xun, J., Cao, F., Wang, Y.H., Soo, S.: Research on energy-efficient driving strategy in Beijing Yizhuang Line. J. Beijing Jiaotong Univ. **40**(4), 19–24 (2016). (in Chinese)
7. Hao, G., Zhang, Y.D., Jin, G., Zhi, Z.: Optimization of energy-efficient train operation based on dynamic programming approach. J. China Railw. Soc. **42**(08), 76–84 (2020). (in Chinese)
8. Liang, Z.C., Wang, Q.Y., He, K.: Energy saving control of electric multiple unit train based on maximum principle. J. China Railw. Soc. **37**(10), 16–25 (2015). (in Chinese)
9. He, Z.Y.: Research on automatic train operation algorithm based on non-parametric iterative learning control. J. China Railw. Soc. **42**(12), 90–96 (2020). (in Chinese)
10. Li, K.P., Gao, Z.Y.: Optimal method of energy saving train operation for railway network. J. Transp. Syst. Eng. Inf. Technol. **9**(04), 90–96 (2009). (in Chinese)
11. Bhanu, P., Krishnanand, K.N., Debasish, G.: Glowworm swarm-based optimization algorithm for multimodal functions with collective robotics applications. Multiagent Grid Syst. **2**(3), 209–222 (2005)
12. Keskin, K., Karamancioglu, A.: Application of firefly algorithm to train operation. In: 2016 IEEE In: 8th International Conference on Intelligent Systems (IS), pp. 692–697 (2016)

13. Huang, Z.X., Zhou, Y.Q.: Using Glowworm swarm optimization algorithm for clustering analysis. J. Convergence Inf. Technol. **6**(2), 78–85 (2011)
14. Kaipa, K.N., Ghose, D.: Glowworm swarm optimization: algorithm development. In: Glowworm Swarm Optimization. SCI, vol. 698, pp. 21–56. Springer, Cham (2017). https://doi.org/10.1007/978-3-319-51595-3_2

Research on Punctual and Energy-Efficient Train Driving Strategy Based on Two-Stage Allocation of Redundant Running Time

Gaofeng Liu[1], Wei Li[1](✉), Kang Li[2], Meiqi Yao[1], Sizhe Zhao[1], and Shijuan Gao[3]

[1] School of Traffic and Transportation Engineering, Central South University, Changsha, Hunan, China
liweijt@csu.edu.cn

[2] School of Electronic and Electrical Engineering, University of Leeds, Leeds, UK

[3] School of Geosciences and Info-Physics, Central South University, Changsha, Hunan, China

Abstract. The energy-efficient train control can effectively reduce energy consumption of railway network by optimizing train driving strategy. This paper proposed a method for optimizing driving strategy by two-stage allocation of redundant running time. The time-optimal train driving strategy requires shortest running time while highest energy consumption. By two-stage allocation of redundant running time, time-optimal train driving strategy can be transferred to energy-optimal train driving strategy under the premise that train arrives on time. In the first stage, part of redundant running time is allocated to long coasting regime before maximum braking. In the second stage, the remaining redundant running time is allocated to the maximum traction-coasting regime pairs, and to further reduce energy consumption. Time complexity of the proposed method is linear. A case study is carried out and the results show that the energy consumption of train operation is reduced by 10.33% under the premise of punctuality.

Keywords: Energy-Efficient Train Control · Redundant running time · Genetic Algorithm · High-speed train

1 Introduction

Railway is currently one of the transportations with the lowest energy consumption per passenger. Railway network has contributed a lot to the economic development, but it inevitably also consumes a lot of energy. From the perspective of train operation, there is a certain room for optimization of the traction energy consumption of train, for the traction energy consumption accounts for more than 70% of the energy consumption of the train operation [1]. The power of auxiliary energy consumption remains basically constant, so the feasibility of optimizing the traction energy consumption in train operation is high.

Since Ichikawa K. [2] proposed the concept of energy-efficient train driving in 1968, and pointed out that an optimal speed profile includes four driving regimes: maximum acceleration, cruising, coasting, and maximum braking. This research field of energy-efficient train control has been developing in the past decades. The complexity of the

K. Li et al. (Eds.): LSMS 2021/ICSEE 2021, CCIS 1468, pp. 133–146, 2021.
https://doi.org/10.1007/978-981-16-7210-1_13

energy-efficient train driving problem modelling developed from level track to varying gradient track, from mechanical braking to regenerative braking which convert kinetic energy into electric energy then feed it back to catenary or on-board energy storage devices. Optimization methods developed from analytical method such as Pontryain's Maximum Principle to numerical method represented by Dynamic Programming. Intelligent optimization methods represented by swarm intelligence algorithms that have developed rapidly in recent decades. The energy-efficient train control problem has attracted many scholars to actively explore because of its positive economic benefits and promotion of low-carbon society. Scholars represented by Milroy I. P., Howlett P., Pudney P., Khmelnitsky E., Albrecht A. R., Liu R. R., Lu S., Su S., etc., extended the problem of energy-efficient train driving to the optimal train operation control, and played a positive role in promoting this research field.

In 1980, Milroy I. P. [3] studied the diesel locomotives at the time in a doctoral dissertation of Loughborough University. Milroy I. P. used Pontryain's Maximum Principle to optimized the continuous variable which was the train acceleration a. Considering that the operation handle in the diesel locomotive's cab at that time was set to discrete positions, the research did not match the actual situation. Howlett P. [4] established a discrete train model considering relatively flat slopes based on Pontryain's Maximum Principle, and pointed out that the optimal train driving strategy on level track include four phases: maximum acceleration, speed holding (cruising), coasting, and maximum braking. Pudney P. and Howlett P. [5] also pointed out in subsequent studies that when the speed limit of line is lower than the optimal cruising speed of the train, the optimal driving strategy is to drive the train at the maximum limited speed. Considering varying gradient, speed limit, and regenerative braking, Khmelnitsky E. [6] established continuous train driving model using distance as independent variable rather than running time. On basis of previous research, Jiaxin C. and Howlett P. [7] firstly proposed the discrete control model with discrete throttle settings, considering varying gradient, that research made the control model more realistic.

In addition to analytical methods, some scholars tried to use other methods to solve the energy-efficient train control problem. Lu S. et al. [8] summarized that train operation control can be categorized as coasting control and general control. Coasting control reduces traction energy consumption by optimizing the coasting margin to use up allowable time, and general control generating speed profile to reduce total energy consumption of train. Lu S. et al. respectively adopted Genetic Algorithm, Ant Colony Optimization and Dynamic Programming, then designed corresponding solution rules based on the characteristics of different algorithms, calculated the energy-efficient speed profile of train. Lu S. et al. pointed out that analytical methods could not solve the energy-efficient train control problem. It is recommended that various methods should be adopted together to improve the global optimality of the optimization in that paper. Haahr J. T. et al. [9] used the time-space graph formulation solved by Dynamic Programming to generate improved train speed profile considering passage point problem. However, Haahr J. T. et al. did not mention the optimization relationship between traction energy consumption and running time.

Modern intelligent optimization algorithms have gradually been widely used in energy-efficient train control problem due to their less requirements for objective functions and the ability to perform global stochastic optimization. In the energy-efficient train control problem, the meta-heuristic algorithm is the most widely used algorithm [10]. Multi-objective Particle Swarm Optimization which had been refined was adopted to optimize the objectives of energy consumption and running time, and taken comfort and operational constraints into consideration, then the improved speed profiles were generated [11].

In view of the excessive simplifications of analytical methods in order to get feasible solutions, the "*curse of dimensionality*" of numerical methods, and the difficulty of ensuring global optimality of one-stage optimization using modern intelligent optimization algorithms, this paper proposed an optimization method based on two-stage allocation of redundant running time using modern intelligent optimization algorithm. Redundant running time, defined as the difference of given running time and shortest running time, can be allocated as time resource to achieve punctual and energy-efficient train driving strategy. The proposed method allocates redundant running time in two stages, and time-optimal speed profile can be transferred to punctual and energy-efficient speed profile after two-stage optimizations.

The optimization methods of train speed profile in the existing literature mostly focus on optimizing the switching points of driving regimes, generating improved speed profiles, or adjusting the allowable running time to reduce energy consumption. The proposed method optimizes two aspects of allocation of redundant running time and switching points of driving regimes. Through the idea of regarding redundant running time as allocable resource, the time resource can also be recycled in the case of slight train delays, so as to alleviate the impact of train delays.

2 Energy-Efficient Train Control Problem

2.1 The Train Dynamics

When the train is running in the section, the basic forces experienced by the train include traction force $F(v)$, braking force $B(v)$, basic resistance $W_0(v)$ and additional resistance $W_i(v, x)$.

The traction force $F(v)$ is determined by the traction characteristics of the train, and is used as the maximum traction force constraint $F_{\max}(v)$ in the optimization process. The traction force from the electric motor under traction regime is related to the running speed of the train, that is, the traction force F is presented as a function of speed v.

$$F = F(v) \tag{1}$$

The braking force $B(v)$ is determined by the braking characteristics of the train, and is used as the maximum braking force constraint $B_{\max}(v)$ in the optimization process. The braking force of the train under braking regime is related to the running speed of the train, that is, the braking force B is presented as a function of speed v.

$$B = B(v) \tag{2}$$

The resistance experienced by the train during the section can be divided into basic resistance $W_0(v)$ and additional resistance $W_i(v, x)$.

The basic resistance can be expressed by Davis equation [12] as:

$$w_0(v) = c_1 + c_2 v + c_3 v^2 \tag{3}$$

$$W_0 = M \cdot g \cdot w_0 \cdot 10^{-3} \tag{4}$$

w_0 represents the unit basic resistance, in *N/kN*. Non-negative empirical coefficient c_1, c_2, and c_3 are according to different model of trains, and are usually given. The constant item c_1 and the linear item $c_2 v$ are the rolling resistance experienced by the train, which is related to the mass of the train. The quadratic item $c_3 v^2$ is the air resistance experienced during the operation of the train, which is mass independent. W_0 is the basic resistance during the actual operation of the train, in *kN*. M is the mass of the train, in *t*. g is the acceleration of gravity, is 9.8 m/s^2.

In addition to the basic resistance, the train running in the section is also affected by the additional resistance. The additional resistance includes the additional resistance of slope, tunnel and curve. This paper only considers the additional resistance of the slope $W_i(v, x)$.

$$w_i(v, x) = M \cdot g \cdot \sin(\theta) \approx \theta \tag{5}$$

$$W_i = M \cdot g \cdot w_i \cdot 10^{-3} \tag{6}$$

The unit additional resistance of slope w_i, in *N/kN*, is a function of the train speed v and the current position x of the train. In practice, w_i can be approximately equal to the train slope value θ in ‰. W_i is the additional resistance of slope experienced by the train in operation, in *kN*.

In this paper, the gradient of slope is set as a function of distance, and the function is called when optimizing.

2.2 Train Driving Regimes

According to the output characteristics of train traction motor during train operation, the driving regimes are divided into four types: maximum acceleration (MA), cruising (CR), coasting (CO) and maximum braking (MB).

The traction regime and braking regime described in this paper are MA and MB. In the following text, CR stands for cruising and CO stands for coasting.

The corresponding force conditions are as follows, and the unit is *KN*.

MA: The train outputs the maximum traction force. This driving regime produces traction energy consumption. The resultant force experienced by the train C_1 is as follows.

$$C_1 = F - W_0 - W_i \tag{7}$$

CR: According to the basic resistance of the train and the additional resistance of the slope, the train outputs partial maximum traction force to offset the reverse force of the basic resistance and the additional resistance of the slope. The resultant force of the train C_2 is zero, and this driving regime produces partial traction energy consumption.

$$C_2 = F - W_0 - W_i = 0 \tag{8}$$

CO: The train does not output traction force, and this driving regime does not produce traction energy consumption. The resultant force on the train C_3 is the sum of the basic resistance and the additional resistance of the slope, and the direction of the resultant force on the train is opposite to the forward direction of the train.

$$C_3 = -W_0 - W_i \tag{9}$$

MB: The train outputs the maximum braking force, and the resultant force experienced by the train C_4 is the sum of the maximum braking force, the basic resistance and the additional resistance of the slope. The direction of the resultant force is opposite to the forward direction of the train.

$$C_4 = -B - W_0 - W_i \tag{10}$$

This paper considers regenerative braking in MB regime. During the MB regime, the train converts part of the kinetic energy into electric energy through a four-quadrant converter and feeds it back to the catenary or on-board energy storage devices.

2.3 Train Energy Consumption Modelling

The single-phase 50 Hz 25 kV AC power obtained from the catenary, part of which is converted from single-phase AC to DC and finally to three-phase AC power, through the main traction converter to drive the traction motor. The other part is converted by the auxiliary converter to complete the same form conversion of electric energy, is used to supply power to non-traction equipment such as air conditioning, lighting and information system. Therefore, the operating energy consumption of the train section consists of traction energy consumption and auxiliary energy consumption. The calculation formulas are as follows.

$$E_{total} = E_{traction} + E_{auxiliary} \tag{11}$$

$$E_{auxiliary} = P_{aulixiary}T \tag{12}$$

The energy consumption of the train section E_{total} is composed of traction energy consumption $E_{traction}$ and auxiliary energy consumption $E_{auxiliary}$. Traction energy consumption $E_{traction}$ is the integral of the running distance of the train's traction force $F(v)$ from the starting point x_0 to the end point x_{end}. Taking the energy consumption efficiency into account, then dividing $F(v)$ by the electric energy utilization coefficient η. The value of auxiliary power $P_{auxiliary}$ is generally known. Therefore, auxiliary energy consumption $E_{auxiliary}$ is the product of auxiliary energy consumption power $P_{auxiliary}$ and the actual running time of the train T.

In this paper, the current position of the train x is used as the independent variable to iterate to calculate the energy consumption of the train. The calculation formulas of traction energy consumption are as follows:

MA:

$$F_{MA} = F(v) \tag{13}$$

$$E_{traction} = \sum_{i=1}^{k} \frac{(b_i - a_i)}{6}(F_{MA}(v_{a_i}) + 4F_{MA}(v_{\frac{a_i+b_i}{2}}) + F_{MA}(v_{b_i})) \tag{14}$$

CR:

$$F_{CR} = -W_0(v) - W_i(v, x) \tag{15}$$

$$E_{traction} = \sum_{i=1}^{k} \frac{b_i - a_i}{6}(F_{CR}(v_{a_i}, x + a_i) + 4F_{CR}(v_{\frac{a_i+b_i}{2}}, x + \frac{a_i + b_i}{2}) + F_{CR}(v_{b_i}, x + b_i)) \tag{16}$$

CO:

$$F_{CO} = 0 \tag{17}$$

$$E_{traction} = 0 \tag{18}$$

MB:

$$F_{MB} = B(v) \tag{19}$$

$$E_{traction} = -\lambda \sum_{i=1}^{k} \frac{b_i - a_i}{6}(F_{MB}(v_{a_i}) + 4F_{MB}(v_{\frac{a_i+b_i}{2}}) + F_{MB}(v_{b_i})) \tag{20}$$

x is the current position of the train, which means the distance the train has traveled from the start point x_0. a_i is the starting position of the minimum iterative interval i. b_i is the end position of the minimum iterative interval i. The minimum iterative interval i of the train operation defined as the distance that the train travels for one second under the acceleration of the current speed of the train. k is the number of minimum iterative intervals of the section. $F(v_{a_i})$ is the dependent variable of the starting velocity v_{a_i} in the current minimum iterative interval i, in kN. In the same way, $F(v_{\frac{a_i+b_i}{2}})$ and $F(v_{b_i})$ are the dependent variables of the midpoint velocity $v_{\frac{a_i+b_i}{2}}$ and the end velocity v_{b_i} in the current minimum iterative interval i respectively. λ is the regenerative braking energy utilization factor.

2.4 Mathematical Model of Energy-Efficient Train Control

The mathematical model of energy-efficient train control is described as follows. By establishing this model, the energy-efficient train driving strategy can be obtained by specific optimization methods.

$$J = \min_{F} \int_{x_0}^{x_{end}} \frac{F(v)}{\eta} dx + P_{auxiliary}T \tag{21}$$

subject to

$$\frac{dt}{dx} = \frac{1}{v} \tag{22}$$

$$\frac{dv}{dx} = \frac{F(v) - B(v) - W_0(v) - W_i(v, x)}{(1 + \rho)M} \tag{23}$$

$$v(x_0) = 0, v(x_{end}) = 0, v(x) \in [0, v_{\max}(x)] \tag{24}$$

$$F(v) \in [0, F_{\max}(v)], B(v) \in [0, B_{\max}(v)] \tag{25}$$

$$T = T_{given} \tag{26}$$

$$x_{end} - x_0 = x_{\sec tion} \tag{27}$$

J is the objective function, that is, the total energy consumption required for the complete operation of the train section, in *kWh*. x_o and x_{end} is the start point and ending point of train operation respectively. $F(v)$, $B(v)$, $W_0(v)$ and $W_i(v, x)$ correspond to the traction force, braking force, basic resistance and additional resistance of slope respectively, all in *kN*. ρ is the dimensionless rotating mass factor [13]. M is the quality of the train, in *t*. $v_{\max}(x)$ is the speed limit in position x. $F_{\max}(v)$ is the maximum traction force constraint. $B_{\max}(v)$ is the maximum braking force constraint. T is the actual train running time. T_{given} is the given running time of the train section which is given by the timetable. $x_{\sec tion}$ is the length of the train section.

3 Two-Stage Allocation of Redundant Running Time

3.1 Time-Optimal Speed Profile

Time-optimal driving strategy of the section is MA-CR-MB, where the cruising speed is the speed limit. The time required for train operation $T_{fastest}$ under this strategy is the shortest, but the traction energy consumption is the highest.

In order to generate the time-optimal driving strategy, three steps need to be taken. Firstly, after the train start running from the starting point, keep the maximum acceleration till the train speed up to speed limit, then keep cruising; Secondly, reversely calculate the maximum braking distance of the train from the end point; Finally, when the train arrives the braking point that calculated before, switching to maximum braking regime for safe parking.

The redundant running time is defined as the difference between the given running time T_{given} and the shortest running time $T_{fastest}$. By generating the time-optimal speed profile, the running time $T_{fastest}$ and energy consumption of the train under this strategy can be obtained. Based on this, the two-stage allocation of redundant running time can be carried out, and the switching points of driving regimes can be optimized simultaneously.

3.2 Energy-Efficient Sequence of Driving Regimes

When the train is running in a section, the switching action between driving regimes must meet certain principles. Unreasonable switching will produce unnecessary energy consumption. The principle of train driving regimes switching as follows, see Fig. 1.

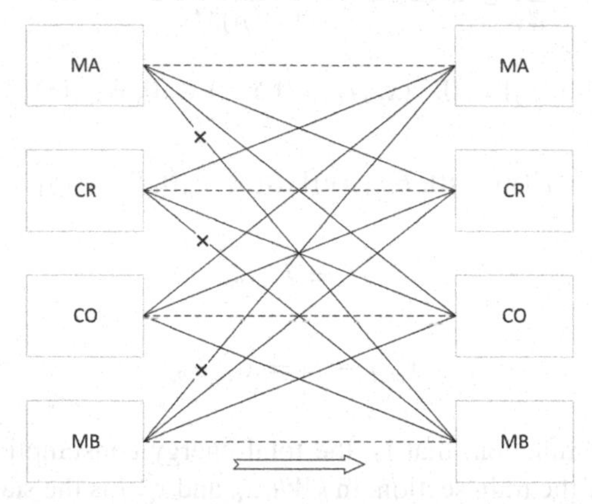

Fig. 1. Switching principles of train driving regimes. Dotted line means that the next driving regime is consistent with the current driving regime. Solid line means that it is allowed to switch from the current driving regime to the next driving regime. Solid line with " ×" means that it is not allowed to switch from the current driving regime to the next driving regime.

The energy-efficient train driving strategy consists of two parts: the energy-efficient sequence of driving regimes and the optimal switching points of driving regimes. The energy-efficient sequence of driving regimes of train operation is composed of four driving regimes mentioned before. If an optimization method is used to find the global energy-efficient sequence of driving regimes directly, it involves the combination and arrangement of the four driving regimes, which is a NP-complete problem. The solution time required for NP-complete problem is unacceptable for energy-efficient train control problem.

Su S. et al. [14] analyzed the energy-efficient sequence of driving regimes of a section, and pointed out the energy-efficient sequence of driving regimes when the section contains one, two, and multiple switching points, as shown in Table 1.

Analyzing the above-mentioned four sections of different switching points of driving regimes, it can be concluded that the energy-efficient sequence of driving regimes is the following three types [14]: MA-CO-MB, MA-(MA-CO pairs)-C-MB, MA-CO-CR-MB.

According to the literature [3], the MA-CO-MB sequence is optimal when train running in short sections, which is suitable for metro and suburban railway system; MA-(MA-CO pairs)-C-MB adds several maximum acceleration-coasting regime pairs, in order to reduce the energy consumption of continuous cruising regime; MA-CO-CR-MB driving regime is suitable for long and steep downhill lines, and the CR means speed holding after the CO regime for the steep downhill gradient. According to the *Code for Design of High-speed Railway* (TB10621–2014) [15], the gradient of the main line of Chinese high-speed railway should not exceed 20‰, and it should not exceed 30‰ under

Table 1. Energy-efficient sequence of driving regimes in a section with various switching points.

Switching points of driving regimes in a section	Energy-efficient sequence of driving regimes
None	MA、CR、CO、MB
One	MA-MB、MA-CO、CO-CR、MA-CR、CO-MB、CR-MB
Two	CO-CR-MB、MA-CO-MB、MA-CO-CR、MA-CR-MB
Multiple	MA-CO、MA-CO-CR-MB、MA-MA-CO-MB

difficult conditions. Therefore, the sequence of driving regimes MA-CO-CR-MB is not consistent with the actual line conditions of Chinese high-speed railways, and is more suitable for freight locomotives in heavy-haul railways.

In summary, the energy-efficient sequence of driving regimes for high-speed train in a section is MA-(MA-CO pairs)-CO-MB. In other words, the energy-efficient sequence of driving regimes of high-speed train in a section is maximum acceleration, several maximum acceleration and coasting pairs, coasting, and maximum braking.

3.3 Two-Stage Allocation of Redundant Running Time

Under the premise of ensuring that the train arrives on time, the redundant running time can be used as the allocable resource to obtain the energy-efficient train driving strategy. By allocating the redundant running time to energy-free coasting regimes, traction energy consumption can be reduced to the lowest under punctuality.

In the first-stage optimization, the redundant time is allocated to the newly added long CR regime before the MB regime. From literature [4], it can be known that the energy-efficient sequence of driving regimes is the MA, CR, CO, and MB. By allocating the redundant running time to CO regime between the CR and the MB regime before the end point, the energy consumption of train in section can be reduced.

In the second stage, MA-CO regime pairs are introduced to replace CR regime in order to further reduce traction energy. Therefore, this stage replaces the optimized CR regime of the first-stage optimized speed profile with several MA-CO regime pairs, and uses Genetic Algorithm to find the optimal allocation of redundant running time and optimal switching points to further reduce the traction energy consumption. The second stage of optimization is to allocate redundant time to MA-CO regime pairs.

Tian Y. et al. [16] used ten optimization algorithms to optimize the solution of one hundred thousand functions. Genetic Algorithm achieved the optimal solutions in the most function experiments than other algorithms. In this paper, Genetic Algorithm is used to optimize the allocation of redundant running time and switching points of driving regimes.

3.4 Relationship Between Allocated Redundant Running Time and Saved Energy

According to literature [17], energy consumption and running time are often in conflict with each other in the optimization process. It is necessary to make a trade-off between objectives to achieve the desired overall optimization effect of each objective.

As shown in the Fig. 2, " + ΔT" means the allocated redundant running time and "-ΔE" means the saved energy by allocation of redundant running time. The minimum running time is the technical shortest running time $T_{\min}$ that constrained by the characteristics of the train and line. The given running time T_{given} is given by the predesigned timetable. It is obvious to know from Fig. 2 that if the traction energy consumption continues to be reduced, the train running time of the section will exceed the given running time which will cause train delays.

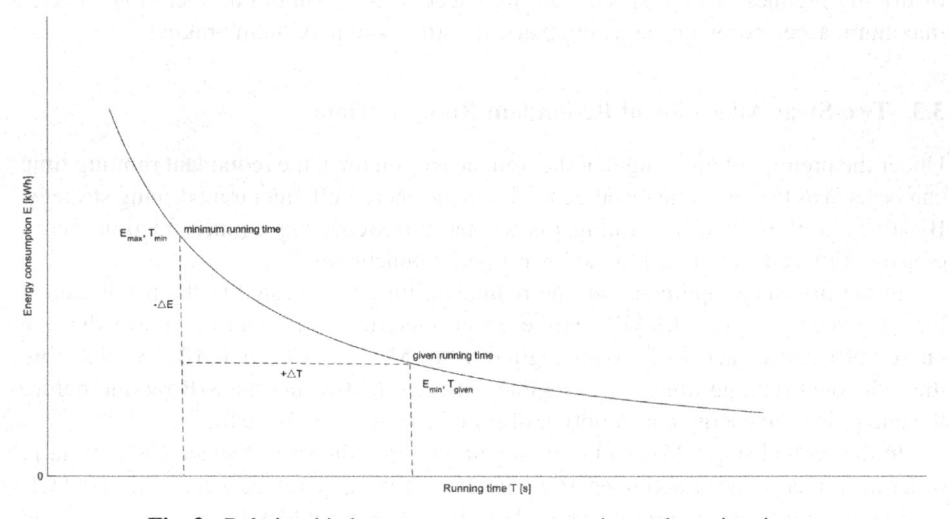

Fig. 2. Relationship between energy consumption and running time.

4 Case Study

4.1 Data of Train and Line

The train model used in the numerical experiment is the CRH3 high-speed train. The main characteristics of the train are shown in Table 2.

Table 2. Main characteristic parameters of CRH3

Characteristics of CRH3	Parameters
Mass (*t*)	563
Dimensionless rotating mass factor	0.06
Maximum traction force (*kN*)	300
Maximum braking force (*kN*)	300
Electric energy utilization coefficient	0.9
Auxiliary energy power	0.05
Regenerative braking energy utilization coefficient	0.85

The line used in the numerical experiment is an actual section of the Shangqiu-Hefei-Hangzhou high-speed railway with a length of 74.8 *km*. The line data is shown in Table 3.

Table 3. Main characteristic parameters of the line.

Characteristics of the line	Parameters
Length of the line (*km*)	74.8
Speed limit (*km/h*)	300
Given running time of section (s)	1200

The gradient of the line is shown in Fig. 3.

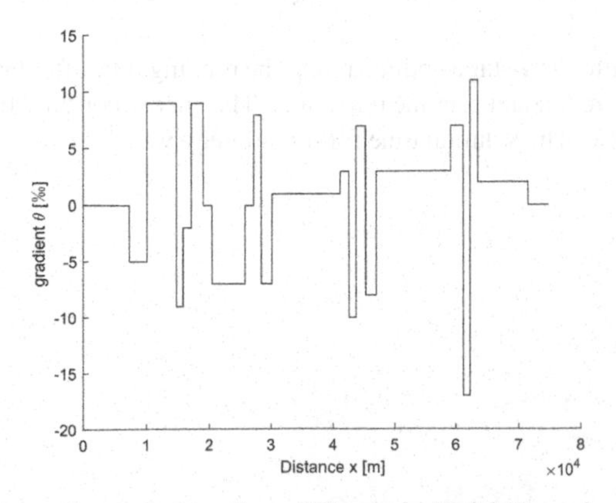

Fig. 3. Varying gradient of an actual section of Shangqiu-Hefei-Hangzhou high-speed railway.

4.2 Time-Optimal Speed Profile

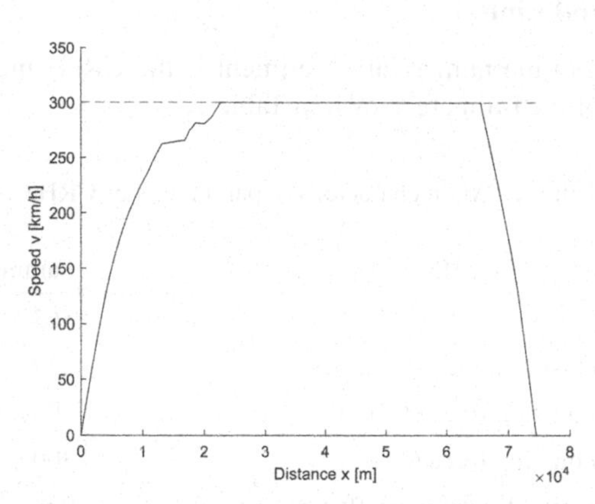

Fig. 4. The time-optimal speed profile considering varying gradient of the line. The running time under this driving strategy was 1105 s while the given running time is 1200 s, so there was 95 s of redundant running time. The energy consumption under this driving strategy was 1443.42 *kWh*. The solution time for this driving strategy was 0.05 s.

4.3 First-Stage Optimized Speed Profile

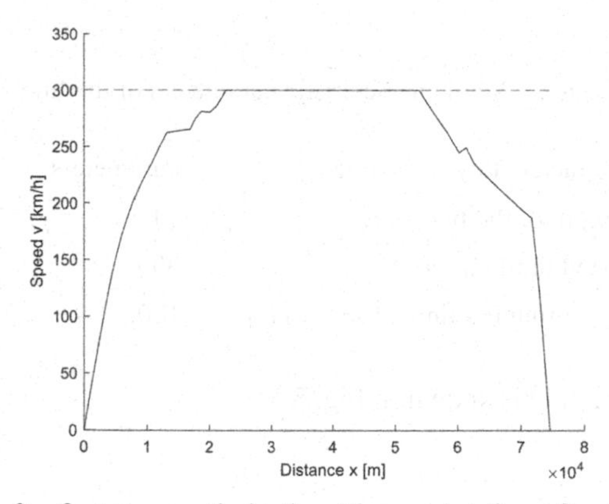

Fig. 5. Speed profile after first-stage optimization. The running time after first-stage optimization was 1144 s, and 56 s of redundant runtime remaining. The energy consumption under this driving strategy increased 0.02%. The solution time for this strategy was 2.69 s.

4.4 Second-Stage Optimized Speed Profile

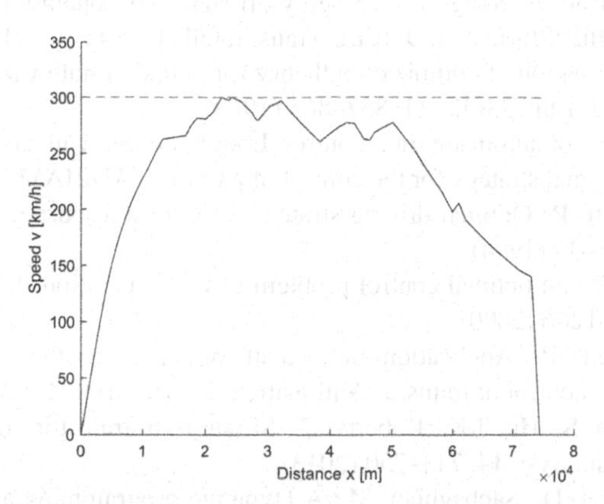

Fig. 6. Speed profile after second-stage optimization. All the redundant running time was allocated. 10.33% energy was saved with the running time consistent with the given running time. The solution time for this strategy was 31.36 s.

5 Conclusion

This paper proposed a two-stage optimization method, which allocates resources of redundant running time to achieve lowest energy consumption under punctuality (Figs. 4, 5 and 6). Under the premise of ensuring punctual arriving, the conversion from the time-optimal speed profile to the punctual and energy-efficient speed profile can be realized. This method combines two optimization ideas of allocation of redundant running time and optimization of switching points. With the given energy-efficient sequence of driving regimes for train operation according to literature, time complexity of the proposed method is linear.

Numerical experiment was carried out. In the first stage, the energy saved by the shorter CR regime and newly added CO regime was less than the regenerative energy produced by the MB regime of the time-optimal speed profile, so the energy consumption increased 0.02%. After two stage optimizations, the proposed method could achieve the energy saving effect of 10.33% compared with the time-optimal speed profile.

Based on the remaining redundant running time, train delays can be alleviated in some extent when the train is not driving under the time-optimal driving strategy. By recycling remaining redundant running time, train delays can be alleviated with higher energy consumption.

Acknowledgements. This research was supported by Major Science and Technology Project of Guangxi Province of China under Grant No. Guike AA20302010.

References

1. Gu, Q., Tang, T., Cao, F., Song, Y.-d.: Energy-efficient train operation in urban rail transit using real-time traffic information. J. IEEE Trans. Intell. Transp. Syst. **15**, 1216–1233 (2014)
2. Ichikawa, K.: Application of optimization theory for bounded state variable problems to the operation of train. J. Bull. JSME. **11**, 857–865 (1968)
3. Milroy, I.P.: Aspects of automatic train control. Loughborough University (1980)
4. Howlett, P.: An optimal strategy for the control of a train. J. ANZIAM J. **31**, 454–471 (1990)
5. Pudney, P., Howlett, P.: Optimal driving strategies for a train journey with speed limits. J. ANZIAM J. **36**, 38–49 (1994)
6. Khmelnitsky, E.: On an optimal control problem of train operation. J. IEEE Trans. Autom. Control. **45**, 1257–1266 (2000)
7. Jiaxin, C., Howlett, P.: Application of critical velocities to the minimisation of fuel consumption in the control of trains. J. Automatica. **28**, 165–169 (1992)
8. Lu, S., Hillmansen, S., Ho, T.K., Roberts, C.: Single-train trajectory optimization. J. IEEE Trans. Intell. Transp. Syst. **14**, 743–750 (2013)
9. Haahr, J.T., Pisinger, D., Sabbaghian, M.: A Dynamic programming approach for optimizing train speed profiles with speed restrictions and passage points. J. Transp. Res. Part B: Methodol. **99**, 167–182 (2017)
10. Fernández, P.M., Sanchís, I.V., Yepes, V., Franco, R.I.: A review of modelling and optimisation methods applied to railways energy consumption. J. Cleaner Prod. **222**, 153–162 (2019)
11. Domínguez, M., Fernández-Cardador, A., Cucala, A.P., Gonsalves, T., Fernandez, A.: Multi objective particle swarm optimization algorithm for the design of efficient ATO speed profiles in metro lines. J. Eng. Appl. Artif. Intell. **29**, 43–53 (2014)
12. Davis, W.J.: The tractive resistance of electric locomotives and cars. Gen. Electr. (1926)
13. Brünger, O., Dahlhaus, E.: Running time estimation. J. Railway Timetable Traffic 58–82 (2014)
14. Su, S., Tang, T., Chen, L., Liu, B.: Energy-efficient train control in urban rail transit systems. J. Proc. Instit. Mech. Eng. Part F: J. Rail Rapid Transit. **229**, 446--454 (2015)
15. Code for Design of High-speed Railway. vol. TB 10621–2014. National Railway Administration of China (2014)
16. Tian, Y., Peng, S., Zhang, X., Rodemann, T., Tan, K.C., Jin, Y.: A recommender system for metaheuristic algorithms for continuous optimization based on deep recurrent neural networks. J. IEEE Trans. Artif. Intell. **1**, 5–18 (2020)
17. Scheepmaker, G.M., Goverde, R.M., Kroon, L.G.: Review of energy-efficient train control and timetabling. J. Eur. J. Oper. Res. **257**, 355–376 (2017)

Computational Intelligence in Utilization of Clean and Renewable Energy Resources, and Intelligent Modelling, Control and Supervision for Energy Saving and Pollution Reduction

Optimization Design of Flexible Solar Cell Interconnect Based on Submodel Method

Xin Wang, Jingjing Xu(✉), Yunpeng Liu, and Liquan Chen

School of Mechatronic Engineering and Automation, Shanghai University, Shanghai, China
{onebluec,xjj125}@shu.edu.cn

Abstract. Due to the different mechanical properties of the components of flexible solar cell, the interconnects will produce large thermal stress under high temperature loading conditions. The connection of spot weld will produce a large pull force. As to this problem, the component system is analyzed using finite element modeling and simulation. The stress distribution of a group of interconnects and their connecting regions was analyzed by using the submodel analysis technique. And this paper carries out optimization design for the structure of the stress-reducing ring of the interconnect, studies the relationship between the design variables and the structural performance of interconnect by response surface technique, and obtains more structural dimension parameters by goal-driven optimization analysis Basing on NSGA-II genetic algorithm.

Keywords: Finite element · Submodel method · Stress · Optimization design · NSGA-II genetic algorithm

1 Introduction

With the energy crisis and the increasingly serious environmental pollution caused by traditional energy, the development of renewable and clean energy has become one of the major strategic issues in the international scope. Solar cells can be used locally to convert the green solar energy into electric energy, which has been widely used as an ideal energy source for spacecraft. The flexible solar cell has the advantages of light weight, high mass ratio power, high flexibility and bendability, good surface coverage, low temperature coefficient of components.

However, flexible materials that constitute solar cell components are easily affected by in-orbit temperature changes. Due to the difference of thermal expansion coefficient and elastic modulus of components, deformation and thermal stress will be generated in space temperature field. This can seriously cause component connection failure.

This paper analyzed the thermal performance of the connection region of the flexible solar cell module under high temperature condition by using global analysis and submodel technology. Taking reducing the equivalent stress of the interconnect and the pull force of the single spot weld as the goal, the design parameters of the stress-reducing ring of the interconnect were designed experimentally. The response surface and objective driven optimization techniques were used to obtain a better interconnect structure, to improve the reliability of connection and the service life of flexible cell components.

K. Li et al. (Eds.): LSMS 2021/ICSEE 2021, CCIS 1468, pp. 149–158, 2021.
https://doi.org/10.1007/978-981-16-7210-1_14

2 Research Object and Physical Model

The structure of flexible thin film GaAs solar cell with the size of 48 mm × 125 mm was selected for full modeling analysis. The solar cell components are composed of flexible printed circuit board with corresponding area and thickness, flexible transparent packaging adhesive, surface packaging film, 4 flexible GaAs solar cells and 10 silver interconnects. The electrodes of the solar cell are connected by the interconnects through welding, and the other materials are bonded by the packaging glue. Dimensional parameters of space flexible thin-film solar cells are shown in Fig. 1.

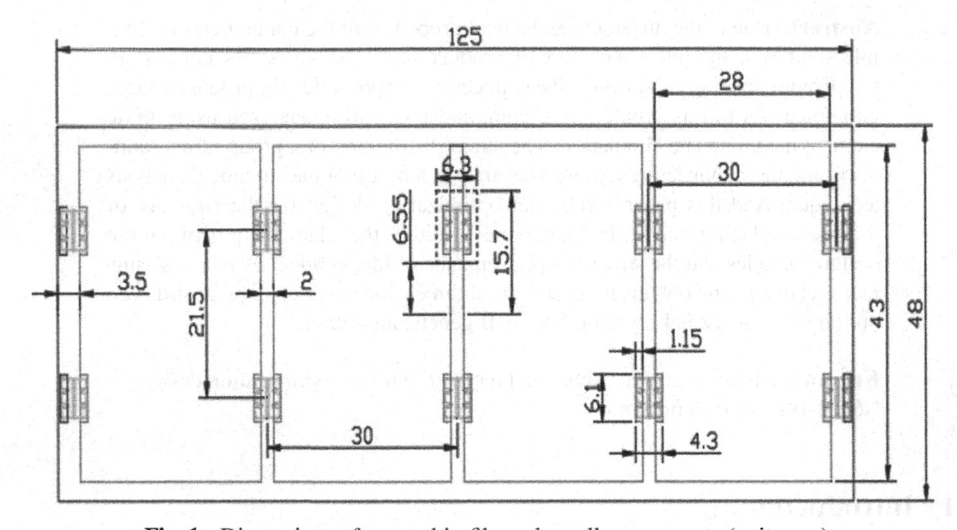

Fig. 1. Dimensions of space thin film solar cell components (unit: mm)

To obtain more accurate mechanical property parameters of interconnect components, the submodel from the full model is taken as shown by the dashed boundary in Fig. 1. The side structure of the components is shown in Fig. 2. From bottom to top, it is: flexible solar cell array substrate, lower package adhesive, flexible GaAs solar cell, upper package adhesive, and surface package film.

Among them, the flexible PCB substrate is a composite material of polyimide and epoxy resin. The flexible thin film solar cell is grown on the rigid GaAs substrate and prepared on the flexible substrate by substrate transfer technology. It consists of polyimide, Au and GaAs. the material parameters corresponding to the thickness of each component are shown in Table 1.

3 Finite Element Analysis of the Model

3.1 FE Analysis of the Full Model

Considering that the plane size of the model is much larger than that in the thickness direction, Choosing shell element type to build finite element model is preferred. But the

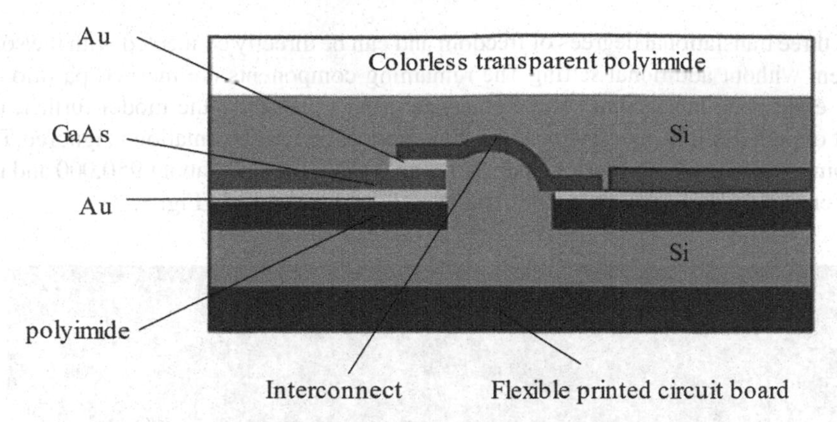

Fig. 2. Submodel component structure

Table 1. Thickness and material parameters of sub-model components

Component	Material	Thickness/μm	Poisson's ratio	Modulus of elasticity /GPa	Coefficient of thermal expansion x10–6 /°C
Surface Package film	Colorless transparent polyimide	50	0.31	3.52	18.4
Upper package adhesive	Space silicone	50	0.37	2e-4	200
Interconnect	Silver	17	0.32	232	25.5
GaAs solar cell	Au	4	0.4	79	14.2
	GaAs	12	0.31	85.26	6.4
	Au	2	0.4	79	14.2
	Polyimide	50	0.34	4	30
Lower package adhesive	Space silicone	50	0.37	2e-4	200
Flexible printed circuit board	Polyimide	25	0.34	4	30
	Epoxy resin	25	0.3	1	70
	Polyimide	25	0.34	4	30

space filled silicone is an irregular entity, it can only be divided by solid element.This will cause the connection between the shell element and the solid element of the silicone is not coordinated.Therefore, the eight-node solid shell element is used to grid the flexible substrate and the solar cell monomer. The topology and characteristics of this element are shown as the continuous medium solid element, which is the same as the solid element.

It has three translational degrees of freedom and can be directly connected with the solid element without additional setting.The remaining components are meshed partitioned using eight-node hexahedral solid element.In order to simplify the model further, the effect of partial filling adhesive near the interconnect on the deformation is ignored.The mapping grid is used for division, and the number of nodes count about 950,000 and the number of element is about 620,000.The overall grid is shown in Fig. 3.

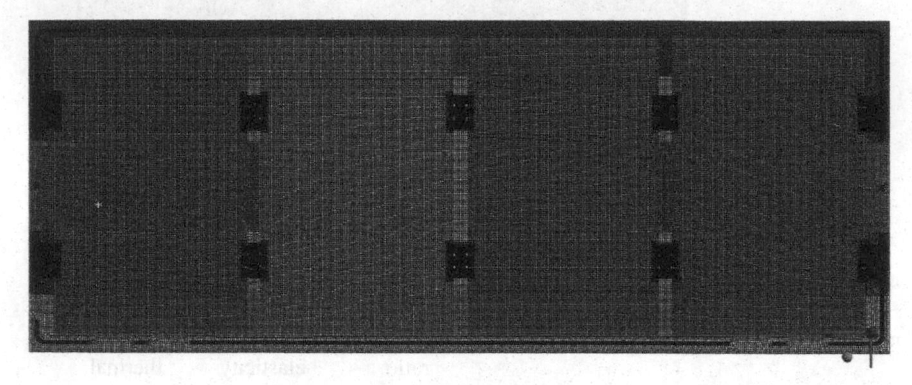

Fig. 3. The FE model of full model

In order to simulate the actual working environment of flexible solar cell component, a normal displacement constraint is applied to the bottom surface of the flexible substrate. Considering the influence of deformation on the global stiffness, the large deformation switch is turned on.At the same time, a weak spring constraint is added to avoid rigid body displacement in the model calculation. As to simulate the thermal deformation of the component under high temperature conditions, a temperature load of 100°C was applied to the whole component.

3.2 FE Analysis of Interconnect Submodel

According to the stress distribution of the full model, the X direction range (−3.15–3.15) mm and Y direction range (6.55–15.7) mm are preliminarily selected, and their positions in the global coordinate system are consistent with those in the full model.

In the submodel, in order to reduce the cost of calculation and the focus is on the interconnect areas, eight-node solid shell element are still used for the solar cell monomer and the flexible substrate, while eight-node hexahedral solid element are used for the rest.However, to obtain accurate thermodynamic properties and pull forces of spot weld on interconnect.Choose to refine the grid on the interconnect and the positive and negative electrodes of the cell.The 0.05 mm size is used to divide the positive and negative electrodes of the solar cell, and the 0.02 mm size is used to divide the interconnect part.The rest part is divided by 0.15 mm.The mesh and interpolation boundary are shown in Fig. 4.

Under the high temperature load, the interconnect components produce a large deformation, and the local region produces a large stress, so that the local area of the interconnect section has entered a plastic state.Therefore, in order to precisely simulate the

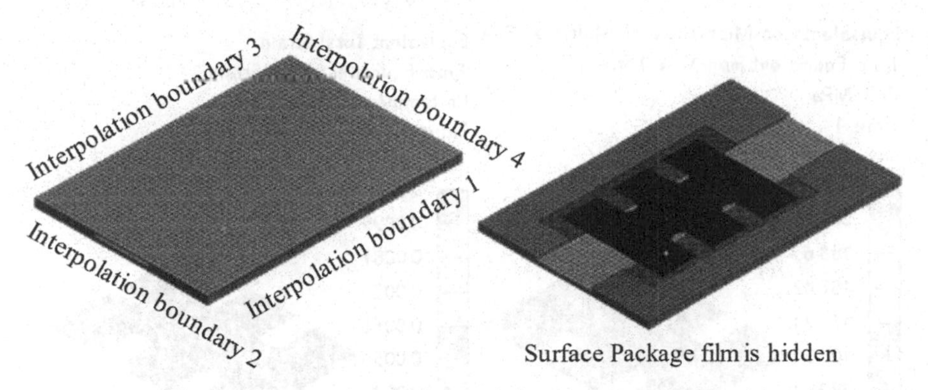

Fig. 4. The FE model and interpolation boundary of submodel

real stress state of the components, the elasto-plastic relationship of the interconnect is considered to solve the problem.Since the stiffness of interconnects contributes little to the stiffness of the integral component, the states of the interconnect cannot affect the deformation of the integral component, so the interpolation displacement boundary conditions of the integral model under linear elasticity can be used as the displacement boundary conditions of the elasto-plastic submodel more accurately.The interpolation boundary conditions derived from the global model are shown in Fig. 5.

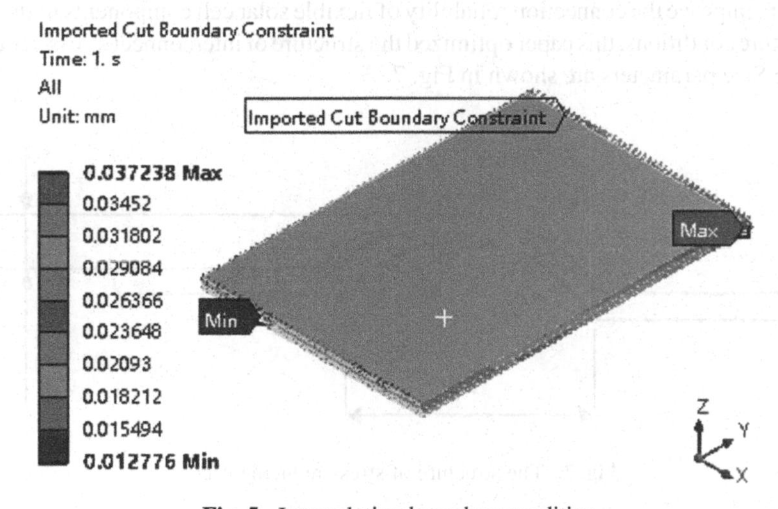

Fig. 5. Interpolation boundary condition

The stress and strain of the interconnect obtained from the solution is shown in Fig. 6.

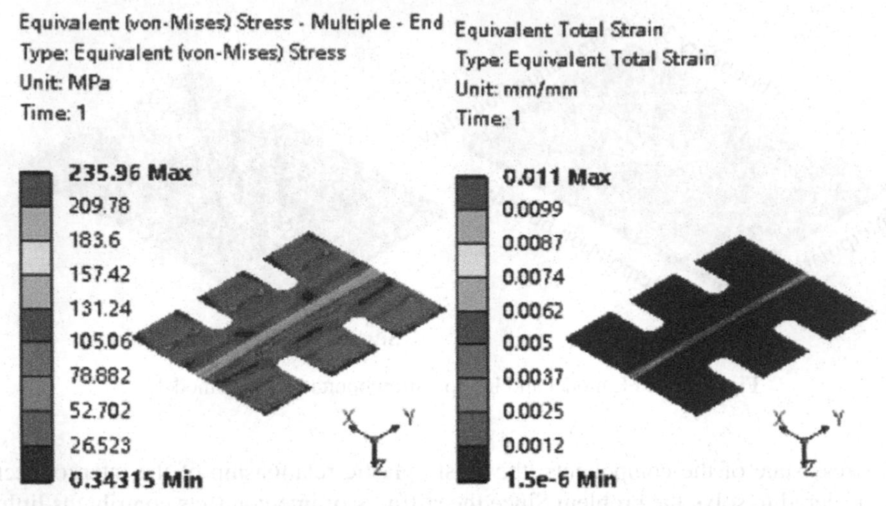

Fig. 6. Stress and strain of submodel

4 Optimal Design of Stress-Reducing Ring

4.1 Mathematical Model of Optimization

In order to improve the connection reliability of flexible solar cell components under high temperature conditions, this paper optimized the structure of interconnect stress-reducing ring.The Size parameters are shown in Fig. 7.

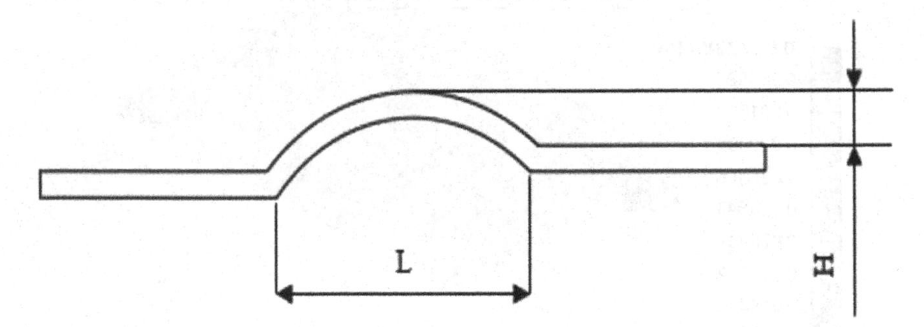

Fig. 7. The structure of stress-reducing ring

As can be seen from the Fig. 7, the parameters of the stress-reducing ring of the interconnect are determined by the height H and the length L,so set height H and length L as the design variables, and the design space of the design variables is shown in Table.2, considering the limitation of the space location of the interconnect.With the maximum deformation of interconnect as constraint, the maximum equivalent strain of interconnect and the maximum pull force of a single spot weld as optimization objectives.

Table 2. Design variable initial values and design space

Design variable	Initial value/mm	Design space/mm
Height H	0.02	0.01–0.028
Length L	0.8	0.4–1.4

4.2 Optimization Algorithm

In the multi-objective optimization problem, it is difficult to find a real optimal solution because of the conflict between these objectives.But there is a set of solutions in which there is at least one goal better than all the other solutions, such a solution is called a non-dominant solution, or a Pareto solution.The process of solving the multi-objective optimization problem is the process of finding the Pareto solution set.

In this paper, NSGA-II algorithm is used to optimize the solution.This algorithm was first proposed by Srinivas and Deb [4]. They introduced the idea of non-dominated sorting into the genetic algorithm and converted the calculation of multiple objective functions into the calculation of virtual fitness for solving multi-objective optimization problems. The core process of this algorithm is shown in Fig. 8.

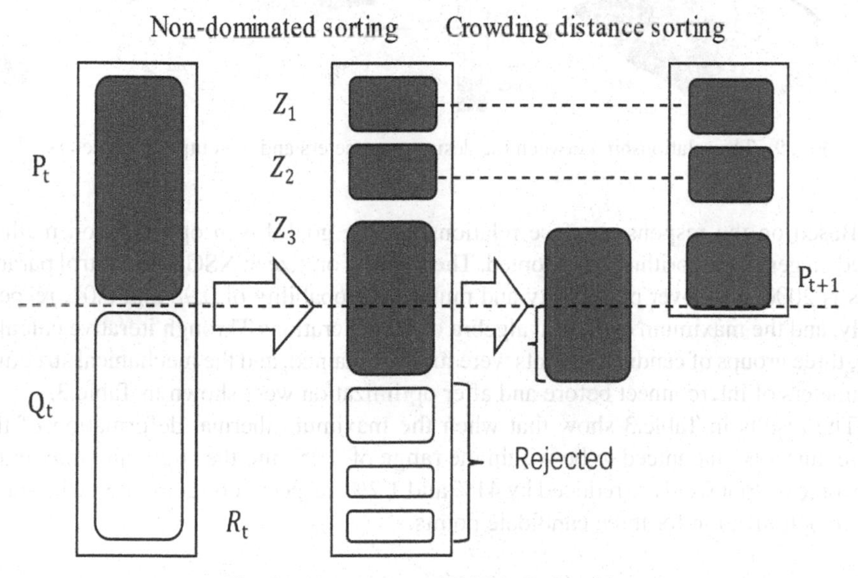

Fig. 8. The process of NSGA-II algorithm

Firstly, the new population Q_t generated in the t generation was merged with the parent generation P_t to form R_t, and the population size was 2 N.Then a non-dominated sort is performed on R_t to generate a series of non-dominated sets Z_i and the crowding degree is calculated.Since both offspring and parent individuals are included in R_t, the individuals contained in the non-dominant set Z_1 after the non-dominant sorting are the best in R_t, so Z_1 is first put into the new parent population P_{t+1}.If the size of is less than N, the next level of non-dominated set Z_2 is continued to be filled into P_{t+1}, until the population size exceeds N when Z_3 is added, and the number of individuals in P_{t+1} reaches N by crowding distance sorting.Then a new progeny population Q_{t+1} is generated by genetic operators (selection, crossover, mutation) [5].

4.3 Optimization Result

The experimental design was used to obtain 17 groups of samples for the simulation calculation of the submodel, and the relationship between the design parameters and the output parameters was obtained based on response surface technology.The response relationship between parameters is shown in Fig. 9.

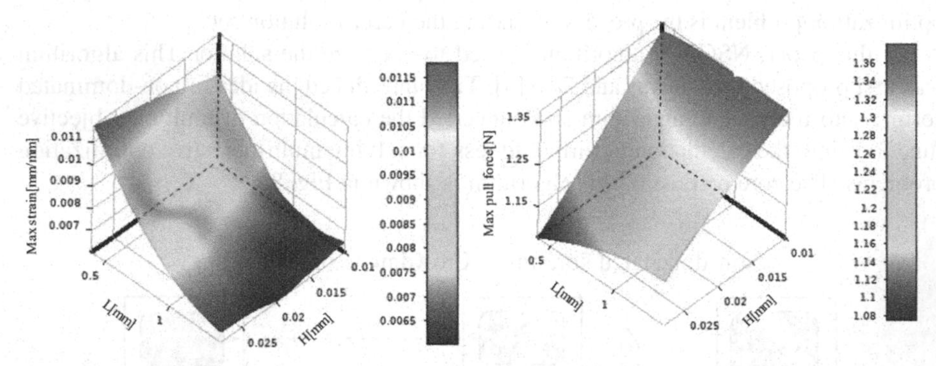

Fig. 9. The relationship between the design parameters and the output parameters

Based on the response surface relationship, the goal-driven optimization method based on genetic algorithm was adopted. The population size in NSGA-II control parameters is 2000, crossover probability and mutation probability of 0.98 and 0.01, respectively, and the maximum evolution algebra of 20 generations.Through iterative calculation, three groups of candidate points were finally obtained, and the mechanical structure parameters of interconnect before and after optimization were shown in Table.3.

The results in Table.3 show that when the maximum thermal deformation of the interconnect is guaranteed to be within the range of 0.05 mm, the strain and maximum pull force of spot weld are reduced by 41% and 1.2%, respectively, compared with those before optimization for three candidate points.

Table 3. Comparison before and after optimization

Input/output variable	Initial point	Candidate point 1	Candidate point 2	Candidate point 3
Height H/mm	0.02	0.028	0.027998	0.027999
Length L/mm	0.8	1.3962	1.3991	1.3909
Max equivalent strain/mm/mm	0.011	0.006459	0.0064567	0.0064632
Max deformation/mm	0.04	0.040695	0.040699	0.040688
Pull force of single spot weld/N	1.1942	1.1797	1.1798	1.1795

5 Conclusion

In this paper, the thermal deformation and thermal stress of flexible solar cell components at a high temperature is analyzed by finite element simulation.The submodel method is used to optimize the structure of interconnect, and the following conclusions are drawn:

1. Under the condition of high temperature, the stress-reducing ring region of the interconnect enters into plastic deformation partially, and the larger stress position occurs in the interconnect and its connection region.This means that this is where the failure of the components is likely to occur.Therefore, in the process of the components design, the type selection and structure design of interconnect are very important.
2. In the aspect of obtaining more accurate mechanical parameters and properties of interconnect.Submodel method is a low cost, high precision and high efficiency method.Therefore, the application of submodel method should be broadened in the field of stress precise analysis.
3. Under the condition of satisfying the target constraint, the structural design of the stress-reducing ring can be changed by using the optimization design method, which can effectively improve the stress and strain state of the interconnect and reduce the pull force in the connection area.It can provide some meaning for the design of interconnect.

References

1. Xia, X., Hu, C., Qu, E.: Applications of ANSYS submodeling technology in dealing with stress concentration. J. Eng. Constr. **20**(2), 92–94 (2006)
2. Wang, H., Li, A., Guo, T., et al.: Accurate stress analysis on rigid central buckle of long-span suspension bridges based on submodel method. J. Sci. China Ser. E-Technol. Sci. **52**, 1019–1026 (2009)
3. Chen, L., Zhou, L., Liang, Y.: Multi-objective optimization design of printed electronics ink-jet printer workbench based on isight (in Chinese). J. Mach. Design Manuf. Eng. **49**(11), 11–16 (2020)

4. Srinivas, N., Deb, K.: Multiobjective optimization using nondominated sorting in genetic algorithms. J. Evol. Computat. **2**(3), 221–248 (1994)
5. Pratap, D.K.A., Agarwal, S., Meyarivan, T.A: Fast andelitist multi-objective genetic algorithm: NSGA-II. J. Evol. Comput. **6**(2), 182–197 (2002)
6. Xu, Y., Qian, L.: Multi-objective optimization of composite barrel based on the improved non-dominated sorting genetic algorithm (in Chinese). J. Acta Armam. **4**, 617–621 (2006)

Hybrid Neural Network Method of the Lithium-Ion Battery Remaining Useful Life Prediction

Dexin Gao[1(✉)], Xin Liu[1], and Qing Yang[2]

[1] College of Automation and Electronic Engineering, Qingdao University of Science and Technology, Qingdao 266061, China

[2] College of Information Science and Technology, Qingdao University of Science and Technology, Qingdao 266061, China

Abstract. In this paper, a hybrid neural network is proposed by combining one dimensional convolutional neural network (1D CNN) and bidirectional long short-term memory (BiLSTM) neural network to improve the accuracy and stability of remaining useful life (RUL) prediction. 1D CNN-BiLSTM hybrid neural network has the ability to extract deep features in data and to save the memory of historical input information. To verify the authority of the proposed method, the lithium-ion battery data of the Center for Advanced Life Cycle Engineering (CALCE) are utilized to make some comparisons among the 1D CNN model, LSTM model and other neural network models. The results show that the hybrid one has higher prediction accuracy than the others.

Keywords: Lithium-Ion battery · Hybrid neural network · Remaining useful life · Long short-term memory (LSTM)

1 Introduction

Compared with traditional power battery, lithium-ion battery has the advantages of green environmental protection, small size, high energy density and so on. No matter in production, use and retirement, lithium-ion batteries will not produce lead, mercury and other toxic substances. Therefore, lithium-ion batteries as green energy are widely used in communication equipment and electric vehicles [1] and other fields. In the charge discharge cycle [2] of lithium-ion battery, there are some phenomena such as irreversible dissolution of positive and negative materials and oxidation decomposition of electrolyte. The capacity of lithium-ion battery [3, 4] decreases gradually until it reaches the failure threshold. Therefore, the stable and accurate estimation of RUL [5–7] is of great significance for improving the safety of lithium-ion battery.

At present, the prediction methods of lithium-ion battery RUL are mainly divided into physical-chemical model and deep learning data-driven. Physical-chemical [8–10] models are constructed to simulate the process of the battery falling from the state of health to the failure threshold. [11] proposes a method to estimate the state of health

K. Li et al. (Eds.): LSMS 2021/ICSEE 2021, CCIS 1468, pp. 159–167, 2021.
https://doi.org/10.1007/978-981-16-7210-1_15

(SOH) by establishing a linear relationship between ohmic internal resistance and internal resistance attenuation. [12] proposes a new capacity decay model based on the single particle model, combined with chemical and mechanical degradation mechanisms. The model can quickly predict the capacity attenuation and voltage distribution changes with the number of cycles and temperature, and predict the lithium-ion battery SOH and RUL with high accuracy. Deep learning algorithms [13–15] mine the hidden health information from the historical monitoring data (current, voltage, temperature, etc.) of lithium-ion battery, and predict the RUL of lithium-ion battery. Considering that the capacity of lithium-ion battery has regeneration phenomenon, [16] proposes a lithium-ion battery RUL prediction method that combines wavelet decomposition technology and nonlinear autoregressive neural network model. This method can effectively capture the regeneration phenomenon and has a higher prediction accuracy. [17] proposes an online estimation method of lithium-ion battery RUL based on feed forward neural network and importance sampling.

Most of the above methods ignore the time series characteristics of lithium-ion battery data. Therefore, this paper proposes to use 1D CNN-BiLSTM hybrid neural network to predict the RUL of lithium-ion battery, which can effectively use the existing historical data to predict and analyze the future state of lithium-ion battery.

2 Hybrid Neural Network

1D CNN-BiLSTM hybrid neural network (as shown in Fig. 1) has the ability to extract deep features in data and to save the memory of historical input information. In the 1D convolution [18] layer, the input data needs to be convolution operation and activation function calculation, as shown in formula 1. Max pooling is used to downsampling to minimize network parameters and obtain deep features.

$$x_t = \sigma_{cnn}(W_{cnn} * n_t + b_{cnn}). \tag{1}$$

where W_{cnn} represents the weight coefficient of the filter, that is, the convolution kernel; n_t represents the t-th input data; $*$ represents the discrete convolution operation; b_{cnn} represents the bias parameter, σ_{cnn} represents the activation function SELU, and x_t represents the output data after the 1D convolution ends.

BiLSTM [19] is composed of forward LSTM (composed of input gates, forgetting gates, output gates and memory cells) hidden layer and backward LSTM hidden layer. It reads lithium-ion battery data in ascending order ($t = 1, 2, 3, \ldots, T$) and descending order ($t = T, \ldots, 3, 2, 1$) respectively. The outcomes of the two hidden layers are then combined into the same output layer. The formula is as follows.

$$i_t = \sigma(W_{xi}x_t + W_{hi}h_{t-1} + W_{ci}c_{t-1} + b_i). \tag{2}$$

$$f_i = \sigma(W_{xf}x_t + W_{hf}h_{t-1} + W_{cf}c_{t-1} + b_f). \tag{3}$$

$$c_t = f_t c_{t-1} + i_t \tanh(W_{xc}x_t + W_{hc}h_{t-1} + b_c). \tag{4}$$

$$o_t = \sigma(W_{xo}x_t + W_{ho}h_{t-1} + W_{co}c_t + b_o). \tag{5}$$

$$h_t = o_t \tanh(c_t). \tag{6}$$

where σ is the sigmoid function, c is the memory cell state, f, i and o are the forget gate, input gate and output gate respectively, all b are biases, and all W are weights.

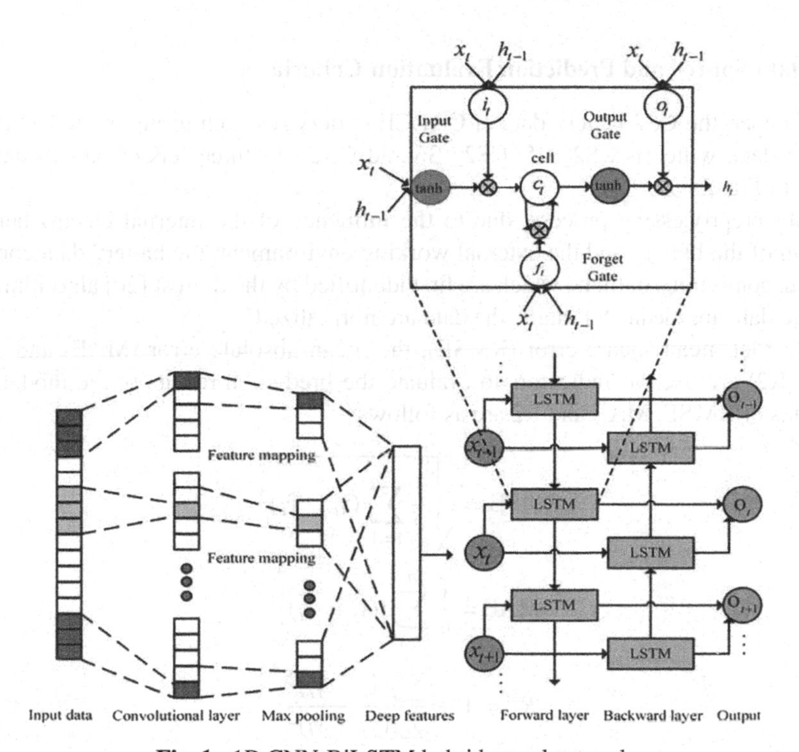

Fig. 1. 1D CNN-BiLSTM hybrid neural network.

3 Prediction Experiment and Analysis

3.1 Research Contents

The RUL of lithium-ion battery refers to the number of charge and discharge cycles required from the current battery capacity storage ability to decay to the failure threshold (the value of SOH = 0.8) after a certain charge and discharge process. The number of cycles of lithium-ion battery means that when the battery reaches a complete charge and discharge cycle, the number of cycles increases once.

SOH can characterize the current battery performance by parameters such as capacity, battery power and battery cycle. In contrast, the capacity of lithium-ion battery is easier to obtain, and the interference factors are relatively less. Therefore, we choose the capacity of lithium-ion battery as the definition standard of SOH, and the formula is as follows:

$$SOH(t) = \frac{C_t}{C_0} \times 100\%. \tag{7}$$

where C_t denotes the lithium-ion battery capacity when fully charged in the cycle, and C_o denotes the rated capacity of the lithium-ion battery.

3.2 Data Source and Prediction Evaluation Criteria

In this paper, the CS2 battery data of CALCE battery research group is used as experimental data, which is CS2_ 35, CS2_ 36 and CS2_ 37 three sets of battery data, as shown in Fig. 2.

Data preprocessing process: due to the influence of the internal electrochemical reaction of the battery and the external working environment, the battery data contains some discontinuous outliers, which are first identified by the iForest [20] algorithm, and then the data are cleaned. Finally, the data are normalized.

The root mean square error (RMSE), the mean absolute error (MAE) and the r-square(R2) are used as indicators to evaluate the prediction results of the model. The formulas of RMSE, MAE and R2 are as follows:

$$\text{RMSE} = \sqrt{\frac{1}{n}\sum_{i=1}^{n}(y_i - \hat{y}_i)^2}. \tag{8}$$

$$\text{MAE} = \frac{1}{n}\sum_{i=1}^{n}\left|(y_i - \hat{y}_i)\right|. \tag{9}$$

$$R^2 = 1 - \frac{\sum_i(\hat{y}_i - y_i)^2}{\sum_i(\overline{y}_i - y_i)^2}. \tag{10}$$

where y_i denotes the actual data of lithium-ion battery, $\hat{y}_i$ denotes the predicted value, and $\overline{y}_i$ denotes the average of the actual data of lithium-ion battery.

3.3 Model Prediction Experiment

In order to verify the prediction accuracy and generalization ability of 1D CNN-BiLSTM model, the charge and discharge data of CS2_36 lithium-ion battery before 200 cycles, 250 cycles, 300 cycles, 350 cycles and 300 cycles are used as training data to predict the change of battery capacity in the future. The prediction results are shown in Table 1.

It can be seen from Table 1 that the prediction results of the five prediction points: the value of MAE is stable within 0.0042, the value of RMSE is stable within 0.0058, the value of R2 is stable above 0.9781, and the error between the predicted fault threshold point and the actual fault threshold point of each prediction point is less than 3.

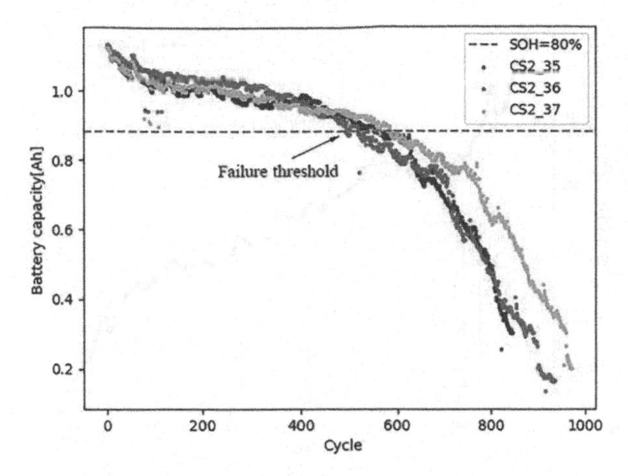

Fig. 2. Capacity curve of CALCE lithium-ion battery.

Table 1. The experimental results of different prediction points.

Prediction point	Real RUL	Prediction RUL	MAE	RMSE	R^2
200	320	322	0.0025	0.0030	0.9972
250	270	273	0.0030	0.0035	0.9939
300	220	222	0.0023	0.0029	0.9968
350	170	173	0.0042	0.0054	0.9844
400	120	122	0.0041	0.0058	0.9781

3.4 Comparisons of Different Prediction Methods

It is compared with the experimental results of 1D CNN model and LSTM model. The first 200 charge and discharge data of CS2_37 lithium-ion battery are selected as the training data set, and the data after 200 charge and discharge are selected as the test data set to predict the residual capacity of lithium-ion battery after 200 charge and discharge. The experimental results of different prediction models are shown in Table 2 and Fig. 3.

It can be seen from Table 2 and Fig. 3 that the fitting degree between the predicted curve of 1D CNN model and the actual curve is the lowest. The value of R2 is 0.9244, the value of MAE is 0.0092, the value of RMSE is 0.0115, and the error between predicted failure threshold point and actual failure threshold point is 27, and the training time is 10.68 s. 1D CNN model has the ability to analyze local features, but the long-term prediction accuracy is low and the training speed is fast.

The prediction results of LSTM model: the value of R2 is 0.9583, the value of MAE is 0.0066, the value of RMSE is 0.0086, and the error between predicted failure threshold point and actual failure threshold point is 20, and the training time is 92.79 s. LSTM model has memory function, but it only analyzes the data in one direction, so the prediction accuracy is low.

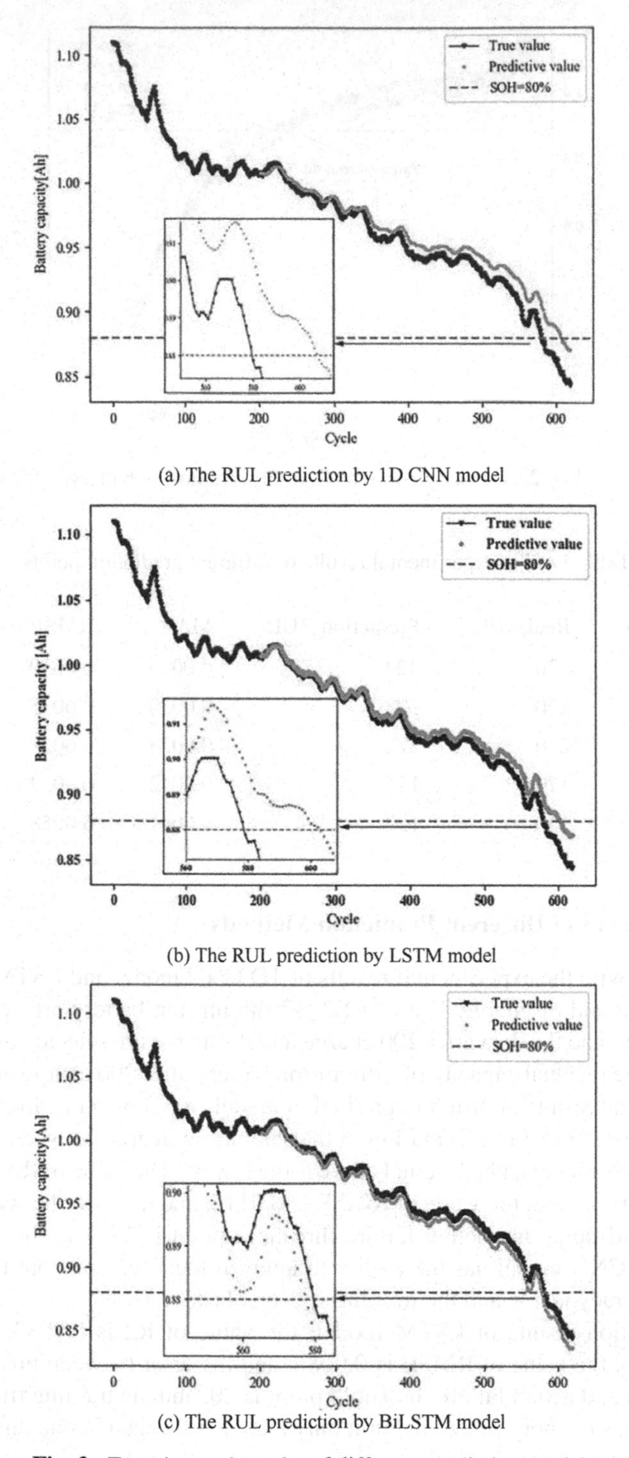

(a) The RUL prediction by 1D CNN model

(b) The RUL prediction by LSTM model

(c) The RUL prediction by BiLSTM model

Fig. 3. Experimental results of different prediction models.

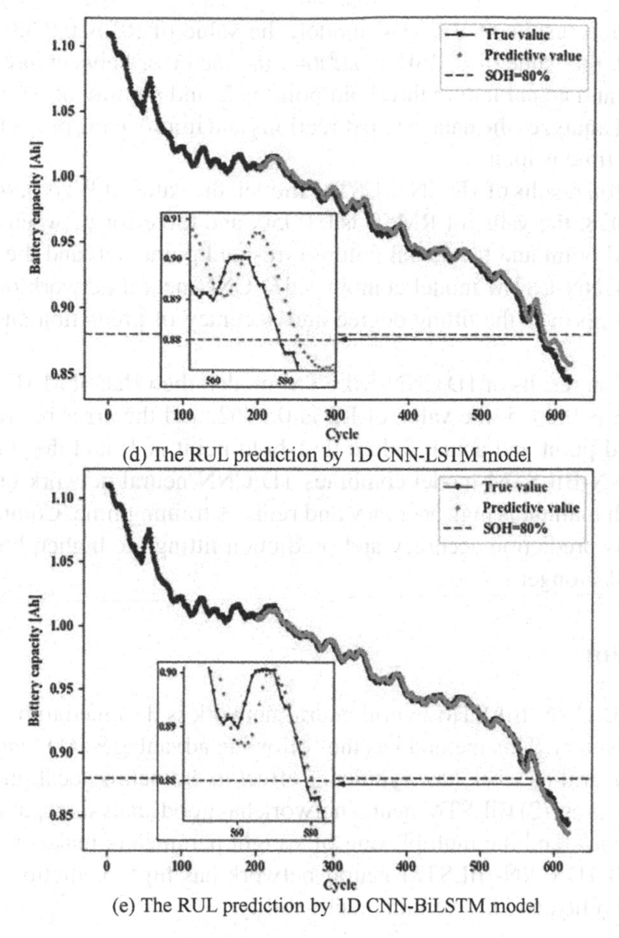

(d) The RUL prediction by 1D CNN-LSTM model

(e) The RUL prediction by 1D CNN-BiLSTM model

Fig. 3. continued

Table 2. The experimental results of different RUL prediction models.

Method	Real RUL	Prediction RUL	MAE	RMSE	R^2	Training Time/s
1D CNN	380	407	0.0092	0.0115	0.9244	10.68
LSTM	380	400	0.0066	0.0086	0.9583	92.79
BiLSTM	380	378	0.0049	0.0064	0.9766	147.18
1D CNN-LSTM	380	384	0.0038	0.0050	0.9857	52.32
1D CNN-BiLSTM	380	381	0.0019	0.0025	0.9962	84.97

The prediction results of BiLSTM model: the value of R2 is 0.9766, the value of MAE is 0.0049, the value of RMSE is 0.0064, and the error between predicted failure threshold point and actual failure threshold point is 2, and the training time is 147.18 s. BiLSTM model analyzes the data in two directions and improve the prediction accuracy, but the training time is long.

The prediction results of 1D CNN-LSTM model: the value of R2 is 0.9857, the value of MAE is 0.0038, the value of RMSE is 0.0050, and the error between the predicted failure threshold point and the actual failure threshold point is 4, and the training time is 52.32 s. 1D CNN-LSTM model combines 1D CNN neural network on the basis of LSTM, which improves the fitting degree and accuracy of prediction and reduces the training time.

The prediction results of 1D CNN-BiLSTM model: the value of MAE is 0.0019, the value of RMSE is 0.0025, the value of R2 is 0.9962, and the error between predicted failure threshold point and actual failure threshold point is 1, and the training time is 84.97 s. 1D CNN-BiLSTM model combines 1D CNN neural network on the basis of BiLSTM, which maintains high accuracy and reduces training time. Compared with 1D CNN-LSTM, its prediction accuracy and prediction fitting are higher, but the training time is about 30 s longer.

4 Conclusion

In this paper, 1D CNN-BiLSTM hybrid neural network is designed to predict the RUL of lithium-ion battery. This method has the following advantages: (1) One dimensional convolutional neural network has significant effect in extracting local information and mining deep features; (2) BiLSTM neural network has good analysis capabilities for continuous time series, and the multiplexing of weight parameters makes it less demanding on data; (3) 1D CNN-BiLSTM neural network has high prediction accuracy and generalization ability, and has certain practicability.

References

1. You, G.W., Park, S., Oh, D.: Real-time state-of-health estimation for electric vehicle batteries: a data-driven approach. J. Appl. Energy **176**, 92–103 (2016)
2. Wang, F.K., Huang, C.Y., Mamo, T.: Ensemble model based on stacked long short-term memory model for cycle life prediction of lithium-ion batteries. J. Appl. Sci. **10**, 3549 (2020)
3. Sui, X., Świerczyński, M., Teodorescu, R., et al.: The degradation behavior of LiFePO4/C batteries during long-term calendar aging. J. Energies **14**, 1732 (2021)
4. Dasari, H., Eisenbraun, E.: Predicting capacity fade in silicon anode-based Li-ion batteries. J. Energies **14**, 1448 (2021)
5. Gao, D., Huang, M.: Prediction of remaining useful life of lithium-ion battery based on multi-kernel support vector machine with particle swarm optimization. J. Power Electron. **17**, 1288–1297 (2017)
6. Yun, Z., Qin, W., Shi, W., et al.: State-of-health prediction for lithium-ion batteries based on a novel hybrid approach. J. Energies **13**, 4858 (2020)
7. Chen, X., Liu, Z., Wang, J., et al.: An adaptive prediction model for the remaining life of an Li-ion battery based on the fusion of the two-phase wiener process and an extreme learning machine. J. Electron. **10**, 540 (2021)

8. Kim, H.K., Lee, K.J.: Scale-up of physics-based models for predicting degradation of large lithium-ion batteries. J. Sustain. **12**, 8544 (2020)
9. Yang, J., Xia, B., Huang, W., et al.: Online state-of-health estimation for lithium-ion batteries using constant-voltage charging current analysis. J. Appl. Energy **212**, 1589–1600 (2018)
10. Zhang, C., Zhang, Y., Li, Y.: A novel battery state-of-health estimation method for hybrid electric vehicles. J. IEEE/ASME Trans. Mechatron. **20**, 2604–2612 (2015)
11. Chen, L., Lü, Z., Lin, W., et al.: A new state-of-health estimation method for lithium-ion batteries through the intrinsic relationship between Ohmic internal resistance and capacity. J. Meas. **116**, 586–595 (2018)
12. Li, J., Adewuyi, K., Yagin, N.L., et al.: A single particle model with chemical/mechanical degradation physics for lithium-ion battery state of health (SOH) estimation. J. Appl. Energy **212**, 1178–1190 (2018)
13. Kim, T.K., Moon, S.C.: Novel practical life cycle prediction method by entropy estimation of Li-ion battery. J. Electron. **10**, 487 (2021)
14. Yin, A., Tan, Z., Tan, J.: Life prediction of battery using a neural Gaussian process with early discharge characteristics. J. Sensors **21**, 1087 (2021)
15. Gao, D., Zhou, Y., Wang, T.: A method for predicting the remaining useful life of lithium-ion batteries based on particle filter using Kendall rank correlation coefficient. J. Energies **13**, 4183 (2020)
16. Pang, X., Huang, R., Wen, J., et al.: A lithium-ion battery RUL prediction method considering the capacity regeneration phenomenon. J. Energies **12**, 2247 (2019)
17. Wu, J., Zhang, C., Chen, Z.: An online method for lithium-ion battery remaining useful life estimation using importance sampling and neural networks. J. Appl. Energy **173**, 134–140 (2016)
18. Jiang, J.R., Lee, J.E., Zeng, Y.M.: Time series multiple channel convolutional neural network with attention-based long short-term memory for predicting bearing remaining useful life. J. Sensors **20**, 166 (2020)
19. Zhao, C., Huang, X., Li, Y., et al.: A double-channel hybrid deep neural network based on CNN and BiLSTM for remaining useful life prediction. J. Sensors **20**, 7109 (2020)
20. Li, X.P., Gao, X., Yan, B., et al.: Outlier detection method of power dispatching flow data based on isolated forest algorithm. J. Power Syst. Technol. **43**, 1447–1456 (2019)

A New Kind of Cross Line Based Crossover Used for Energy Saving and Security Enhancement of Industrial Wireless System

Mingtao Qiu(✉) and Dakui Wu

School of Mechatronic Engineering and Automation,
Shanghai University, Shanghai 200444, China

Abstract. A new type of compact crossover with a center frequency of 4.9 GHz which used for 5G wireless industrial application power reduction and security performance enhancement was proposed in this paper. It is composed of two identical compact cross line-based patch couplers and is characterized by small size, low loss, and easily integrated with other elements for its four ports are distributed in different directions. This new crossover can be used in beam-forming part of 5G industrial wireless systems for energy saving and security enhancement. Good performance of this cross over can be observed both in simulation results which are calculated by different tools.

Keywords: Energy saving · Wireless industrial systems · Crossover · 5G · Security enhancement

1 Introduction

Recently, wireless system is widely used in industrial scenario such as smart factory, intelligent vehicle system, power system, chemical industry etc. On the other hand, compared with 4G communication, the transmission rate of 5G is increased by 10 to 100 times, the peak rate reaches 10 Gbps, the delay is as low as 1 ms, and it can achieve 1 million massive connections per square kilometer. Although 5G is characterized with wide bandwidth, low delay, etc., the power consumption is becoming another problem for its mass application. The power consumption of 5G station is 2.5 to 3.5 times more than 4G station. The power consumption of the base station mainly comes from the radio frequency unit, and the power amplifier has the highest power consumption in the radio frequency unit. The power amplifier can linearly amplify the input signal within a certain range, and the input signal that exceeds this range will be amplified by distortion, resulting in a decrease in the efficiency of the power amplifier. Secondly, the power consumption of the base station is also closely related to the operating temperature. Generally, the higher the operating temperature, the higher the leakage current of the device and the higher the power consumption of the base station. In addition, the data processing part of the base station also occupies a lot of power consumption overhead, and there is also a large room for optimization in this part. 5G is responsible for the

K. Li et al. (Eds.): LSMS 2021/ICSEE 2021, CCIS 1468, pp. 168–179, 2021.
https://doi.org/10.1007/978-981-16-7210-1_16

transmission of confidential data, such as personal information, financial data, confidential formulas, medical records, etc. Therefore, security issues in 5G communications have also attracted much attention [1]. In the industrial Internet, wireless technology has greater advantages in terms of cost and flexibility than wired connections. However, due to the broadcasting characteristics of wireless channels and limited bandwidth, industrial wireless networks are more vulnerable to be attacked and eavesdropped than wired networks [2].

Artificial noise can improve channel security by restraining the eavesdropper's received signal power and fixed beam AP is usually used for artificial noise generating [3]. In real scenarios, fixed beam AP usually results in extra power consumption and is harmful to human body according to IEEE C95.1–2005. If the man-made noise and beam scanning technology are combined to make the beam scan at a certain frequency, it can not only improve the hazard of long-term fixed beam irradiation, but also can reduce transmission power of industrial wireless terminal consumption through beamforming technology. In addition, man-made noise can also be used to improve security performance.

The beamforming network (BFN) is the most important part for the beam-scanning terminal, which is generally composed of couplers, crossover, and phase shifters. Compared with normally BFN, such as Blass matrix, Nolen matrix and Rot-man lens, the Butler matrix (BM) is characterized with easily manufacturing, compactness, and low cost. In order to expand the data capacity and cover the base station sector simultaneously, many kinds of BMs are usually applied in wireless system, but rare application are found in wireless industrial security enhancement scenario. A low-loss three-beam broadband BM was designed in [4]. In [5], a broadband 4×4 BM which can be equivalent to different feed network in different frequency ranges was presented. And a 4×4 BM for 5G terminals was designed in [6], which was implemented on a substrate integrated wave-guide. As a planar micro-strip circuit, BM consumes a great amount of space. Many scholars focus on the size optimization theory of BM. A method to reduce the size of BM through left-handed media was proposed in [7], using the phase-delay characteristics of the transmission line of the load complementary split resonant ring (CSRR) to achieve miniaturization. With this method, compared with the traditional BM, the area is reduced by 48%.

2 Signal Propagation Loss Analysis

Radio wave propagation in wireless communication system will be affected by free space loss, topography blocking, multipath effect and weather factors. The so-called free space refers to the uniform and ideal propagation space in which radio wave propagation in wireless communication system will not be blocked, and refraction, scattering, absorption and other phenomena will not occur. Assuming that d represents the propagation distance of radio waves in the wireless communication system and f represents the propagation frequency, so the loss of radio waves in the free space propagation of the wireless communication system can be:

$$L_f = 32.44 + 20logf + 20logd. \tag{1}$$

However, there are usually obstacles in the plant to produce propagation loss. The ray tracking method takes the electromagnetic wave emitted from the transmitting point as the rays propagating in all directions and tracks each ray [8]. When it encounters obstacles, the field intensity can be calculated according to reflection, transmission or diffraction. The field intensity of each ray reaching the point is superimposed at the receiving point, so as to realize propagation prediction. The total distance that the light travels from the transmitting antenna to the receiving antenna is D. The acceptance point is taken as the center of the receiving sphere, and the included angle between adjacent rays is α, then the effective receiving radius is $\alpha D/\sqrt{3}$, and only the rays within the effective receiving radius can be received by the receiver. The correct radius of the receiving sphere allows the receiving sphere to receive only one ray at a time. But if the radius of the receiving sphere is larger than $\alpha D/\sqrt{3}$, then the receiving sphere can receive two rays at the same time, resulting in the rays on the same path being counted twice and the received power will be in an error of 3 dB. Also, if the radius of the receiving sphere is too small and no light intersects the receiving sphere, the multipath information will be filtered out. According to the above calculation, it can be seen that the propagation loss of radio waves in the wireless communication system in free space is related to the propagation distance d and the propagation frequency f. As the propagation distance of radio waves gradually increases, the transmission loss of radio waves also increases.

Assuming that the reflecting surface is S and the reflecting point is Q_R, and then the intensity of the reflected receiving electric field at the distance from the reflecting point S can be:

$$E^r(S) = E^i(Q_R)R\sqrt{\frac{p_1^r p_2^r}{(p_1^r + s)(p_2^r + s)}}e^{-j\beta s}. \tag{2}$$

In the formula, p_1^r and p_2^r represent the radius of the curvature of the reflected wave front at the reflection point, which are related to the radius of the curvature of the incident wave front p_1^i and p_2^i, and they can be expressed as:

$$\frac{1}{p_1^r} = \frac{1}{2}\left\{\frac{1}{p_1^i} + \frac{1}{p_2^i}\right\} + \frac{1}{f_1}, \frac{1}{p_2^r} = \frac{1}{2}\left\{\frac{1}{p_1^i} + \frac{1}{p_2^i}\right\} + \frac{1}{f_2}. \tag{3}$$

In the formula, f_1 and f_2 respectively represent the focal length in the zy and zx directions.

$$\frac{1}{f_1} = \frac{1}{cos\theta_i}\left(\frac{sin^2\theta_2}{R_1} + \frac{sin^2\theta_1}{R_2}\right) + \sqrt{\frac{1}{cos^2\theta_i}\left(\frac{sin^2\theta_2}{R_1} + \frac{sin^2\theta_1}{R_2}\right)^2 - \frac{4}{R_1 R_2}}$$

$$\frac{1}{f_2} = \frac{1}{cos\theta_i}\left(\frac{sin^2\theta_2}{R_1} + \frac{sin^2\theta_1}{R_2}\right) - \sqrt{\frac{1}{cos^2\theta_i}\left(\frac{sin^2\theta_2}{R_1} + \frac{sin^2\theta_1}{R_2}\right)^2 - \frac{4}{R_1 R_2}}. \tag{4}$$

In the formula, R_1 and R_2 represent the radius of curvature of the reflecting surface, θ_1 and θ_2 represent the angle between the incident wave and zx, zy, and θ_i represents the incident angle. For p_1^i and p_2^i, the radius of curvature of the incident wavefront (s' refers to the radius of the spherical wave) can be:

$$\begin{cases} p_1^i = p_2^i = s\prime & \text{Spherical incident wave} \\ p_1^i = p\prime, p_2^i = \infty & \text{Cylindrical incident wave} \\ p_1^i = p_2^i = \infty & \text{Plane incident wave} \end{cases}$$

Assuming that the spherical wave is the incident wave and the reflecting surface is a plane, the focal length $f_1 = f_2 = \infty$, in the direction of zx and zy, then the receiving electric field intensity in formula (1) can be simplified as:

$$E^r(S) = E^i(Q_R)R\frac{s'}{s+s'}e^{-j\beta s}. \tag{5}$$

Since each received signal has to go through multiple reflections, the final received electric field can be:

$$E_i = E_0 G_{ti} G_{ri} \prod_j R_j \frac{s'}{s+s'} \frac{e^{-j\beta(d+d_0)}}{d_0}. \tag{6}$$

In the formula, G_{ti} and G_{ri} respectively are the antenna gains of the transmitter Tx and the receiver Rx, d_0 represents the distance between the source signal point and the first interaction point, and d represents the distance between the first interaction point and the acceptance point. Considering the multipath effect in the channel, that is, there is more than one reaching the acceptance point, and then the total received electric field can be:

$$E_{Tot} = \sum_{i=1}^{N} E_i + E_{LOS}. \tag{7}$$

In the formula, E_{LOS} represents the electric field intensity of the direct component. The path loss can be calculated according to formula (7), and its unit is dB.

$$PL = 20lg\left(\frac{\lambda}{4\pi}\frac{E_{Tot}}{E_0}\right). \tag{8}$$

From the above formula, it can be concluded that the noise signal will have a certain degree of loss in the propagation process. Assuming that there are no obstacles and losses between the transmitting point and the receiving point, the Fries formula can be expressed as:

$$P_R = \frac{P_T G_T G_R \lambda^2}{(4\pi R)^2}. \tag{9}$$

In this formula, P_R and P_T represent the power level of the receiving antenna and the transmitting antenna, respectively, and G_T and G_R represent the antenna gain of the receiving antenna and the transmitting antenna, respectively. λ is the operating wavelength, and R is the distance between the antennas. The formula (9) shows that when P_R, R, and λ remain unchanged, the transmit power will decrease as the antenna gain increases. So as a conclusion, it can be seen that when using the BFN to increase the antenna gain and obtain the same receiving power level, the transmitting power can be lowered. In that case, the power loss can be reduced, compared with the case without using beam-forming technology (Fig. 1).

3 Beamforming Theory Analysis

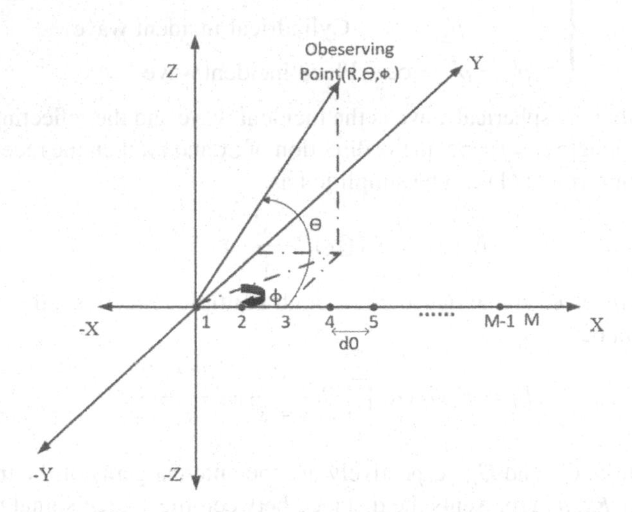

Fig. 1. Theory of beamforming array antenna.

Array is a system in which a number of discrete radiating elements are arranged according to a certain rule and connected to each other to form an antenna system to generate highly directional radiation. Antenna array is composed of many antenna cells which are so called as elements. The radiation characteristics of the array, such as beam width, beam position, gain and front-to-rear ratio, depend on the type, number, arrangement, spacing of array elements, and the amplitude and phase distribution of the current on the array elements. For any equidistant linear array, the required pattern can be generated by changing the current amplitude and phase of each element of the antenna. Assuming that the number of array antenna elements is odd, that is, $m = 2n + 1$, and then the array factor can be expressed as:

$$f_{2n+1}(\varphi) = A_0 + A_1 e^{j\varphi} + A_2 e^{j2\varphi} + \cdots + A_{n-1} e^{j(n-1)\varphi} + A_n e^{jn\varphi}$$

$$+ A_{n+1} e^{j(n+1)\varphi} + \ldots + A_{2n} e^{j2n\varphi}. \tag{10}$$

In the above formula, the complex number A represents the amplitude and phase of the current, and d represents the spacing between adjacent elements, and the phase can be expressed as $\phi = \beta d cos\theta, \theta$ is the azimuth from the axis of the antenna array.

Dividing the above formula by $e^{jn\phi}$ without changing the expression of the array factor, we can get:

$$f_{2n+1}(\varphi) = A_0 e^{-jn\varphi} + A_1 e^{-j(n-1)\varphi} + \ldots + A_{n-1} e^{-j\varphi} + A_n$$

$$+ A_{n+1} e^{j\varphi} + \ldots + A_{2n} e^{jn\varphi}. \tag{11}$$

If the current amplitude of each element of the array antenna is symmetrically distributed about the midpoint, and the phase of the left element lagging behind the midpoint is equal to the phase of the right element before the midpoint, then A_{n-k} and A_{n+k} are in a conjugate relationship. That is:

$$A_{n-k} = a_k - jb_k, A_{n+k} = a_k + jb_k. \tag{12}$$

Substitute the above formula into the exponential form and simplify, then we can get:

$$A_{n-k}e^{-jk\varphi} = A_{n+k}e^{jk\varphi} = a_k\left(e^{jk\varphi} + e^{-jk\varphi}\right) + jb_k\left(e^{jk\varphi} - e^{-jk\varphi}\right)$$

$$= 2a_k cosk\varphi - 2b_k sink\varphi. \tag{13}$$

Let $A_m = a_0$, and substitute into the above equation to simplify, then we can get:

$$f_{2n+1}(\varphi) = 2\left[\frac{1}{2}a_0 + (a_1 cos\varphi + a_2 cos2\varphi + \ldots + a_n cosn\varphi) - (b_1 sin\varphi + b_2 sin2\varphi + \ldots + b_n sinn\varphi)\right]$$

$$= 2\left\{\frac{a_0}{2} + \sum\nolimits_{k=1}^{n} [a_k cosk\varphi + b_k sink\varphi]\right\}. \tag{14}$$

The above formula is the first $2n + 1$ terms of the Fourier series. Any directional graph is an angle θ, and therefore is also a periodic function $f(\varphi)$ of φ, so it can be expanded into Fourier series with infinite terms.

For a linear array with constant amplitude and linear change in phase, the array factor can be expressed as:

$$f(\theta) = \frac{\sin(u - v)}{u - v}. \tag{15}$$

In the formula, $u = \frac{\beta l}{2}cos\theta$, $v = \frac{1}{2}\beta_1$, $\beta_1 = \frac{\alpha}{d}$ is the phase change per unit length.

As for the specified directional graph, some points can be selected on it, which are respectively satisfied by the above main lobe that deviates from zero. Then, the total directional graph factor thus constituted is:

$$f(\theta) = \sum\nolimits_m C_m \frac{\sin(u - v_n)}{u - v_n}. \tag{16}$$

In the formula above, C_m is the amplitude of the main lobe of each unit.

According to the obtained total directional pattern array factor, the specific array spacing and other parameters can be synthesized.

4 Structure of the Crossover

Crossover is the most important element for the BFN, which usually consists of two cascaded 90° hybrid couplers and plays an important role in power distribution. As shown in Fig. 2, we proposed a novel kind of crossover in this paper which consists of two similar special patch coupler cells. Each single cell is constructed by etching asymmetrical cross grooves on a square patch. This structure can realize same amplitude output and a 90° difference at the two output ports. Meanwhile, it has a good isolation effect at the output port 4. [9] introduces the simulation results of different parameters of the single coupler. The crossover formed by the two couplers can produce full output at the output port 3 through cancellation. The dielectric constant of the substrate used for simulation is 10.8 mm, the thickness is 0.635 mm, and the length and width of other wires and grooves are show in Table 1.

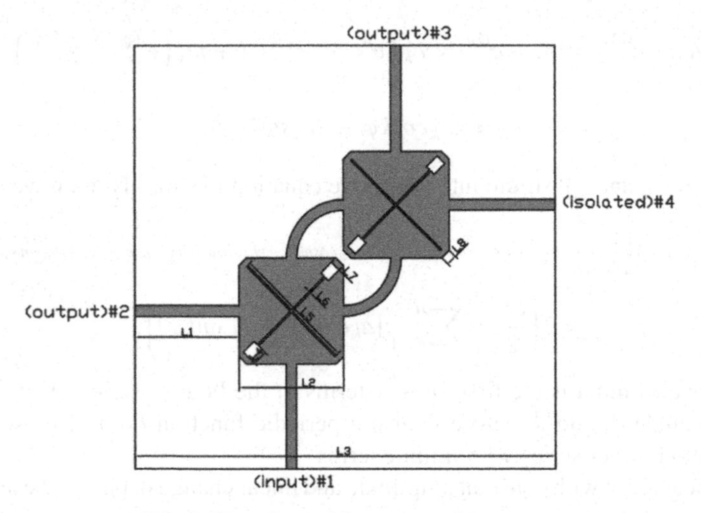

Fig. 2. Top view of the proposed crossover.

Table 1. The parameters of the crossover.

Parameters	Dimension(mm)
L1	6.00
L2	6.00
L3	24.00
L4	0.86
L5	7.20
L6	0.10
L7	0.62
L8	0.60

For a single hybrid coupler, the characteristic impedances Z_{01}, Z_{02}, and Z_{03} can be expressed by the following formulas:

$$Z_{01} = \frac{Z_{in}}{k}. \tag{17}$$

$$Z_{02} = \sqrt{\frac{Z_{out} Z_{in}}{1 + k^2}}. \tag{18}$$

$$Z_{03} = \frac{Z_{out}}{k}. \tag{19}$$

In the formula, *Zin* and *Zout* respectively represent the input impedance and the output impedance. *k* refers to the coupling factor and is defined as $\left|\frac{S_{21}}{S_{21}}\right|$. Under the premise of an equal power split coupler, that is, $k = 1$ and $z_{in} = z_{out}$ and then the formulas above can be simplified as:

$$Z_{01} = Z_{03} = Z_0. \tag{20}$$

$$Z_{02} = \frac{Z_0}{\sqrt{2}}. \tag{21}$$

In the formula, *Z0* refers to the characteristic impedance.

The even-odd mode can be used to analyze the branch line coupler [10]. The initial characteristic impedance of the coupler, which is normalized to Z0, is shown in Fig. 3.

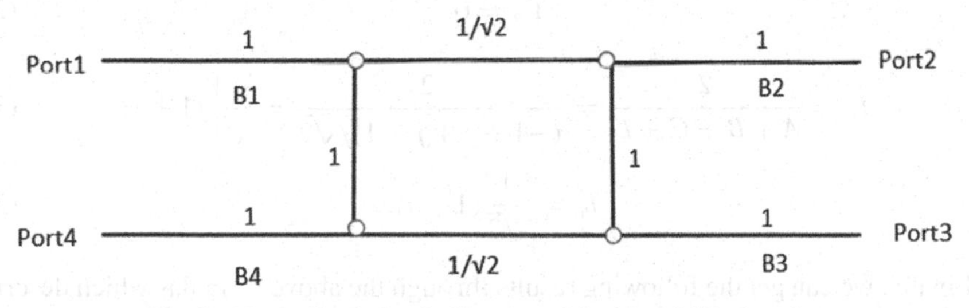

Fig. 3. Normalized circuit diagram of the branch line coupler.

It is assumed that an incident wave of uniform amplitude reaches port 1. It's because of the symmetrical structure of the coupler that the circuit can be decomposed into the superposition of even-mode excitation and odd-mode excitation. The original excitation of Fig. 3 can be produced by adding both even–odd mode excitations and the actual response is the sum of the responses to the even and odd mode excitations.

For these two–port networks, the amplitudes of the incident waves are respectively + 1/2 and –1/2. The amplitudes of the emerging wave at each port of the quadrature hybrid are:

$$B_1 = \frac{1}{2}\Gamma_e + \frac{1}{2}\Gamma_o. \tag{22}$$

$$B_2 = \frac{1}{2}T_e + \frac{1}{2}T_o. \tag{23}$$

$$B_3 = \frac{1}{2}T_e - \frac{1}{2}T_o. \tag{24}$$

$$B_4 = \frac{1}{2}\Gamma_e - \frac{1}{2}\Gamma_o. \tag{25}$$

In the formula above, T_e and T_o respectively represent the even and odd mode transmission coefficients, and Γ_e and Γ_o the even and odd mode reflection coefficients.

Taking into account the even mode two-port circuit, T_e and Γ_e can be found by obtaining the scattering parameters.

$$\begin{bmatrix} A & B \\ C & D \end{bmatrix}_e = \begin{bmatrix} 1 & 0 \\ j & 1 \end{bmatrix} \begin{bmatrix} 0 & \frac{j}{\sqrt{2}} \\ \frac{j}{\sqrt{2}} & 0 \end{bmatrix} = \frac{1}{\sqrt{2}} \begin{bmatrix} -1 & j \\ j & -1 \end{bmatrix}. \tag{26}$$

$$\begin{bmatrix} A & B \\ C & D \end{bmatrix}_o = \frac{1}{\sqrt{2}} \begin{bmatrix} 1 & j \\ j & 1 \end{bmatrix}. \tag{27}$$

The *ABCD* parameter can be transformed into S-parameter equivalent to the reflection coefficient and transmission coefficient of the even mode.

$$\Gamma_e = \frac{A + B - C - D}{A + B + C + D} = \frac{(-1 + j - j + 1)/\sqrt{2}}{(-1 + j + j - 1)/\sqrt{2}} = 0. \tag{28}$$

$$\Gamma_o = 0. \tag{29}$$

$$T_e = \frac{2}{A + B + C + D} = \frac{2}{(-1 + j + j - 1)/\sqrt{2}} = \frac{-1}{\sqrt{2}}(1 + j). \tag{30}$$

$$T_o = \frac{1}{\sqrt{2}}(1 - j). \tag{31}$$

Finally, we can get the following results through the above formulas which describe the fundamental operations of the branch line coupler.

$$B_1 = 0$$

$$B_2 = \frac{-j}{2}$$

$$B_3 = \frac{-1}{\sqrt{2}}$$

$$B_4 = 0$$

The result shows that when port 1 is in good condition, the phase difference between port 1 and port 2 is −90°, the phase difference between port 1 and port 3 is −180°, and port 4 has no power. The power split between the output ports is −3 dB, and the phase difference is 90°.

5 Validation

ANSOFT HFSS is used to validate this new kind of crossover. As shown in Fig. 4. Since 4.9 GHz is one of the important frequency bands of 5G communication, the crossover designed in this paper performs well at 4.9 GHz. Figure 5 shows the S11 and S41, and the simulation results of S11 at 4.9 GHz are −12.5 dB. Figure 6 demonstrates the results of S21 and S31, showing that S21 is −9 dB and S31 is −2 dB at 4.9 GHz. The simulation of the phase response of the crossover prototype is shown in Fig. 7, and it can be seen that the phase difference between S21 and S31 is 90°. According to above figures, good performance of this novel kind of crossover in 4.9 GHz is observed.

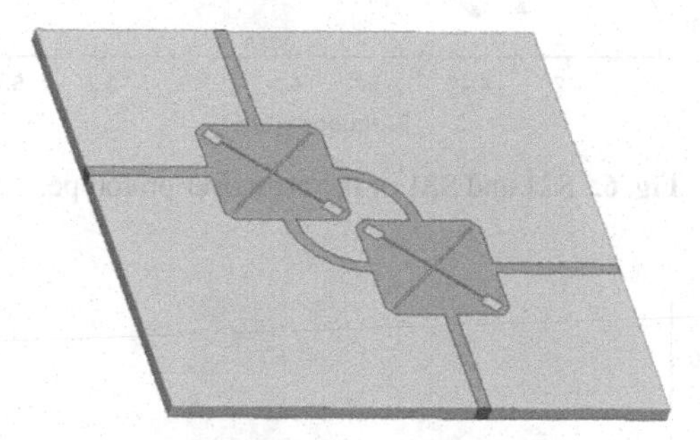

Fig. 4. 3D model for Crossover.

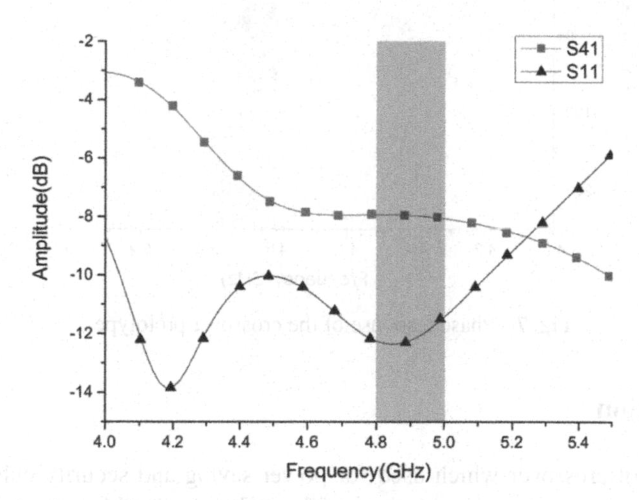

Fig. 5. S11 and S41 of the crossover prototype.

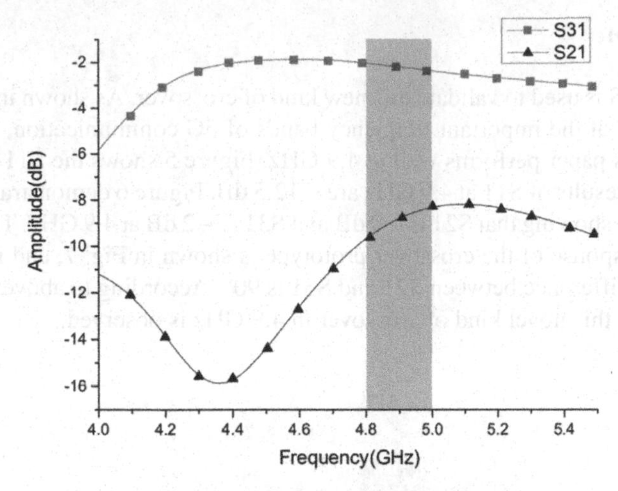

Fig. 6. S21 and S31 of the crossover prototype.

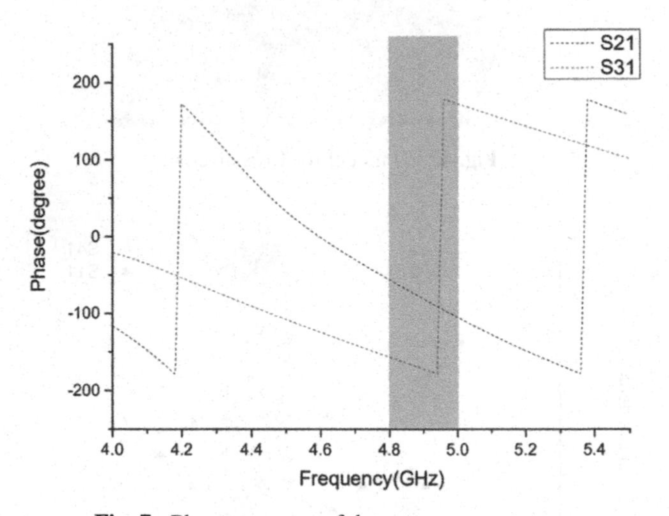

Fig. 7. Phase response of the crossover prototype.

6 Conclusion

A novel kind of crossover which used for power saving and security enhancement is proposed in this paper. It is characterized with small size, flexible output direction and can easily integrate with other element of wireless system. Also, it performs well in 4.9 GHz operating frequency and can be used in 5G and other industrial wireless systems.

References

1. Cheng, D., Vij, S., Jain, A.: 5G: evolution of a secure mobile technology. In: 2016 3rd International Conference on Computing for Sustainable Global Development (INDIACom), New Delhi, pp. 2192–2196. Springer, Heidelberg (2016)
2. Wang, H.: The application and research of wireless technology in industrial network. In: 2008 4th International Conference on Wireless Communications. pp. 1–4, Networking and Mobile Computing, Dalian (2008)
3. Chen, A., Zhao, K.: Fixed-frequency beam-steerable leaky-wave antenna operating over a wide band. In: 2017 International Symposium on Antennas and Propagation (ISAP) (2017)
4. Zhu, H., et al.: Butler matrix based multi-beam base station antenna array. In: 2019 13th European Conference on Antennas and Propagation (EuCAP) IEEE (2019)
5. Ran, X.K., Chen, F.C.: 4×4 broadband butler matrix and its application in antenna arrays. In: 2019 IEEE International Symposium on Antennas and Propagation and USNC-URSI Radio Science Meeting IEEE (2019)
6. Yang, Q.L., et al.: Butler matrix beamforming network based on substrate integrated technology for 5G mobile devices. In: Antennas and Propagation IEEE (2016)
7. Rao, P.H., Sajeevan, S.J., Kushal, K.: Miniaturisation of switched beam array antenna using phase delay properties of CSRR-loaded transmission line. J. IET Microw. Antennas Propag. **12**, 1960–1966 (2018)
8. Chen, X., Pan, Y.T., Wu, Y.M., Zheng, G.X.: Statistical characteristics of multipath channels in subway tunnels. J. Appl. Sci. **33**(4), 389–398 (2015)
9. Sun, S., Zhu, L.: Miniaturised patch hybrid couplers using asymmetrically loaded cross slots. J. Microw. Antennas Propag. IET **4**(9), 1427–1433 (2010)
10. Rahimian, A.: Steerable antennas for automotive communication systems. J. Comput. Sci. (2014)

Review of Machine Learning for Short Term Load Forecasting

Aijia Ding[1], Huifen Chen[2](✉), and Tingzhang Liu[1]

[1] School of Mechatronic Engineering and Automation, Shanghai University, Shanghai 200444, China

[2] Toulouse School of Management, The University of Toulouse 1, Capitole, Toulouse, France

Huifen.chen@tsm-education.fr

Abstract. Short term load forecasting is among the most significant issues of research for efficient, reliable and safe operation of power system. It takes significant part in the field of power dispatch, system extension, power flow analysis, scheduling and preservation of power system. This work contains a review on the various excellent reported literatures of machine learning methodologies utilized in the domain of short-term load forecasting. Specifically, hybrid models which combine machine learning with statistic methods, evolutionary algorithm and non-parameter methods achieve continuous attention. Recurrent Neural Network, Long Short-Term Memory, Gated Recurrent Unit, Convolutional Neural Network and variant of residual network were introduced to reveal the improved performance of comprehensive methods.

Keywords: Short term load forecasting · Machine learning · Recurrent neural network · Long short term memory · Gated recurrent unit · Convolutional neural network · Residual network

1 Introduction

Short term load forecasting (STLF) is among the most significant issues which involves electricity planning and scheduling, dispatching, and reconstruction in power system [1] especially during the national electricity market reform. Precise load forecasting can reduce the cost of power plant, decrease power transmission loss, cut down the probability of power interruption events and avoid unessential energy waste.

Roadblocks in STLF are of measurement of load data, missing necessary data, acquisition of meteorological data and appropriate load forecasting models. The former issues have been addressed with the development of various meters. Research of STLF began around 1970s and great amount of methodologies have been put forward to sharpen the accuracy of the forecasting results. The forecasting methodologies can be categorized as two types: statistical methodologies and artificial intelligence (AI) ones. With the development of AI approaches in various fields, medical area, agriculture, service industry, finance and electricity, AI methodologies can be hardly distinguished from machine learning which far outweighs statistical methods due to the non-stationary of load data,

K. Li et al. (Eds.): LSMS 2021/ICSEE 2021, CCIS 1468, pp. 180–190, 2021.
https://doi.org/10.1007/978-981-16-7210-1_17

while statistical approaches involving Multiple linear regression [2 , 3], Exponential Smoothing method [4], Kalman Filter with wavelet [5], Auto Regressive approach [6], Auto Regressive Moving Average approach [7], show superior performance in handling stationary time series. An obvious tendency can be observed that more and more researchers apply machine learning methodologies to STLF. The benchmarks of machine learning involve support vector machine (SVM) [8], fuzzy logic [9], genetic approaches such as further developed Ant colony [10], Particle Swarm Optimization (PSO) [11] and Artificial Bee Colony [12], artificial neural network (ANN) [13–15] and cloud computing [16]. Different algorithms share different advantage; therefore, researchers combine different methods to achieve better performance. This paper focuses on machine learning applied in STLF. Section 2 describes various combined methodologies and elaborates models together with inputs, creation points and performance of each methods. Section 3 states the shortcomings and the trend of STLF methods.

2 A Brief Introduction to Machine Learning for LSTM

Machine Learning is currently main technique to resolve AI issues which attract AI researchers and technology corporation home and abroad. A bulk of machine learning based models have been proposed and achieve remarkable performance.

2.1 ANN

ANN is one of the most critical methodologies along with the development of STLF approaches. The theory of ANN is derived from human brain neurons. The fundamental structure of ANN contains three layers including input layer, output layer and hidden layer which exists between input layer and output layer. Besides, there are weights between neurons in different layers along with bias in each layer and activation function in each neuron except input ones. The weights, bias and parameters in activation function can be trained and tuned to make the network fulfill deterministic task. Various kinds of ANN structures have been developed to address diverse issues including image recognition, facial recognition speech processing, auto-driving. Full-connected neural network, recurrent neural network (RNN), convolutional neural network (CNN), and Hopfield network are among the typical multi-layer neural networks with various structures. A two layer of feed forward NN is illustrated in Fig. 1.

2.2 SVM

Support vector machine (SVM) achieved the best results in classification problems until deep learning technology was developed. The theory of SVM lies in the basis of maximizing the total distance of samples from various classification. The formula $\omega \bullet x + b = 0$ determines the separating hyperplane. There are endless hyperplanes with regards to linearly separable data sets while the hyperplane with maximum geometric distance is unique. As can be seen from Fig. 2, the red spots are support vector.

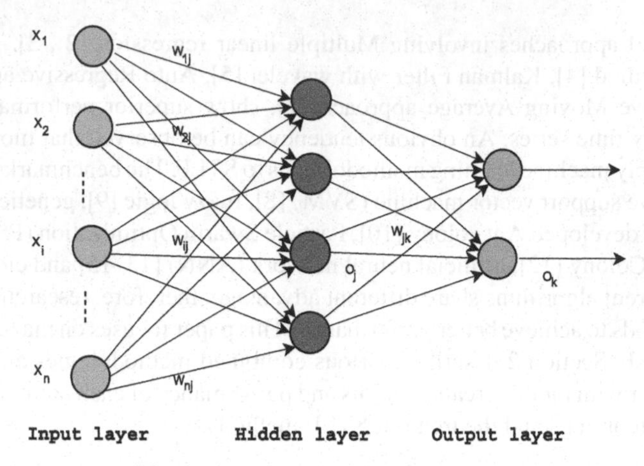

Fig. 1. A two layer of feed forward NN

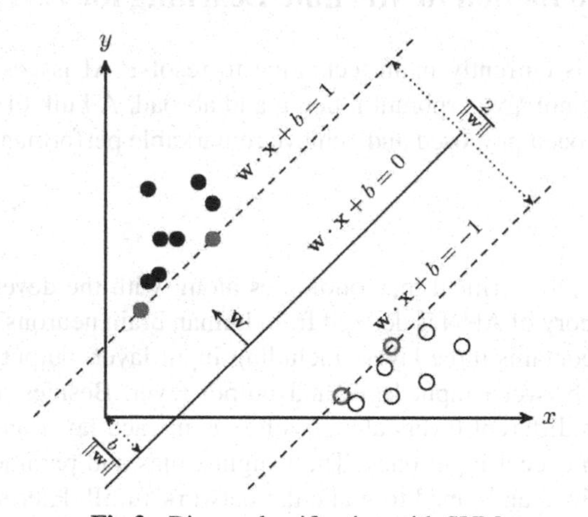

Fig.2. Binary classification with SVM

2.3 KNN

K-nearest neighbor (KNN) aims to discover k training samples which are nearest the distinction object in the training set, where k, the number of training set, is a hyperparameter and should be determined before classification. The dominant category can be recognized from the k training samples and be considered as the target object's category. Since the KNN methodology strongly depends on the surrounding limited adjacent samples, while many other classification approaches rely upon the approach of discriminant domain mechanism to determine the classes. Therefore, the KNN method is more satisfactory than other approaches. The idea of KNN method is demonstrated in Fig. 3. Xu in the figure goes with the category (ω_1) as four neighboring samples go with ω_1, only one neighboring sample goes with ω_3 and no sample belongs to ω_2 [18].

The forecasting algorithm of KNN contains four procedures,

(1) Determine the hyperparameter value k;
(2) discover k samples which are nearest the distinction object and have been classified into different categories;
(3) Calculate the number of chosen samples belong to each category;
(4) Choose the category that has the largest number.

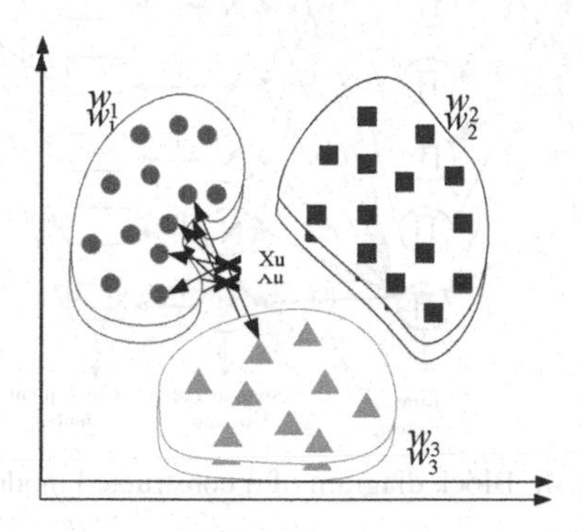

Fig. 3. KNN algorithm map

2.4 Related Work

ANNs have been abundantly used in recently years, while overfitting and curse of dimensionality effects may occur in many cases. Overfitting happens when noise and random error is learned by the NN, while the curse of dimensionality is caused by the complexity of unnecessary deep NN structure.

Support vector regression has been proposed to overcome fore-mentioned problems [19], but the method proposed is time consuming. To address this issue, error correction algorithm is proposed to train radial basic function (RBF) neural network with less RBF units, less training time and lower mean absolute percentage error (MAPE) [1]. The high dimension of input variables is one of the factors causing the curse of dimensionality. Rough set is introduced into BP neural network and reduces attributes based on variable accuracy. The impact of data with noise and inconspicuously interdependent data is weakened and time cost for training procedure is reduced due to less input data [20].

Historical load, temperature and weather condition including humidity, wind speed, wind degree and cloud conditions, are used partly or totally in most of the published papers, while the system operation state is usually ignored. The system state (node voltage, phase angle and power flow) is taken into consideration and accurately estimated by weighted least square (WLS) [21]. The precise estimate system state together with preliminary forecast load generated from neural network is introduced to adaptive neuro-fuzzy inference system (ANFIS). The constructed model is shown in Fig. 4.

It is inevitable that deep learning is dominating the STLF methods due to its accurate performance. Deep residual network (ResNet) is utilized to predict short term load

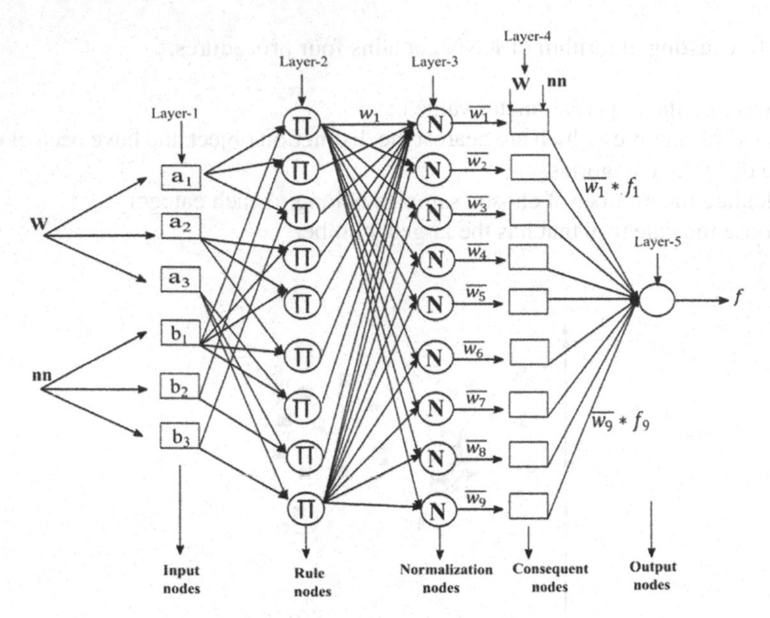

Fig. 4. Block diagram of a constructed model

and load interval prediction is conducted with Monte Carlo (MC) dropout based on ResNet. Three widely acknowledged datasets are applied to evaluate the validity and generalization ability of the raised method. This literature adopted a two-stage ensemble strategy to improve the generalization ability and this strategy is shown in Fig. 5. In the first stage, a few of snapshots are taken during the period of training procedure of a separate model and Adam (abbreviated from adaptive moment estimation), which has the default parameters, is used as the optimizer. In other words, a few of snapshots of the same model are taken during its training procedure. The second stage is implemented where the parameters of the same model are simply reinitialized, which is among the common practices of acquiring good integration models [22]. To further estimate the robustness of the raised approach, the literature added different Gaussian noise to the observed temperature.

Case1: the literature added Gaussian noise with mean standard deviation 1 °F and mean 0 °F to the observed temperature before normalization.

Case2: the literature added Gaussian noise with mean standard deviation 2 °F and mean 0 °F to the observed temperature before normalization.

Case3: the literature added Gaussian noise with mean standard deviation 3 °F and mean 0 °F to the observed temperature before normalization.

It can be concluded from Fig. 6 that model with ensemble strategy of two stage decreases the growth of MAPE. The prediction result of a summer week is shown in Fig. 7 and the 90% probabilistic load forecasting of a winter week is shown in Fig. 8. The actual coverage from the proposed model compared with expected coverage with regards to various z-score is implied in Table 1. Load interval prediction results of the same day with regards to various z-score are shown in Fig. 9. However, the proposed model with ensemble strategy is time consuming.

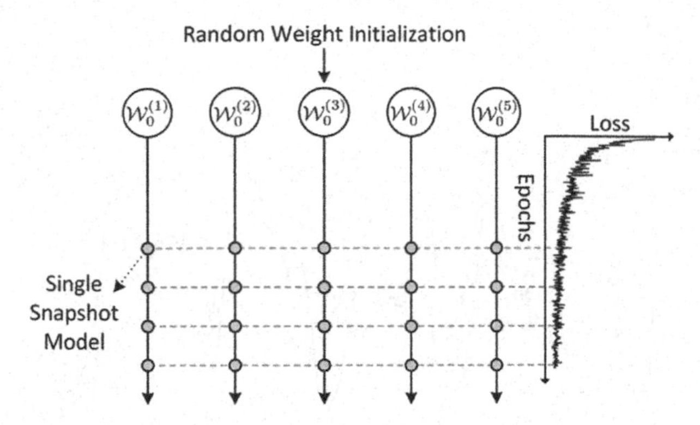

Fig. 5. The ensemble strategy with random weight initialization

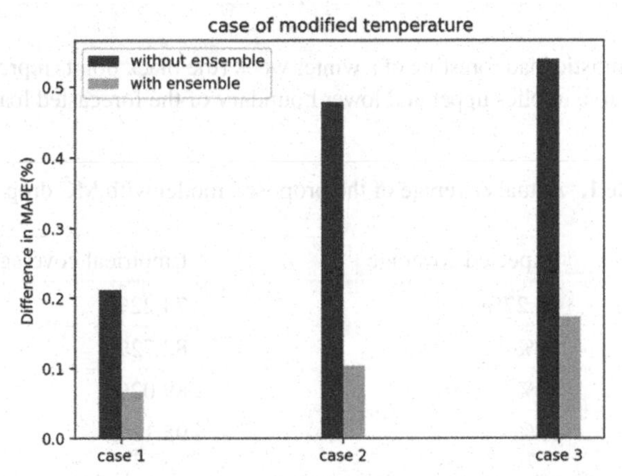

Fig. 6. The comparison of the raised approach with the ensemble strategy and the raised approach without ensemble when various cases of adjusted temperature are adopted.

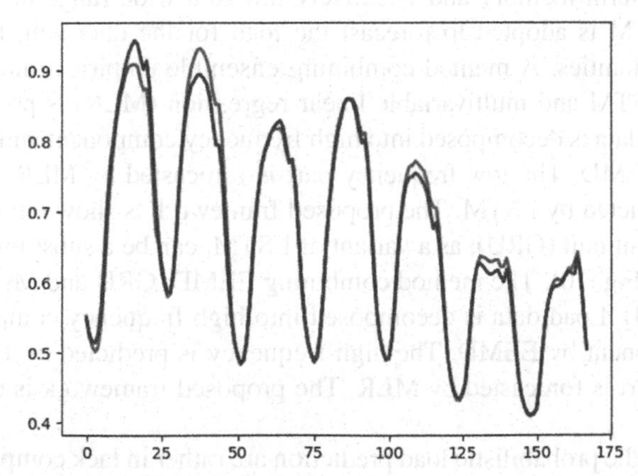

Fig. 7. Prediction result of a summer week (the green curve implies the observed load, the blue curve implies the forecasted load) (Color figure online)

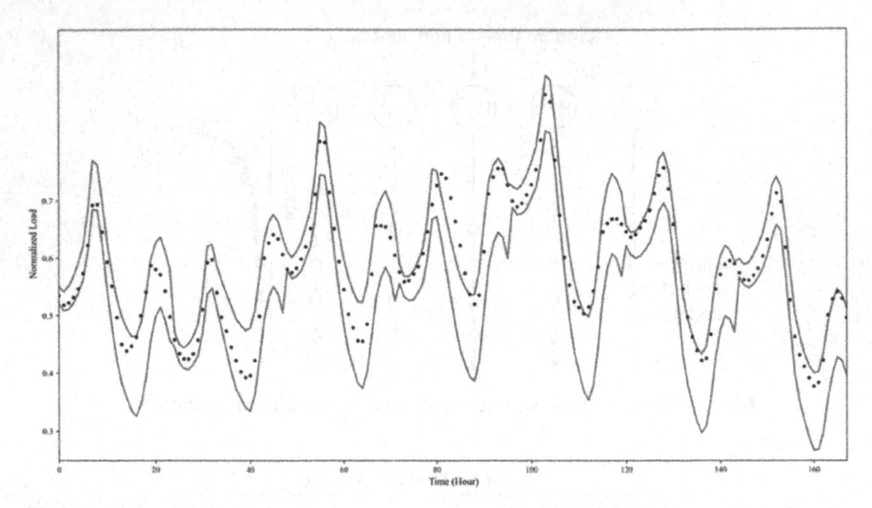

Fig. 8. 90% probabilistic load foresting of a winter week (the black points represent the observed load and the grey curve implies upper and lower boundary of the forecasted load respectively)

Table 1. Actual coverage of the proposed model with MC dropout

z-score	Expected coverage	Empirical coverage
1.000	68.27%	74.22%
1.280	80%	82.72%
1.645	90%	89.02%
1.960	95%	95.11%

Long short term memory (LSTM) is another deep learning methodology which can extend the short term memory and effectively utilize a wide range of historical data. Specifically, LSTM is adopted to forecast the load for the data with high frequency and strong uncertainties. A method combining ensemble empirical-mode decomposition (EEMD), LSTM and multivariable linear regression (MLR) is proposed, namely ELM [23]. Load data is decomposed into high frequency component and low frequency component by EEMD. The low frequency part is forecasted by MLR, while the high frequency is predicted by LSTM. The proposed framework is shown in Fig. 9.

Gated recurrent unit (GRU), as a variant of LSTM, can be a substitute for LSTM in many scenarios (Fig. 10). The method combining EEMD, GRU and MLR is proposed, namely EGM [24]. Load data is decomposed into high frequency component and low frequency component by EEMD. The high frequency is predicted by GRU, while the low frequency part is forecasted by MLR. The proposed framework is quite similar to Fig. 9.

Essays about the probabilistic load prediction are rather in lack compared with point forecasting. [25] raised a hybrid approach based on two-stage NN to calculate the average value and fiducial intervals of daily peak load prediction. [26] evaluated the fluctuation

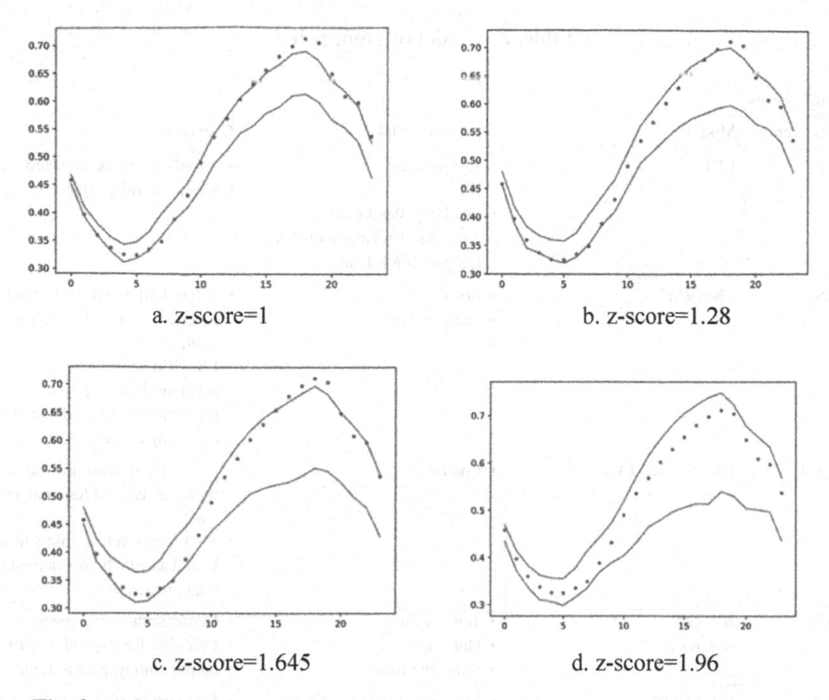

a. z-score=1 b. z-score=1.28

c. z-score=1.645 d. z-score=1.96

Fig. 9. Load interval prediction of the same day with respect to different z-score

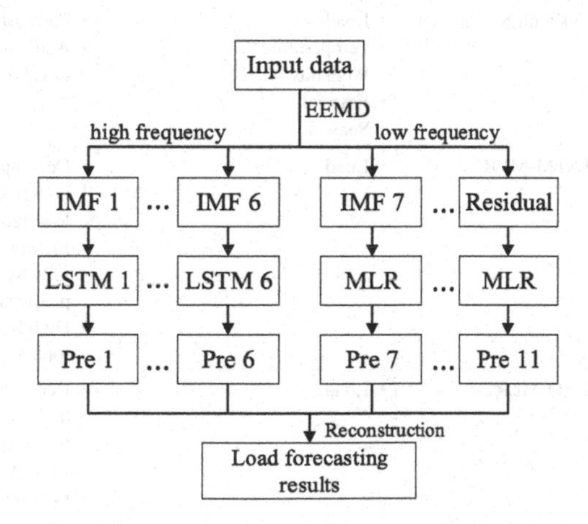

Fig. 10. Proposed model framework of ELM

of forecasting deviations and prediction intervals (PIs) using load prediction based on meteorology variable integration. [27] put forward a sparse heteroskedastic methodology for predicting the load of energy-intensive corporation. There is another research field in short-term probabilistic load prediction is about fuzzy interval load prediction. [28] adopted fuzzy interaction regression (FIR) to yield short-term load prediction with fuzzy

Table. 2. Model references list

Model references				
No	Reference	Model	Inputs to model	Overview
1	[1]	RBF	• Temperature • Hour • Weekday/Weekend/ • Average of former 24 h-load • Lagged 168 h load	• Simpler network structure • Shorter training time
2	[8]	LS-SVM	• Load • Temperature	• Adopt RBF function as a kernel function for a stable and reliable method • LS-SVM extrapolates relationship between the temperature and load data then obtaining the predicted load
3	[11]	PSO for ARMAX	• Load data	• Prediction of load prior to more than one day and less than one week • PSO points to few points in the load data which are claimed for load prediction
4	[20]	BP with rough set	• Temperature • Humidity • Sunshine time	• Decrease the input space • Calculate the dependency of output with respect to input
5	[21]	WLANFIS	• Forecasted load from NN • System state	• Forecasting of short term and medium term load
6	[22]	ResNetPlus with MC dropout	• Load • Temperature • Weekday/weekend • Festival • Season	• Forecasting day-ahead load • Add noise into the temperature to validate the model's robust
7	[23]	EEMD-LSTM-MLR	• Load	• Decompose the load data into high frequency component and low frequency component by EEMD • Low frequency component is predicted by MLR • High frequency component is forecasted by LSTM
8	[24]	EEMD-GRU-MLR	• Load	• Decompose the load data into high frequency component and low frequency component by EEMD • Low frequency component is predicted by MRL • High frequency component is forecasted by GRU

intervals. [29] also put forward a novel model to implement fuzzy load prediction in microgrids region (Table 2).

3 Conclusion

In this essay, we have presented the recent released literatures on various hybrid methods that have been resoundingly utilized to predict short term load. From the literatures published by various research workers, we can come to the conclusion that the machine learning, especially deep learning, based forecasting methodologies are proved to be promising approaches for this challenging work of nonlinear time series forecasting. Various random optimization technologies, such as PSO, AIS and BFO, which have the ability of global search ability have also been emphasized and have been combined with basic network of ANN to address this difficult and meaningful issue. The argued methodologies present their strong capability in electrical load prediction which fundamentally decreased the maintenance cost of power system and improves the system operation efficiency.

It is notable that the combination of different methods can bring each method's superiority into the process of STLF. The deep learning would have promising future in forecasting field, while they are unstable and time consuming in some case. The mechanism for determination of hyperparameter is not clear yet. There still much work can be done to promote the development of STLF.

References

1. Cecati, C., Kolbusz, J., Różycki, P., Siano, P., Wilamowski, B.M.: A novel RBF training algorithm for short-term electric load forecasting and comparative studies. J. IEEE Trans. Ind. Electron. **62**(10), 6519–6529 (2015)
2. Amral, N., Ozveren, C.S., King, D.: Short term load forecasting using multiple linear regression. In: 42nd Universities Power Engineering Conference (UPEC), pp. 1192–1198. IEEE Press, New York (2007)
3. Papalexopoulos, A.D., Hesterberg, T.C.: A regression-based approach to short-term system load forecasting. In: IEEE Trans. Power Syst. **5**(4), pp. 1535–1547 (1990)
4. Christiaanse, W.R.: Short-term load forecasting using general exponential smoothing. J. IEEE Trans. Power Appar. Syst. PAS **90**(2), 900–911 (971)
5. Zheng, T., Girgis, A.A., Makram, E.B.: Hybrid wavelet-Kalman filter method for load forecasting. J. Electr. Power Syst. Res. **54**(1), 11–17 (2000)
6. Liu, K., et al.: Comparison of very short-term load forecasting techniques. IEEE Trans. Power Syst. **11**(2), 877–882 (1996)
7. Huang, C.-M., Yang, H.-T.: A time series approach to short term load forecasting through evolutionary programming structures. In: International Conference on Energy Management and Power Delivery, 2, pp. 583–588 (1995)
8. Yuancheng, L., Tingjian, F., Erkeng, Y.: I.: introduction: short-term electrical load forecasting using least squares support vector machines. In: International Conference on Power System Technology IEEE, New York, pp. 230–233. IEEE Press (2002)
9. Mori, H., Kobayashi, H.: Optimal fuzzy inference for short-term load forecasting. J. IEEE Trans. Power Syst. **11**(1), 390–396 (1996)
10. Li, W., Niu, D., Han, Z.: An improved ant colony clustering for power load forecasting problem. J. Soc. 1441–1445 (2008)
11. Huang, C.-M., Huang, C.-J., Wang, M.-L.: A particle swarm optimization to identifying the ARMAX model for short-term load forecasting. J. IEEE Trans. Power Syst. **20**, 1126–1133 (2005)

12. Uguz, H., Hakli, H., Baykan, O.K.: A new algorithm based on artificial bee colony algorithm for energy demand forecasting in Turkey. In: 4th International Conference on Advanced Computer Science Applications and Technologies, pp. 56–61 (2015)
13. Baumann, T., Germond, A.J.: Application of the Kohonen network to short-term load forecasting. In: 2nd International Forum on Applications of Neural Networks to Power System, Yokohama, Japan, pp. 407–412 (1993)
14. Kim, B., Velas, J.P., Lee, J., Park, J., Shin, J., Lee, K.Y.: Short-term system marginal price forecasting using system-type neural network architecture. In: 2006 IEEE PES Power Systems Conference and Exposition, PSCE 2006 - Proceedings, pp. 1753–1758 (2006)
15. Ding, N., Benoit, C., Foggia, G., Besanger, Y., Wurtz, F.: Neural network-based model design for short-term load forecast in distribution systems. J. IEEE Trans. Power Syst. **31**(1), 72–81 (2016)
16. Ota, K., Dong, M., Li, L.: When weather matters: IoT-based electrical load forecasting for smart grid. J. IEEE Commun. Mag. **55**(10), 46–51 (2017)
17. Singh, P., Dwivedi, P: Integration of new evolutionary approach with artificial neural network for solving short term load forecast problem. J. Appl. Energy **217**, 537–549 (2018)
18. Fan, G.-F., Guo, Y.-H., Zheng, J.-M., Hong, W.-C.: Application of the weighted K-nearest neighbor algorithm for short-term load forecasting. J. Energies **12**(5) (2019)
19. Mohandes, M.: Support vector machines for short-term electrical load forecasting. J. Int. J. Energy Res.**26**(4), 335–345 (2002)
20. Xiao, X., Ye, S.-J., Zhong, B., Sun, C-X.: BP neural network with rough set for short term load forecasting. J. Expert Syst. Appl. **36**(1), 273–279 (2009)
21. Ali, M., Adnan, M., Tariq, M., Poor, H.V.: Load forecasting through estimated parametrized based fuzzy inference system in smart grids. J. IEEE Trans. Fuzzy Syst. **29**(1), 156–165 (2021)
22. Chen, K., Chen, K., Wang, Q., He, Z., J. Hu, J. He.: Short-Term Load Forecasting With Deep Residual Networks. J. IEEE Transactions on Smart Grid **10**(4), 3943–3952 (2019)
23. Li, J., et al.: A novel hybrid short-term load forecasting method of smart grid using MLR and LSTM neural network. J. IEEE Trans. Ind. Inf. **17**(4), 2443–2452 (2021)
24. Deng, D., Li, J., Zhang, Z., Teng, Y., Huang, Q.: Short-term electric load forecasting based on EEMD-GRU-MLR. J. Power Syst. Technol. **44**(2), 593–602 (2020)
25. Ranaweera, D.K., Karady, G.G., Farmer, R.G.: Effect of probabilistic inputs on neural network-based electric load forecasting. J. IEEE Trans. Neural Netw. **7**(6), 1528–1532 (1996)
26. Taylor, J.W., Buizza, R.: Neural network load forecasting with weather ensemble predictions. J. IEEE Trans. Power Syst. **17**(3), 626–632 (2002)
27. Kou, P., Gao, F.: A sparse heteroscedastic model for the probabilistic load forecasting in energy-intensive enterprises. J. Int. J. Elect. Power Energy Syst. **55**, 144–154 (2014)
28. Hong, T., Wang, P.: Fuzzy interaction regression for short term load forecasting. Fuzzy Optim. Decis. Making **13**(1), 91–103 (2013). https://doi.org/10.1007/s10700-013-9166-9
29. Sáez, D., Ávila, F., Olivares, D., Cañizares, C., Marín, L.: Fuzzy prediction interval models for forecasting renewable resources and loads in microgrids. J. IEEE Trans. Smart Grid **6**(2), 548–556 (2015)

Design of an Online Detection System for High-Power Submerged Arc Furnace with Magnetic Field Difference

Wei-Ling Liu[1], Zi-Kai Yang[2], Ling-Zhen Yang[3](✉), and Xiao-Ming Chang[4]

[1] School of Data Science, Taiyuan University of Technology, Taiyuan 030024, Shanxi, China
liuweiling@tyut.edu.cn
[2] School of Microelectronics, Tianjin University, Tianjin 300072, China
zikai_yang@tju.edu.cn
[3] School of Physics and Optoelectronics, Taiyuan University of Technology, Taiyuan 030024, Shanxi, China
office-science@tyut.edu.cn
[4] School of Information and Computer Science, Taiyuan University of Technology, Taiyuan 030024, Shanxi, China
lab@xiaoming-lab.com

Abstract. An online measurement system to effectively detect electrode tip position of high-power submerged arc furnace (SAF) based on magnetic field difference were proposed and demonstrated. According to the current distribution characteristics near the electrode tip position and the theory of electromagnetism, a magnetic field radiation model was established, and the differential array sensing system was explored and developed based on sensing mechanism of differential magnetic field. The electrode tip position is obtained through collecting magnetic field information outside the furnace and analyzing the current difference characteristics. The experiment and simulation results showed that the peak point of the magnetic field distribution curve on the electrode line is near the electrode tip position of the furnace. The differential on-line measuring system can accurately detect the electrode tip position of the high-power SAF, which has important significance in optimizing the smelting process, improving the quality of products and reducing energy consumption.

Keywords: Submerged arc furnace · Electrode tip position · Magnetic field · Difference · Array sensing

1 Introduction

Submerged arc furnace (SAF) has become the main equipment in ferroalloy production. However, most of the SAFs are still in the rough production mode, so it brings a lot of hidden dangers to the environment, such as high unit power consumption, serious environmental pollution, low resource utilization rate [1]. In order to transform the rough production mode and improve the smelting efficiency of the SAF, an electrode

K. Li et al. (Eds.): LSMS 2021/ICSEE 2021, CCIS 1468, pp. 191–200, 2021.
https://doi.org/10.1007/978-981-16-7210-1_18

lifting control system is added to improve the heat distribution in furnace and balance degree of three-phase power in molten bath. Therefore, research of the electrode control system of SAF has important economic and social benefits. Electrode lifting control system is a complex MIMO system [2], and the electrode tip position is one of the most important input parameters of the system. In order to obtain the input parameter, the electrode tip position needs to be detected. At present, researchers have proposed some detection methods of the electrode tip position.

A Japanese company proposes a gas temperature-composition analysis method [3]. The electrode tip position affects the temperature and the concentration of CO or CO_2 in the furnace, so the electrode tip position was calculated by analyzing the temperature and concentration of CO or CO_2. The advantage of this method is convenient operation. However, there are errors in the measurement of the temperature and the concentration of CO or CO2, and the errors will be greater when the model is used for indirect calculation and iteration. Ref [4] proposed an electrode weighing system to monitor the electrode tip position. The system not only displays the electrode tip position and trend curves, but also gives inputs for controlling the carbon balance and the electrode operation. The mathematical model of the electrode weighing system is established under the assumption that the electrodes are ideal cylinders. However, the electrode is not regular and uniform, because electrode tip will be lost in the smelting, which could decrease the precision of measure system. Bai et al. [5–7] proposed an indirect measurement method to obtain the electrode tip position and completed the online detection of the relative position of the electrode using the rope position sensor and photoelectric sensor. This method has high detection accuracy. However, the installation of photoelectric and position sensors is not convenient. Wang et al. [11, 12] designed the three-phase electrode position detection system. This method has been improved in the terms of data processing, but only preliminary simulation has been carried out at present.

Liu et al. [8–10] developed a set of non-contact sensing measuring device with electromagnetic radiation. This method is a non-contact magnetic field detection method based on the electromagnetic principle, which identifies the electrode tip position by analyzing the measured magnetic field curve, without iteration operation, so it will not produce iteration error, and the theoretical model of this method is based on the furnace circuit structure and electromagnetics principle, and is independent of the electrode shape and combustion state. Moreover, this method is a non-contact measurement method, which requires no installation of any equipment on the furnace, and only needs to place a magnetic field array at about 1 m away from the furnace. It has advantages of high precision, ease of installation. However, this method is only suitable for small capacity SAF. To solve this problem, the difference method of the magnetic field outside the furnace is proposed. Based on the inner structure of SAF and electromagnetism, the magnetic field have the significant difference in the current distribution between the electrode zone and arc zone [13], and must be reflected outside the furnace. For this reason, if the difference method to process the magnetic field outside the furnace can be used, the magnetic field difference between the electrode zone and the arc zone is the largest and most pronounced where the effective information can be extracted, and the position corresponds to the electrode tip position. Therefore, we use the difference method of the magnetic field outside the furnace to extract the characteristic values. A differential magnetic field array detection system is designed in order to identify the electrode tip position of the high-power furnace. The experiment verifies that the system not only has the advantage of high precision, low cost, ease of installation, but also can be used in high-power and high-capacity SAF.

2 Magnetic Field Differential Array Detection Model

Three electrodes are arranged in an equilateral triangle in the furnace [14]. The powerful currents flow into the furnace through the three electrodes, where the circuit can be regarded as the triangle circuit loop [15]. The magnetic field radiation model [8] is shown in Fig. 1.

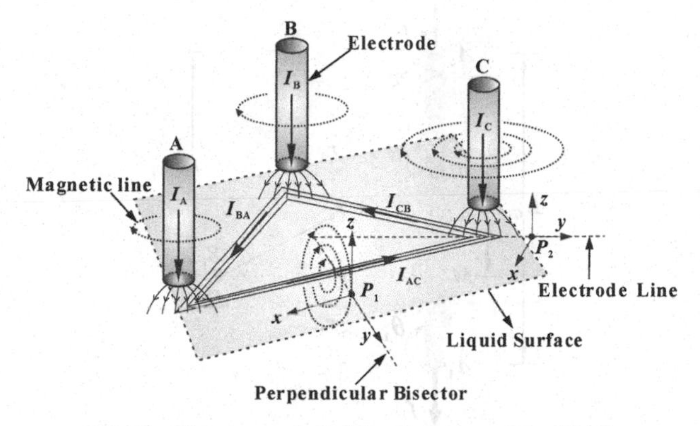

Fig. 1. The magnetic field radiation model of SAF

I_A, I_B, and I_C are the currents of electrode A, B, and C, respectively, and I_{AC}, I_{CB}, and I_{BA} are the currents in the molten bath. I_{AC} is the molten bath current parallel to the x-axis at point P_1; I_C is the electrode current parallel to the z-axis at point P_2. The vertical line of each edge of the regular triangle is called by perpendicular bisector, and the line connecting the electrode and the furnace core is termed as electrode line. The

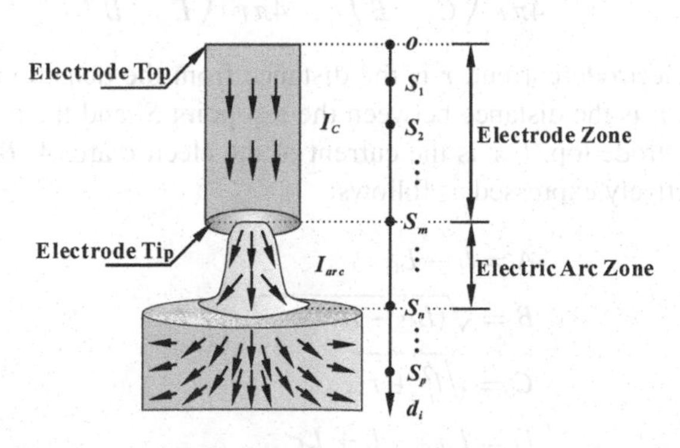

Fig. 2. The current distribution of electrode, electric arc and molten bath

distribution model of electrode current, arc current and the current in molten liquid [16] are shown in Fig. 2.

In Fig. 2, the electrode top is assumed to be the origin *o*, and we take *p* points vertically down as the test points (S_1, …, S_m, …, S_n, …, S_p), among which the point S_m is located at the electrode tip, and the point S_n is at the liquid level, $m < n < p$, and the distances between the point *o* and these test points are (d_1, …, d_m, …, d_n, …, d_p), respectively. The current is uniformly distributed in the electrode above the point S_m in the range of $d_1 \leq d_i \leq d_m$. However, the direction of the arc current appears randomness and inhomogeneity owing to the changing medium between the detected point S_m and S_n ($d_m \leq d_i \leq d_n$) in the arc zone [17]. The electrode is equivalent to a current-carrying straight wire A_1A_2 with the finite length, as shown in Fig. 3.

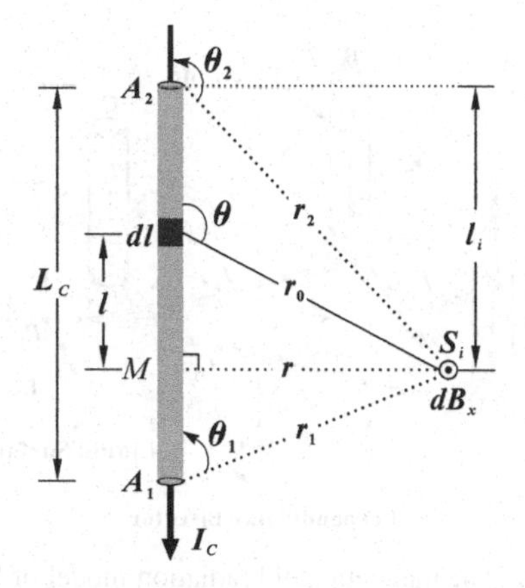

Fig. 3. The magnetic induction intensity of the measuring point S_i in electrode zone

The magnetic induction intensity *dB* which is generated by any current element *Idl* is parallel to the *x*-axis. B_x is the algebraic sum of the magnetic induction intensity in the arc zone current and the electrode zone current, as shown in Eq. (1) [18].

$$B_x \approx \frac{\mu_0 I_C}{4\pi r}\left(\frac{l_i}{C} - \frac{A}{B}\right) + \frac{\mu_0 I_{arc}}{4\pi r}\left(\frac{D}{E} + \frac{A}{B}\right). \tag{1}$$

where I_C is the electrode current. *r* is the distance from the detection point S_i to the vertical line S_iM. l_i is the distance between the test point S_i and the point A_2 which is located at the electrode top. I_{arc} is the current of the electric arc. *A*, *B*, *C*, *D* and *E* in Eq. (1) are respectively expressed as follows:

$$\begin{aligned} A &= l_i - L_C \\ B &= \sqrt{(L_C - l_i)^2 + r^2} \\ C &= \sqrt{l_i^2 + r^2}. \\ D &= L_{arc} - l_i + L_C \\ E &= \sqrt{(L_{arc} - l_i + L_C)^2 + r^2} \end{aligned} \tag{2}$$

where L_C is the total length of electrode, or the distance between the points A_1 and A_2 of the electrode. L_{arc} is the distance between the extreme points A_1 and A_2 of the electric arc.

If l_i and l_{i+1} are the height of the detection point S_i and S_{i+1}, and B_{xi} and $B_{x(i+1)}$ are the magnetic flux density of S_i and S_{i+1}, respectively. Set $\Delta l = l_{(i+1)} - l_i$, the variation of ΔB_x between adjacent points is defined as $\Delta B_x = B_{x(i+1)} - B_x$. Using the method of local linearization of nonlinear functions in differential calculus [19], the output of the array of the differential magnetic field is presented in Eq. (3).

$$|\Delta B_x| \approx dB_x = \left|B_x^{'}(l_i)\right| \Delta l. \tag{3}$$

Thus, we can differentiate the formula for B_x to get the change rate of the total magnetic induction intensity B_x in the electrode line, as shown in Eq. (4).

$$\left|B_x^{'}(l_i)\right| = \frac{\mu_0}{4\pi r}\left|I_C\left(\frac{A^2}{B^3} - \frac{1}{B} + \frac{1}{C} - \frac{l_i^2}{C^3}\right) + I_{arc}\left(\frac{D^2}{E^3} - \frac{1}{E} + \frac{1}{B} - \frac{A^2}{B^3}\right)\right|. \tag{4}$$

Take $\mu_0 = 4\pi \times 10^{-7}$ H/m, $r = 1$ m, $\Delta l = 0.1$ m, $I_C \approx 54.414$ kA [2], $I_{arc} \approx 10$ kA [20], $L_{arc} \approx 0.5$ m [21], $L_C = 2$ m. Therefore, the simulation results of the output ΔB_x are set out in Fig. 4.

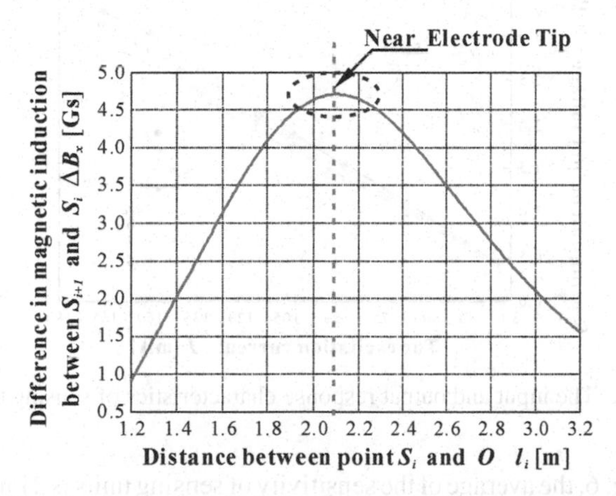

Fig. 4. The simulation result of magnetic induction intensity in the arc zone and electrode zone

From Fig. 4, the output value of the array rises to the peak point at $l_i \approx 2.1$ m, and begins to fall within the range of $l_i \geq 2.1$ m. As the total length of the electrode L_C is equal to 2 m, it can be concluded that the peak point among the output values of the array is near the electrode tip position.

3 System Design and Implementation

The magnetic field array detection system is composed of array sensors, wireless modules (RF), microprocessors (MCU), Internet-enabled device (Raspberry Pi) and an industrial computer. The architecture of the magnetic field array sensing system is shown in Fig. 5.

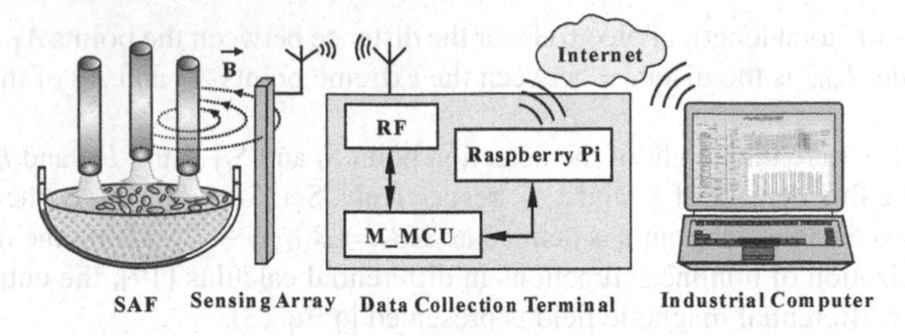

Fig. 5. Schematic diagram of magnetic field array sensing system

In Fig. 5, M_MCU regularly sends instructions to the array sensors and forwards the collected data to the Raspberry Pi. The collected information then is saved to the cloud database, so that the industrial computer can obtain information at any time. The probe output of the differential magnetic field sensor unit is converted into a true RMS signal ΔV_{xi} by the signal amplification circuit, the low-pass filter circuit and the true RMS conversion circuit. The eight sensing units are randomly sampled and energized, and the measured response curve is shown in Fig. 6.

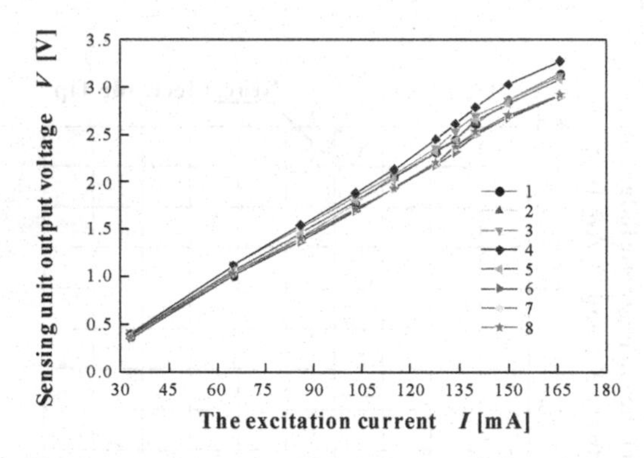

Fig. 6. The input and output response characteristics of sensing units

According Fig. 6, the average of the sensitivity of sensing units is 21 mV/mA. Sensing units have certain differences, so it is necessary to correct the collected data. Suppose the number of sensing units is M, and the slope and intercept of the response curve of the input and output characteristics of the sensing units are respectively (k_1, k_2, ..., k_n, ..., k_M), (b_1, b_2, ..., b_n, ..., b_M). Therefore, the correction curve of the sensing units can be obtained as shown in Eq. (5).

$$V = \frac{\overline{k}}{k_n} \times V_n + \left(\overline{b} - \frac{\overline{k}}{k_n} \times b_n \right). \tag{5}$$

When the excitation current I is 115 mA, the arithmetic mean of the sensing unit output is approximately 2.019 V before correction. A magnetic field array composed of 24 sensing units is corrected. The curves before and after correction are shown in Fig. 7.

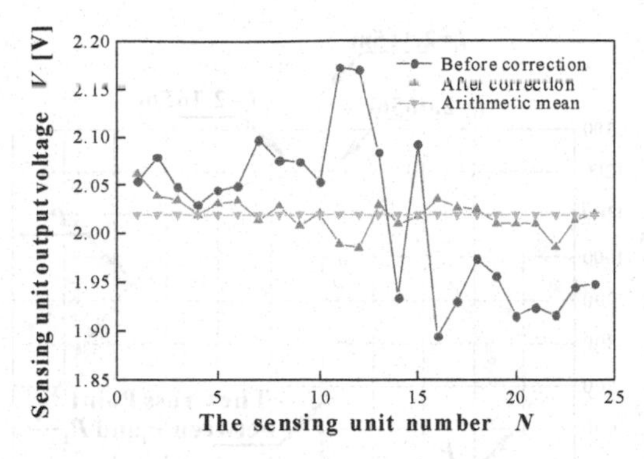

Fig. 7. The correction curve of array sensing units

From Fig. 7, the deviation degree of measurement results and arithmetic mean after correction is smaller than that before correction. Standard deviation [22] is the most commonly used estimate of random error under certain measurement conditions, and it represents the difference between each discrete measurement value and the arithmetic mean of the sensing units. We define the standard deviation of the sensing unit output as σ that can be expressed as Eq. (6).

$$\sigma = \sqrt{\frac{\sum_{i=1}^{n} \left(V_i - \overline{V}\right)^2}{n-1}}. \tag{6}$$

where n is the measurement times of the sensing unit. $\overline{V}$ is the arithmetic mean of the sensing unit output.

According to Eq. (16), when the excitation current I is 115 mA, the standard deviation of the output of 24 sensing units before and after correction are 0.08 V and 0.03 V respectively.

As previously stated, the consistency of the sensing units has been significantly improved after the imbalance of the sensing units has been corrected.

4 System Solution Verification

The measurements in real scene are performed on the 33000KVA SAF at a metallurgical company of Yinchuan city in China. The test site is chosen at the electrode line and is about 1 m away from the lateral wall through a number of trials. The height of the furnace is 5.6 m. The differential magnetic array is vertically placed on the lifting and lowering device. During the process of the test, the array is moved up and down, and its moving range covers the electric tip. The electrode top is set as the coordinate origin O.

The experimental results show that the most ideal sampling result can be achieved when the sampling interval is 10 cm and the distance between the two coils of the system is 3 cm. Especially, the characteristic of the electrode tip position is relatively obvious when the detection point is within the range of 1.1 m–2.2 m. The average values of multiple sampling points are computed. The test results are presented in Fig. 8.

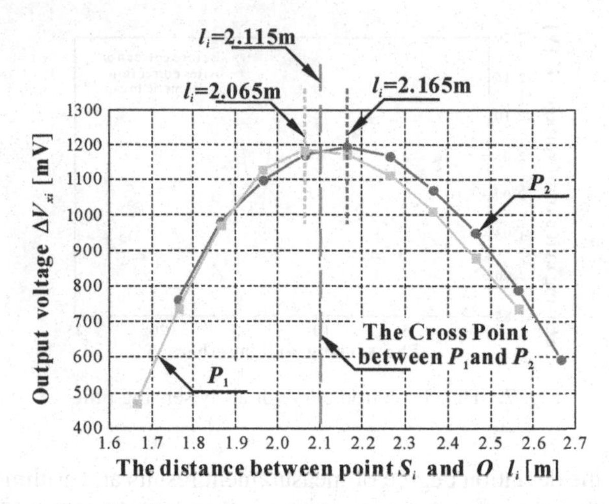

Fig. 8. Measured curves after the moving the array

As can be seen from Fig. 8, the peak point before and after moving the array are 2.065 m and 2.165 m, respectively. The horizontal ordinate of the cross point between the two curves lies at 2.115 m. It can be inferred that the electrode tip position is located near 2.1 m, which is in accordance with the simulation results in Fig. 4.

5 Conclusion

The paper puts forward to a differential magnetic field detection for electrode tip position of large capacity and high power SAF, and the feasibility of this method is proved theoretically and demonstrated practically. The design and implementation of the magnetic field array detection system are introduced, and the unbalance of sensor array characteristics and its correction methods are analysized. The field experiment shows that the system can identify the electrode tip position of the large SAF. The main conclusions are as follows:

(1) It has been proved that the magnetic field difference method of can detect the electrode tip position of large SAF.
(2) The output deviation and characteristic difference of the array sensing units can be reduced by data correction in the later stage.
(3) The experimental results show that the measured magnetic field curve can clearly reflect the electrode tip position when the sampling interval is 10 cm and the system coil interval is 3 cm.

Acknowledgments. This research has been supported by National Natural Science Foundation of China under Grant (61975141 and 61575137).

References

1. Sun, W.Q., Cai, J.J., Ye, Z.: Advances in energy conservation of China steel industry. J. Sci. World **2**, 105–112 (2013)
2. Jiang, X.D., Zhang, J.L., Zuo, H.B.: Application and development of electrode system in mineral furnace. J. Chem. Autom. Instrum. **39**(11), 1404–1408 (2012)
3. kotaritaro.: Ferroalloy electric furnace uses computer to improve energy efficiency and high carbon ferromanganese melt uses direct decarburization method to produce low carbon ferromanganese. In: Proceedings of the Fourth International Conference on Ferroalloys, pp. 33–38 (1986)
4. Gudmundsson, E.B., Halfdanarson, J.: Electrode length measurement with elomos. In: INFACON 7 Proceedings, pp. 423–430 (1992)
5. Liu, Y.Q.: Research and application of automatic control system for three-phase electrode lifting in mineral furnace. Master, Central South University (2010)
6. Bai, Y., Wang, Q., Meng, F.R., Wang, H.Y.: Arc furnace electrode non-contact on-line detection system. J. Changchun. Univ. Technol. (Nat. Sci. Ed.). **33**(4), 383–386 (2012)
7. An, Z.H.: Design of on-line measurement system for optimum position of electrode discharge in mineral furnace. Master, Changchun University of Technology (2010)
8. Liu, W.L., Chang, X.M.: A non-contact detection method for smelting in submerged arc furnace based on magnetic field radiation. J. Magn. **21**(2), 204–208 (2016)
9. Liu, W.L., Han, X.H., Yang, L.Z., Chang, X.M.: Array sensing using electromagnetic method for detection of smelting in submerged arc furnaces. J. Magn. **21**(3), 322–329 (2016)
10. Liu, W.L., Yang, L.Z., Chang, X.M.: Detection for position of electrode tip of submerged arc furnaces with the differential magnetic field radiation. J. Magn. **25**(1), 78–85 (2020)
11. Wang, L., Zhou, T., Niu, Q.F., Cui, J.: Design of three-phase electrode position detection system based on external magnetic field signals of submerged arc furnace. J. Sci. Technol. Eng. **20**(6), 2344–2351 (2020)
12. Zhou, T., Wang, L., Niu, Q.F.: Research on electromagnetic field simulation and electrode position detection method of submerged arc furnace. J. Comput. Simul. **37**(5), 206–212 (2020)
13. Wang, Z.K., Li, T.W., Li, B.K.: An analysis of current density and temperature field for the ferronickel submerged arc furnace. J. Mater. Metall. **12**(3), 177–217 (2013)
14. Zhang, N.N., Wang, Z.J., Zhang, D.J.: The multi-scale forecast of submerged arc furnace energy consumption base on support vector machine. In: Proceedings of 2010 International Conference on Computer, Mechatronics, Control and Electronic Engineering (CMCE 2010), pp. 108–111 (2010)
15. Zhang, N.S.: The concept of electric heat distribution in ferrosilicate furnace. J. Ferro-Alloys **6**, 1–7 (1986)
16. Alexis, J., Ramirez, M., Trapaga, G.: Modeling of a DC electric arc furnace: heat transfer from the arc. J. ISIJ Int. **40**(11), 1089–1097 (2000)
17. Tesfahunegn, Y.A., Magnusson, T., Tangstad, M., Saevarsdottir, G.: Effect of carbide configuration on the current distribution in submerged arc furnaces for silicon production—a modelling approach. In: Nastac, L., Pericleous, K., Sabau, A.S., Zhang, L., Thomas, B.G. (eds.) TMS 2018. TMMMS, pp. 175–185. Springer, Cham (2018). https://doi.org/10.1007/978-3-319-72059-3_17
18. Liu, W.L., Han, X.H., Yang, L.Z., Chang, X.M.: Array sensing using differential magnetic field method for electrode tip position of submerged arc furnaces. J. Magn. **26**(1), 101–110 (2021)
19. Department of Mathematics of Tongji University: Higher Mathematics. Higher Education Press, Beijing (2007)

20. Martell-Chávez, F., Ramírez-Argáez, M., Llamas-Terres, A.: Theoretical estimation of peak arc power to increase energy efficiency in electric arc furnaces. J. ISIJ Int. **53**, 743–750 (2013)
21. Mao, Z.Z., Shang, H.Y., Ma, J.J.: A new method to estimate arc current for electric arc furnace. J. Northeastern. Univ. Nat. Sci. **10**, 1398–1401 (2008)
22. Zhu, X.Q., Ling, Y., Yuan, C.L.: Sensors and Detecting Technology. Tsinghua University Press, Beijing (2014)

Distributed Model Predictive Active Power Control for Large-Scale Wind Farm Based on ADMM Algorithm

Shuai Xue, Houlei Gao(✉), Bin Xu, and Yitong Wu

School of Electrical Engineering, Shandong University, Jinan 250061, China
houleig@sdu.edu.can

Abstract. This paper provided a distributed model predictive control (DMPC) for the active power of a large-scale wind farm. The active power produced by each wind turbine (WT) was regulated by DMPC to reduce the fatigue loads and to track the dispatch command obtained from the transmission system operator (TSO). Model predictive control (MPC) and alternating direction multiplier method (ADMM) were adopted in the proposed DMPC, which can make the computational pressure lower and improve the scalability of the wind farm. This study used a wind farm model containing 80 WTs to test the applicability of DMPC scheme.

Keywords: Large-scale offshore wind farm · Model predictive control · ADMM algorithm · Active power control

1 Introduction

Recently, developing renewable sources has gradually become the consensus of various countries and governments [1]. Among the renewable energy technologies, offshore wind power is developing more and more rapidly due to its advantages such as stable resource conditions and proximity to the load center. According to research by GWEC, by 2025, the worldwide growth of offshore wind power capacity in each year will exceed 20 GW, and by 2030 it will exceed 30 GW. The offshore wind power capacity installed in the next 10 years will exceed 205 GW [2].

As offshore wind farms get larger and larger, its influence on the power system also increases, higher requirements should be put forward for the active power control system of large-scale wind farms. In this context, it is of great significance to propose a more economical and efficient active power control strategy.

In the active power control of a wind farm, a controller is installed to receive the power dispatch command from TSO and distributes it to each WT according to a certain strategy. The main objective of wind farm dispatch schemes is power tracking [3, 4]. In [5], an optimal control method was proposed to maximize outputs of WTs while minimizing the line loss, which is based on the ultra-short-term forecasting. In [6], a bi-level control method was proposed for better active power dispatch by achieving fair active power

K. Li et al. (Eds.): LSMS 2021/ICSEE 2021, CCIS 1468, pp. 201–209, 2021.
https://doi.org/10.1007/978-981-16-7210-1_19

sharing and decreasing the fatigue loads of WTs. In [7], a new auxiliary damping control method was proposed to suppress the subsynchronous resonant oscillation of the nearby turbo-generator. The authors of [8] proposed a closed-loop framework for the active power control while reducing structural loads caused by wakes during the interaction of a fully developed wind farm flow with the atmospheric boundary layer.

In previous studies, the active power control methods of wind farms can be divided into centralized methods and distributed methods according to different topological structures. The performance of centralized control relies heavily on the central controller. As the scale of wind farms grow, the central controller's computational pressure will be greater. In distributed control strategies, each WT is equipped with a separate controller and each controller solves the control problem in parallel, so that the computational burden can be effectively reduced. However, as wind farms grow in size, the number of iterations and the resulting communication delay will greatly influence the convergence time.

Therefore, a DMPC method for large-scale wind farms is proposed with centralized communication and distributed computing. Based on the framework of ADMM, the original optimization problem can be divided into multiple problems that can be solved simultaneously in each WT controller. Part of the problems are solved in the central controller, and others are solved locally in each WT controller. Therefore, the computational pressure of the central controller can be effectively reduced. At the same time, as part of the optimization problem is solved in parallel in each WT controller, the convergence speed is not affected by the increase in the scale of the wind farm.

2 Architecture of DMPC

2.1 Configuration of a Wind Farm

Figure 1 shows the structure of a typical large-scale offshore wind farm, which is connected to the onshore AC grid through the voltage-source-converter high-voltage-direct-current (VSC-HVDC) system. The power output of each WT is collected through collector substations, collected to the high-voltage (HV) transmission cable through the

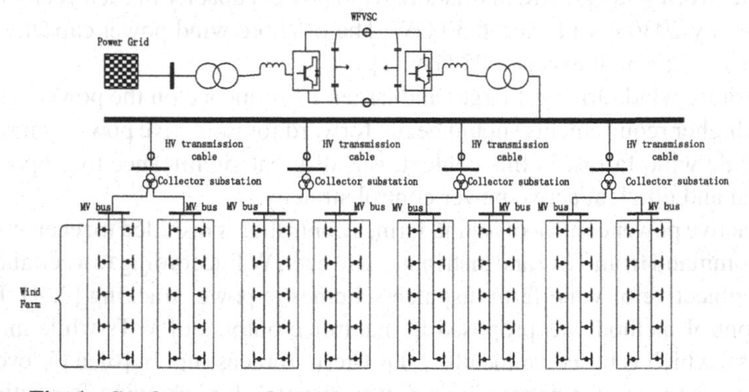

Fig. 1. Configuration of a VSC-HVDC connected offshore wind farm

step-up transformer, and then sent to the onshore AC grid through a wind farm side voltage-source -converter(WFVSC).

2.2 Concept of DMPC

The structure of DMPC proposed in this paper is shown in Fig. 2. A central controller is adopted in the wind farm and several WT controllers are set for each WT.

The proposed active power control scheme is based on the MPC and ADMM. Linearize the mathematical model of the WT at the operating point, and then formulate the optimization problem. The control object of the optimization problem is to track the power dispatch command from the TSO while prolonging the operation life of WTs. Through the ADMM algorithm framework, the original optimization problem can be divided into multiple optimization problems. These sub-problems can be solved in parallel on the central controller and the WT controllers, and the global optimal solution is obtained through continuous iteration between the central controller and the WT controllers.

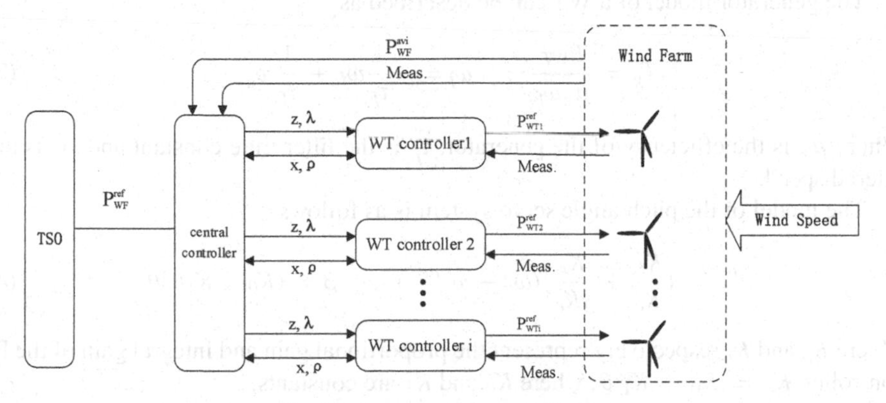

Fig. 2. Control structure of DMPC

3 DMPC Active Power Control of Wind Farm

The DMPC method regulates the output produced by each WT to track the dispatch command from TSO while minimizing the fatigue loads.

In this paper, the WT shaft load and the structural load of the tower are used to measure the fatigue load [9]. The fatigue load can be effectively reduced by reducing the fluctuation of the shaft torque of the WT and the thrust of the tower.

3.1 Predictive Model

The WT model used to study DMPC is based on the 5 MW nonlinear variable speed WT system of NREL [10]. The dynamic characteristics of the pitch angle servo system should be considered when establishing the nonlinear model of the WT because it makes a big difference to the state of the WT.

The aerodynamic model of the WT is as follows

$$T_a = \frac{0.5\pi\rho R^2 V_W^3 C_P(\lambda,\theta)}{\omega_r} \quad , \quad F_t = 0.5\pi\rho R^2 V_W^3 C_t(\lambda,\theta). \tag{1}$$

Where T_{a} is the aerodynamic moment; R is the blade length; ω_r is the rotor speed; V_W is the effective wind speed; θ is the pitch angle; C_P is the power coefficient; C_t is the thrust coefficient and λ is the tip speed ratio.

Model the drive system as follows

$$T_s = T_a - J_r\dot{\omega}_r = \frac{\eta_g^2 J_g}{J_{\mathrm{r}} + \eta_g^2 J_g} T_a + \frac{\eta_g J_r}{J_{\mathrm{r}} + \eta_g^2 J_g}. \tag{2}$$

Where η_g is the transformation ratio of the gearbox; T_g is the torque of the generator; J_r is the rotor mass and J_g is the generator mass.

The generator model of a WT can be described as

$$T_g = \frac{P_{WT}}{\mu_g \omega_f} \quad , \quad \dot{\omega}_f = -\frac{1}{\tau_f}\omega_f + \frac{1}{\tau_f}\omega_g. \tag{3}$$

Where μ_g is the efficiency of the generator, τ_f is the filter time constant and ω_f is the filtered speed.

The model of the pitch angle servo system is as follows

$$\theta^{ref} = \left(\frac{K_P}{K_c} + \frac{K_i}{sK_c}\right)(\omega_f - \omega_g^{rated}) \quad , \quad \beta = (K_0 + K_1\theta)\theta. \tag{4}$$

Where K_p and K_i respectively represent the proportional gain and integral gain of the PI controller. $K_c = K_0 + K_1\,\theta$, where K_0 and K_1 are constants.

Assuming that the operating point is t_0 and the wind speed does not change dramatically over short periods of time, define the state values of the WT at t_0 time as $T_{a,0}$, $T_{g,0}$, θ_0, $\omega_{g,0}$, $\omega_{f,0}$ and $P_{WT,0}$. It can be deduced that the incremental state space model at the operating point is expressed as

$$\Delta\dot{x} = A\Delta x + B\Delta u + E \quad , \quad \Delta y = C\Delta x + D\Delta u. \tag{5}$$

Where $\Delta x = [\Delta\omega_g, \Delta\omega_f, \Delta\beta]^{\mathrm{T}}$, $\Delta u = P_{\mathrm{ref}W}$ and $\Delta y = [\Delta T_s, \Delta F_t]^{\mathrm{T}}$. The state space matrix is as follows

$$A = \begin{pmatrix} \frac{\eta_g}{J_t}\frac{\partial T_a}{\partial \omega_g} & \frac{\eta_g^2 P_{WT,0}}{\mu_g J_t \omega_{f,0}^2} & \frac{\eta_g}{J_t}\frac{\partial T_a}{\partial \beta} \\ \frac{1}{\tau_f} & -\frac{1}{\tau_f} & 0 \\ \frac{K_p}{\tau_f} & -\frac{K_p}{\tau_f + K_i} & 0 \end{pmatrix} B = \begin{pmatrix} \frac{\eta_g^2 P_{WT,0}}{\mu_g J_t \omega_{f,0}^2} \\ 0 \\ 0 \end{pmatrix}$$

$$C = \begin{pmatrix} \frac{\eta_g^2 J_g}{J_t}\frac{\partial T_a}{\partial \omega_g} & \frac{\eta_g J_r P_{W,0}}{\mu_g J_t \omega_{f,0}^2} & \frac{\eta_g^2 J_g}{J_t}\frac{\partial T_a}{\partial \beta} \\ \frac{\partial F_t}{\partial \omega_g} & 0 & \frac{\partial F_t}{\partial \beta} \end{pmatrix} D = \begin{pmatrix} \frac{\eta_g J_r}{\mu_g J_t \omega_{f,0}} \\ 0 \end{pmatrix} E = \begin{pmatrix} \frac{\eta_g}{J_t}(T_{a,0} - \eta_g T_{g,0}) \\ 0 \\ K_i(\omega_{f,0} - \omega_g^{rated}) \end{pmatrix}.$$

Therefore, the discretized state space equation of the WT can be expressed as

$$\Delta x(k+1) = A_d \Delta x(k) + B_d \Delta u(k+1) + E_d. \tag{6}$$

$$\Delta y(k) = C_d \Delta x(k) + D_d \Delta u(k). \tag{7}$$

3.2 Optimization Problem Formulation

MPC is essentially a model-based finite time-domain optimal control algorithm, which is dedicated to decomposing the optimization control problem of a longer time span into several shorter time spans. The method of optimizing the control problem, and still pursuing the optimal solution to a certain extent [11].

When designing the objective function, first consider making the shaft torque and the thrust of the tower as small as possible to reduce fatigue load, thereby prolonging the operating life of WTs; secondly, it is also necessary to consider fair distribution of the active power between WTs in the wind farm. Therefore, the objective function can be expressed as

$$\min \sum_{i=1}^{N_T} \left\| P_{WT,i}^{ref}(k) - P_{pd,i} \right\|_{Q_P}^2 + \left\| \Delta T_{s,i}(k) \right\|_{Q_T}^2 + \left\| \Delta F_{t,i}(k) \right\|_{Q_F}^2. \tag{8}$$

Where $P_{pd,i}$ is the power reference for WT-i when proportional dispatch (PD) method is adopted, Q_P is the weighting coefficient to ensure the fair distribution of active power, Q_T and Q_F are the weighting coefficients to minimize the changes in the shaft torque and tower thrust of WTs. $P_{pd,i}$ can be obtained by the following formula

$$P_{pd,i} = P_{WF}^{ref} \cdot \frac{P_{WT,i}^{avi}}{P_{WF}^{avi}}. \tag{9}$$

The constraints of the optimization problem should be considered from both a single WT and the wind farm as a whole. For the wind farm, the sum of the outputs produced by each WT should track the dispatch command given by TSO, namely

$$\sum_{i=1}^{N_T} P_{WT,i}^{ref} = P_{WF}^{ref}. \tag{10}$$

For a WT, its power output should be within the range of available power of the WT, namely

$$0 \le P_{WT,i}^{ref} \le P_{WT,i}^{avi}, \quad \forall i \in N_T. \tag{11}$$

3.3 Solution Based on ADMM

The ADMM algorithm is a widely used method for solving distributed optimization problems. ADMM algorithm has many excellent characteristics, such as simple form, decomposability, nice convergence and high robustness [12, 13]. The standard form of ADMM is as follows

$$\begin{cases} \min f(x) + g(z) \\ s.t.\ \ Ax + Bz = c \end{cases}. \tag{12}$$

In the formula: f and g are convex functions; $x \in R^n$; $z \in R^m$; $A \in R^{p\times n}$; $B \in R^{p\times m}$; $c \in R^p$.

The steps of the ADMM algorithm can be expressed as follows

$$\begin{aligned} step1 : x^{k+1} &= \arg\min_z L(x, z^k, \lambda^k) \\ step2 : z^{k+1} &= \arg\min_z L(x^{k+1}, z, \lambda^k). \\ step3 : \lambda^{k+1} &= \lambda^k + \rho(Ax^{k+1} + Bz^{k+1} - c) \end{aligned} \tag{13}$$

Where ρ is the iteration step size of the ADMM algorithm.

Due to the special structure of the active power control problem, that is, the objective function is the sum of multiple decoupled objective functions, and the constraints are coupled with each other, so the ADMM algorithm can be used to disassemble the problem into multiple sub-optimization problems. Through iterative calculation with the central controller, the constraints are gradually satisfied. In order to deal with the constraints, variables Δz are introduced and the original problem is rewritten into ADMM form.

$$\begin{cases} \min f(\Delta u) + g(\Delta z) \\ \quad s.t.\ \ \Delta u = \Delta z \\ \qquad f(\Delta u) = \sum_{i=1}^{N_T} \left\| P_{WT,0}^{(i)} + \Delta u_i - P_{pd,i} \right\|_{Q_P}^2 \\ \qquad g(\Delta z) = \sum_{i=1}^{N_T} \left(\left\| \frac{\partial T_{s,i}}{\partial P_{WT,i}^{ref}} \Delta z_i \right\|_{Q_T}^2 + \left\| \frac{\partial F_{t,i}}{\partial P_{WT,i}^{ref}} \Delta z_i \right\|_{Q_F}^2 \right) \\ \qquad \sum_{i=1}^{N_T} \Delta u_i = P_{WF}^{ref} - \sum_{i=1}^{N_T} P_{WT,0}^{i}, \quad \forall i \in N_T \\ \qquad 0 - P_{WT,0}^{(i)} \le \Delta u_i \le P_{WT,i}^{avi} - P_{WT,0}^{(i)}, \quad \forall i \in N_T \end{cases} \tag{14}$$

Among them, $f(\Delta u)$ needs the information of TSO, and its optimization can be carried out in the central controller; the part $g(\Delta z)$ is only related to the information of each WT and can be completed locally in each WT controller. The two parts of optimization problems are processed in parallel, and then iterative calculations are performed according to the ADMM algorithm, and finally converges to the optimal solution.

4 Case Study

In order to test the performance of the DMPC scheme proposed in this paper, a wind farm model with 80 5 MW-WTs was established based on the MATLAB/Simulink platform according to the wind farm structure in Fig. 1. The SimWindFarm toolbox is used to dynamically simulate the wind conditions. The performance of DMPC is compared with the simulation result of the PD con scheme based on the available power.

The control performance of DMPC method is analyzed by selecting WT-1 as the representative WT. The simulation results of WT-1 are shown in Fig. 3.

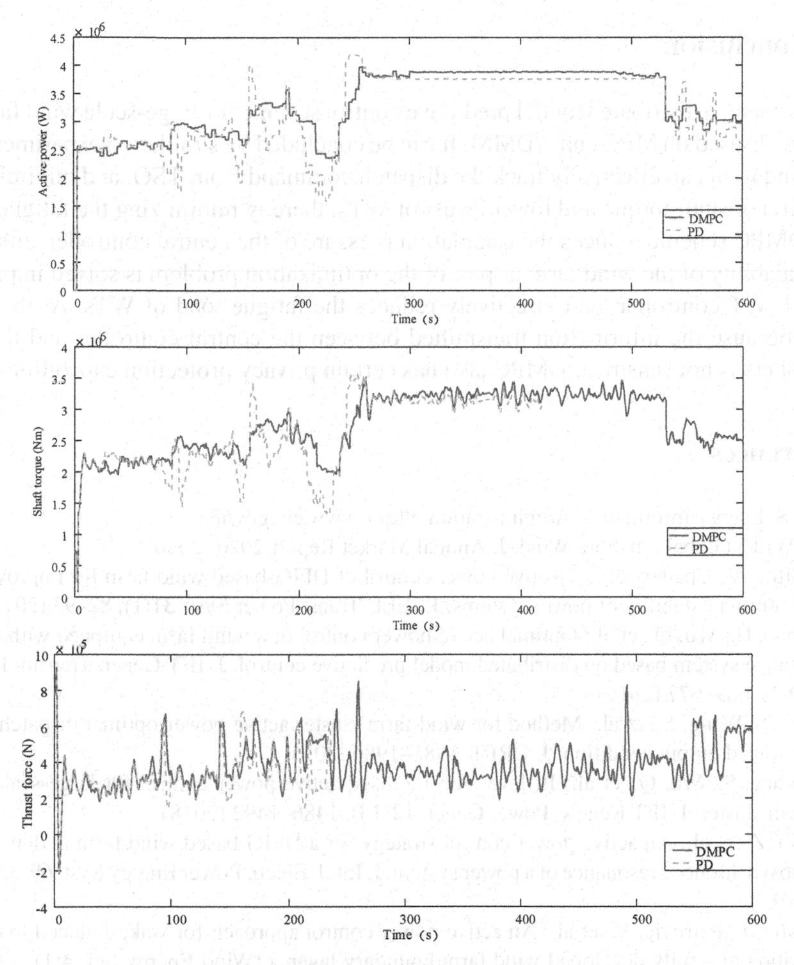

Fig. 3. Simulation results

As can be seen from the figure above, the proposed DMPC scheme has good convergence, and the output of the optimized WT is relatively close to that of PD scheme. At the same time, compared with PD scheme, under the DMPC scheme, the variation of the shaft torque of WTs and the thrust of tower is smaller, and the output of the WT is smoother. Therefore, the DMPC scheme in this paper can minimize the fatigue load of WTs while prolonging the service life under the premise of meeting the requirements of active power dispatching.

In addition, the optimization problem of the DMPC scheme is disassembled into two parts, one of which is solved locally in WTs in parallel, the other part is solved in the central controller. As part of the data is solved in WT controllers, this control can relieve the calculation pressure of the central controller. At the same time, because the information transmitted between the central controller and the WT controllers is not sensitive, DMPC also has certain privacy protection capabilities.

5 Conclusion

In this paper, a distributed model predictive control scheme for large-scale wind farms is proposed based on MPC and ADMM. It can be concluded by simulation experiment that the wind farm can effectively track the dispatch command from TSO, and minimize the variation of shaft torque and tower thrust of WTs, thereby minimizing the fatigue load. The DMPC scheme reduces the calculation pressure of the central controller, enhances the scalability of the wind farm as part of the optimization problem is solved in parallel in each WT controller, and effectively reduces the fatigue load of WTs. At the same time, because the information transmitted between the central controller and the WT controllers is not sensitive, DMPC also has certain privacy protection capabilities.

References

1. U.S. Energy Information Administration. http://www.eia.gov/ieo
2. GWEC: Global Offshore Wind. J. Annual Market Report 2020 (2020)
3. Mitra, A., Chatterjee, D.: Active power control of DFIG-based wind farm for improvement of transient stability of power systems. J. IEEE Trans. Power Syst. **31**(1), 82–93 (2015)
4. Zhao, H., Wu, Q., et al.: Optimal active power control of a wind farm equipped with energy storage system based on distributed model predictive control. J. IET Gener. Transm. Distrib. **10**(3), 669–677 (2016)
5. Li, D., Wang, S., et al.: Method for wind farm cluster active power optimal dispatch under restricted output condition. J. DRPT. 1981–1986 (2015)
6. Huang, S., Wu, Q., et al.: Bi-level decentralised active power control for large-scale wind farm cluster. J. IET Renew. Power Gener. **12**(13), 1486–1492 (2018)
7. Bin, Z., et al.: An active power control strategy for a DFIG-based wind farm to depress the subsynchronous resonance of a power system. J. Int. J. Electr. Power Energy Syst. **69**, 327–334 (2015)
8. Vali, M., Petrović, V., et al.: An active power control approach for wake-induced load alleviation in a fully developed wind farm boundary layer. J. Wind Energy Sci. **4**(1), 139–161 (2019)
9. Zhao, H., Wu, Q., et al.: Fatigue load sensitivity-based optimal active power dispatch for wind farms. J. IEEE Trans. Sustain. Energy. **8**(3), 1247–1259 (2017)
10. Jonkman, J., Butterfield, S., et al.: Definition of a 5-MW reference wind turbine for offshore system development. In: National Renewable Energy Lab.(NREL). Golden, CO, USA (2009)

11. Chen, H., Zhang, J., et al.: Asymmetric GARCH type models for asymmetric volatility characteristics analysis and wind power forecasting. Prot. Control Mod. J. Power Syst. **4**(1), 1–11 (2019)
12. Chang, T.-H., Liao, W.-C., et al.: Asynchronous distributed ADMM for large-scale optimization—Part II: Linear convergence analysis and numerical performance. J. IEEE Trans. Signal Process **64**(12), 3131–3144 (2016)
13. Boyd, S., Parikh, N., et al.: Distributed optimization and statistical learning via the alternating direction method of multipliers (2011)

Modelling and Practical Stability Analysis of Wind Turbines with Battery Energy Storage System Based on Dwell Time

Guangrui Zhang, Cheng Peng, and Yang Song(✉)

School of Mechatronic Engineering and Automation,
Shanghai University, Shanghai 200072, People's Republic of China
y_song@shu.edu.cn

Abstract. This paper studies the modelling and practical stability analysis of the wind turbine linked with a battery energy storage system (WT/BESS). An offset strategy is proposed for BESS to keep the state of charging (SOC) at the middle value. And a model of switched system with multiple equilibriums is established to describe the dynamics of the WT/BESS under the SOC offset strategy. The equilibriums of the model switches according to a time-dependent switching signal. Based on this, the sufficient condition for the practical stability of the switched system with multiple equilibriums is given, and the feasibility of the proposed results is certificated by a case.

Keywords: Wind turbine · Battery energy storage system · Switched system with multiple equilibriums · SOC offset strategy · Practical stability

1 Introduction

Nowadays, wind energy as a renewable energy has been used universally [1], and he control of WT has attracted much interest [2]. In generally, the control strategies of WT can be divided into maximum power point tracking (MPPT) control and variable pitch control. When the maximum wind energy captured by the WT is less than the expected output power. In order to capture the maximum wind energy, MPPT is often used [3]. In addition, when the maximum wind energy captured by the WT is more than the expected output power, the variable pitch control is usually used to decrease the wind energy captured, so that the output power of the WT maintains near the demand power [4].

Although the output power undulation of wind turbine could be mitigated by the above strategies of control. When the intermittent latency and fluctuation of wind energy is very severe, it is not adequate to suppress the power's undulation of WT just through those approaches [5]. In the situation, a feasible method that links the WT with BESS to recompense the power undulation of WT [6]. The battery energy storage system can dynamically absorb the excess output power of the wind turbine, and can also supplement the insufficient output power of the wind turbine when needed.

K. Li et al. (Eds.): LSMS 2021/ICSEE 2021, CCIS 1468, pp. 210–220, 2021.
https://doi.org/10.1007/978-981-16-7210-1_20

For the case variable wind speed, [7, 8] propose some state of charging (SOC) regulate approaches of battery by utilizing a prediction model. Noticed that during the battery is in the charging (or discharging), the power output to the BESS will be raised (or reduced). This is to say, when the battery state switches between charging and discharging, the operation points of the WT/BESS will transform as well, even if the wind speed is fixed. From the point of switched systems, the switched system with multiple equilibriums can characterizes such model of WT/BESS. For a complete WT/BESS model, there are complex logic switching and arbitrary switching in the model [9]. There is a close coupling relationship between WT and BESS. The existing switching system theory cannot analyze its stability. The existing research often needs to simplify the system model, ignoring part of the dynamics between the WT and BESS. For example, [10] does not consider the affect of the battery state on the wind turbine, and the system model is a single equilibrium point switching system because it is linearized only at the junction of the under power region and the over power region. [11] considers the influence of battery state on wind turbine, and the system model is a multiple equilibriums switching system under the influence of arbitrary switching rules, i.e. the change of battery state is modeled as arbitrary switching.

This paper considers two equilibriums relating to the two different operation points of WT. The switching rules in the WT/BESS depends on the initial SOC and the energy scheduling methods of the BESS. The battery state switching satisfies the constraint of dwell time, and the model of the WT/BESS is a time-dependent switched system with multiple equilibriums. The sufficient condition for the practical stability of switched system with multiple equilibriums is derived and be applied to the practical stability of WT/BESS.

The remainder of the paper is organized as follow. Section 2 depicts the dynamics of WT and the SOC scheming approach of BESS. In Sect. 3, practical stability based on the dwell time for WT/BESS is studied. Finally, the simulation result is given in Sect. 4, and Sect. 5 gives the conclusion.

2 Modelling of WT/BESS

2.1 Modelling of WT

Figure 1 illustrates the schematic diagram of a WT/BESS, where P_g and P_o are the instantaneous output power of WT and WT/BESS respectively, P_g^* is the expected output power, the demand power of power grid P_d is usually constant, P_c is the compensation power of BESS and is determined by P_d and P_g, it is clear that $P_c = P_d - P_g$.

The dynamic model of the WT can be expressed as

$$\dot{\omega}(t) = \frac{1}{J}\left(\tau_{aero}(t) - \tau_{\sigma(t)}\right). \tag{1}$$

$$P_g = \tau_{\sigma(t)}\omega. \tag{2}$$

where ω is the rotor angular velocity, J is the combined rotational inertia of the rotor, gearbox, generator and shafts, $\tau_{aero}(t)$ and $\tau_{\sigma(t)}$ denote the aerodynamic torque and the

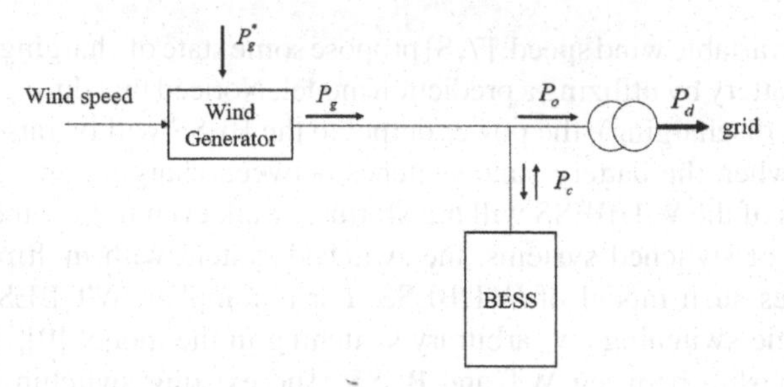

Fig. 1. WT/BESS system

load torque of the turbine respectively, which are given by

$$\tau_{aero}(t) = \frac{1}{2}\rho_{aero}\pi R^2 v_{\omega}^3(t)\frac{C_P(\lambda, \beta)}{\omega(t)}. \tag{3}$$

$$\tau_{\sigma(t)} = \frac{P_g^*}{\omega_e}. \tag{4}$$

where ρ_{aero} is the air density, R is the rotor radius of WT, v_ω is the input wind speed, and ω_e is the rotor speed at the equilibrium. The function $C_p(\lambda, \beta)$ denotes the power coefficient of WT, which represents the capture efficiency of the wind energy. $C_p(\lambda, \beta)$ is a non-linear function of blade pitch angle β and tip speed ratio λ as

$$C_P(\lambda, \beta) = 0.4654(\frac{116}{\lambda'} - 0.4\beta - 5)e^{\frac{-20.24}{\lambda'}}. \tag{5}$$

where $\frac{1}{\lambda'} = \frac{1}{\lambda + 0.08\beta} - \frac{0.035}{\beta^3+1}$, tip speed ratio $\lambda = \frac{\omega R}{v_\omega}$.

Generally, a proportional integral controller is used to control the pitch angle quickly and accurately. Select K_P, K_I as the PI controller parameters, therefore

$$\beta(t) - \beta_e = K_P(\omega - \omega_e) + K_I\int_0^t (\omega - \omega_e)d\tau$$

where β_e is the expected pitch angle at the equilibrium. Equivalently,

$$\dot{\beta} = K_P\dot{\omega} + K_I(\omega - \omega_e). \tag{6}$$

The approach of linearization is applied to investigate the stability of WT [12]. At the point (ω_e, β_e, v_e) about (1) and (6), the following equations can be yield by using the Taylor expansion.

$$\dot{\omega} = \frac{1}{J}\left(\frac{\partial\tau_{aero}}{\partial\omega}|_{\omega=\omega_e}(\omega - \omega_e) + \frac{\partial\tau_{aero}}{\partial\beta}|_{\beta=\beta_e}(\beta - \beta_e) + \frac{\partial\tau_{aero}}{\partial v}|_{v=v_e}(v - v_e)\right). \tag{7}$$

$$\dot{\beta} = \frac{K_P}{J}\left(\frac{\partial\tau_{aero}}{\partial\beta}|_{\beta=\beta_e}(\beta - \beta_e) + \frac{\partial\tau_{aero}}{\partial v}|_{v=v_e}(v - v_e)\right) + \left(\frac{K_P}{J}\frac{\partial\tau_{aero}}{\partial\omega}|_{\omega=\omega_e} + K_I\right)(\omega - \omega_e). \tag{8}$$

2.2 BESS Model

The random fluctuation of wind speed results that wind power grid-connected power fluctuates largely, thus a BESS is generally connected with WT system in practice to ease the undulation of the output power P_g by the recompense power P_c of BESS. P_c is usually taken to be the difference between the demand power P_d and the instantaneous power P_g. But in practice, the charging and discharging power of BESS have the limit $P_{\lim it}$, so the dynamics of the SOC of BESS is

$$S\dot{O}C = Sa_{[-P_{\lim it}, P_{\lim it}]}\left(\frac{P_g - P_d}{nV_b C_b}\right). \tag{9}$$

where, C_b denotes the nominal capacity, n denotes the amount of batteries, and V_b denotes the terminal voltage of battery. The $S_a(\cdot)$ denotes a saturation function.

It should be noted that too low SOC will reduce the battery life dominantly. To deal with the problem, we propose an SOC offset strategy of BESS by changing the expected output power P_g^*. If the SOC of battery below SOC = SOCL, i.e. the expected output power of WT P_g^* will increase to $P_d + P_{\lim it}$ for charging BESS till SOC = SOCH, when the SOC is very lower. Figure 2 shows the control process of the SOC recompense strategy.

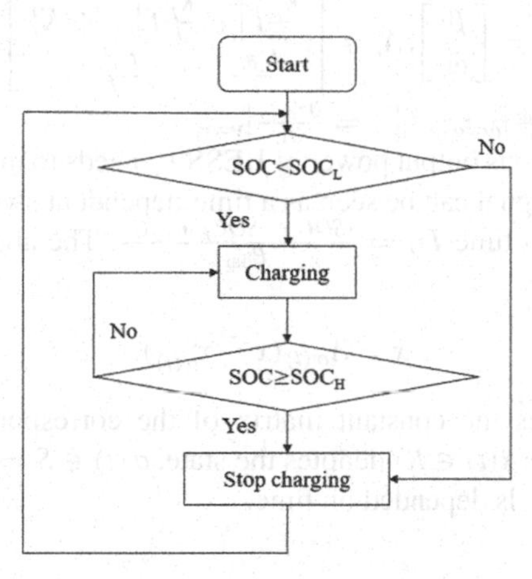

Fig. 2. The schedule of battery charging

2.3 Modelling of WT/BESS

Figure 3 demonstrates the change of power curve caused by this SOC offset strategy, which directly changes the number and distribution of operation points. The operation point will transform while the expected output power switches from P_1 to P_2. That is to say, around the fixed wind speed, the expected pitch angle β_s will transform from β_1 to β_2 at the operation point, and the expected rotor speed ω_s will also change from

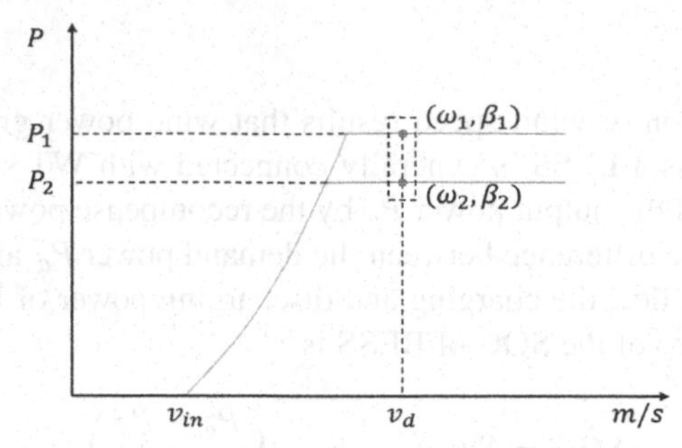

Fig. 3. Operation points of WT/BESS

ω_1 to ω_2 as well. Hence, the model of WT/BESS is linearized at (β_1, ω_1) and (β_2, ω_2) respectively.

This paper takes a series of proportional integral controller parameters K_P^i, K_I^i for every subsystem. The following shows the model of switched system with multiple equilibriums by linearization around the two equilibriums

$$\dot{x} = A_i(x - x_i^e) + B_i(v - v_i). \tag{10}$$

where $x = \begin{bmatrix} \beta \\ \omega \end{bmatrix}, x_i^e = \begin{bmatrix} \beta_i \\ \omega_i \end{bmatrix}, A_i = \begin{bmatrix} \frac{K_P^i}{J} f\,|_{\beta_i} & \frac{K_P^i}{J} f\,|_{\omega_i} + K_I^i \\ \frac{f\,|_{\beta_i}}{J} & \frac{f\,|_{\omega_i}}{J} \end{bmatrix}, B_i = \begin{bmatrix} \frac{K_P^i f\,|_{v_i}}{J} \\ \frac{f\,|_{v_i}}{J} \end{bmatrix}, f\,|_{\beta_i} = \frac{\partial \tau_{aero}}{\partial \beta}\,|_{\beta=\beta_i}, f\,|_{\omega_i} = \frac{\partial \tau_{aero}}{\partial \omega}\,|_{\omega=\omega_i}, f\,|_{v_i} = \frac{\partial \tau_{aero}}{\partial v}\,|_{v=v_i}$

Since the instantaneous output power of BESS P_c needs to meet the limit, i.e. $|P_c| \leq P_{\lim it}$. The switching signal can be seen as a time-dependent switching signal, which is restricted through dwell time $T_D = \frac{(SOC_H - SOC_L)nV_bC_b}{P_{\lim it}}$. The above system (10) can be expressed as

$$\dot{x} = A_{\sigma(t)}(x - x_{\sigma(t)}^e). \tag{11}$$

where $A_{\sigma(t)}$ denotes the constant matrix of the corresponding dimension, $x_{\sigma(t)}^e$ denotes the equilibrium, $x(t) \in R^n$ denotes the state, $\sigma(t) \in S = \{1, \ldots, m\}$ is segment constant switching signals depended on time.

3 Practical Stability Based on the Dwell Time for WT/BESS

Due to the limitation of the battery's own charging and discharging power, the switching of the battery state needs to meet the dwell time constraint. This section will consider the sufficient condition for practical stability under the dwell constraint, and the stability region will also be given.

Definition 3.1. The continuous time switched linear system with multi-equilibriums (11) is practical stability, if the trajectory of switched system starting from initial state guided by switching signal $\sigma(t)$ satisfies

$$x(t) \in \Omega_2, \ \ \forall t \in [0, T]. \tag{12}$$

where $x(t_0) \in \Omega_1$, Ω_1,Ω_2 is the set of positive real numbers, and $\Omega_1 \subset \Omega_2$, real number $T > 0$, switching sequence $\sigma(t) \in S = \{1, \ldots, m\}$.

Definition 3.2. For $\sigma(t) = i_k \in S$, if any two consecutive switching times t_k and t_{k+1}, satisfies the following condition:

$$t_{k+1} - t_k \geq T_D. \tag{13}$$

where T_D is a positive real number. Then T_D is named dwell time of switching sequence $\sigma(t)$.

Theorem 3.3. Considering a switched linear system with multi-equilibriums (11), if there exists $\mu \geq 1$, and positive definite matrix $\overline{Q}_{\sigma(t)}$=$R^{-1/2}Q_{\sigma(t)}R^{-1/2}$,$Q_{\sigma(t)}$,$H$, and positive real number γ, for $\forall i, j \in S$, such that the following inequalities hold:

$$A_i^T Q_i + Q_i A_i \leq 0. \tag{14}$$

$$\begin{bmatrix} e^{A_j^T T_D} Q_j e^{A_j T_D} - \mu P_i & e^{A_j^T T_D} Q_j \\ Q_j e^{A_j T_D} & Q_j - \gamma H \end{bmatrix} \leq 0. \tag{15}$$

$$c_2 \geq \frac{\lambda_1}{\lambda_2} \mu^{N(0,T)} c_1 + \frac{\gamma d \lambda_{\max}(H)}{\lambda_2} \sum_{q=0}^{N(0,T)-1} \mu^q. \tag{16}$$

Then, for the continuous time system with multi-equilibrium (11) satisfying the dwell time T_D, it is practical stability on $[0, T]$ with respect to Ω_1 and Ω_2, where

$$\Omega_1 = \bigcap_{i=1}^{M} \left\{x \in \mathbb{R}^n : \left(x - x_{e_i}\right)^T R\left(x - x_{e_i}\right) < c_1\right\},$$

$$\Omega_2 = \bigcup_{i=1}^{M} \left\{x \in \mathbb{R}^n : \left(x - x_{e_i}\right)^T R\left(x - x_{e_i}\right) < c_2\right\},$$

$$\lambda_1 = \lambda_{\max}\left(\overline{Q}_i\right), \lambda_2 = \lambda_{\min}\left(\overline{Q}_i\right), N(0, T) = \left\lceil \frac{T}{T_D} \right\rceil, d = \max_{i,j \in S}\left(\left\|x_{e_i} - x_{e_j}\right\|^2\right).$$

Proof: Construct the following Lyapunov function $v(x(t)) = V_{\sigma(t)}(x(t)) = \left(x(t) - x_{e_{\sigma(t)}}\right)^T Q_{\sigma(t)}\left(x(t) - x_{e_{\sigma(t)}}\right)$, where $\sigma(t) \in S = \{1, 2\}$.

The proof of the Theorem 3.3 is divided in three steps.

Step 1. Let t_k represents the last switching time closest to the current time t, for any $t \in \left[t_k, t_{k+1}\right)$, assume $\sigma(t_k^-)$=i, $\sigma(t_k)$=j, where $i, j \in S$. According to the constructed Lyapunov function and continuous time multiple equilibriums switching system (13), then

$$\dot{v}(x(t)) = \dot{x}^T Q_j\left(x(t_k) - x_{e_j}\right) + \left(x(t_k) - x_{e_j}\right)^T Q_j \dot{x}$$

$$= \left(x(t_k) - x_{e_j}\right)^T \left(A_j^T Q_j + Q_j A_j\right)\left(x(t_k) - x_{e_j}\right). \tag{17}$$

If condition (14) holds, according to (17) could obtain

$$\dot{v}(x(t)) \leq 0. \tag{18}$$

Step 2. Consider switching time t_k, suppose $\sigma(t_k^-)=\sigma(t_{k-1}) = i$, so the Lyapunov function becomes:

$$v(x(t_k)) = V_j(x(t_k)) = \left(x(t_k) - x_{e_j}\right)^T Q_j\left(x(t_k) - x_{e_j}\right). \tag{19}$$

According to system (11) can know

$$x\left(t_k^-\right) - x_{e_i} = e^{A_j(t_k - t_{k-1})}\left(x(t_{k-1}) - x_{e_i}\right). \tag{20}$$

where, since the system state have not jump, then

$$x(t_k) = x\left(t_k^-\right) = e^{A_j(t_k - t_{k-1})}\left(x(t_{k-1}) - x_{e_i}\right) + x_{e_i}. \tag{21}$$

Substitute it into Eq. (19) to obtain

$$V_j(x(t_k)) = \left(e^{A_j(t_k - t_{k-1})}\left(x(t_{k-1}) - x_{e_i}\right) + x_{e_i} - x_{e_j}\right)^T Q_j\left(e^{A_j(t_k - t_{k-1})}\left(x(t_{k-1}) - x_{e_i}\right) + x_{e_i} - x_{e_j}\right)$$

Let $T_{k-1} = t_k - t_{k-1}$, Substitute it into the above equation

$$\begin{aligned} V_j(x(t_k)) &= \begin{bmatrix} x(t_{k-1}) - x_{e_i} \\ x_{e_i} - x_{e_j} \end{bmatrix}^T \begin{bmatrix} e^{A_j^T T_{k-1}} Q_j e^{A_j T_{k-1}} & e^{A_j^T T_{k-1}} Q_j \\ Q_j e^{A_j T_{k-1}} & Q_j \end{bmatrix} \begin{bmatrix} x(t_{k-1}) - x_{e_i} \\ x_{e_i} - x_{e_j} \end{bmatrix} \\ &= \begin{bmatrix} e^{A_j(T_{k-1} - T_D)}\left(x(t_{k-1}) - x_{e_i}\right) \\ x_{e_i} - x_{e_j} \end{bmatrix}^T \begin{bmatrix} e^{A_j^T T_D} Q_j e^{A_j T_D} & e^{A_j^T T_D} Q_j \\ Q_j e^{A_j T_D} & Q_j \end{bmatrix} \begin{bmatrix} e^{A_j(T_{k-1} - T_D)}\left(x(t_{k-1}) - x_{e_i}\right) \\ x_{e_i} - x_{e_j} \end{bmatrix}. \end{aligned} \tag{22}$$

If condition (15) holds, according to (22) could obtain

$$\begin{aligned} V_j(x(t_k)) &\leq \mu\left(x(t_{k-1}) - x_{e_i}\right)^T e^{A_j^T (T_{k-1} - T_D)} Q_i e^{A_j(T_{k-1} - T_D)}\left(x(t_{k-1}) - x_{e_i}\right) \\ &+ \gamma\left(x_{e_i} - x_{e_j}\right)^T H\left(x_{e_i} - x_{e_j}\right) \\ &\leq \mu\left(x(t_{k-1}) - x_{e_i}\right)^T e^{A_j^T (T_{k-1} - T_D)} Q_i e^{A_j(T_{k-1} - T_D)}\left(x(t_{k-1}) - x_{e_i}\right) + \gamma d \lambda_{\max}(H) \end{aligned} \tag{23}$$

where, $d = \max\limits_{i,j \in S}\left(\left\|x_{e_i} - x_{e_j}\right\|^2\right)$, since the dwell time is T_D, then $t_k - T_D \geq t_{k-1}$, i.e. $\sigma(t_k - T_D) = i$, so

$$x(t_k - T_D) - x_{e_i} = e^{A_j(T_{k-1} - T_D)}\left(x(t_{k-1}) - x_{e_i}\right). \tag{24}$$

Substitute the above equation into (23)

$$\begin{aligned} V_j(x(t_k)) &\leq \mu\left(x(t_k - T_D) - x_{e_i}\right)^T Q_i\left(x(t_k - T_D) - x_{e_i}\right) + \gamma d \lambda_{\max}(H) \\ &\leq \mu V_i(x(t_k - T_D)) + \gamma d \lambda_{\max}(H). \end{aligned} \tag{25}$$

Step 3. For any $t \in \left[t_k, t_{k+1}\right)$, combine (18) and (25)

$$v(x(t)) = V_j(x(t)) \leq V_j(x(t_k))$$

$$\leq \mu^{N(0,t_k)} V_{\sigma(0)}(x(0)) + \sum_{q=0}^{N(0,t_k)-1} \mu^q \gamma d \lambda_{\max}(H). \tag{26}$$

where $N(0, t_k) = \left\lceil \frac{t_k}{\tau_a} \right\rceil$ is maximum value of switches on the time zone $[0, t_k]$, according to the definition of $\overline{Q}_i$, then

$$v(x(t)) = V_{\sigma(t)}(x(t)) = (x(t) - \overline{x})^T Q_{\sigma(t)}(x(t) - \overline{x})$$

$$\geq \lambda_2 (x(t) - \overline{x})^T R(x(t) - \overline{x}). \tag{27}$$

And

$$V_{\sigma(0)}(x(0)) = (x(0) - \overline{x})^T Q_{\sigma(0)}(x(0) - \overline{x})$$

$$\leq \lambda_1 (x(0) - \overline{x})^T R(x(0) - \overline{x}). \tag{28}$$

So, according to Eq. (26) to (28)

$$\left(x(t) - x_{e_{\sigma(t)}}\right)^T R\left(x(t) - x_{e_{\sigma(t)}}\right) \leq \frac{\lambda_1}{\lambda_2} \mu^{N(0,t_k)} \left(x(0) - x_{e_{\sigma(0)}}\right)^T R\left(x(0) - x_{e_{\sigma(0)}}\right)$$

$$+ \frac{\gamma d \lambda_{\max}(H)}{\lambda_2} \sum_{q=0}^{N(0,t_k)-1} \mu^q. \tag{29}$$

Notice that $\left(x(0) - x_{e_{\sigma(0)}}\right)^T R\left(x(0) - x_{e_{\sigma(0)}}\right) \leq c_1$, the condition (16) holds, then

$$\Omega_2 = \bigcup_{i=1}^{M} \left\{ x \in \mathbb{R}^n : \left(x - x_{e_i}\right)^T R\left(x - x_{e_i}\right) < c_2 \right\}. \tag{30}$$

So, the multi-equilibrium switching system (11) is practical stable with respect to Ω_1 and Ω_2.

4 Simulation Result

The simulation of this paper is based on the WT/BESS model considered in [6]. where the amount of batteries $n = 2$, the terminal voltage of battery $V_b = 3V$, the nominal capacity $C_b = 1\,Ah$, the rotor inertia $J = 26000\,\text{kg} \cdot \text{m}^2$, the rotor radius $R = 9\,\text{m}$, $C_{p\max} = 0.4101$, and $P_{\lim it} = 2.5\,kw$.

Through the design method expressed in Sect. 3, the PI parameters as well as the relative stable region can be given under the circumstance of swift rate of wind. When the expected output power P_1=27500 w and P_2=25000 w, the dwell time $T_D = 2.6\,s$, and the state matrices of two subsystems osf the wind turbine at the equilibriums are

$$A_1 = \begin{bmatrix} -0.0613 & 14.8957 \\ -0.0077 & -0.0130 \end{bmatrix},\ A_2 = \begin{bmatrix} -0.0878 & 19.8704 \\ -0.0073 & -0.0108 \end{bmatrix}$$

According to Theorem 3.3, it can be obtained by using the MATLAB LMI toolbox.

$$Q_1 = \begin{bmatrix} 0.0232 & -0.1038 \\ -0.1038 & 40.6074 \end{bmatrix},\ Q_2 = \begin{bmatrix} 0.0246 & -0.0003 \\ -0.0003 & 0.0008 \end{bmatrix}$$

Q_1 and Q_2 are positive define.

Under the above conditions, the simulation results are given below.

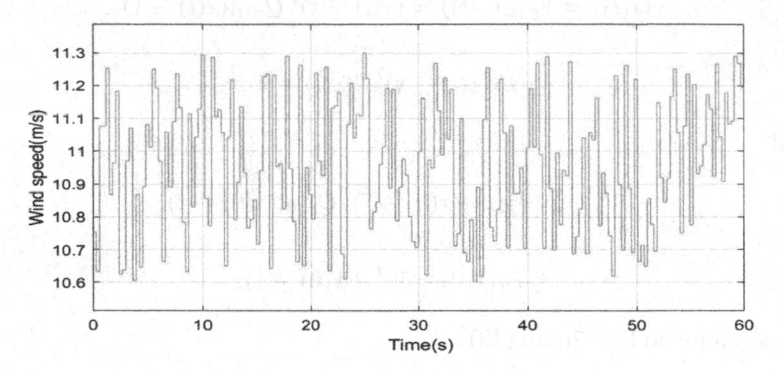

Fig. 4. The curve of wind speed

The above Fig. 4 is the curve of wind speed, and it is random.

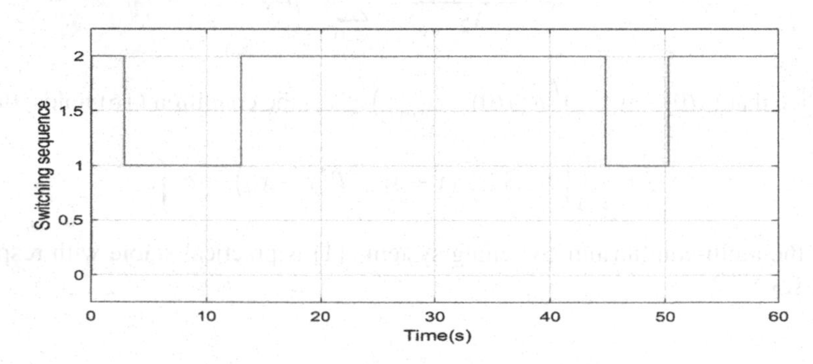

Fig. 5. The switching sequence of battery

The Fig. 5 is the switching sequence of battery. According to the schematic of Fig. 3, it shows that the switching of modes between charging and discharging of the battery is coincident with the change of operation points. The Fig. 6 is the trajectories of SOC of battery. When the SOC < 0.2, the wind turbine increases the output power to the BESS; When the SOC > 0.5, the wind turbine decreases the output to the BESS. This is to say, when the battery state switches between charging and discharging, the operation points of the WT/BESS will transform as well.

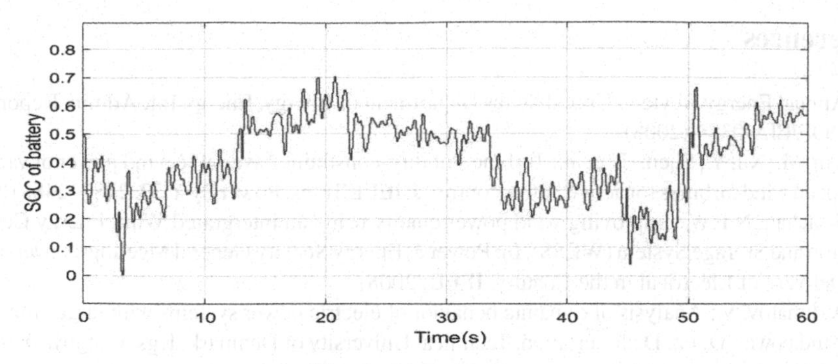

Fig. 6. SOC of battery

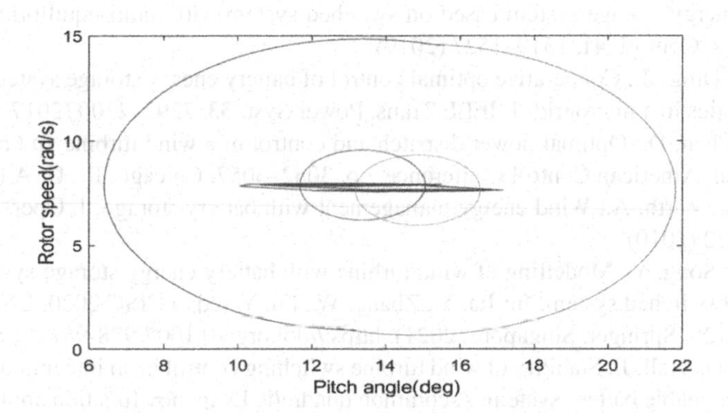

Fig. 7. Stability region of switched systems of WT/BESS

As shown in Fig. 7, the stability region for the two switched systems of WT/BESS is plotted. The trajectories started from initial point always keep in the stability region at the equilibriums.

5 Conclusion

In this paper, a new SOC offset strategy is proposed to reduce the output power fluctuation of the WT/BESS and prolong the battery life. Based on this, a switched model with multiple equilibriums is established to describe the dynamics of the WT/BESS under the dwell time constraint. The sufficient condition for the practical stability of the switched system with multiple equilibriums is given and applied to the practical stability analysis of the WT/BESS. Finally, the simulation example shows that the theory is feasible.

Acknowledgments. This work was supported by the National Natural Science Foundation of China (61573237, 61600016), NSFC/Royal Society Cooperation and Exchange Project (62111530154), 111Project (D18003), Project of Science and Technology Commission of Shanghai Municipality (19500712300).

References

1. Annual Energy Review: United States Department of Energy. Energy Inf. Admin. Report No. DOE/EIA-0384 (2008)
2. Yin, M., Xu, Y., Shen, C., et al.: Turbine stability-constrained available wind power of variable speed wind turbines for active power control. J. IEEE Trans. Power Syst. **32**, 2487–2488 (2016)
3. Strachan, N.P.W.: Improving wind power quality using an intergrated Wind Energy Conversion and Storage System (WESS). In: Power & Energy Society General Meeting-conversion & Delivery of Electrical in the Century. IEEE (2008)
4. Akhmatov, V.: Analysis of dynamic behavior of electric power systems with large amount of wind power. D. Ph. D. dissertation, Technical University of Denmark, Kgs. Lyngby, Denmark (2003)
5. Dai, X., Song, Y., Yang, T.: Modelling and region stability analysis of wind turbines with battery energy storage system based on switched system with multi-equilibriums. J. Trans. Inst. Meas. Control. **41**, 1519–1527 (2019)
6. Zhao, T., Ding, Z.: Cooperative optimal control of battery energy storage system under wind uncertainties in a microgrid. J. IEEE Trans. Power Syst. **33**, 2292–2300 (2017)
7. Ma, Z., Chen, D.: Optimal power dispatch and control of a wind turbine and battery hybrid system. In: American Control Conference, pp. 3052–3057. Chicago, IL, USA (2015)
8. Zhang, L., Wirth, A.: Wind energy management with battery storage. J. Oper. Res. Soc. **61**, 1510–1522 (2010)
9. Peng, C., Song, Y.: Modelling of wind turbine with battery energy storage system based on two-level switched system. In: Jia, Y., Zhang, W., Fu, Y. (eds.) CISC 2020. LNEE, vol. 706, pp. 117–125. Springer, Singapore (2021). https://doi.org/10.1007/978-981-15-8458-9_13
10. Palejiya, D., Hall, J.: Stability of wind turbine switching control in an integrated wind turbine and rechargeable battery system: A common quadratic Lyapunov function approach. J. Dyn. Syst., Measur. Control **135** (2013)
11. Guo, R.W., Wang, Y.Z.: Region stability for switched nonlinear systems with multiple equilibria. J. **15**, 1–8 (2010)
12. Patel, M.: Wind and solar power systems design, analysis, and operation. J. Taylor & Francis, Boca Raton (2006)

Intelligent Methods in Developing Electric Vehicles, Energy Conversion Devices and Equipment

Design of DC Electric Parameter Measurement System

Juan Li[1(✉)], Zhongde Su[2], and Dongxue Yang[2]

[1] Department of Electrical and Electronic Engineering, Anhui University of Information Engineering, Wuhu, China

[2] Key Laboratory of Electric Drive and Control of Anhui Higher Education Institutes, Anhui Polytechnic University, Wuhu, China

Abstract. For electric parameters measurement and facilitation, through the study of control and theory of power parameter measurement, design a kind of take alarm device can measure the voltage and current at the same time power three electricity parameters of simple measuring instrument, the measurement system uses single-chip microcomputer as the main control unit, using ADC0832 as voltage acquisition chip, through multiple harmonic oscillation circuit resistance is converted into frequency, again by resistance transformation to calculate the current, finally calculated by single chip microcomputer power displayed on the LCD screen. Alarm circuit by relay all the way and the light-emitting diode (led), as well as of the horn, when an ADC chip, the detection of voltage, current, or power more than setting the maximum threshold, can make the I/O mouth into a low level, Triode and relay conduct, led and horn line pressure drop resulting alarm, through simulation and experiment validate the correctness and feasibility of the design.

Keywords: Electrical energy measurement · Voltage acquisition · Multiple resonant oscillation circuit

1 Introduction

Electric parameter measurement in industrial control, electric power systems, electrical measurement and various kinds of electromechanical integration device is of great significance [1], the traditional voltage current power parameters measuring instruments are separate, for the measurement of voltage and current is used mostly voltage transformer and current transformer [2–4], low accuracy and measurement of trouble, the literature [5] for electric mechanical and electrical parameters measurements is ignoring the motor copper loss and core loss and magnetic flux leakage of measurement, though this measuring method can measure multiple parameters measurement circuit structure is complex, algorithm complexity, poor reliability, The traditional instrument for measuring electrical parameters is large in volume and weight, and the measurement accuracy is not high because of the large power loss. At present, there are many kinds of electric parameter measuring instruments in the market, among which most of

K. Li et al. (Eds.): LSMS 2021/ICSEE 2021, CCIS 1468, pp. 223–231, 2021.
https://doi.org/10.1007/978-981-16-7210-1_21

the domestic measuring instruments have low accuracy and large test error. Measuring method in literature [6, 7] of voltage sampling and current sampling method is adopted to measure the voltage and current, and the precision of this method due to the sampling resistor different will affect the final measured values, the literature [8] proposed the use of advanced precise electronic current transformer and voltage transformer power measurement study, also is the use of current transformer and voltage transformer for the study of the power, circuit structure and the algorithm is complex, poor reliability, and foreign production of digital measuring instrument is expensive, not suitable for general industrial applications; Therefore, it is of great significance to design a kind of electrical parameter measuring instrument which can measure a variety of electrical parameters at the same time. In this paper, an intelligent measurement instrument for electric parameters of resistance current and power is designed for the control system based on 51 MCU.

2 System Design

This electric parameter measurement system design of DC voltage measurement range is 1~30 V, the maximum set to 5 A loop current, the measurement module mainly consists of single chip microcomputer minimum system, D/A conversion circuit, input and output circuit, alarm, display circuit, etc., MCU as the main control unit, ADC0832 as the voltage acquisition chip, parameters of resistance by harmonic oscillation circuit to make it into parameters frequency, into single chip microcomputer, then by the voltage and resistance transformation to calculate the current value, finally through the calculation of voltage and current measured load circuit power consumption. The maximum value of voltage, current and power can be set when the system is measured. When the voltage, current or power value flowing through the measurement loop exceeds the set maximum threshold, the LED and the buzzer alarm, and the LCD display displays the actual value currently measured. Each part of the circuit is described in the following (Fig. 1).

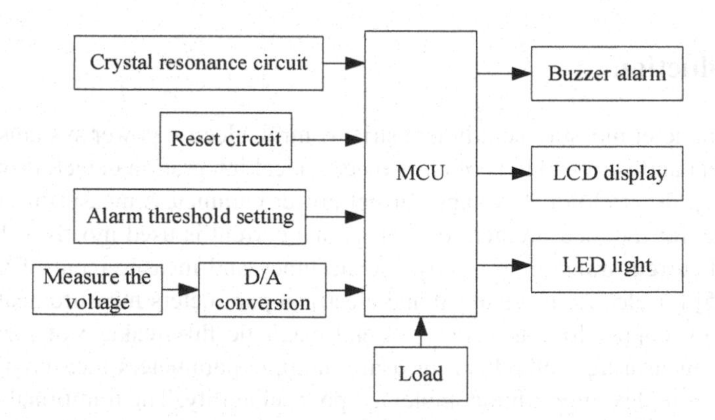

Fig. 1. System overall design drawing

2.1 MCU Minimum System

Figure 2 shows the minimum system of MCU, that is, can let the microcontroller work up the smallest unit, it is composed of a single chip microcomputer, the reset circuit includes power reset and manual reset, want to reset the MCU, the need to MCU pin 9 2 machine cycle or more high level, System just before power on, C2 capacitor is no electricity, started to electricity power supply through the R1 to charging capacitor C2, charging time is more than two machine cycles, so when a electricity microcontroller reset, if there is no R1, the power supply will be directly grounded by capacitance C2, which is equivalent to short circuit, so the R1 is necessary, then pin 9 into a low level, manual reset is presses the button reset K4, when K4 key is pressed, the reset pin from the MCU to connect power directly, just press down the cycle time is greater than two machines, microcontroller will reset; Crystals of these circuits is also called the clock circuit, Y1 is crystals, crystal vibration is similar to the heart of the single chip microcomputer, provide single chip microcomputer frequency, generally uses 12 M or 11.0592 M crystals, here are 12 M of crystals, crystal vibration is connected to the MCU pin 18 and pin19, then from the two pins to connect two 30 pf capacitance to the ground, used to stabilize the normal work of the crystals.

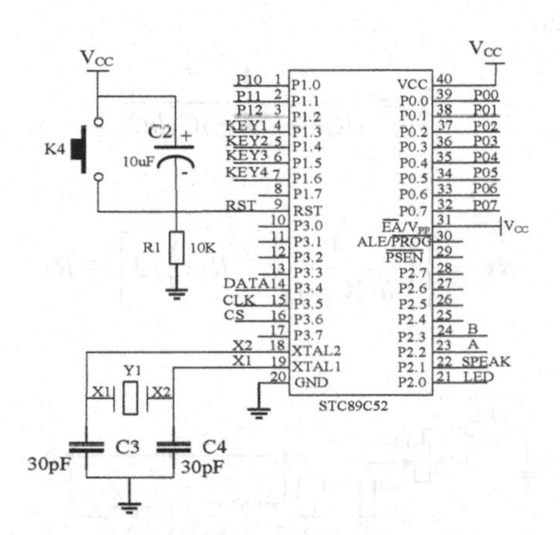

Fig. 2. MCU minimum system

2.2 Voltage Acquisition Module

Adopt ADC0832 as voltage acquisition chip, because ADC0832 maximum input voltage of 5 V, so need to compare A/D measuring voltage and ADC0832 maximum input voltage, if greater than 5 V first with partial pressure resistance, make the test voltage range into 0 to 5 V, conform to the requirements of the ADC0832 chip of input voltage, and then through single chip processing, on the basis of the original voltage quality in partial pressure coefficient can restore the previous voltage under test.

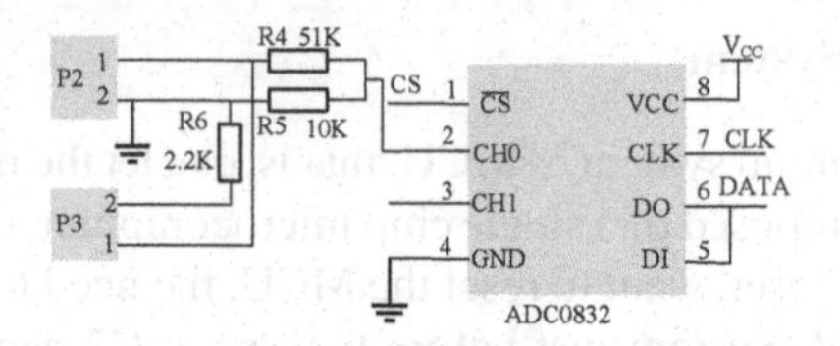

Fig. 3. Voltage acquisition module

2.3 Resistance Measurement Module

The P1 port in Fig. 4 is the port for measuring the load resistance. The part of measuring the resistance in this system adopts the pulse counting method. The multi-resonant swing circuit composed of 555 circuit calculates the magnitude of the measured resistance by calculating the frequency of the oscillation output.

According to the parameters configured in Fig. 3, the oscillation period formula of the multi-resonant oscillation circuit can be obtained as follows:

$$T = (R_{10} + R_X) * C_1 * ln2 + R_X * C_1 * ln2 = (R_{10} + 2R_X)C_1 * ln2 \tag{1}$$

It is concluded that:

$$f_X = \frac{1}{(R_{10} + 2R_X)C_1 ln2} \tag{2}$$

That is:

$$R_X = \left[\left(\frac{1}{ln2C_1 * f_X} - R_{10}\right)/2\right] - R_7 \tag{3}$$

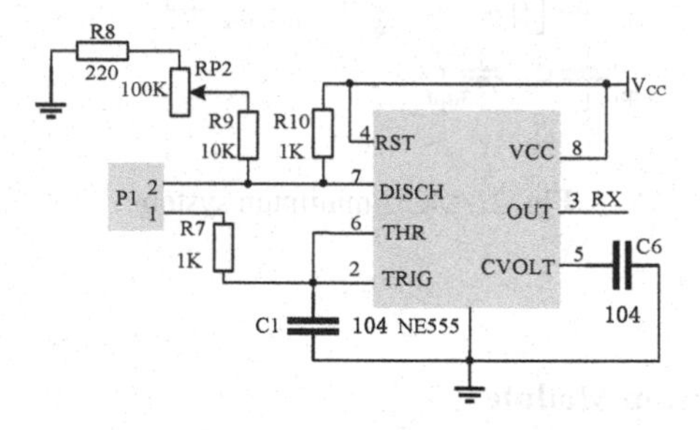

Fig. 4. Resistance measuring module

2.4 Alarm Module

As shown in Fig. 5, the alarm module circuit mainly includes the alarm threshold setting circuit, as shown in Fig. 5 (a), as well as the alarm display circuit, as shown in Fig. 5 (b).

Figure 5 (a) is used to set the maximum voltage, current and power, once measured values more than the set value, a buzzer alarm, led flashing, K1 in Fig. 5 (a) is used to switch parameter setting mode, keys K1 press next to enter the maximum voltage, current and power to set the mode switch interface, press will in turn according to the maximum voltage, maximum current and maximum power sequence of loop switch, K2 in Fig. 5 (a) is the confirmation of the currently selected maximum parameters, K3 in Fig. 5 (a) is the currently selected parameters for data increase,K4 in Fig. 5 (a) is to reduce the data of the currently selected parameters, in which the unit of voltage change is set as 0.1 V, the unit of current change is set as 1 mA, and the unit of power change is set as 0.1 W, where the unit change value can be changed through the program.

The voltage alarm circuit consists of relay, LED and buzzer. When the voltage, current or power detected by the ADC chip exceeds the set maximum value, the specific IO port will be turned into a low level, and the PNP triode will be switched on, so that the relay will be switched on, and the luminous LED and buzzer will alarm.

The microcontroller compares the parameters collected with the alarm threshold. When any acquisition value of voltage, current or power is higher than the alarm threshold voltage, the buzzer alarm and light-emitting diode alarm will be started.

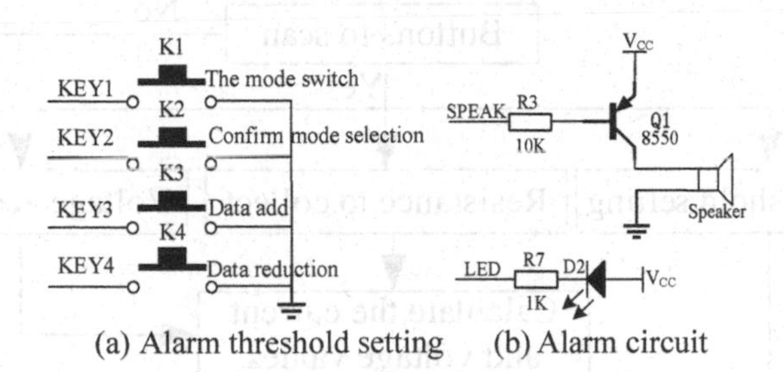

(a) Alarm threshold setting (b) Alarm circuit

Fig. 5. Alarm module

2.5 Display Module

The display circuit, as shown in Fig. 6 is controlled by the P0 port of the single chip microcomputer for the communication between the single chip microcomputer and the liquid crystal. It contains two sets of VCC and GND. The third pin is used to adjust the display contrast and is connected with a 3 K resistance.

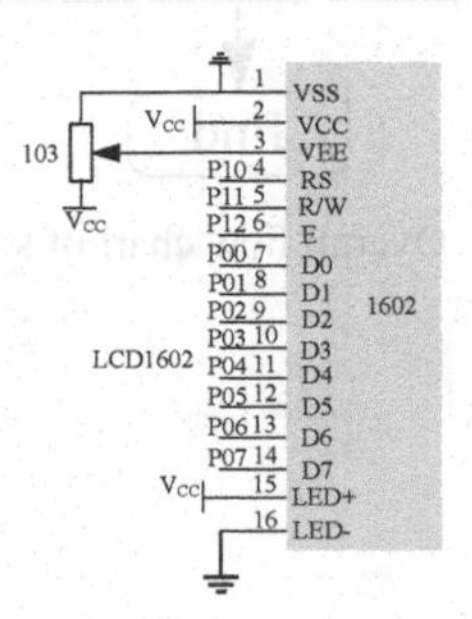

Fig. 6. Display module

3 The Program Design

Using keil programming, as shown in Fig. 7 is the main program main process diagram, after the system is powered on, first initialize screen, and then an analysis of the scanning, each press the button press the button to automatically determine which step should be performed, then through the MCU interrupt to timer is set to open channel and light Settings and calculation to the final value measuring the parameters of the liquid crystal display, these functions are controlled by single chip microcomputer program.

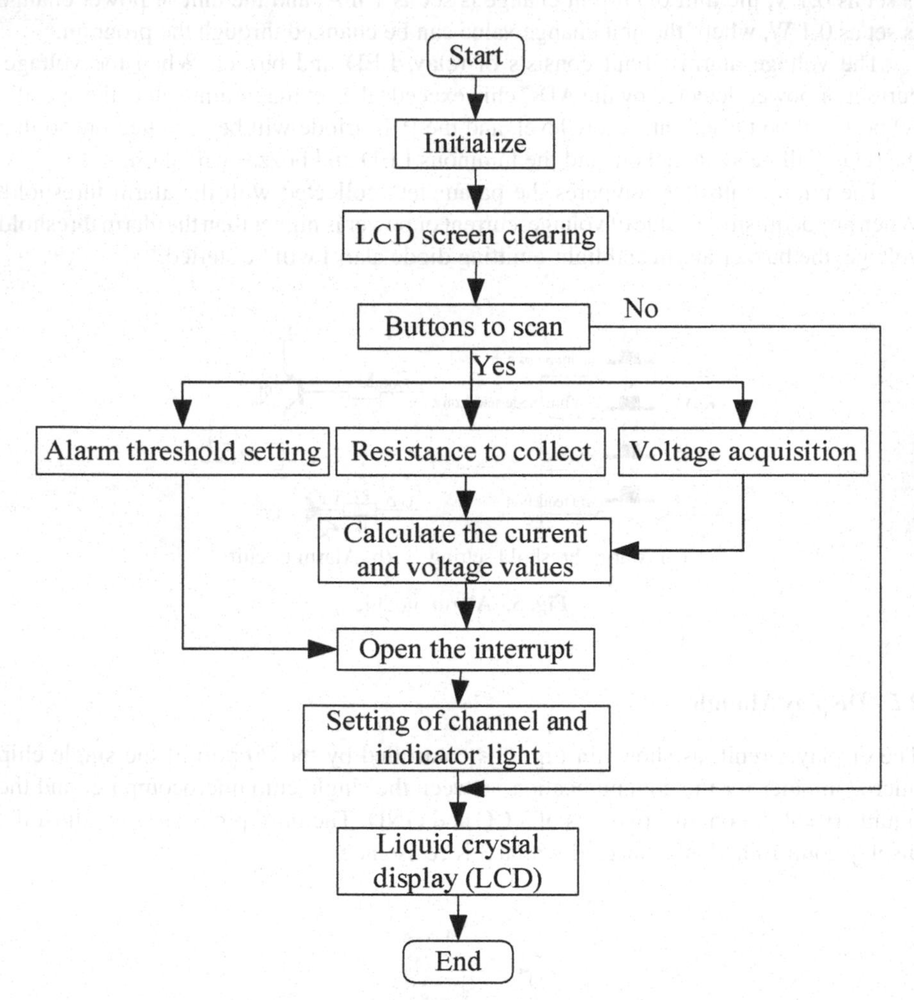

Fig. 7. Overall flow chart of software

4 Circuit Simulation

Figure 8 shows the simulation results of the circuit. U, I and P in the LCD liquid crystal display refer to the measured voltage, current and power values, in which the minimum increment or decrease of voltage U is 0.1 V, the minimum increment or decrease of current I is 1 mA, and the minimum increment or decrease of power P is 0.1W.R14 in the simulation circuit is two ports of external access to the load resistance. Through simulation, the correctness of the program and the feasibility of the hardware circuit are verified.

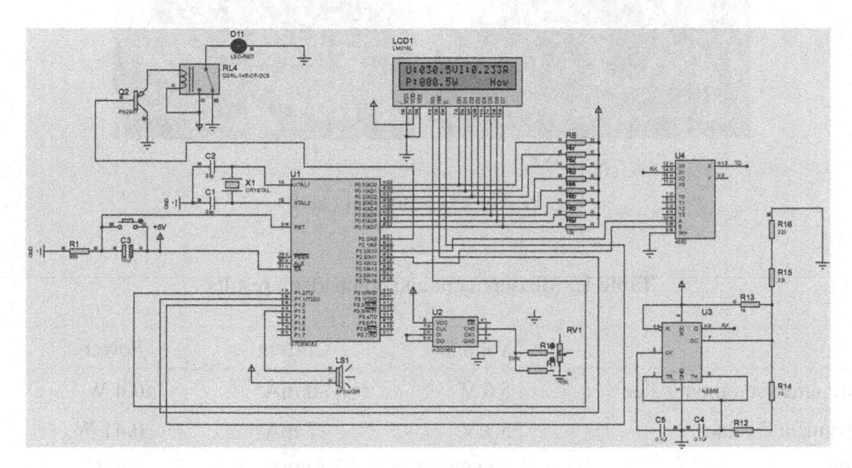

Fig. 8. The simulation results

5 Experimental Verification

Figure 9 shows the experimental results. Figure 9 (a) shows the hardware circuit diagram with a load fan. Figure 9 (b) shows the voltage, current and power values when the load is not connected. Are shown in Fig. 9 (c) set the voltage, current and power of the largest alarm threshold, is currently the most powerful alarm threshold setting, can press the switch button to set the biggest alarm threshold voltage or current, current setting said the biggest alarm threshold voltage of 2.2 V, current largest alarm threshold is 38 mA, the biggest alarm threshold power of 2.2 W; Fig. 9 (d) measures the voltage, current and power values after load access. Since the measured voltage and current values are greater than the set alarm threshold, the LED and buzzer will alarm at this time.

In order to verify the accuracy of the measurement results, the voltage measured by the system was measured with a multimeter in the experiment, and the current of the loop was measured with a current clamp. The errors of the measured values were calculated to be less than 6%. The errors of the components themselves were not considered here. The error calculated from the experimental data shows that the measuring accuracy of the electrical parameter measuring system is good (Table 1).

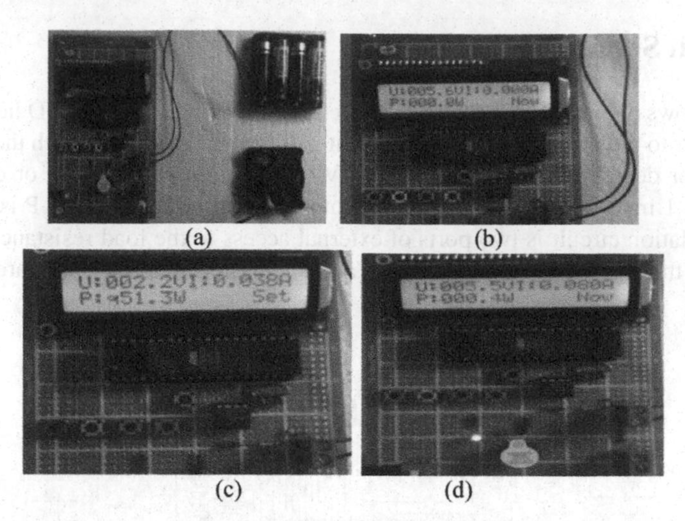

(a) (b) (c) (d)

Fig. 9. The experimental results

Table 1. Measurement and comparison results

	Voltage	Current	Power
System measurement value	5.6 V	80 mA	0.4 W
Instrumental value	5.3 V	77 mA	0.41 W
Error	5.66%	3.89%	2.44%

6 Conclusion

To sum up, the DC parameter measurement system adopts AT89C52 single chip microcomputer as the main control chip, which can measure the voltage, current and power values at the same time. The measurement range and maximum alarm threshold can be adjusted and controlled through the program. The measured parameter values are displayed by liquid crystal display, and equipped with LED and buzzer alarm device. In this design, the resistance parameters need to be converted into frequency by multi-resonant oscillation circuit, and then the converted frequency is processed by single chip microcomputer, which makes the hardware circuit structure simple and the system reliability high. The correctness and feasibility of the design are verified by Keil program compilation, Proteus simulation and experiment. The measurement values of inductance and capacitance parameters can be embedded in the system and displayed simultaneously.

References

1. Ruuskanen, V., et al.: Power quality estimation of water electrolyzers based on current and voltage measurements. J. Power Sources. **450** (2020)

2. Shao, X., Xie, M., Wang, T., Yu, B., Li, X.: Design of current measurement sensor with self-powered function. J. Power Syst. Prot. Control **48**(16), 155–162 (2020)
3. Wang, J.: Design of power control system and chip in smart electricity meter. J. Electr. Technol. **49**(06), 13 (2020)
4. Alubodi, A.O. Mashhadani, A.Al.I.B.N.: Mahdi, S.S.: Design and Implementation of a Zigbee, Bluetooth, and GSM-based smart meter smart grid. J. IOP Conf. Series: Mater. Sci. Eng. **1067**(1), (2021)
5. Li, H., Xu, X., Yu, X., Wu, Z.: Design of intelligent generator parameter measurement device based on STM32. J. Electr. Measur. Instru. **53**(02), 125–128 (2016)
6. Chen, Y.: Design of Automatic Power Measurement Based on STM32. J. Electronic Technology and Software Engineering. **20**, 130–131 (2019)
7. Yiqing, Y., Wei, Z., Lei, C.Q., Wang, S.H.: Power measurement accuracy analysis in the presence of interharmonics. J. Meas. **154** (2020)
8. Sirshendu, S., Bera, S.K., Mandal, H., Sadhu, P.S., Bera, S.C.: Study of an accurate electronic power measurement technique using modified current transformer and potential transformer. J. Trans. Inst. Meas. Control. **41**(13), 3663678 (2019)

Research on V2G Reactive Power Compensation and Optimization Strategy Based on Genetic Algorithm Optimization

Qi Zhu(✉), Hui Cai, Yue Xie, and Qian Guo

College of Mechanical and Electrical Engineering, China Jiliang University, Hangzhou, China

Abstract. The design capacity of electric vehicle charging piles is increasing, but the problems of its own utilization rate and low capacity utilization rate still exist, and the V2G technology for grid reactive power compensation is a potential solution to this issue. This paper proposes a V2G reactive power compensation mode suitable for power grid and user hierarchical dispatch. First, the genetic algorithm (GA) is used to optimize the orderly charging method to improve the load peak-valley difference rate. Then, the remaining capacity of the charging pile is used to compensate the power grid to reduce the voltage deviation. Finally, according to the simulation results, the proposed orderly tuning method can effectively reduce the fluctuation of the charging load; the reactive power compensation can increase the low-voltage node voltage while effectively suppress the extra network loss caused by electric vehicle charging.

Keywords: GA · Orderly tuning · Reactive power compensation

1 Introduction

The development of electric vehicles has benefited from the continuous attention to non-renewable energy and environmental pollution issues worldwide. The rapid growth of its number has brought challenges to grid control and management, but also brought great opportunities [1]. Literature [2] uses the quadratic programming method to optimize the control of vehicle charging and discharging to achieve the goal of saving charging costs and increasing discharge benefits. Literature [3] uses linear programming to analyze and study the smart charging scheme of electric vehicles according to the degree of electricity price response.

With the further opening of the electricity market, the future reactive power market will develop in a distributed manner. Devices with distributed locations such as electric vehicle charging stations, flexible site selection, and reactive power compensation capabilities will have potential utilization values. Literature [4, 5] demonstrates through experiments the reactive power capacity that the charging pile can provide at different times during the charging process of the power battery; Literature [6] applies voltage feedback control to make the charging pile group follow the voltage change for reactive power compensation. The main disadvantages of the above-mentioned compensation

K. Li et al. (Eds.): LSMS 2021/ICSEE 2021, CCIS 1468, pp. 232–241, 2021.
https://doi.org/10.1007/978-981-16-7210-1_22

methods are: First, the individual charging pile cannot and does not have the authority and ability to collect enough grid data to support its own calculations, which may cause over-compensation.

This paper proposes a method that adopts GA to optimize the peak-valley difference rate to meeting the charging needs of electric vehicle users, which can effectively guide and regulate the charging behavior of electric vehicles, and reduce the impact and load impact on the power grid and save power transmission. The investment cost of power supply and distribution construction is conducive to the coordinated development of power grids and vehicles to achieve the effect of peak shaving and valley filling; Combined with a reactive power compensation mode based on user charging habits, obtained by Monte Carlo load model combined with node power flow calculation. According to the data, the compensation amount is calculated according to the voltage situation of the vehicle network, and the global optimized reactive power compensation is realized through the instruction interaction with the charging user.

The structure of this article is as follows. Section 1 describes the current application of electric vehicle V2G development, the utilization of reactive power and the impact of reactive power compensation on the power grid. Section 2 introduces the Monte Carlo load forecasting method, and combines the user's charging habits, IEEE100 node to predict the real-time load curve and power quality. Section 3 introduces the GA, which divides the load curve into peak and valley periods; Establishes an optimization model with the goal of minimizing the peak-valley difference of the power grid, and uses the GA to solve the model. Section 4 analyzes the feasibility of the basic algorithm and the proposed method from the algorithm accuracy and voltage optimization. Finally, conclude in Sect. 5.

1.1 The Feasibility of Adjusting Power Factor of V2G Charging Pile System

With the advancement of electric vehicle charger technology, in addition to the role of an energy storage system based on active power, electric vehicles can also be used as an energy storage system for reactive power sources. Reactive power compensation is achieved through electric vehicle chargers [7], which is different from using electric vehicles to provide active power to the grid. In order to improve the power density and charging and discharging efficiency of the two-way converter, and to realize the large-capacity energy storage system and modularity, it is meaningful to study the new topology structure and control strategy of the energy conversion system (or called the charge-discharge bidirectional converter) composed of power electronic devices. Figure 1 is the basic topology diagram of the bidirectional charger.

Figure 2 is the single phase equivalent circuit diagram of the grid-side converter link. In the figure, U_S is the unit value of the fundamental phasor of the grid voltage u_S; U_{Ls} is the unit value of the fundamental phasor of the voltage u_{Ls} across the inductor L_S; U_{AB} is the DC bus voltage after sinusoidal pulse width modulation (SPWM); The unit value of the fundamental phasor of the AC side voltage u_{AB}; I_S is the unit value of the fundamental phasor of the grid-connected current.

Under steady-state conditions, ignoring the harmonic components of the grid side voltage, there is

$$\dot{U}_s = j\omega L_s \dot{I}_s + \dot{U}_{AB}. \tag{1}$$

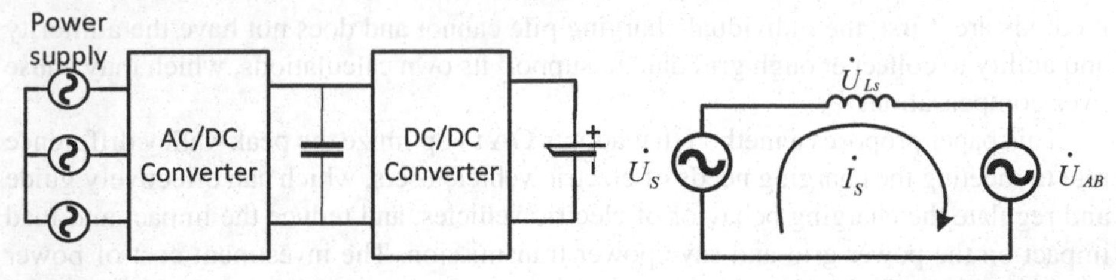

Fig. 1. Basic topology diagram of two-way charger.

Fig. 2. The equivalent circuit diagram of the grid-side converter link.

In Eq. (1), $\dot{U}_s$ is the fundamental wave vector of grid voltage; $\dot{U}_{AB}$ is the fundamental wave vector of DC bus voltage, L_s is the inductance, and $\dot{I}_s$ is the fundamental wave vector of grid-connected current.

Assuming that the magnitude of the grid-connected current is fixed $|\dot{I}_s|$, then the voltage magnitude at both ends of the inductor $|\dot{U}_{Ls}| = \omega L|\dot{I}_s|$, which can be obtained from the formula, is also a fixed value, and the grid voltage $\dot{U}_s$ remains unchanged, the rectifier bridge The voltage on the AC side will move on a circle, and the resulting voltage vector diagram is shown in Fig. 3. The reactive power compensation method in this paper selects this point C as the operating state of the grid-side converter link.

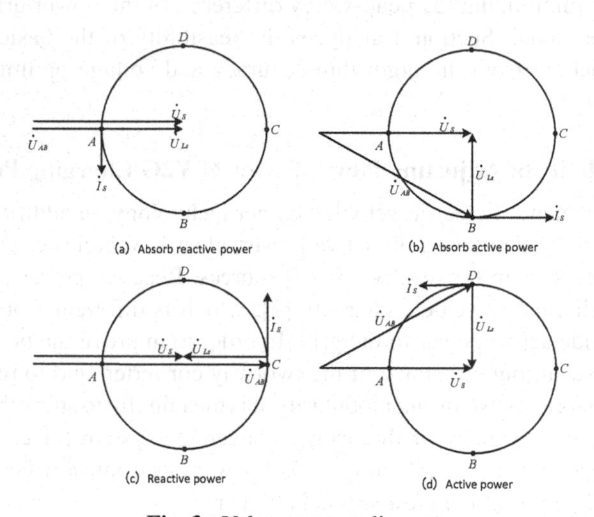

Fig. 3. Voltage vector diagram.

2 Monte Carlo Modeling of Charging Stations

Monte Carlo method, also known as statistical simulation method, is a stochastic simulation method based on probability theory and statistical theory. The implementation steps are generally divided into three steps, as shown in Fig. 4:

1) Analyze the possible variation range and probability distribution of each variable.
2) According to the value range and probability distribution of each parameter, a single sample is obtained by random sampling, including random numbers and computer random simulation.
3) Repeat step 2) to obtain multiple samples. After all samples are superimposed, an overall model that satisfies the characteristics of the probability distribution is obtained.

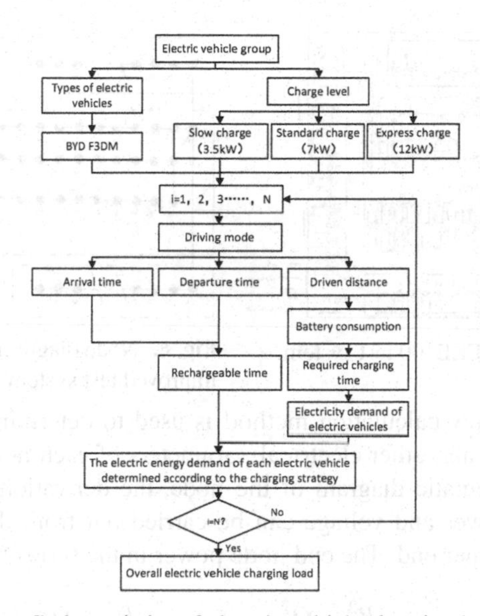

Fig. 4. Monte Carlo statistics of electric vehicle charging load flow chart.

2.1 Modeling of User Charging Habits Based on Multi-purpose Charging Stations

At present, there are three ways to charge electric vehicles in multi-purpose charging stations: residential area, public area, and commercial area. Different charging methods and charging habits of electric vehicles will bring different results to the distribution network [9].

The statistical results of the operation data of electric vehicles are as follows: the morning and evening peak hours correspond to the commuting time (6:30–9:00, 17:00–19:00), and the residual electric battery in household charging stations follows the distribution of N(0.55,0.62). The electric vehicle in the parking lot needs to replenish the electric quantity twice a day, which is 11:30~14:30 and 19:00~21:00. Parked in malls, electric-car travel and charging habits are concentrated during rush hour and holidays spread throughout the day.

2.2 Combining IEEE100 Node and Power Flow Calculation Modeling

The test system uses an improved IEEE100 bus system, as shown in Fig. 5. The figure shows three type of areas: residential area, public area, and commercial area. In order to simply allocate the charging power, three type of chargers with rated powers of 3.5, 7.5 and 12 kW are considered in the simulation.

Perform simple node matching, and the corresponding node diagram is shown in Fig. 6.

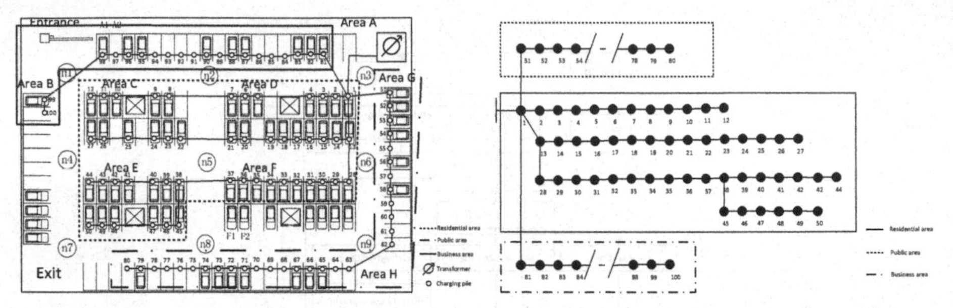

Fig. 5. Improved IEEE100 test system.

Fig. 6. Node diagram corresponding to the improved test system.

Then the power flow calculation method is used to determine the voltage, active power, reactive power and other electrical parameters of each node in the test system. According to the schematic diagram of the node, the derivation from the initial end voltage to the end power and voltage can be carried out from the known power and voltage at the transformer end. The end node power in the series branch is:

$$S_2 = P_1 - \frac{P_1^2 + Q_1^2}{U_1^2} R_1 + j(Q - \frac{P_2^2 + Q_2^2}{U_1^2} X_1). \tag{2}$$

The terminal node voltage in the series branch is:

$$U_2 = \sqrt{(U_1 - \frac{P_1R_1 + Q_1X_1}{U_1})^2 + (\frac{P_1X_1 - Q_1R_1}{U_1})^2} \tag{3}$$

In Eq. (2) & (3), S_2 is the end node power, U_2 is the end node voltage. After bringing in the node parameters, the voltage calculation results of each node are shown in Fig. 7.

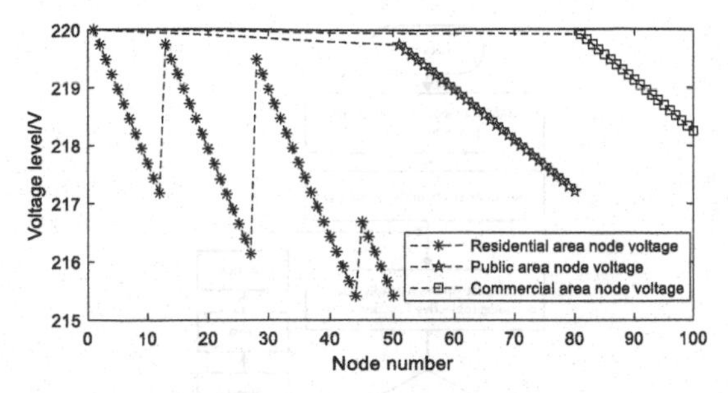

Fig. 7. Calculation results of node voltage at each point.

3 Process Design of Load Peak-Valley Difference Rate Based on Genetic Algorithm Control

GA is an intelligent search method that simulates natural biological selection and genetic mechanism. It uses the coded form to simulate the gene chain of biological chromosomes, and through random and orderly genetic operations, recombination produces a more adaptable population. It has a wide range of feasible solutions and strong applicability; At the same time, it has significant global search and optimization capabilities for a large number of individual parallel processing; Heuristic search and optimization capabilities make it have better optimization processing capabilities. The basic process of GA to solve the optimization problem is shown in Fig. 8. The basic operations of GA include coding, selection, crossover and mutation. Among them, the coding completes the conversion from the solution space data to the genetic space gene structure type. The commonly used coding methods are binary and decimal; The selection follows the principle of survival of the fittest in nature, sets the selection algorithm, and selects individuals with higher fitness as the parent. The next generation of reproduction; Crossover is the most important genetic operation. A new generation of individuals integrates the characteristics of their parents and effectively realizes inter-generational variation. This variation occurs in a very small number of individuals and reflects variability.

In order to effectively reduce the load fluctuation of the distribution network and optimize the efficiency of the system, The problem solution is formulated as the adjustable range of three charging methods, the objective function value is formulated as the load peak-valley difference rate. The formula is as follows:

$$LPV = 1 - \frac{\min(P_{l+ev})}{\max(P_{l+ev})} \tag{4}$$

In Eq. (4), LPV is the load peak-valley difference rate, P_{l+ev} is the total load, that is, the sum of the charging load and the original load.

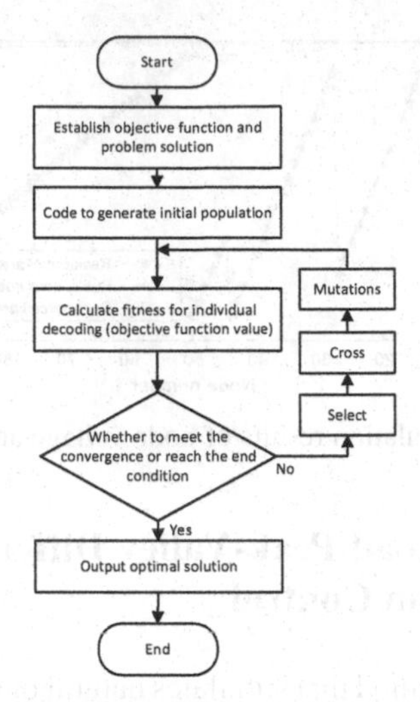

Fig. 8. The basic flow chart of GA to solve optimization problems.

4 Simulation Analysis

Use Monte Carlo method to simulate the charging demand of users, and the conventional electric load power curve and the vehicle disorderly charging load power superimposed curve are obtained; Combined with the GA and the objective function, the vehicle orderly charging load power superimposed curve is obtained. Compile a standard GA to optimize the load peak-valley difference rate based on the 0~50% of the user's charging habits change range. The results of the optimal range of change are shown in Fig. 9, and the result of 300 iterations to find the optimal target value is shown in Fig. 10.

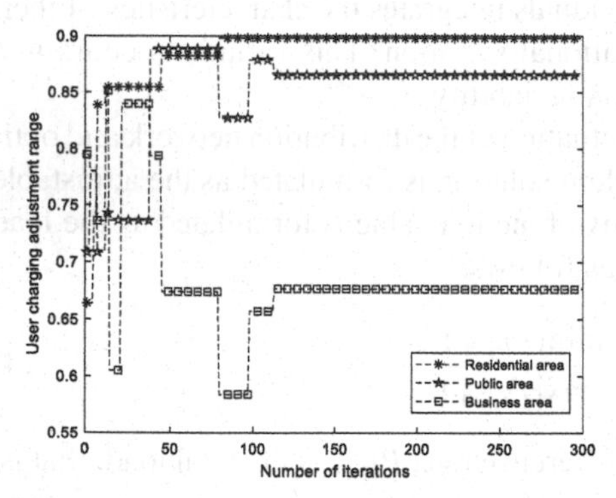

Fig. 9. Results of 300 iterations of optimal range of change.

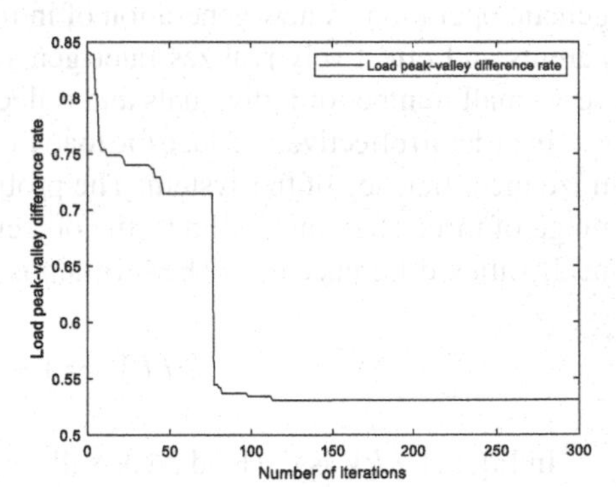

Fig. 10. 300 iterations to find the optimal the optimal target value target value.

From the comparison of the curves in Fig. 11, it can be seen that the disorderly charging behavior of electric vehicles will have a certain impact on the load of the distribution network. The charging behavior of car owners in accordance with their own usage habits makes the charging load coincide with the peak period of daily electricity load, and the consequence of "peak on top of peak" occurs. In this calculation example, during the first peak period, the disorderly charging behavior of vehicles caused the peak load of the distribution network to rise from 13.69 kW to 88.83 kW, an increase of 75.15 kW; During the second peak period, the peak load of the distribution network increased from 23.00 kW. It rises to 128.77 kW, an increase of 76.41 kW, and the peak-valley difference is 82.14%. After tuning, the peak load of the distribution network increased from 29.96 kW to 70.95 kW, an increase of 40.99 kW; During the second peak period, the peak load of the distribution network increased from 56.25 kW to 119.74 kW, an increase of 63.49 kW, peak and valley The difference rate is reduced to 53.02%, which is a 29.12% reduction in the peak-valley difference of the distribution network load curve compared with before the optimization. The utilization rate of power supply equipment is improved, and the economic cost of power grid operation is reduced.

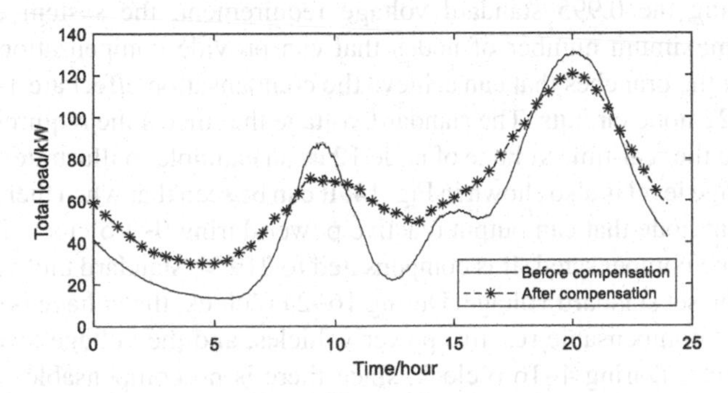

Fig. 11. Comparison chart before and after tuning of total charging load.

Figure 12 shows the process of reactive power compensation in the above model according to different voltage standards. Taking into account the user's mobility, only residential areas have users participating in reactive power compensation coordination for the time being. Therefore, the area where reactive power compensation can be performed is a residential area (the first 50 nodes). There, i is the number of electric vehicles that can be compensated; N is the number of nodes to be compensated; t is the time.

After simulation, the real-time number of charging vehicles and the number of vehicles that can provide reactive power compensation are shown in Fig. 13.

After obtaining the number of vehicles that can provide reactive power compensation, the above reactive power compensation strategy is adopted to carry out reactive power compensation, and the average voltages within one day of each node are obtained, as shown in Fig. 14.

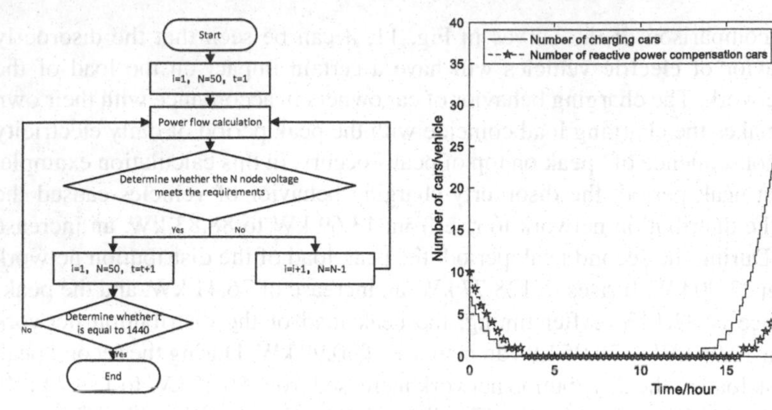

Fig. 12. Flow chart of reactive power Compensation.

Fig. 13. The number of charging vehicles and the number of vehicles that can provide reactive power compensation.

After setting the 0.995 standard voltage requirement, the system compensates. Because the maximum number of nodes that can provide compensation at a certain moment is 20, the branches that can achieve the compensation effect are 1–12 node circuits and 13–27 node circuits. The standard voltage that meets the requirements can be obtained. Take the real-time voltage of node 12 as an example to illustrate. The 12-node real-time voltage level is also shown in Fig. 14. It can be seen that when there are vehicles before the front node that can output reactive power during 0~4 o'clock, the voltage at this node can be compensated. It is compensated to 219 V (standard unit value 0.9954), being above the set standard voltage. During 16~24 o'clock, the voltage is compensated when there are compensable reactive power vehicles, and the voltage level also meets the requirements; During 4~16 o'clock, since there is no compensable vehicle in the network, the voltage cannot be compensated, which is no increasing (Fig. 15).

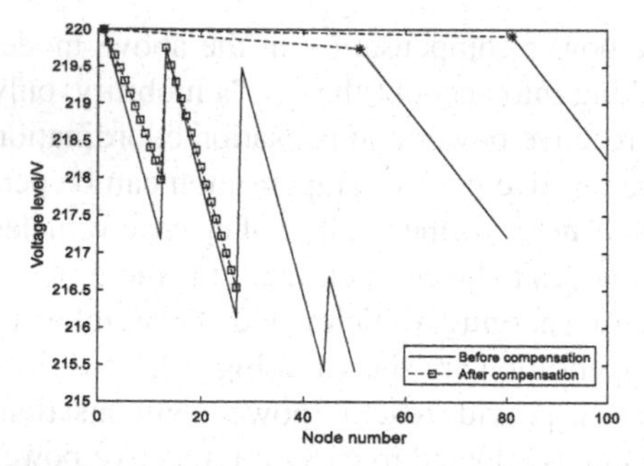

Fig. 14. Reactive power compensation result graph.

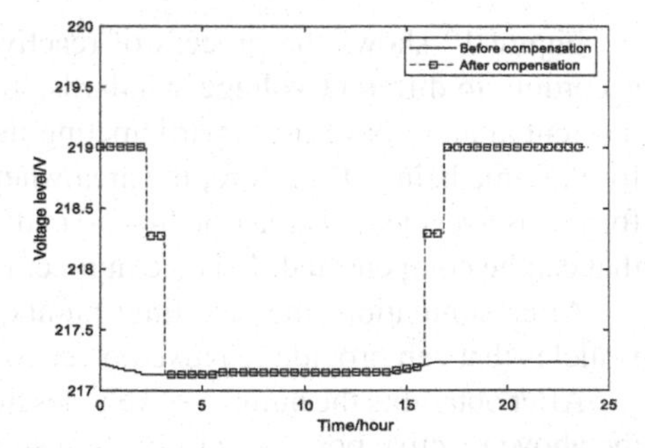

Fig. 15. 12-node real-time voltage level diagram.

5 Conclusion

This article mainly studies the reactive power compensation strategy of V2G to optimize the design and application of orderly charging, combined with the Monte Carlo model based on the user's charging habits, and the main results obtained are as follows:

1) This compensation mode can effectively reduce the load fluctuation caused by the additional charging load, and the fluctuation of the original load is reduced by 29.12% under the working conditions in the article.
2) This compensation mode can effectively reduce the voltage deviation during the whole day. Compared with the original load situation, it can reduce the average voltage deviation under the orderly charging mode.
3) This compensation mode can effectively improve the voltage quality of the vehicle network on the basis of effective coordination with orderly charging. Compared with the working conditions before reactive power compensation, it can bring an improvement to the lowest voltage node during peak power consumption

It can be seen from this paper that this compensation mode has obvious effects in improving the voltage quality of the vehicle network, meanwhile, it can restrain the violent fluctuation of the load to a certain extent. It is a potential V2G utilization method with good application value and commercial value.

Acknowledgements. This research was supported by Zhejiang Provincial Natural Science Foundation of China under Grant No. LGG20E070003.

References

1. Liu, X., Zhang, Q., Cui, S.: Overview of electric vehicle V2G technology. J. Trans. Chin. Soc. Electr. Eng. **27**(2), 121–127, Appendix: Springer-Author Discount (2012)
2. Hu, z., Song, Y., Xu, Z., et al.: The impact and utilization of electric vehicles connected to the grid. J. Proc. Chin. Soc. Electr. Eng. **4**, 1–10 (2012)
3. Chenggang, D., Zhang, H., Li, J., et al.: Application of electric vehicle access technology in smart grid. J. East China Electric Power **38**(4), 557–560 (2010)
4. Kisacikoglu, M.C., Ozpineci, B., Tolbert, L.M.: EV/PHEV bidirectional charger assessment for v2g reactive power operation. J. IEEE Trans. Power Electr. **28**(12), 5717–5727 (2013)
5. Buja, G., Bertoluzzo, M., Fontana, C.: Reactive Power Compensation Capabilities of V2G-Enabled Electric Vehicles. In: IEEE Transactions on Power Electronics, vol. 32, pp. 9447–9459 (2017)
6. Fan, Y., Zhang, L., Xue, Z., et al.: Reactive power compensation technology based on V2G. J. Power Syst. Technol. **2**, 307–311 (2013)
7. Taghizadeh, S., Hossain, M.J., Lu, J., et al.: A unified multi-functional on-board EV charger for power-quality control in household networks. J. Appl. Energy **215**, 186–201 (2018)
8. Zhou, C., Qian, K., Allan, M., et al.: Modeling of the cost of EV battery wear due to V2G application in power systems. J. IEEE Trans. Energy Convers. **26**(4), 1041–1050 (2011)
9. Sortomme, E., El-Sharkawi, M.A.: Optimal charging strategies for unidirectional vehicle-to-grid. J. IEEE Trans. Smart Grid **2**(1), 131–138 (2011)

Power Battery Reverse Logistics Network Optimization Model Under Tax Relief Policy

Qian Guan and Yuxiang Yang(✉)

China Jiliang University, Hangzhou 310018, China

Abstract. In this paper, a two-stage mixed integer linear programming model is established based on 1) the uncertainty of quantity and quality of used power battery recycling and 2) the government's tax reduction policy. A solution strategy combining SAA with genetic algorithm is proposed. The effectiveness of the stochastic model is verified in numerical experiments, where the model results are compared and analyzed under settings with both certain and uncertain parameters. In addition, this paper investigates the impact of tax relief policy on the layout decision and cost of logistics network.

Keywords: Network optimization · Power battery · Tax relief policy · Genetic algorithm · Sample average approximation

1 Introduction

With the continuous increase of the total production of new energy vehicles, it is estimated that a large number of power batteries will enter the centralized scrapping stage. However, at the present stage, China's recycling capacity of used power batteries is limited, the enthusiasm of recycling enterprises is not high, and the relevant recycling policy is not perfect, which lead to the low efficiency of power battery recycling in China. Therefore, it is of great significance to construct an effective recycling network, to plan the node facilities reasonably, and to optimize the design of the network.

At present, there are many researches on the optimization of electronic waste reverse logistics network. Fu et al. [1] aimed at the reverse logistics of recycling waste electronic products, and constructed a game model for the selection of recycling channels for processors under different modes. Duan et al. [2] constructed a mixed integer programming model with e-waste as the research object, and designed a heuristic solution algorithm. Dutta et al. [3], Gao et al. [4] also studied the optimization of reverse logistics network in uncertain environment. However, in reality, the effect of uncertain factors on the reverse logistics network is very complicated, and the values solved under certain and uncertain conditions are different.

However, the recycling and processing of used power batteries has its own particularities. If it is not handled properly, it will cause serious environmental pollution. In order to solve the problem of low recovery rate, Yao et al. designed the recycling mode of power batteries under the extended responsibility system of producers [5]. Gu et al.

K. Li et al. (Eds.): LSMS 2021/ICSEE 2021, CCIS 1468, pp. 242–252, 2021.
https://doi.org/10.1007/978-981-16-7210-1_23

[6] discussed the impact of government subsidies and electric battery recycling on production strategies in the case of uncertain market demand. Xie et al. [7] focused on the multi-level supply chain network of power battery recycling, and designed the revenue-sharing contract model of supply chain under different recycling modes. Zhu et al. [8] take the power battery of SAIC as the research object, consider the uncertainty factors, then use the fuzzy comprehensive evaluation method to select the reverse logistics mode of the used power battery.

In addition, the tax policies issued by the government play an important role in promoting the work of power battery recycling. In the document of "Notice on Organizing the Pilot Work of Recycling New Energy Vehicle Power Batteries", it is mentioned that we should increase policy support. In addition, the state has also introduced a series of preferential tax policies to promote battery recycling. It can be seen that the tax relief policy will have an inevitable impact on the recycling of used power batteries. Based on this, the paper constructs a multi-level reverse logistics network of used power battery recycling. Considering the Uncertain factors in network, and taking into account the tax breaks for power battery recycling center, a two-stage stochastic programming model was established, then a hybrid algorithm combining SAA and genetic algorithm is designed. Based on the case study, the paper compares the results of the model in the certain environment and the uncertain environment, then discusses the impact of tax relief policy on the reverse logistics network structure.

2 Model Establishment

2.1 Problem Description

The network includes consumers, electric vehicle dealers, car dismantling companies, power battery recycling points, second-hand markets, waste metal recycling centers and waste processing points. Consumers transport power batteries to electric vehicle dealers and used car dismantling companies through "vehicle scrap" and "old-for-new", then the batteries are recycled through the recycling center. The used power batteries are processed at the recycling point, the power batteries which meet the recycling standards are sent to the second-hand market for cascade utilization. Other power batteries are disassembled. the metals that can be used as raw materials are sent to the metal recycling center, and the remaining part with no economic value is sent to the waste treatment station for landfill or incineration. The network structure is shown in Fig. 1.

To describe the model clearly, the following assumptions are made for the model:

(1) The capacity of potential nodes is all limited within their assigned value;
(2) The alternative locations of used power battery recycling centers, electric vehicle dealers and used vehicle dismantling companies are known, and the locations of other facilities are known;
(3) The government provides subsidies to used power battery recycling enterprises according to the battery capacity. The tax rate charged by the government on used power battery is set as δ.
(4) The recovery quantity of used power battery is a random parameter;

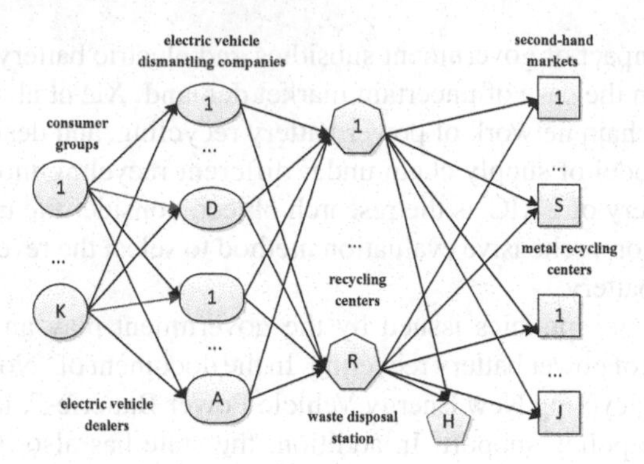

Fig. 1. Network structure of Power battery reverse logistics.

(5) The quality level of waste batteries recycled in the recycling center was measured by the random parameter quality index.
(6) The transportation cost of used power batteries from consumers to collection points is not considered.

For the convenience of description, the following symbols are introduced:

(1) Indices and sets:

$k \in K$, set of consumer group; $d \in D$:set of electric vehicle dismantling company;$a \in A$: set of electric vehicle dealer; $r \in R$: set of used power battery recycling center;$s \in S$: set of second-hand market; $j \in J$: set of waste metal recycling center; $h \in H$: set of waste disposal station.

(2) Model parameter

f_d: fixed cost of car dismantling company d.(f_a, f_r are the same).u_d: unit variable costs of dismantling company d.u_a: unit variable cost of electric vehicle dealer a.c_{dr}: unit product transportation cost from dismantling company d to recycling center r.(c_{ar}, c_{rs}, c_{rh}, c_{rj} are the same).rf: unit disassembly cost of used power battery.sf: unit dismantling cost of electric vehicle.hf: unit weight processing cost of waste treatment point.ws: the weight of the parts of the power battery after being tested and disassembled.c_{apa}: maximum storage capacity of electric vehicle dealer a(Similarly, c_{apd}, c_{apr} are the same).δ: the tax rate charged by the government.P: unit used power battery recycling price.PS: unit sales price of power battery transported to the secondary market.Pj: unit weight sales price of power battery to metal recovery station.θ_p^1: reutilization of used power batteries.θ_p^2: remanufacturing rate of used power battery.θ_p^3: scrap rate of power battery parts after disassembly.

(3) Decision variables

X_{dr}^r: quantity of product delivered from EV dismantling company d to power battery recycling center r.X_{ar}^r: quantity of product deliveries from electric vehicle dealers a

to power battery recycling centers r; (X_{ka}^{a}, X_{kg}^{g}, X_{rh}^{h}, X_{rs}^{s}, X_{rj}^{j} are the same).y_r, y_d, y_a: whether to build recycling center 、dismantling center and electric vehicle distributor in the alternative address, when it is equal to one, it means selection.

2.2 Mathematical Formulation

In the reverse logistics network, the total cost includes the following parts: fixed investment costs of alternative manufacturers; transportation cost; normal operation cost of recycling center and other costs, which can be represented by TF_1, TF_2, respectively. In other words, TF_1 represents the fixed investment cost of node facilities, TF_2 represents the others.

Let vector $X = (X^a, X^d, X^r, X^s, X^j, X^h)_{a+d+r+s+j+h}$, $\xi = (R_r, \theta_p^1, \theta_p^2, \theta_p^3)$, Ω is the random result space extracted from the probability density function of ξ. The costs set are as follows:

$$TF_1 = \sum_r f_r y_r + \sum_a f_a y_a + \sum_d f_d y_d. \tag{1}$$

$$\begin{aligned} TF_2 = & \sum_a \sum_r X_{ar}^a c_{ar} + \sum_d \sum_r X_{dr}^d c_{dr} + \sum_r \sum_s X_{rs}^r c_{rs} + \sum_r \sum_j X_{rj}^r c_{rj} + \sum_h X_{rh}^h c_{rh} + \delta * (\sum_r \sum_s X_{rs}^s PS + \\ & \sum_r \sum_j X_{rj}^j Pj) + \sum_r P * R_r(m) + \sum_r sf * R_r(m) + \sum_r (1 - \theta_p^1(m))\theta_p^2(m) * R_r(m) rf * ws + \\ & \sum_r R_r(m)(1 - \theta_p^1(m))(1 - \theta_p^2(m))\theta_p^3(m) ws * hf + \sum_k \sum_a X_{ka}^a u_a + \sum_k \sum_d X_{kd}^d u_d. \end{aligned} \tag{2}$$

Considering the uncertainty of the recovery quantity and the quality of used power batteries, two-stage model is established. The first stage is to select the alternative address before the random vector ξ is implemented. Equation (3) represents the minimization of the total cost of the objective function in the first stage, in which is a linear programming model with random parameters in the second stage o, and Eq. (4) represents binary variable. The model is as follows:

$$\min F = TF_1 + E[G(Y, \xi)]. \tag{3}$$

$$s.t.$$

$$y_a, y_d, y_r \in \{0, 1\}, \forall a, d, r. \tag{4}$$

The stochastic model of the second stage is the logistics distribution of used power batteries based on the given decision of the first stage. For the $m \in \Omega$, m are the realization values of any group of random vectors. It can be expressed as the following stochastic model:

$$G(Y, Z, \xi) = TF_2 \tag{5}$$

$$s.t.$$

$$\sum_k X_{ka}^a \leq c_{apa} y_a, \quad \forall a \tag{6}$$

$$\sum_{k} X_{kd}^{d} \leq c_{apd} y_d, \ \ \forall d \tag{7}$$

$$\sum_{d} X_{dr}^{r} + \sum_{a} X_{ar}^{r} \leq c_{apr} y_r, \ \ \forall r \tag{8}$$

$$R_r(m) = \sum_{d} X_{dr}^{r} + \sum_{a} X_{ar}^{r}, \ \ \forall r \tag{9}$$

$$\sum_{r} X_{rs}^{s} = R_r(m) * \theta_p^1(m), \forall s \tag{10}$$

$$\sum_{r} X_{rj}^{j} = \sum_{r} R_r(m) * \theta_p^2(m)(1 - \theta_p^1(m)), \forall j \tag{11}$$

$$\sum_{r} X_{rh}^{h} = \sum_{r} R_r(m) * \theta_p^3(m)(1 - \theta_p^1(m))(1 - \theta_p^2(m))ws, \forall h \tag{12}$$

$$\theta_p^1(m) + \theta_p^2(m) + \theta_p^3(m) = 1 \tag{13}$$

$$X_{ka}^{a}, X_{kd}^{d}, X_{rs}^{s}, X_{rj}^{j}, X_{rh}^{h}, X_{ar}^{r}, X_{dr}^{r} \geq 0, \forall k, a, r, s, j, h, d \tag{14}$$

Constraint (6), (7), (8), represent the constraint of the maximum processing capacity, constraint (9) - (12) represent the flow limit, constraint (13) represents the linear relationship between the recovery quality, and constraint (14) represents the value constraints of decision variables.

3 Model Solution

From the research of previous scholars, we know scholars usually use deterministic model transformation method [9], Expectation Programming Method [10] and chance constrained programming method [11] in stochastic programming to deal with the random parameters in the mode. However, in reality, uncertain factors of waste products recycling will have an impact on the selection of alternative address and the material allocation between nodes, which makes the value solved different from the actual value.

In the uncertain environment, with the increase of the size of the random variables to solve the problem, the calculation pressure will increase, and the Sample Average Approximation (SAA) [12] is an important method to solve such problems, which overcomes the defect that the sample parameterization method does not conform to the reality [13]. This paper uses the hybrid algorithm of SAA and genetic algorithm to solve the model. it is assumed that N samples are generated under the condition of random simulation. Among the N samples, there are n samples that can make the model have the optimal solution, and the probability of the feasible scheme is $\alpha = n/N$. Assuming the confidence is γ, when $\alpha \geq \gamma$, it can be considered that the location model in the first stage is feasible, and then the estimated value of F is obtained [14]. The estimated value of F is:

$$\widetilde{\mathrm{F}} = \frac{1}{n}\sum_{i=1}^{n} G(Y, \xi) + TF1. \tag{15}$$

3.1 Genetic Algorithm

(1) Coding: binary coding.
(2) Fitness calculation: among N random samples selected, n samples have feasible solutions. if and only if $n/N \geq \gamma$, the chromosome is a feasible solution, and then the estimated value of F is calculated. On the contrary, the sufficient value is set as the fitness value of the chromosome.
(3) Selection: use roulette algorithm to select population.
(4) Crossover: use a single-point crossover strategy, according to the cross probability, the adjacent chromosomes are genetically exchanged at the cross position to form a new chromosome.
(5) Mutation: select the position of chromosome based on the probability and decide whether to mutate.

3.2 Hybrid Algorithm Design Optimization Process

Due to the randomness of the samples generated by the system, the target value may be smaller than the actual value, so that the result may not be objective. In order to solve this problem, we can consider the samples as a whole and extract multiple groups of samples for optimization [14]. At the same time, a large number of samples are selected to test the results of the selected scheme, and make the final function estimation will be closer to the actual value. The specific optimization steps are as follows.

1) Randomly select the independent sample with the capacity of G, and record it as $\xi^g = (\xi_1^g, \xi_2^g, \ldots \xi_G^g)$.
2) The independent and identically distributed individual samples of group A with capacity of N were randomly selected and recorded as $\xi = (\xi_1^t, \xi_2^t, \ldots \xi_N^t)$, $t = 1, 2, \ldots A$ as the initial sample.
3) The initial population of group A randomly selected above is transferred to the model of the second stage, then solved by linear programming, assuming that there are feasible solutions and feasible schemes for group A'. Let the feasible solution of one group of samples is $v_{A,N}^t$, and the feasible scheme is $F_{A,N}^t$. the upper bound estimator is as follows:

$$\tilde{v}_{A,N} = \frac{1}{A'} \sum_{t=1}^{A'} v_{A,N}^t. \tag{16}$$

$$\delta_{v_{A,N}}^2 = \frac{1}{A'(A'-1)} \sum_{t=1}^{A'} (v_{A,N}^t - \tilde{v}_{A,N})^2. \tag{17}$$

4) According to the solution idea of reference [19], Calculate the variance of the estimate:

$$\delta_F^2 = \frac{1}{G'(G'-1)} \sum_{g=1}^{G'} (F - \tilde{F})^2 \tag{18}$$

5) Relative error of calculation estimate:

$$Gap = \tilde{F} - \tilde{v}_{A,N} \tag{19}$$

$$\sigma_{Gap}^2 = \delta_{\tilde{v}_{A,N}}^2 + \delta_F^2 \tag{20}$$

Calculate the error of the objective function, if the error does not meet the requirements, the sample size should be increased to continue the optimization operation.

4 Numerical Examples

4.1 Validation Model

In order to verify the validity of the model, a power battery recycling network is considered. Suppose There are 3 consumer clusters, 3 s-hand markets, 3 metal recycling centers and one waste treatment station, 6 alternative electric vehicle dismantling companies, 6 alternative power battery recycling centers and 8 alternative electric vehicle sellers. It is known that the number of recovered power batteries in the recovery center obeys normal distribution, which are N (2800, 280), N (2400, 240), N (2000, 200), N (1500, 150), N (2500, 250), N(2700, 270), N (2000, 200), N (1700, 170), respectively. The quality index of the recovered power batteries follows the uniform distribution. the fixed investment cost of electric vehicle dismantling company is 280000 yuan, that of power battery dealer is 230000 yuan, and the fixed investment cost of the power battery recycling center is 250,000 yuan. The unit variable cost of the electric vehicle dismantling company is 400 yuan, and that of electric vehicle dealer is 500 yuan. In addition, the maximum storage capacity of electric vehicle dealers is 3500 tons, the maximum recovery capacity of electric vehicle dismantling company is 3500 tons, the maximum recovery capacity of the recovery center is 4500 tons. The government charges a tax rate of 15% on the recycling center. Other parameters are shown in Table 1.

Table 1. Values of model parameters.

Parameter	Value	Parameter	Value
c_{ar}	(0.5, 3.5)	c_{rh}	(10, 50)
c_{dr}	(0.5, 3.5)	rf	(200, 500)
c_{rs}	(0.5, 3.5)	sf	(0.2, 0.9)
c_{rj}	(10, 50)	δ	15%

Assuming the confidence level is 0.8, the calculation program is compiled based on MATLAB R2019a platform. The experimental parameters are as follows: the initial population is 100, the crossover rate is 0.8, the variation rate is 0.05, and the number of iterations is 100 generations. Then N = 20, A = 20, G = 3000, e = 0.01.The calculation results are shown in Table 2–4.

From Table 2, seven different solutions can be obtained in a random environment. When checking under the condition of large samples, it can be seen that the total cost of candidate solution 3 is the minimum. so the total cost of candidate solution 3 can be calculated as F as an upper bound. It can be seen from Table 3, in the random environment, the error of satisfactory solution 3 is E = (713.17−686.51)/686.51 = 0.04 > 0.01, which is beyond the allowable range. Therefore, it is necessary to increase the capacity of random samples and carry out optimization again. The calculation results are shown in Table 4. The error is E = (713.17−707.31)/707.31 = 0.008 < 0.01, the solution 3 can be selected, so it's better to select alternative addresses 2 and 6 as electric vehicle seller, and choose No. 1, 6 alternative automobile disassembly company, No. 4, 7 and No. 8 alternative addresses are selected for the power battery recycling center.

Table 2. Comparison of satisfactory solutions in two environments with confidence of 0.8.

Environment	Solution	a						d						r							
		1	2	3	4	5	6	1	2	3	4	5	6	1	2	3	4	5	6	7	8
Ran-dom	1	1	0	0	1	0	1	1	0	0	0	1	0	0	0	0	1	0	0	1	1
	2	0	0	0	1	0	1	0	0	1	0	0	1	0	0	0	1	0	0	1	1
	3	0	1	0	0	0	1	1	0	0	0	0	1	0	0	0	1	0	0	1	1
	4	0	0	1	0	0	1	0	0	1	0	1	0	0	0	0	1	0	0	1	1
	5	1	0	0	1	0	1	1	0	0	0	1	0	0	0	0	1	0	0	1	1
	6	0	1	0	0	1	1	1	1	0	0	0	0	0	0	1	1	0	0	0	1
	7	1	0	1	0	0	0	0	0	1	0	0	1	0	0	1	1	0	0	0	1
Deter-mine	8	1	1	0	1	0	0	0	0	0	1	0	1	0	0	0	1	0	0	1	1

Table 3. Test results of large samples (10000 yuan).

Environment	Solution	Average total cost	Standard deviation	Applicable probability	Gap	Estimated variance
Random	1	722.11	0.78	0.92	35.60	1.11
	2	714.66	0.76	0.88	28.15	1.09
	3	713.17	0.79	0.91	26.66	1.12
	4	715.39	0.74	0.93	28.88	1.06
	5	723.60	0.78	0.84	37.09	1.14
	6	724.10	0.75	0.93	27.59	1.07
	7	714.18	0.77	0.92	27.67	1.13
Determine	8	754.16	-	1.00	67.65	0.50

Table 4. Calculation results after expanding sample size.

N	Average total cost	Gap	The estimated variance of gap
80	707.31	7.35	0.89
100	718.48	5.31	0.78
150	710.83	2.34	0.73

4.2 Analysis on the Influence of Tax Policy

Sensitivity analysis method is used to study the influence of government tax relief policy on the location planning and cost of reverse logistics network. Assuming that the random parameter is a fixed value, the tax rate of government reduction and exemption is increased from 0 to 0.05 by 0.01 step. According to the genetic algorithm, the decision-making schemes of each facility can be obtained as shown in Table 5 and Table 6 below.

Table 5. Facility layout decisions at different levels.

δ	a						d						r							
	1	2	3	4	5	6	1	2	3	4	5	6	1	2	3	4	5	6	7	8
0.15	1	1	0	1	0	0	0	0	0	1	0	1	0	0	0	1	0	0	1	1
0.14	0	1	1	1	1	0	0	0	1	1	0	0	0	0	0	1	0	0	1	1
0.13	0	1	0	1	1	1	1	1	0	0	0	0	0	0	0	1	0	0	1	1
0.12	1	0	0	0	0	1	0	1	1	0	0	1	0	0	1	1	0	0	0	1
0.11	1	1	0	1	1	0	0	0	0	1	0	1	0	0	1	1	0	0	0	1
0.10	0	0	1	1	1	0	0	1	0	1	0	0	0	0	1	1	0	0	1	0

Table 6. The rate of tax cost in the total cost of recycling center under different levels.

δ	Proportion	δ	Proportion
15%	3.62%	12%	2.92%
14%	3.38%	11%	2.68%
13%	3.15%	10%	2.44%

It can be seen from Table 5, with the change of tax policy, the layout of alternative manufacturers will change. From the Table 6, the ratio of tax cost to the total cost of the recovery center in the reverse logistics structure increases with the increase of tax rate. This means that when the government adopt the corresponding tax relief policy, which will not only reduce the total cost of the supply chain, but also change the location of stations. Therefore, the government should adopt appropriate tax policies according to the current situation, which can encourage more enterprises to recycle used power batteries, so as to make the whole system of power battery recycling more perfect.

5 Conclusion

Consider the uncertainty in the supply chain network, in this paper, we design a reverse logistics network of power battery recycling. A two-stage mixed integer linear programming model is established, and a hybrid algorithm based on SAA and genetic algorithm is designed. The results show the mixed-integer linear programming model considering the stochastics conditions is more effective, and the sensitivity analysis of tax reduction and exemption policies is carried out. It is found that the model is more sensitive to the tax policies adopted by the government. Therefore, the model in this paper can reduce the decision-making risk of uncertain factors on used power battery recycling network, and provide good reference value for rational planning of the used power battery recycling network structure.

In the future, we can expand the network structure, introducing forward logistics on the basis of reverse logistics, so as to study the design of closed-loop network, and focus on multi-transportation methods of power battery recycling.

Acknowledgements. The authors acknowledge the support from Philosophy and Social Sciences Planning Foundation of Zhejiang Province (no. 20NDJC114YB), National Natural Science Foundation of China" (no. 71972172; no. 71801199), Humanity and Social Science Planning Foundation of Ministry of Education of China (no.19YJA630101).

References

1. Fu, X.Y., Zhu, Q.H., Zhao, T.L.: Selection of the recovery channel based on recovery price competition between the reverse supply chains. J. Chin. J. Manage. Sci. **22**, 72–79 (2014)
2. Duan, Y.T., Zhao, M.X., Li, T.X.: Study on location model and algorithm of electronic waste reverse logistics network. J. Logistics Technol. **21**, 217220+350 (2013)
3. Dutta, P., Mishra, A., Khandelwal, S., Katthawala, I.: A multiobjective optimization model for sustainable reverse logistics in Indian E-commerce market. J. J. Cleaner Prod. **249**, 1–13 (2020)
4. Gao, J.H., Wang, R., Wang, H.Y.: Closed-loop supply chain network design under carbon subsidies. J. Comput. Int. Manuf. Syst. **21**, 3033–3040 (2015)
5. Yao, H.L., Wang, X., Huang, J.B.: Mode of new energy automotive battery reclamation with restriction of extended producer responsibility. J. Sci. Technol. Manage. Res. **18**, 84–89 (2015)
6. Gu, H., Liu, Z., Qing, Q.: Optimal electric vehicle production strategy under subsidy and battery recycling. J. Energy Policy. **109**, 579–589 (2017)
7. Xie, J.P., Li, J., Yang, F.F., Liang, L.: Decision-making and coordination optimized for multi-stage closed-loop supply chain of new energy vehicle. J. J. Ind. Eng. Eng. Manage. **34**, 180–193 (2019)
8. Zhu, L.Y., Chen, M.: Research on reverse logistics model and network for used electric vehicle batteries. J. China Mech. Eng. **30**, 1828–1836 (2019)

9. Sun, Q., Shen, Y.Z., Li, S.J.: Robust optimization and design for WEEE reverse logistics network under recovery uncertainty. J. Comput. Eng. Appl. **54**, 263–270 (2017)
10. Chen, G., Wang, F., Li, X.M.: Optimal design of reverse logistics network for third-party logistics in uncertain environment. J. J. Railway Sci. Eng. **10**, 86–91 (2008)
11. Zhou, X.H., Cheng, S.J., Cheng, P.F.: Multi cycle an multi objective location planning for remanufacturing reverse logistics network under self recovery mode. J. Syst. Eng. **36**, 146–153 (2018)
12. Trochu, J., Chaabane, A., Ouhimmou, M.: A two-stage stochastic optimization model for reverse logistics network design under dynamic suppliers' locations. J. Waste Manage. **95**, 569–583 (2019)
13. Lam, H., Zhou, E.: The empirical likelihood approach to quantifying uncertainty in sample average approximation. J. Oper. Res. Lett. **45**, 301–307 (2017)
14. Li, X., Zhang, K.: A sample average approximation approach for supply chain network design with facility disruptions. J. Comput. Ind. Eng. **126**, 243–251 (2018)

The Research on Ventilation Cooling System for DLP Laser Projector

Guiqing Li[1], Zijie He[1], Jun Wang[1], and Peter Mitrouchev[2](✉)

[1] Shanghai Key Laboratory of Intelligent Manufacturing and Robotics, Shanghai University, Shanghai 200072, China
leeching@shu.edu.cn
[2] University Grenoble Alpes, G-SCOP, 38031 Grenoble, France
peter.mitrouchev@grenoble-inp.fr

Abstract. Laser projection is the mainstream projection technology in the market, but compared with other projection technologies, the laser source has higher demand on the stability of operating temperature. Therefore, the ventilation system is one of the most important components in the laser projector and its performance restricts the lifespan and stability of the laser projector. In this paper, a forced ventilation cooling system is proposed. Firstly, in order to achieve the best cooling effect, the types of ventilators, the effect of air leak, the pressure loss and wind resistance characteristics were analyzed. Then, the forward air inlet and the bottom air inlet structures were compared by the computer simulation. The results showed that the bottom air inlet structure is better than the forward one. Finally, the cooling system was applied to a laser projector to test the heat dissipation effect. The result showed that the ventilation cooling system design can meet the cooling demand.

Keywords: Laser proctor · Forced ventilation cooling system · Thermal analysis

1 Introduction

Projection display is a technology that displays image information on a screen through photoelectric conversion [1]. The Digital Light Processing (DLP) is the mainstream projection display technology in the market for its good image color, simple light path, uniform contrast and brightness, light weight, small size, and long life [2]. Therefore, many companies have increased their investment in DLP laser projector. However, laser projectors still have many technical difficulties, and thermal design is one of them.

Laser source has higher demand on the stability of operating temperature, compared with other projection technologies, because temperature fluctuations will influence the image of the laser projection. Consequently, when the temperature is higher than 50°C, the laser light source cannot work normally. The heat dissipation of the laser projector has become recognized as one of the major obstacles, which restricts the lifespan and stability of the laser projector [3]. Therefore, many domestic and foreign scholars have conducted a lot of research and achieved many results:

K. Li et al. (Eds.): LSMS 2021/ICSEE 2021, CCIS 1468, pp. 253–262, 2021.
https://doi.org/10.1007/978-981-16-7210-1_24

Younes Shabany [4] studied the thermal management of electronic devices and analyzed electronic cooling problems. Schmidt R.R, Notohardjono B.D [5] introduced the design of high-end server low-temperature cooling system and expounded some key parameters in the design. Yuancheng Xie, Zhonghong Ou [6] analyzed the heat dissipation status and new heat dissipation methods of electronic devices. Bin Zhang, Xiaoyang Tian, Guolian Song [7] studied a heat dissipation design scheme and thermal performance simulation of a forced ventilation cooling electronic chassis, which provided a reference for the optimization design of the heat dissipation system of the chassis.

In this paper, a ventilation cooling system for DLP laser projector is proposed. The cooling system is designed to reduce the ventilation wind resistance and control and contribute the airflow. With the help of ANSYS software, both the forward air inlet structure and bottom air inlet structure are simulated to obtain the better wind resistance characteristics. Consequently, the bottom air inlet structure is applied to the laser projector. Finally, a temperature test is performed to test the performance of the forced ventilation cooling system. The result shows that the temperature of components can meet the demand of temperature. This study can be further used in the cooling system of other electronic chassis.

2 The Design of Forced Ventilation Cooling System

In this section, the structure of ventilation cooling system is discussed and the pressure loss is calculated.

2.1 The Design of Ventilation Structure

Forced Air Cooling
There are three types of forced air cooling [8]:

(1) Forced air cooling with single air duct: The airflow flows along the axis of a single heating device. It has good ventilation effect and it is suitable for components with large local heat generation.
(2) Forced air cooling by exhaust fan: exhaust cooling is mainly suitable for the whole chassis with relatively dispersed heat and the heat can be discharged into the atmosphere through a special air duct. The exhaust cooling has large air flow and low air pressure.
(3) Forced air cooling by air-blower: The blow cooling is usually suitable for uneven heat in the units, and each unit needs special air duct cooling. The blow cooling has small air flow and high air pressure.

In laser projectors, the printed circuit board and laser light source are the main heating components, which means that the heat is dispersed. Therefore, exhaust fans are selected.

The Selection of Ventilator
There are two types of fans: axial fans and centrifugal fans [9] (Fig. 1).

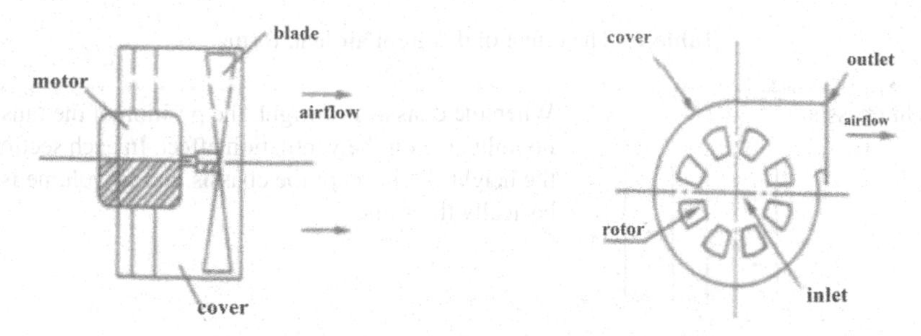

Fig. 1. Axial fans and centrifugal fans

The working principle of axial fans is that when the air enters the ventilator from the inlet, the rotation of fans is generating a pressure difference, which makes air flow out in the axial direction. Axial fans generate larger air flow and lower air pressure. The working principle of centrifugal fans is that when the air enters the ventilator from the inlet, the rotation of fans is generating a centrifugal force, which makes air flow out in the radial direction. Centrifugal fans generate smaller air flow and higher air pressure.

When choosing the ventilator, there are many factors that have to be taken into consideration, such as: air flow, air pressure, efficiency, space, noise, volume and weight, etc. The main parameters are air flow and air pressure of the fan. Because the laser projector requires higher air pressure for cooling system, centrifugal fans are selected.

The Effect of Air Leak

The effect of air leak on the air volume is shown in Table 1.

In Table 1, when there is air leak, the outlet air volume of the exhaust chassis is larger than the inlet air volume, and the outlet flow volume is increased. The outlet air volume of the blow chassis is smaller than the inlet air volume, and the outlet flow volume is decreased. Therefore, the ventilation effect of the exhaust chassis is better than that of the blow chassis.

The Arrangement of Components

In the design of the forced ventilation cooling system, the arrangement of components has a great influence on the cooling effect. When the cooling air flow rate is not large (the Reynolds number is not large), the components should be arranged in a cross pattern, which can increase the turbulence of the air flow and increase the capacity of heat dissipation capacity. For printed boards with many integrated components, turbulators can be added between the integrated components to increase the degree of air turbulence.

When the cooling airflow passes through multiple printed boards, the gap between the printed boards should be controlled at about 13mm. In order to prevent the airflow from forming a boundary layer on the surface of the printed board and affecting the heat exchange effect, turbulators should be added at an appropriate location to destroy the boundary layer, increase the degree of turbulence, and improve the performance of heat transfer.

Table 1. The effect of different air leak forms

Airtight chassis	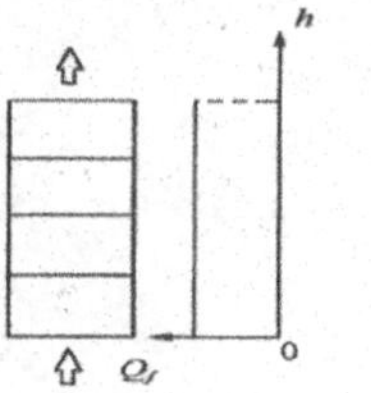	When the chassis is airtight, the position of the fans has no influence on the ventilation effect. In each section in the height direction of the chassis, the air volume is basically the same
Exhaust chassis	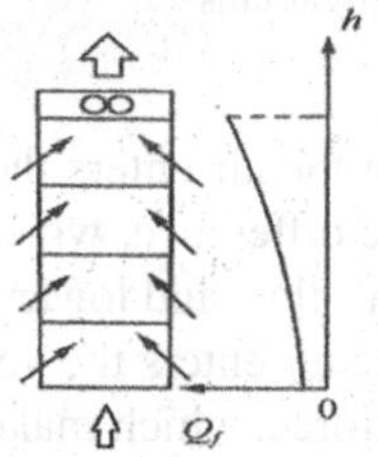	When the chassis is not airtight and an exhaust fan is installed at the air outlet, the air pressure in the chassis decreases and air flows from the gap, so the air volume gradually increases from the air inlet to the air outlet
Blow chassis	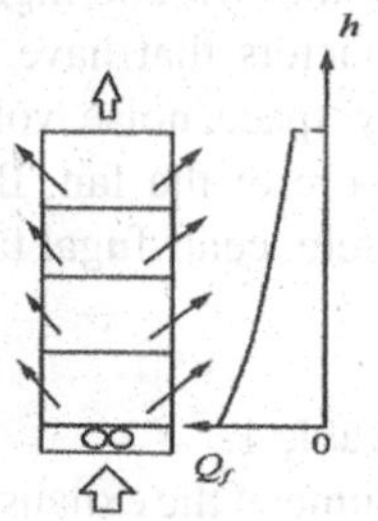	When the chassis is not airtight and an air-blower is installed at the air outlet, the air pressure in the chassis increases and air leaks from the gap, so the air volume gradually decreases from the air inlet to the air outlet

2.2 Ventilation System Calculation

The pressure loss of the ventilation system includes two types of losses: drag loss and local loss. Drag loss is caused by the resistance formed by the movement of the airflow and the friction between the airflow and the system (or pipe). The local loss is caused by the change of the airflow direction or the sudden change of the pipe section.

Drag Loss

Most gases are turbulence [10] and experiments have shown that the drag loss is proportional to the density of the fluid (ρ) and the square of its velocity (ω).

$$\Delta p_t = f \frac{l}{d_c} \cdot \frac{\omega^2 \rho}{2} (Pa) \quad (1)$$

Where: f is the coefficient of frictional Drag, which is related to Reynolds (Re) and relative roughness ε/d;l is the length of pipe (m); *ω is the velocity of air (m/s); ρ is the density of air (kg/m3); dc is the equivalent diameter (m).*

Local Loss

Local loss ΔPc is calculated by the following formula:

$$\Delta p_c = \sum \zeta \frac{\omega^2 \rho}{2} (Pa) \tag{2}$$

Where: ζ is the coefficient of local resistance; ω is the velocity of air (m/s); ρ is the density of air (kg/m^3).

In some cases, it is difficult to calculate the resistance of electronic equipment, such as integrated electronic equipment. The resistance characteristics of exhaust fans and air-blower can only be determined with the help of test [11]. Figure 2 shows the resistance curve of several chassis. Curve A is the resistance characteristic curve of a typical case, workbench or frame, where the air circulation area is 100%; curve B is the resistance characteristic curve of case (a) with a printed circuit board or micro-component, where the air circulation area is reduced by 30%; curve C is the resistance characteristic curve of case (a) with a printed circuit board or micro-component, where the air circulation area is reduced by 60%; Curve D is the resistance characteristic curve of a typical vertical case, where the air circulation area is 100%; curve E is the resistance characteristic curve of case (d) with a printed circuit board and micro-component, where the air circulation area is reduced by 75%. The resistance of various cabinets can be calculated on the corresponding curve in Fig. 2 according to the air volume.

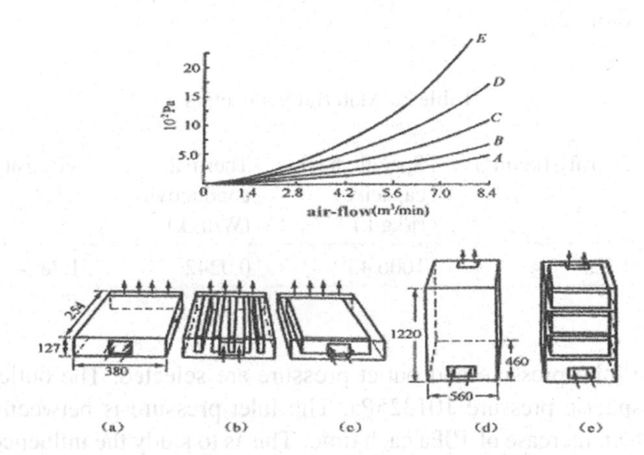

Fig. 2. Different chassis and the relationship between air-flow and resistance loss

3 Simulation of Ventilation System of Laser Projector

3.1 Model and Boundary Condition

Due to the complicated structure of the laser projector and the irregular arrangement of the ducts, it is difficult to calculate the wind resistance loss according to the existing formula, so computer simulation is used to measure the wind resistance loss of this turbulence ventilation system. According to the internal structure, two different air inlets are designed (Fig. 3) and the better one will be adopted through the comparison of the simulation. In Fig. 3(a), the inlet is in the front of the laser projector, in Fig. 3(b), the inlet is at the bottom of the laser projector.

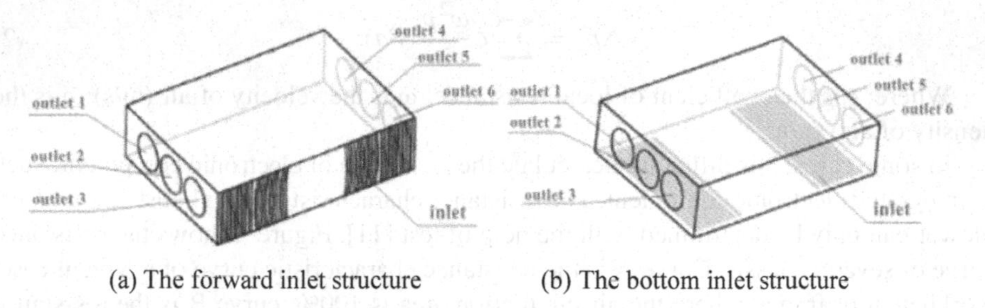

(a) The forward inlet structure (b) The bottom inlet structure

Fig. 3. Two different inlet structures

Firstly, the established physical model of the laser projector is imported into ANSYS Icem and the adoptive tetrahedral mesh is selected. Then the mesh model is imported into ANSYS Fluent and the Reynolds number is calculated ($Re = \frac{\rho v d}{\eta} \approx 7.54 \times 10^4 > 10^4$) which shows that the air-flow is turbulent. Therefore, we choose the k-ε standard equation to simulate in the Fluent.

The material properties of the air fluid are set. Some specific material parameters can be seen in Table 2.

Table 2. Material parameters

Names	Density (kg/m^3)	Specific heat capacity (j/kg·k)	Thermal conductivity (W/m·k)	Viscosity (kg/m·s)
Parameters	1.225	1006.43	0.0242	1.7894 × 10–5

Finally, the inlet pressure and outlet pressure are selected. The outlet pressure is standard atmospheric pressure 101325Pa. The inlet pressure is between 101325 and 101425Pa, with an increase of 10Pa each time. This is to study the influence of different inlet pressures on the flow field, that is, the influence of different fans on heat dissipation.

3.2 The Result of the Simulation of Ventilation Duct System

During the simulation of ventilation duct system, different inlet pressure boundary conditions were applied to obtain the velocity vector diagram of the air. The case where the outlet pressure is 101325Pa and the inlet pressure is 101335Pa (ΔPa = 10Pa) is shown in Fig. 4 and Fig. 5.

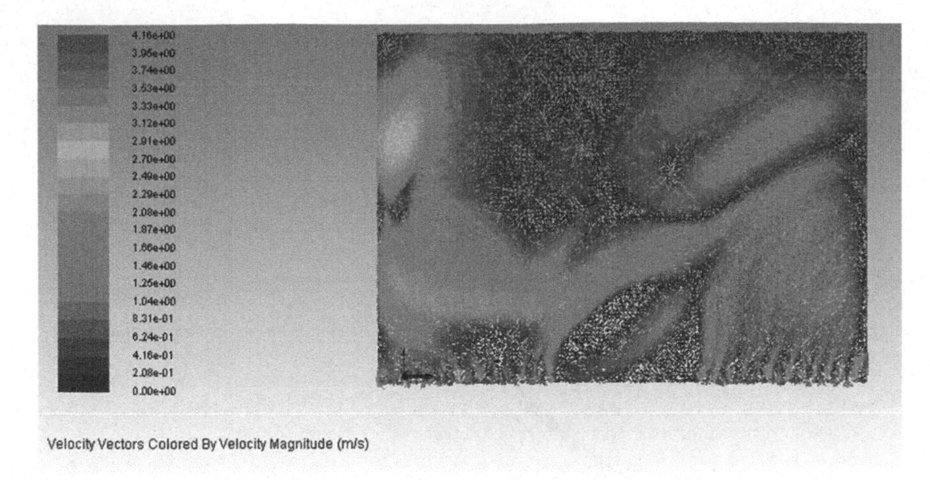

Fig. 4. The velocity vector diagram of the forward inlet structures at ΔPa = 10Pa

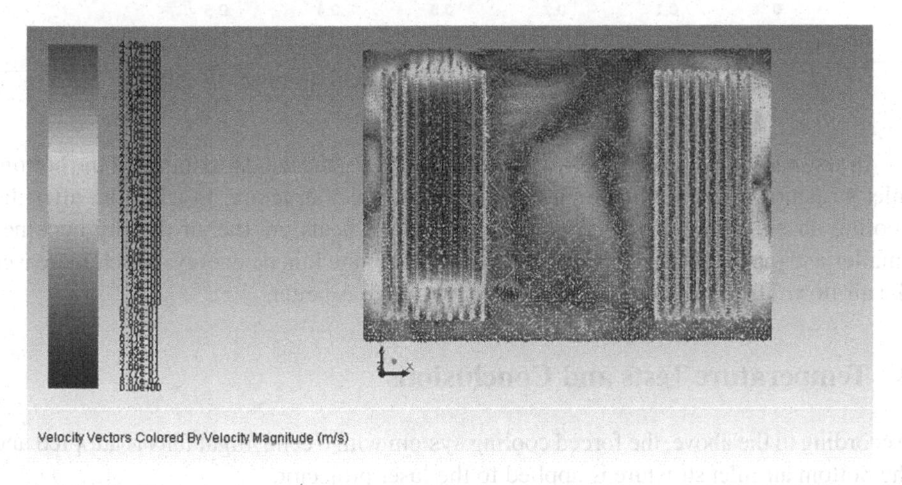

Fig. 5. The velocity vector diagram of the bottom inlet structures at ΔPa = 10Pa

Because there is lens in the middle of the projector, in Fig. 4, the flow velocity in the middle part is lower and the velocity loss in in the middle part is larger. In Fig. 5, the flow velocity in the middle part is larger than both sides. Compared with the forward wind structure, in the chassis the velocity vector diagram of the bottom wind structure is denser, which indicates that the flow velocity inside the cassis has increased. When the inlet ΔPa increases, this phenomenon becomes more obvious.

The wind resistance characteristic curve obtained by simulation is shown in Fig. 6. In the two different air inlet structures, three fans are installed in parallel on both sides. It can be seen that the wind resistance characteristic curve of the bottom inlet structure is below the wind resistance characteristic curve of the forward inlet structure. The main reason is that the bottom air inlet is larger than the forward air inlet, and the drag loss is smaller in the bottom air inlet structure.

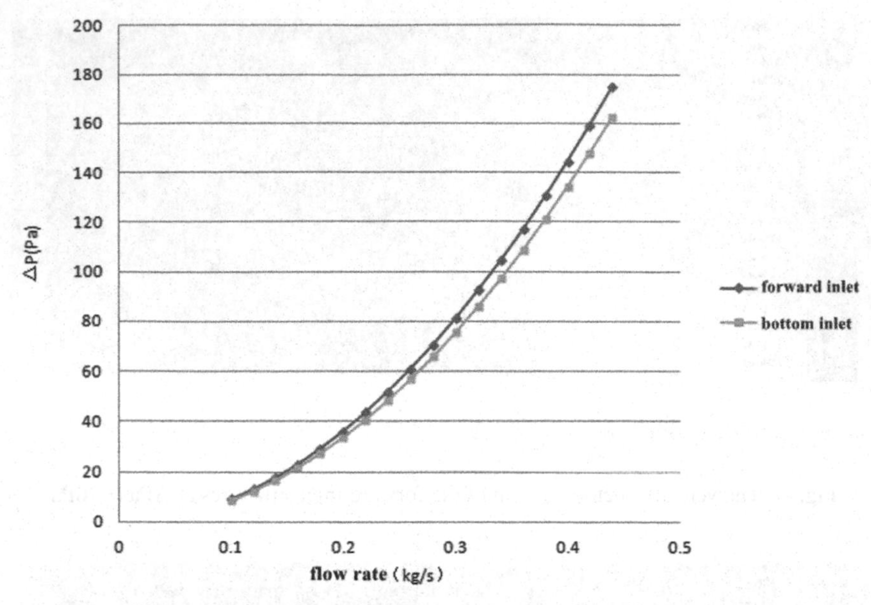

Fig. 6. The wind resistance characteristic curve of two structures

In summary, when ΔPa is between 10Pa and 100Pa, the wind resistance of the bottom inlet structure is smaller than that of the forward inlet structure. In addition, after the cooling air enters the chassis from the bottom and heats up, the air density becomes smaller and part of the heat energy is converted into the kinetic energy, which increases the air flow. Therefore, the bottom air inlet structure is better.

4 Temperature Tests and Conclusions

According to the above, the forced cooling system with 6 centrifugal fans is adopted and the bottom air inlet structure is applied to the laser projector.

The temperature of each component of the projector in Fig. 7 is tested and shown in Table 3.

The large horizontal transformer is the working power source of the laser projector, and its heat is the largest. It is placed at the left air outlet (downstream of the cooling airflow) so that the heat can be quickly discharged to the outside, which improves the heat dissipation efficiency. The purpose of the ventilation cooling system of the laser projector is to ensure the temperature of the laser source. The laser source is installed on the base of the light source, so the temperature of the left and right bases is the most important. When the ambient temperature is 25°C, the temperature of the left and right light source bases are 43.4°C and 46.2°C. When the ambient temperature the temperature of the left and right light source bases are 44.2°C and 46.8°C.

According to the temperature test, the temperature of light source is lower than 50°C, which indicates that the ventilation cooling system can meet the requirement of temperature.

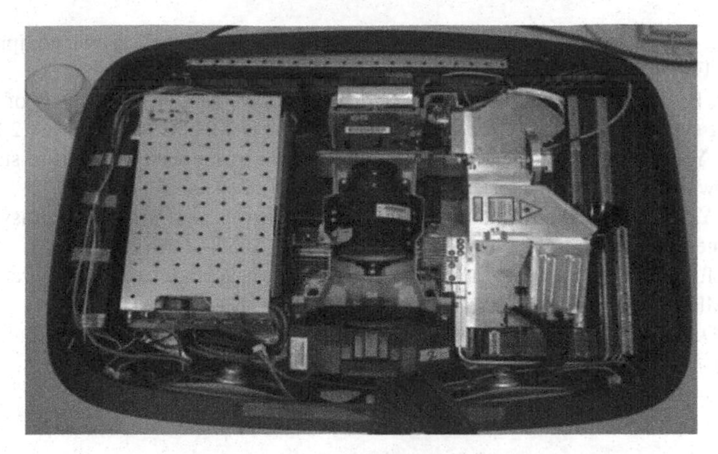

Fig. 7. The DLP laser projector

Table 3. The temperature of each component

Number	Components	Ambient temperature: 25°C	Ambient temperature: 35°C
1	Large horizontal transformer	70.0	75.0
2	Cooling plate for DMD	50.7	51.7
3	Longitudinal transformer	53.2	55.6
4	The exterior of the right light source	45.3	48.2
5	The bottom of the right light source	46.2	46.8
6	The bottom of the left light source	43.4	44.2
7	The copper of left light source	44.3	46.5
8	Horizontal transformer	49.0	52.0
9	capacitor	49.6	50.0
10	lens	35.2	36.5

References

1. Lee, J., Bagheri, B., Kao, H.-A.: A cyber-physical systems architecture for industry 4.0-based manufacturing systems. J. Manuf. Lett. **3**, 18–23 (2015)
2. Chen, H., Yuan, D., Zhang, F.: Projection display technology. J. AV Technol. (010), 59–63 (2008)
3. Tingbiao, H.: Introduction and application of digital light processor projection display technology. J. Telecommun. Electr. Power Syst. **24**(11), 32–34 (2003)
4. Shabany, Y.: Heat Transfer: Thermal Management of Electronics Beijing: China Machine Press (2013)
5. Schmidt, R.R., Notohardjono, B.D.: High-end server low-temperature cooling. J. IBM J. Res. Dev. **46**(6), 112–115 (2002)

6. Xie, Y., Ou, Z.: Development of heat dissipation technology for electronic equipment. Ship Electr. Eng. (2019)
7. Zhang. B., Tian, X., Song, G.: Design and performance simulation analysis of forced air cooling system for electronic chassis. J. Technol. Innov. Appl. **319**(27), 87–89 (2020)
8. Wang, Y., Vafai, K.: Transient characterization of flat plate heat pipe during star tup and shutdown operation. J. Int. J. Heat Mass Trans. **43**(15), 2641–2655 (2000)
9. Wang, Y., Vafai, K.: Experimental investigation of the thermal performance of an asymmetrical flat plate heat pipe. J. Int. J. Heat Mass Trans. **43**(15), 2657–2668 (2000)
10. Keng, T.B.: Study of multiple heat sources on a flat plate heat pipe using point source approach. J. ASME. Fluids Eng. Div. **250**, 175–182 (1999)
11. Zhao, D.: Thermal Design of Electronic Equipment. Publishing House of Electronics Industry (2009)

Velocity Prediction Based Model Predictive Control for Energy Management in Land and Air Vehicle with Turboshaft Engine

Zhengchao Wei[1], Yue Ma[1,2](✉), Changle Xiang[1], and Dabo Liu[2]

[1] Beijing Institute of Technology, Beijing 100081, China
[2] Beijing Institute of Technology Chongqing Innovation Center, Chongqing 401120, China

Abstract. In this paper, a land and air vehicle equipped with hybrid power unit with turboshaft engine was introduced, and a velocity prediction-based model predictive control (VPMPC) for the energy management strategy (EMS) was proposed for the vehicle. Firstly, based on the experiment data, the modeling approach based on data driven method was adopted to obtain the mathematical model of turboshaft engine. Besides, the models of generator, battery, motors were established. The deep learning method was adopted to design power predictor by training with random integrated driving cycles to improve the accuracy of the prediction model for the model predictive control (MPC). Subsequently, the EMS based on MPC using the power predictor was introduced to regulate the state of charge of battery and the exhaust gas temperature of turboshaft engine. The compared simulation results of different weight coefficients for proposed EMS were also discussed. The simulation results showed the high accuracy of mathematical model of turboshaft engine, and the effectiveness of the proposed EMS is demonstrated.

Keywords: Hybrid Power Unit (HPU) · Turboshaft engine · Land and air vehicle · Velocity prediction based on deep learning · Velocity prediction based Model Predictive Control (VPMPC)

1 Introduction

The land and air vehicle is a promising solution orienting the future transportation, which is capable of running on the land by wheels like conventional vehicle, and fly in the air by rotary-wing or fixed-wing aircraft in air mode. Consequently, the integration of the multiple functions of land and air vehicle contributes to relieving the traffic pressure and improving traffic efficiency [11]. While, more driving power is required for the land and air vehicle to meet the demand from the multiple situations compared to the conventional vehicles. Generally it is believed that the ongoing use of fossil fuel resources leads to the main environmental problems [1]. If the full driving power of land and air vehicle is still provided by the conventional combustion engine which use fossil fuel without any improvement, it makes no contribution to energy conservation and pollution reduction. Therefore, it is crucial to finding alternatives to fossil fuels in aviation now than ever. Hybrid and pure electric driving solutions have been applied in the automotive industry

K. Li et al. (Eds.): LSMS 2021/ICSEE 2021, CCIS 1468, pp. 263–273, 2021.
https://doi.org/10.1007/978-981-16-7210-1_25

for many years and the great success have been gained in the land-based vehicles, Recently these environmentally friendly solutions are being exploited for unmanned aerial vehicles and general aviation aircraft and some research results about electric and hybrid power unit power demand estimation, reduction of required power and battery sizes have been obtained [2, 3, 12]. Moreover, before the energy density of storage system can be increased greatly, the hybrid power unit (HPU) is the excellent solution and it has more advantages in the endurance mileage.

Compared to the piston engine, the turboshaft engine has the characteristics of high power to weight ratio especially when it concerns high power requirement, and these advantages make the turboshaft engine has preferable performance in HPU for the land and air vehicles. An accurate engine model which can reflects the characteristics is necessary for research on HPU. Compare to the modelling by mechanism or neutral network [6, 7], the data driven method can use the experiment data to acquire the steady-state and dynamics parameters, and it is effective for the modeling of turboshaft engine, which has the advantage of fast calculation speed and high accuracy when the internal parameters of engine are not concerned.

The appropriate energy management strategies can greatly develop the potential of HPU [2, 8, 9]. Among optimization-based methods, model predictive control (MPC) method can be solved online by predicting the required parameters in the finite time horizon of the future unlike dynamic programming (DP). However, the conventional MPC method ignores possible changes to the power demand during the prediction horizon, and it can lead to inaccurate prediction model. The power demands closely relate to the velocity of the vehicles, Hence the velocity prediction method is focused on in this paper. Among several speed prediction methods [4, 5], the deep learning prediction method is a potential candidate due to its high accuracy. This paper proposes a power demand predictor with the deep learning neural network model to forecast the future power demands over the prediction horizon MPC.

This paper is organized as follows. In Sect. 2, the distributed structure of land and air vehicle is introduced and the models of components including turboshaft engine are established. In Sect. 3, energy management problem is introduced, the velocity prediction method based on deep learning for MPC is proposed. In Sect. 4, the accuracy of the turboshaft engine model based on data driven method is shown compared to experiment data, and the performance of the proposed deep learning prediction based MPC is presented. Besides, the affects of different weight coefficient in multi-objective function of the proposed EMS on control performance are also discussed. Some conclusions are drawn in the final Section.

2 Configuration and Modeling

2.1 Configuration

The distributed structure of land and air vehicle with HPU in Fig. 1 (a) is designed and produced by Vehicle Research Center in Beijing Institute of Technology. The configuration of HPU and the topological structure of vehicle are shown in Fig. 1 (b). The turboshaft engine and the generator are connected mechanically as the main energy source of HPU, and the power battery is used as an auxiliary energy source for voltage stabilization and power compensation. The electric power from HPU is transmitted to driving motors for wheels and rotor wings to meet the power requirement.

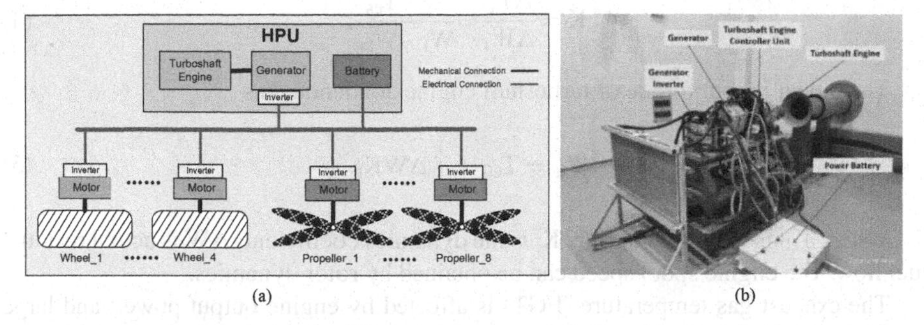

Fig. 1. (a) Is description of the land and air vehicle with HPU. (b) is configuration of HPU

2.2 Modeling

Because of the feature of application environments, the vehicle usually runs in land mode at most of working time. The research about EMS of the vehicle with HPU under the complex road working condition is concerned, therefore the models including turboshaft engine, dynamic battery, electric generator, distributed drive motors should be established for land mode.

Modeling of Turboshaft Engine Based on Data Driven Method. The turboshaft engine starting experiment and load experiment have been conducted, and the experiment results are collected. According to turboshaft engine theory, at the constant atmospheric temperature and pressure, the output rotation speed, torque, power, EGT of the engine are approximately determined by the fuel flow when the vane angles of compressor and turbine do not change. According to experiment data, when the engine output torque is stable at around certain value, the fuel flow also stabilizes at around certain value. Based on the experiment data, the steady-state fuel flow is described by

$$W_{fs} = f_{load}(T_e). \tag{1}$$

Where T_e is the engine output torque in engine load process. W_{fs} are the steady-state fuel flow. When the actual fuel flow changes from the "steady-state point", the actual outputs such as engine rotate speed also change from the "steady-state point" into dynamic state. The difference between actual fuel flow and steady-state fuel flow usually is known as the actual residual fuel flow. The actual residual fuel flow can be expressed as

$$\Delta W_f = W_f - W_{fs}. \tag{2}$$

Where W_f is actual fuel flow. The corresponding change of output torque of turboshaft engine can be expressed as

$$\Delta T_e = T_e - T_{es}. \tag{3}$$

Where T_e is actual output torque, T_{es} indicates steady-state output torque. The dynamics coefficient K_t can be described respectively as follows

$$K_t = \frac{\Delta T_e}{\Delta W_f} = \frac{T_e - T_{es}}{W_f - W_{fs}}. \tag{4}$$

The model output torque of turboshaft engine are denoted as

$$T_{et} = T_{e0} + \int_0^t \Delta W K_t. \tag{5}$$

Where T_{e0} is the initial torque, K_t is the dynamics coefficient, ΔW is actual residual fuel flow. The engine spool speed can be obtained by rotor dynamics.

The exhaust gas temperature (EGT) is affected by engine output power, and large power means high EGT. The dynamics model of EGT is regarded as first order inertia plant and the parameters of model can be obtained from the experiment data.

Modeling of Electric Generator and Motor. The generator or drive motor control the output torque on the motor rotor according to the reference torque command, and the model can be is regarded as first order inertia plant [8].

Modeling of Power Battery. Considering the practicability of model, the internal resistance model is applied to simulate the characteristics of the battery [8].

3 Velocity Prediction Based MPC (VPMPC) for EMS

3.1 Problem Statement

According to the power demand analysis during vehicle operation, the power demands P_{de} of the vehicle can be expressed as:

$$P_{de} = P_{resist\_env} + P_{acc} = P_{e-g} + P_{batt}. \tag{6}$$

Where P_{resist_env} is environment resistance power, P_{acc} is acceleration resistance power, P_{batt} is the power of battery, $P_{e\text{-}g}$ is the power of generator. To achieve the low fuel consumption and good performance, how to effectively distribute the power between $P_{e\text{-}g}$ and P_{batt} is focused on. The operation process of the proposed VPMPC-based EMS is displayed in Fig. 2. The proposed EMS method adopts the power predictor with deep neural network model to obtain the future change of power demands. Subsequently the predictive power demands and the current vehicle states including the state of charge (SOC) and exhaust gas temperature (EGT) of turboshaft engine are delivered into MPC controller to achieve the optimal control command. Different from the conventional MPC which assumes the power demands to be constant over the prediction horizon, the accurate power demands from the power predictor can improve the accuracy of the prediction model in MPC controller for HPU.

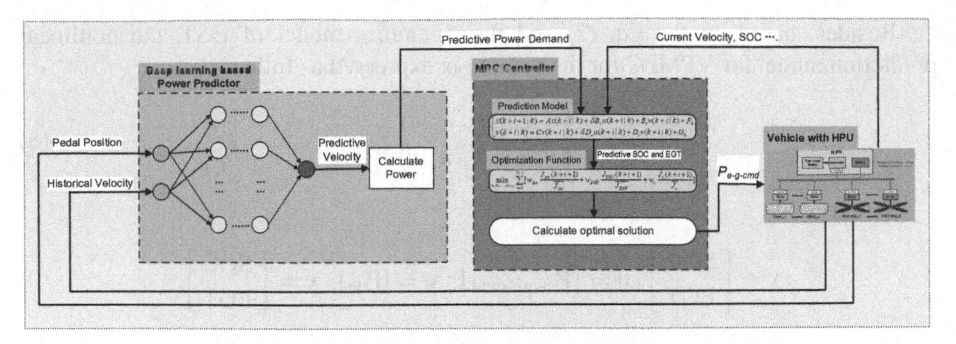

Fig. 2. Velocity prediction-based MPC for energy management scheme

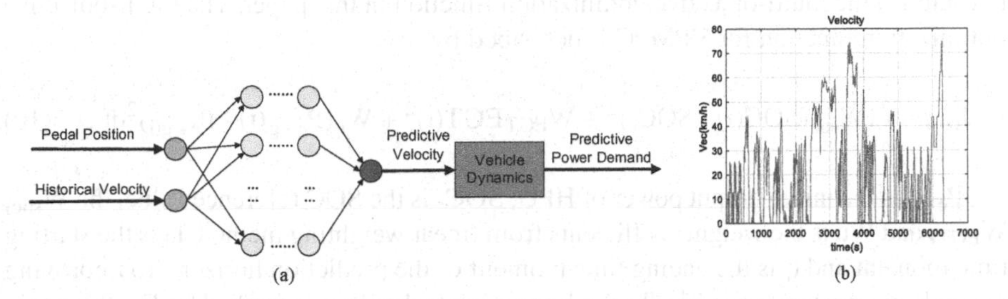

Fig. 3. (a) Is Deep learning based velocity prediction scheme and (b) is training sample

3.2 Deep Learning Based Velocity Prediction

According to Sect. 3.1, the future change of power demand P_{de} plays a vital roles on MPC method, and it is usually determined by the vehicle velocity. In this paper, the deep neural networks learning is adopted to predict the velocity due to its high accuracy. The deep neural networks predictor (DNNP) uses the historical velocity and the acceleration/brake pedal position as inputs, to predict the future velocity in upcoming period of time. The predictive power demands in upcoming period of time can be obtained by the vehicle dynamics calculation. The proposed power predictor scheme is shown in Fig. 3 (a). Training sample data for DNNP are composed of the 5 driving cycles power of CSHVR, NEDC, 1015_6PRIUS, HWFET and UNIF01, and these 5 driving cycles are arranged randomly in each training sample to increase training information, as shown in Fig. 3 (b).

3.3 Prediction Model and Objective Function for VPMPC

According to Sect. 2.2, the dynamic characteristics of the battery SOC is described by

$$\dot{SOC}(t) = \frac{-U_{ocv}(t) + \sqrt{U_{ocv}^2(t) - 4R_{int}(t)(P_{de}(t) - P_{e-g}(t))}}{2R_{int}(t)Q_{batt}}. \quad (7)$$

Besides, based on the Eq. (7) and the dynamics model of EGT, the nonlinear prediction model for VPMPC for the vehicle is expressed as follows:

$$\begin{cases} \dot{x} = f(x, u, v) \\ y = g(x, u, v) \end{cases}. \tag{8}$$

Where

$$x = \begin{bmatrix} SOC \\ EGT \end{bmatrix}, u = [P_{e-g-cmd}], v = [P_{de}], x = \begin{bmatrix} SOC \\ EGT \end{bmatrix}. \tag{9}$$

x is state variable, u is control variable, v is measured input variable, and y is output variable. Different from the works in references [8, 9], linear weighting method is applied to establish the multi-objective optimization function in this paper. The multi-objective optimization function for VPMPC is described by

$$J_{cost} = \int_{t_0}^{t_f} (W_{soc}(SOC(t) - SOC_r)^2 + W_{EGT}EGT(t)^2 + W_u(P_{e-g}(t) - P_{e-g0})^2 dt. \tag{10}$$

$P_{e\text{-}g0}$ is the initial output power of HPU, SOC_r is the SOC reference value, the W_{soc}, W_{EGT}, and Wu is the weight coefficients from linear weighting method. t_0 is the starting time moment and t_f is the ending time moment of the prediction horizon. To improving the real-time performance, The Taylor Expansion technology is applied for Eq. (8) obtain the linear model which is expressed as

$$\begin{cases} \dot{x} = Ax + B_u u + B_v v + F_0 \\ y = Cx + D_u u + D_v v + G_0 \end{cases}. \tag{11}$$

Where

$$G_0 = g(x_0, u_0, v_0) - Cx_0 - D_u u_0 - D_v v_0, \; F_0 = f(x_0, u_0, v_0) - Ax_0 - B_u u_0 - B_v v_0. \tag{12}$$

Here A, B_u, B_v C, D_u, D_v are the derivative of Eq. (8) in current state variables (u_0, x_0, v_0). And the linear prediction model can be discretized, and the corresponding discrete objective cost function is expressed as

$$J = \sum_{i=0}^{N-1} [W_{soc}(SOC(k+i+1) - SOC_r)^2 + W_{EGT}EGT(k+i+1)^2 + W_u(P_{e-g}(k+i+1) - P_{e-g0})^2]. \tag{13}$$

Where k represents the current step, k + i represents i-th step in prediction horizon. The interior-point method can be applied to solve the Eq. (13) to obtain the control sequences.

4 Results and Discussion

4.1 Simulation Validation of Turboshaft Engine Dynamics Model

For verifying the accuracy of the turboshaft engine model, the simulation output results of turboshafte engine model is compared to turboshaft engine experiment data in Fig. 4.

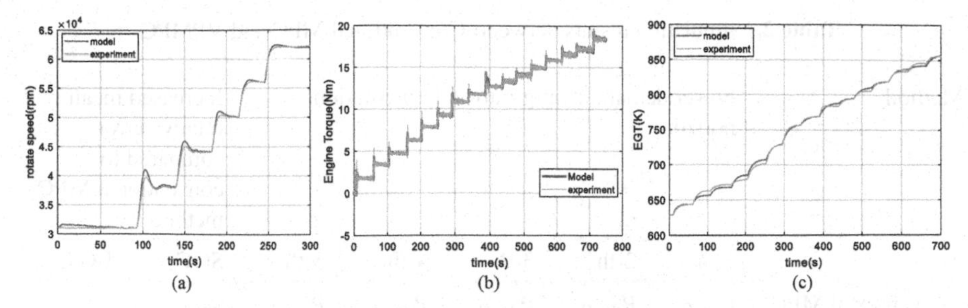

Fig. 4. (a), (b), (c) are the rotate speed, torque results and EGT results respectively

The simulation input of turboshaft engine model is experiment fuel flow data. As shown in Fig. 4 (a), the maximum rotate speed relative error is 4.2% in the engine starting process; As shown in Fig. 4 (b), the maximum output torque relative error is 5% in engine load process. Furthermore in Fig. 4 (c), the maximum EGT relative error is 1.7%. The simulation results shows that the proposed dynamics coefficient modeling method in this paper has good accuracy and is very efficient and practical for the turboshaft engine dynamics model.

4.2 Simulation of EMS Based on VPMPC

To verify the performance of the proposed EMS based on VPMPC, the simulation for the vehicle model with HPU in land mode is established in MATLAB/SIMULINK based on the subsystem models including turboshaft engine, generator, battery, distributed motors. Besides, The LA92 driving cycle is used to offer power demand. The main parameters of model are shown in the following Table 1.

Table 1. Main parameters of HPU

Parameters	Value
Rated power of turboshaft engine (kW)	120
Rated speed of turboshaft engine (rpm)	62000
Rated power of generator (kW)	116
Rated speed of generator (rpm)	5800
Rated Voltage of battery (V)	504
Rated capacity of battery (Ah)	6

Comparison between Conventional MPC and VPMPC. In this paper, the prediction horizon is set as the future 5 time moments. P_{dep_x} is supposed to be the predictive power demand in x-th time moment in the prediction horizon. The simulation mean relative deviation of SOC and EGT are compared in Table 2. Compared to conventional MPC, the mean relative error of SOC from VPMPC is smaller than that for conventional MPC.

Table 2. Simulation results between Conventional MPC and VPMPC

Method	Power demand configuration in prediction horizon					Increased mean relative error compared to conventional MPC method (%)	
	1-th	2-th	3-th	4-th	5-th	SOC	EGT
Conventional MPC	P_{de_0}	P_{de_0}	P_{de_0}	P_{de_0}	P_{de_0}	–	–
VPMPC	P_{dep_1}	P_{dep_2}	P_{dep_3}	P_{dep_4}	P_{dep_5}	-0.0949	-0.0026

Besides, the mean relative error of EGT from VPMPC is also smaller. The results show that compared to conventional MPC, the higher control performance can be obtained by proposed VPMPC.

4.3 Weight Coefficient of VPMPC

In order to compare the affect of the different weight coefficient in the multi-objective function in Eq. 25, firstly W_{soc} and W_u, the W_{EGT} is set as the following Table 3, and the simulation tests are carried out, the results are also shown in Table 3. In this paper, for comparing the performance of the control targets, the following 5 indexes are introduced: the maximum relative error (MARE) of SOC from the reference value, the mean relative error (MERE) of SOC from the reference value, the MARE of EGT from the reference value, the MERE of EGT from the reference value.

Table 3. Simulation results comparison

No.	W_{soc}	W_u	W_{EGT}	MARE of SOC (%)	MERE of SOC (%)	MARE of EGT (%)	MERE of EGT (%)
1	0.1	0.8	0.1	–	–	–	–
2	0.3	0.6	0.1	–	–	–	–
3	0.5	0.4	0.1	95.0221	52.9644	18.0913	3.2712
4	0.7	0.2	0.1	44.8893	22.6797	20.5868	3.7511
5	0.1	0.6	0.3	–	–	–	–
6	0.1	0.4	0.5	–	–	–	–
7	0.1	0.2	0.7	–	–	–	–

As shown in Table 3, that the simulation tests which adopt the No. 1, No. 2, No. 5,No. 6 and No. 7 configuration of weight coefficients cannot complete the entire driving cycle due to the too low SOC which reach the low limit. The simulation tests which adopt the No. 3, and No. 4 configuration suggests that the small weight coefficient W_{soc} and large

Table 4. Simulation results comparison

No.	W_{soc}	W_u	W_{EGT}	MARE of SOC (%)	MERE of SOC (%)	MARE of EGT (%)	MERE of EGT (%)
1	0.1	0.85	0.05	–	–	–	–
2	0.3	0.65	0.05	–	–	–	–
3	0.5	0.45	0.05	–	–	–	–
4	0.6	0.35	0.05	79.79	43.36	17.87	3.41
5	0.7	0.25	0.05	56.50	29.48	19.18	3.62
6	0.8	0.15	0.05	35.56	17.59	20.68	3.80
7	0.9	0.05	0.05	18.67	6.60	22.81	3.96

W_{EGT} means the restriction on SOC is reduced and more power from power battery is used to satisfy the requirements from driver, which contributes to the low EGT. However, and too small W_{soc} leads to the over-discharging of power battery easily. Although the larger W_{soc} can lead to higher EGT, the increase in relative error of EGT is lower than the decrease in relative error of SOC, Therefore large W_{soc} and small W_{EGT} should be applied for VPMPC-based EMS of vehicle.

Therefore, in order to determine the final weight coefficient, the W_{soc}, W_{EGT}, and W_u are set as following Table 4. The simulation results are also shown in Table 4. When the $W_{soc} = 0.90$, $W_{EGT} = 0.05$, and $W_u = 0.05$, the control performance is acceptable. Compared to the results of other configurations, there are a few increase in the MARE and MERE of EGT, however there are more decrease in the MARE and MERE of SOC. Therefore, the weight coefficients of multi-object cost function of VPMPC for EMS are set as $W_{soc} = 0.90$, $W_{EGT} = 0.05$, and $W_u = 0.05$.

Comparison between Rule-based MPC and VPMPC. To better demonstrate the performance achieved by the proposed EMS, the power following rule-based EMS is introduced for comparisons. The simulation results are shown in Fig. 5 and Table 5. The maximum EGT of turboshaft engine by the rule-based strategy is approximate 20 degrees Kelvin higher than the maximum EGT by MPC. And the FC of engine power in rule-based EMS is larger than that in VPMPC-based EMS. Besides, the Fig. 5 (c) also shows the turboshaft engine power in rule-based EMS changes more sharply than that in VPMPC-based EMS, which can lead to the unstable working state of turboshaft engine. The larger deviation of SOC in VPMPC based EMS is tolerable considering maintaining the turboshaft engine in stable working state. The results show VPMPC based EMS can offer the better control performance for aerial vehicle with HPU.

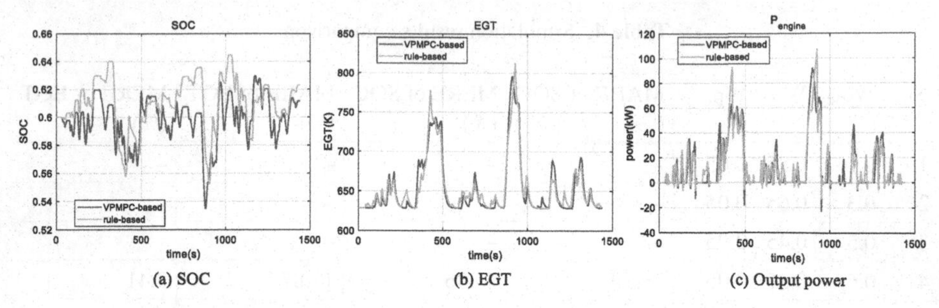

(a) SOC (b) EGT (c) Output power

Fig. 5. Simulation results comparison between rule-based control and MPC

Table 5. Simulation results comparison

Method	MARE of SOC (%)	MERE of SOC (%)	MARE of EGT (%)	MERE of EGT (%)
Rule-based	7.9236	2.8751	29.3705	4.1464
VPMPC-based	10.8278	2.0424	26.6842	4.0562

5 Conclusion

In this paper, the land mode model of a new land and air vehicle is established and the VPMPC based EMS is proposed. Firstly, a modeling method of turboshaft engine based on data driven method is proposed. And the great accuracy of turboshaft engine model compared to experiment data is demonstrated by simulation results. A power demand predictor with historical velocity and pedal position as input is designed via deep neural network learning to obtain the future power demands for MPC. The prediction model of MPC including the SOC and EGT as state variables is established. The effects of the different weight coefficients on VPMPC based EMS are discussed, and a series of simulation results are conducted to determine the appropriate weight coefficients in optimization function. The simulation results between rule-based EMS and VPMPC based EMS demonstrate the advantages of the proposed EMS.

References

1. Guillem, M.B., Nurgeldy, P., Arpad, V.: Performance analysis of hybrid electric and distributed propulsion system applied on a light aircraft. J. Energy. **241**, 118823 (2021)
2. Wangai, A.W., Nguyen, D., Rohacs, D.: Forecast of electric, hybrid-electric aircraft. J. Int. Symposiumon Electr. Aviation Auton. Syst. **44**, 26--29 (2019)
3. Economou, J.T., Tsourdos, A., Wang, S.: Design of a distributed hybrid electric propulsion system for a light aircraft based on genetic algorithm. J. AIAA Propulsion Energy (2019)
4. Mei, Y., Menglin, L., Hongwen, H., Jiankun, P.: Deep learning for vehicle speed prediction. J. Energy Procedia. **152**, 618–623 (2018)
5. Jaewook, S., Myoungho, S.: Vehicle speed prediction using a markov chain with speed constraints. J. IEEE Trans. Intell. Trans. Syst. **20**, 3201–3211 (2019)

6. Hanlin, S., et al.: Research on dynamic modeling and performance analysis of helicopter turboshaft engine's start-up process. J. Aerospace Sci. Technol. **106**, 106097 (2020)
7. Petrochenkov, A.B., Romodin, A.V., Leizgold, D.Y., Semenov, A.S.: Modeling power-supply systems with gas-turbine units as energy sources. Russ. Electr. Eng. **91**(11), 673–680 (2020). https://doi.org/10.3103/S1068371220110103
8. Xiang, C., Ding, F., Wang, W., He, W.: Energy management of a dual-mode power-split hybrid electric vehicle based on velocity prediction and nonlinear model predictive control. J. Appl. Energy **189**, 640–653 (2017)
9. Xuechao, W., Jinzhou, C., Shengwei, Q., Ya-Xiong, W., Hongwen, H.: Hierarchical model predictive control via deep learning vehicle speed predictions for oxygen stoichiometry regulation of fuel cells. J. Appl. Energy **276**, 115460 (2020)
10. Xing, Y., Huang, J., Lu, F., Yao, W.: Research on simplified real-time model of turboshaft engine. J. Aeroengine **38**, 15–18 (2012)
11. Ai, T., Xu, B., Xiang, C., Fan, W., Zhang, Y.: Modeling and multimode analysis of electrically driven flying car. In: 2020 International Conference on Unmanned Aircraft Systems, pp. 1565–1571 (2020)
12. Wangai, A., Nguyen, D., Rohacs, D.: Forecast of electric, hybrid-electric aircraft. J. Int. Symposiumon Electr. Aviation Auton. Syst. **44**, 26–29 (2019)

Networked Active Suspension Control of In-Wheel Motor Driven Electric Vehicles Under Aperiodic Sampling and Transmission Delays

Iftikhar Ahmad, Xiaohua Ge(✉), Qing-Long Han, and Zhenwei Cao

School of Science, Computing and Engineering Technologies,
Swinburne University of Technology, Hawthorn, VIC 3122, Australia
{iftikharahmad,xge,qhan,zcao}@swin.edu.au

Abstract. This paper addresses the networked H_∞ control problem for active suspension systems equipped with dynamic dampers. First, a networked active suspension control model of a 6-DOF half in-wheel motor driven electric vehicle is proposed to account for the effects of time-varying transmission delays, aperiodic sampling periods, and uncertain masses. Then, a delicate networked H_∞ state-feedback controller is developed to guarantee the asymptotic stability as well as simultaneously preserving the expected suspension performance and H_∞ disturbance attenuation performance requirements of proposed closed-loop suspension system. Finally, simulation results are provided to validate the effectiveness of the designed networked active suspension control method in both time and frequency domains.

Keywords: Active suspension system · Networked control · Electric vehicles · Dynamic dampers · Aperiodic sampling · Transmission delays

1 Introduction

Being an essential source of communication between cities, vehicles are evolving day by day. The automotive industries and researchers are continuously striving for incorporating the latest technologies in the vehicles to make them user-comfortable and environment-friendly. A promising solution to such a vehicle feature requirement is the replacement of traditional passive suspension systems and internal combustion engine operated vehicles with active suspension systems for increased comfortability and with electric motor operated vehicles for environmental friendliness, respectively.

In the in-wheel motor (IWM) driven electric vehicles, electric motors are installed into each wheel separately to generate accurate and fast torques. However, the addition of the electric motor increases the wheel mass, which greatly deteriorates the vehicle ride comfort. To reduce the influence of the IWM on the ride comfort, the motor is isolated from the wheel by a dynamic damping

K. Li et al. (Eds.): LSMS 2021/ICSEE 2021, CCIS 1468, pp. 274–284, 2021.
https://doi.org/10.1007/978-981-16-7210-1_26

system called the advanced dynamic damper (ADM) model [1]. The installation of active suspension can further improve the ride comfort of the vehicle. Substantial research has been performed to enhance the active suspension system performance by utilizing different control methods, e.g., robust H_∞ control [2], T-S fuzzy control [3,4] and linear optimal control [5]. Also, the combined effects of the IWM and active suspension system on the electric vehicle performance have been extensively studied in the literature; see, e.g., [6–9].

An in-vehicle controller area network (CAN) connects different components of the vehicle suspension system such as sensors, actuators and control units. In majority of the existing active suspension control methods, a continuous data transmission paradigm has been widely considered among system components. However, the sensor data and control commands essentially need to be scheduled and sampled before possible transmissions over digital communication networks [10]. Therefore, considerable research effort has been carried out to address the networked-active suspension system data scheduling and control problem. For example, data scheduling and control problem over CAN for uncertain vehicle active suspension system was investigated under dynamic event triggering in [11]. Several important issues, such as network-induced delays, data losses, and event-triggered transmissions, were addressed. A fuzzy control approach was utilized to resolve the problem of sampled-data control for uncertain vehicle active suspension system by Li et al. [12]. By using Lyapunov functional method, sampled-data dynamic output feedback and state feedback controllers were designed. A robust sampled-data H_∞ control approach has been adopted for vehicle active suspension system by Gao et al., [13]. Quarter vehicle active suspension in IWM equipped with ADM under dynamic event-triggered control was investigated in an in-vehicle networked environment in [9]. It is noteworthy that the results above consider only quarter vehicle model such that vehicle's body vertical motion is involved during analysis and design of the control techniques. In contrast, the half vehicle model is more practical because it empowers both the pitch angular and vertical motions of the body to be accommodated. On the other hand, it should be mentioned that only the periodic (or constant) sampling paradigm has been considered in the aforementioned results for the active suspension systems. The periodic sampling paradigm simplifies the analysis and design procedures for the vehicle active suspension systems, that further allows the classical sampled-data control theory to be easily applied. Nevertheless, sampling all the sensor signals at a constant rate may be impractical because practical samplers/sensors often possess the restricted energy and/or the system may already approach its steady state with little fluctuations. Therefore, time-varying and uncertain sampling is more preferable in a networked control context.

In this paper, we investigate an in-wheel-motor driven electric half vehicle with ADM and active suspension system, where uncertain sampling periods and time-varying transmission delays occur during vehicle data transmissions in an in-vehicle CAN setting. An uncertain dynamic model of the half vehicle active suspension (HVAS) system in a networked environment will be derived first. A

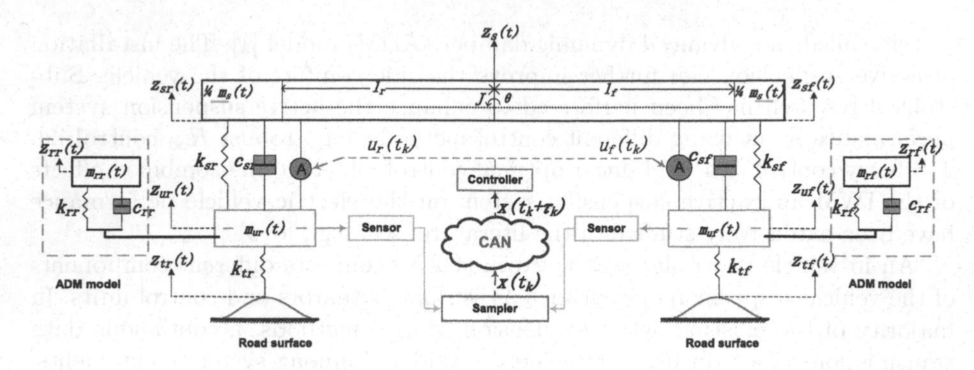

Fig. 1. Schematic of networked IWM electric half vehicle with dynamic dampers and active suspension control

fuzzy network-based state feedback controller will be designed that can guarantee the asymptotic stability of the developed active suspension model and simultaneously achieving the desired suspension performance and H_∞ disturbance attenuation performance requirements. Finally, the usefulness of the proposed networked active suspension control approach will be validated by providing some simulation results.

2 Problem Formulation

2.1 An Uncertain Half-Vehicle Active Suspension Dynamic Model

A 6-DoF in-wheel-motor driven electric half vehicle equipped with dynamic dampers and active suspension systems is considered in this study, as shown in Fig. 1. The equations of motion of the proposed HVAS can be derived as

$$m_s(t)\ddot{z}_s(t) = -f_{sf} - f_{sr}; \qquad J\ddot{\theta}(t) = -l_f f_{sf} + l_r f_{sr} \tag{1}$$

$$m_{uf}(t)\ddot{z}_{uf}(t) = -f_{uf} + f_{sf} + f_{rf}; \quad m_{ur}(t)\ddot{z}_{ur}(t) = -f_{ur} + f_{sr} + f_{rr} \tag{2}$$

$$m_{rf}(t)\ddot{z}_{rf}(t) = -f_{rf}; \qquad m_{rr}(t)\ddot{z}_{rr}(t) = -f_{rr} \tag{3}$$

where $f_{sf} = k_{sf}(z_{sf}(t) - z_{uf}(t)) + c_{sf}(\dot{z}_{sf}(t) - \dot{z}_{uf}(t)) - u_f(t)$, $f_{sr} = k_{sr}(z_{sr}(t) - z_{ur}(t)) + c_{sr}(\dot{z}_{sr}(t) - \dot{z}_{ur}(t)) - u_r(t)$, $f_{rf} = k_{rf}(z_{rf}(t) - z_{uf}(t)) + c_{rf}(\dot{z}_{rf}(t) - \dot{z}_{uf}(t))$, $f_{rr} = k_{rr}(z_{rr}(t) - z_{ur}(t)) + c_{rr}(\dot{z}_{rr}(t) - \dot{z}_{ur}(t))$, $f_{uf} = k_{tf}(z_{uf}(t) - z_{tf}(t))$, $f_{ur} = k_{tr}(z_{ur}(t) - z_{tr}(t))$, and $m_s(t)$, $m_{ur}(t)$, $m_{uf}(t)$, $m_{rr}(t)$ and $m_{rf}(t)$ are the sprung, rear and front unsprung and ADMs masses, $z_s(t)$, $z_{sr}(t)$, $z_{sf}(t)$, $z_{ur}(t)$, $z_{uf}(t)$, $z_{rr}(t)$, $z_{rf}(t)$, $z_{tr}(t)$ and $z_{tf}(t)$ are the vehicle's body, rear and front vehicle's body, suspensions, ADMs and tyres vertical displacements, k_{sr}, k_{sf}, k_{tr}, k_{tf}, k_{rr} and k_{rf} are rear and front suspensions, tyre and ADMs stiffness coefficients, c_{sr}, c_{sf}, c_{rr} and c_{rf} are rear and front suspensions and ADMs damping coefficients, J is the moment of inertia, $\theta(t)$ is the vehicle's body angular displacement at centre of gravity, l_r and l_f, are the longitudinal distances of rear and front tyres to the centre of gravity, $u_r(t)$ and $u_f(t)$ are rear and front actuators inputs, respectively.

The relationship between the vehicle's body pitch angular and vertical motions with respect to rear and front sprung masses displacements is given by $z_{sr}(t) = z_s(t) - l_r\theta(t)$ and $z_{sf}(t) = z_s(t) + l_f\theta(t)$ [14]. Denote the new state variables as $x_1(t) = z_{sf}(t) - z_{uf}(t)$, $x_2(t) = \dot{z}_s(t)$, $x_3(t) = z_{sr}(t) - z_{ur}(t)$, $x_4(t) = \dot{\theta}(t)$, $x_5(t) = \dot{z}_{uf}(t)$, $x_6(t) = \dot{z}_{ur}(t)$, $x_7(t) = z_{rf}(t) - z_{uf}(t)$, $x_8(t) = z_{uf}(t) - z_{tf}(t)$, $x_9(t) = \dot{z}_{rf}(t)$, $x_{10}(t) = z_{rr}(t) - z_{ur}(t)$, $x_{11}(t) = z_{ur}(t) - z_{tr}(t)$, $x_{12}(t) = \dot{z}_{rr}(t)$. Then we can obtain the following state-space model

$$\dot{x}(t) = A(t)x(t) + B(t)u(t) + Ew(t), \tag{4}$$

where $x(t) = [x_1(t), x_2(t), \cdots, x_{12}(t)]^T$ is the state vector, $u(t) = [u_f^T(t), u_r^T(t)]^T$ is the actual control input, $w(t) = [\dot{z}_{tf}^T(t), \dot{z}_{tr}^T(t)]^T$ is the road input, and

$$A(t) = \begin{bmatrix} a_{11} & a_{12} & a_{13} \\ a_{21} & a_{22} & a_{23} \\ a_{31} & a_{32} & a_{33} \end{bmatrix}, a_{11} = \begin{bmatrix} 0 & 1 & 0 & l_f \\ -\frac{k_{sf}}{m_s(t)} & -\frac{c_{sf}+c_{sr}}{m_s(t)} & -\frac{k_{sr}}{m_s(t)} & \frac{-l_f c_{sf}+l_r c_{sr}}{m_s(t)} \\ 0 & 1 & 0 & -l_r \\ -\frac{l_f k_{sf}}{J} & \frac{-l_f c_{sf}+l_r c_{sr}}{J} & \frac{l_r k_{sr}}{J} & -\frac{l_f^2 c_{sf}+l_r^2 c_{sr}}{J} \end{bmatrix}$$

$$a_{21} = \begin{bmatrix} \frac{k_{sf}}{m_{uf}(t)} & \frac{c_{sf}}{m_{uf}(t)} & 0 & \frac{l_f c_{sf}}{m_{uf}(t)} \\ 0 & \frac{c_{sr}}{m_{ur}(t)} & \frac{k_{sr}}{m_{ur}(t)} & -\frac{l_r c_{sr}}{m_{ur}(t)} \\ 0 & 0 & 0 & 0 \\ 0 & 0 & 0 & 0 \end{bmatrix}, a_{23} = \begin{bmatrix} \frac{c_{rf}}{m_{uf}(t)} & 0 & 0 & 0 \\ 0 & \frac{k_{rr}}{m_{ur}(t)} & -\frac{k_{tr}}{m_{ur}(t)} & \frac{c_{rr}}{m_{ur}(t)} \\ 1 & 0 & 0 & 0 \\ 0 & 0 & 0 & 0 \end{bmatrix}$$

$$a_{22} = \begin{bmatrix} -\frac{c_{sf}+c_{rf}}{m_{uf}(t)} & 0 & \frac{k_{rf}}{m_{uf}(t)} & -\frac{k_{tf}}{m_{uf}(t)} \\ 0 & -\frac{c_{sr}+c_{rr}}{m_{ur}(t)} & 0 & 0 \\ -1 & 0 & 0 & 0 \\ 1 & 0 & 0 & 0 \end{bmatrix}, a_{32} = \begin{bmatrix} \frac{c_{rf}}{m_{rf}(t)} & 0 & -\frac{k_{rf}}{m_{rf}(t)} & 0 \\ 0 & -1 & 0 & 0 \\ 0 & 1 & 0 & 0 \\ 0 & \frac{c_{rr}}{m_{rr}(t)} & 0 & 0 \end{bmatrix}$$

$$a_{12} = \begin{bmatrix} -1 & 0 & 0 & 0 \\ \frac{c_{sf}}{m_s(t)} & \frac{c_{sr}}{m_s(t)} & 0 & 0 \\ 0 & -1 & 0 & 0 \\ \frac{l_f c_{sf}}{J} & -\frac{l_r c_{sr}}{J} & 0 & 0 \end{bmatrix}, a_{33} = \begin{bmatrix} -\frac{c_{rf}}{m_{rf}(t)} & 0 & 0 & 0 \\ 0 & 0 & 0 & 1 \\ 0 & 0 & 0 & 0 \\ 0 & 0 & -\frac{k_{rr}}{m_{rr}(t)} & -\frac{c_{rr}}{m_{rr}(t)} \end{bmatrix}$$

$$a_{13} = a_{31} = \mathbf{0_{4\times4}}, B(t) = \begin{bmatrix} 0 & \frac{1}{m_s(t)} & 0 & \frac{l_f}{J} & -\frac{1}{m_{uf}(t)} & 0 & 0_{1\times6} \\ 0 & \frac{1}{m_s(t)} & 0 & -\frac{l_r}{J} & 0 & -\frac{1}{m_{ur}(t)} & 0_{1\times6} \end{bmatrix}^T$$

$$E = \begin{bmatrix} 0_{1\times7} & -1 & 0 & 0 & 0 & 0 \\ 0_{1\times7} & 0 & 0 & 0 & -1 & 0 \end{bmatrix}^T.$$

2.2 A T-S Fuzzy Model of the Uncertain HVAS System

The masses (including sprung, unsprung, and ADMs) of the HVAS system may vary according to the changes in the number of the passengers occupying the vehicle. Specifically, these masses are uncertain but bounded, i.e., $m_s(t) \in [\check{m}_s, \hat{m}_s]$, $m_{r(f,r)}(t) \in [\check{m}_{r(f,r)}, \hat{m}_{r(f,r)}]$ and $m_{u(f,r)}(t) \in [\check{m}_{u(f,r)}, \hat{m}_{u(f,r)}]$, where (f, r) stands for the front and rear masses, respectively. Applying similar T-S fuzzy rules as defined in [9], we obtain the blended T-S fuzzy model as given

$$\dot{x}(t) = \sum_{i=1}^{8} \eta_i(t)(A_i x(t) + B_i u(t) + Ew(t)), \tag{5}$$

where $\eta_1(t) = M_1(\phi_1(t)) \cdot M_2(\phi_2(t)) \cdot M_3(\phi_3(t))$, $\eta_2(t) = M_1(\phi_1(t)) \cdot M_2(\phi_2(t)) \cdot N_3(\phi_3(t))$, $\eta_3(t) = M_1(\phi_1(t)) \cdot N_2(\phi_2(t)) \cdot M_3(\phi_3(t))$, $\eta_4(t) = M_1(\phi_1(t)) \cdot N_2(\phi_2(t)) \cdot N_3(\phi_3(t))$, $\eta_5(t) = N_1(\phi_1(t)) \cdot M_2(\phi_2(t)) \cdot M_3(\phi_3(t))$, $\eta_6(t) = N_1(\phi_1(t)) \cdot M_2(\phi_2(t)) \cdot N_3(\phi_3(t))$, $\eta_7(t) = N_1(\phi_1(t)) \cdot N_2(\phi_2(t)) \cdot M_3(\phi_3(t))$, $\eta_8(t) = N_1(\phi_1(t)) \cdot N_2(\phi_2(t)) \cdot N_3(\phi_3(t))$ are fuzzy weighting functions satisfying $\sum_{i=1}^{8} \eta_i(t) = 1$ and $\eta_i(t) \geq 0$.

2.3 A Networked Controller

The system state $x(t)$ is measured and sampled at sampling points $\{t_k | k \in \mathbb{N}, t_k \in \mathbb{R}\}$ before transmitting over the in-vehicle communication network. Thus, the sampled data $x(t_k)$ is available at the networked controller side only over the interval $t \in [t_k, t_{k+1})$. During the in-vehicle network transmission, the sampled vehicular data packet $(t_k, x(t_k))$ experiences a time-varying transmission delay $\tau_k \in [\tau_m, \overline{\tau}]$ from the sensor to the controller, with $\tau_m \geq 0$ and $\overline{\tau}$ denoting the lower and upper bounds of the delay, respectively. Without loss of generality, the time-varying intervals between two consecutive sampling instants are assumed to be bounded as

$$0 \leq \underline{h} \leq t_{k+1} - t_k \leq \overline{h} < \infty, \quad t_0 = 0, \quad \forall\, k \geq 0, \tag{6}$$

where $\underline{h}$ and $\overline{h}$ are the lower and upper bounds of the time-varying sampling interval, respectively. Clearly, if the sampling interval $t_{k+1} - t_k = h$, then the traditional periodic sampling paradigm is retrieved.

We are interested in constructing and designing a networked state feedback controller of the following form

$$u(t) = u(t_k) = Kx(t_k), \quad \forall\, t \in [t_k + \tau_k, t_{k+1} + \tau_{k+1}), \tag{7}$$

where K is the control gain matrix to be designed. To facilitate the subsequent analysis and design, an artificial time delay function $d(t) = t - t_k$, $t \in [t_k + \tau_k, t_{k+1} + \tau_{k+1})$ is introduced, which satisfies

$$\begin{cases} \tau_m \leq \tau_k \leq d(t) = t - t_k < t_{k+1} - t_k + \tau_{k+1} \leq \overline{\tau} + \overline{h} \triangleq \tau_M \\ \dot{d}(t) = 1 \text{ at } t \neq t_k, \end{cases} \tag{8}$$

where τ_M denotes the maximum time span to update the instant at the ZOH.

Then, the overall fuzzy controller can be written as

$$u(t) = \sum_{j=1}^{8} \eta_j(t) K_j x(t - d(t)), \quad t \in [t_k + \tau_k, t_{k+1} + \tau_{k+1}). \tag{9}$$

2.4 Control Objectives of the HVAS System

We aim to achieve three vital control objectives as defined in [9], during the analysis and design processes to improve the suspension system performance with modified third objective as,

$$k_{tf}(z_{uf}(t) - z_{tf}(t)) = k_{tf}x_8 \leq (\frac{l_r}{l_f + l_r}\check{m}_s + \check{m}_{uf} + \check{m}_{rf})g, \tag{10}$$

$$k_{tr}(z_{ur}(t) - z_{tr}(t)) = k_{tr}x_{11} \leq (\frac{l_f}{l_f + l_r}\check{m}_s + \check{m}_{uf} + \check{m}_{rf})g. \tag{11}$$

Substituting (9) into the system (5), we obtained the following closed-loop model

$$\dot{x}(t) = \sum_{i=1}^{8}\sum_{j=1}^{8}\eta_i(t)\eta_j(t)(A_i x(t) + B_i K_j(x(t - d(t)) + Ew(t)) \tag{12}$$

for any $t \in [t_k + \tau_k, t_{k+1} + \tau_{k+1})$. The initial condition for the state $x(t)$ is defined on $[-\tau_M, 0]$ as $x(e) = \Phi(e)$ in $e \in [-\tau_M, 0]$ with $\Phi(e)$ being a continuous function.

Hence, the aim of this paper is to design a networked state feedback controller of the form (9) for the half vehicle active suspension system such that similar conditions given in [9] are guaranteed.

3 Main Results

3.1 Stability and H_∞ Performance Analysis

A sufficient condition is provided in the following theorem to guarantee the asymptotic stability of the developed active suspension system (12) under the assumed suspension performance. The proof is similar to the one provided in [9,11], and thus omitted due to space limitation.

Theorem 1. *Given positive scalars α, γ, λ, τ_m, τ_M, and control gain matrices K_j, $j = 1, \cdots, 8$, the resulting active suspension system* (12) *is asymptotically stable under the H_∞ performance index γ and assumed suspension performance, if there exist matrices $P > 0$, $Q > 0$, $R > 0$, $S > 0$ and U of appropriate dimensions such that the following inequalities*

$$\begin{bmatrix} -P & \sqrt{\lambda}C_n^T \\ \sqrt{\lambda}C_n & -\mathbf{I} \end{bmatrix} < 0\ , \quad n = 1, 2 \tag{13}$$

$$\Pi_{ij} = \begin{bmatrix} \Gamma_1 & \Gamma_2 & R & U^T B_i K_j & U^T E & C_i^T \\ * & \Gamma_3 & 0 & \alpha U^T B_i K_j & \alpha U^T E & 0 \\ * & * & \Gamma_4 & \frac{\pi^2}{4}S & 0 & 0 \\ * & * & * & -\frac{\pi^2}{4}S & 0 & (D_i K_j)^T \\ * & * & * & * & -\gamma^2 I & 0 \\ * & * & * & * & * & -I \end{bmatrix} < 0 \tag{14}$$

hold, where $\Gamma_1 = Q - R + U^T A_i + A_i^T U$, $\Gamma_2 = P - U^T + \alpha A_i^T U$, $\Gamma_3 = \tau_m^2 R + (\tau_M - \tau_m)^2 S - \alpha U - \alpha U^T$, and $\Gamma_4 = -\frac{\pi^2}{4}S - Q - R$.

Table 1. Parameters and values of the HVAS system

Parameter	Value	Unit	Parameter	Value	Unit	Parameter	Value	Unit
J	1222	kg·m^2	k_{sf}	18	kN/m	k_{rr}	32	kN/m
z_{max}	0.055	m	k_{sr}	20	kN/m	$c_{sf} = c_{rf}$	1	kN·s/m
l_f	1.3	m	$k_{tf} = k_{tr}$	200	kN/m	c_{sr}	1.1	kN·s/m
l_r	1.5	m	k_{rf}	30	kN/m	c_{rr}	1.08	kN·s/m

3.2 Networked H_∞ Controller Design

Theorem 2. *Given positive scalars α, γ, λ, τ_m and τ_M, the resulting active suspension system* (12) *is asymptotically stable under the H_∞ performance index γ and assumed suspension performance, if there exist matrices $\bar{P} > 0$, $\bar{Q} > 0$, $\bar{R} > 0$, $\bar{S} > 0$, $\bar{K}$ and $\bar{U}$ of appropriate dimensions such that the following inequalities*

$$\begin{bmatrix} -\bar{P} & \sqrt{\lambda}\bar{U}^T C_n^T \\ * & -\mathbf{I} \end{bmatrix} < 0\,, \quad n = 1, 2 \tag{15}$$

$$\bar{\Pi}_{ij} = \begin{bmatrix} \bar{\Gamma}_1 & \bar{\Gamma}_2 & \bar{R} & B_i\bar{K}_j & E & \bar{U}^T C_i^T \\ * & \bar{\Gamma}_3 & 0 & \alpha B_i \bar{K}_j & \alpha E & 0 \\ * & * & \bar{\Gamma}_4 & \frac{\pi^2}{4}\bar{S} & 0 & 0 \\ * & * & * & -\frac{\pi^2}{4}\bar{S} & 0 & \bar{K}_j^T D_i^T \\ * & * & * & * & -\gamma^2 I & 0 \\ * & * & * & * & * & -I \end{bmatrix} < 0 \tag{16}$$

hold, where $\bar{\Gamma}_1 = \bar{Q} - \bar{R} + A_i\bar{U} + \bar{U}^T A_i^T$, $\bar{\Gamma}_2 = \bar{P} - \bar{U} + \alpha\bar{U}^T A_i^T$, $\bar{\Gamma}_3 = \tau_m^2\bar{R} + (\tau_M - \tau_m)^2\bar{S} - \alpha\bar{U} - \alpha\bar{U}^T$, $\bar{\Gamma}_4 = -\frac{\pi^2}{4}\bar{S} - \bar{Q} - \bar{R}$. Then the required control gain matrices can be obtained from $K_j = \bar{K}_j\bar{U}^{-1}$.

Proof. The proof follows a similar line of that in [11], and is omitted for brevity.

4 Simulation Results

The effectiveness of the developed networked active suspension control approach is validated by examining an HVAS model of an IWM electric vehicle in this section. The parameters for the HVAS are provided in the Table 1.

The masses are considered as uncertain and assumed to vary in specific ranges, i.e., $m_s(t) = [621\,\text{kg}, 759\,\text{kg}]$, $m_{u(f,r)}(t) = [39.6\,\text{kg}, 40.4\,\text{kg}]$ and $m_{r(f,r)}(t) = [19.9\,\text{kg}, 20.9\,\text{kg}]$. The simulation results are examined under a bump road disturbance input as given in [6].

Solving the inequalities in Theorem 2, it is found that the problem is feasible for $\alpha = \lambda = 0.001$ and prescribed H_∞ performance level $\gamma = 9$. Four different networked transmission cases are evaluated to demonstrate the resulting

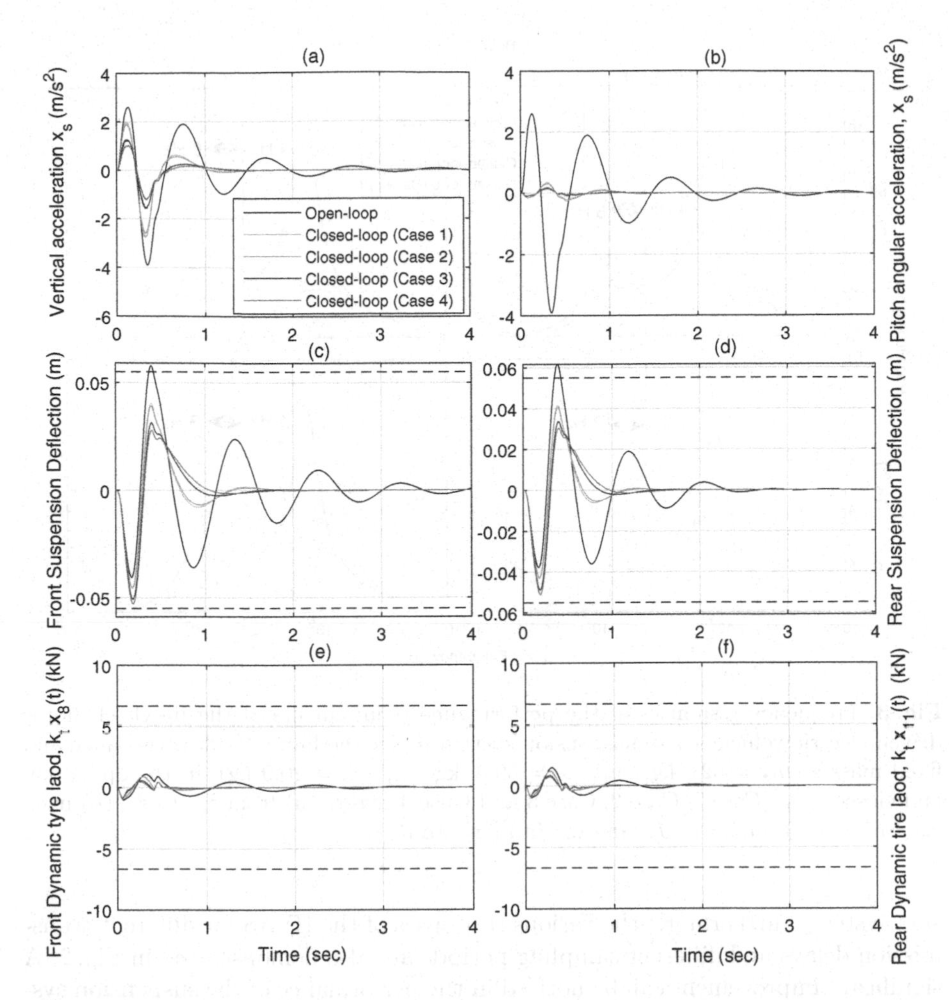

Fig. 2. Time responses of the performance requirements of the in-wheel motor driven electric vehicle active suspension system under the bump road disturbance in the open-loop and closed-loop (Case 1, Case 2, Case 3 and Case 4) cases: (a) Vehicle body vertical acceleration (b) Pitch angular acceleration (c) Front suspension deflection (d) Rear suspension deflection (e) Front dynamic tyre load and (f) Rear dynamic tyre load

suspension system performance under the effect of the time-varying transmission delays and uncertain sampling periods, i.e., **Case 1**: $\tau_m = \bar{\tau} = 8\,\text{ms}$ and $\underline{h} = 0\,\text{ms}, \overline{h} = 5\,\text{ms}$; **Case 2**: $\tau_m = \bar{\tau} = 8\,\text{ms}$ and $\underline{h} = 0\,\text{ms}, \overline{h} = 3\,\text{ms}$; **Case 3**: $\tau_m = \bar{\tau} = 2\,\text{ms}$ and $\underline{h} = 0\,\text{ms}, \overline{h} = 5\,\text{ms}$; and **Case 4**: $\tau_m = \bar{\tau} = 2\,\text{ms}$ and $\underline{h} = 0\,\text{ms}, \overline{h} = 3\,\text{ms}$.

Simulation results for the HVAS performance requirements in the time domain are provided in Fig. 2. It is shown in the Fig. 2(a)–(b) that both the vehicle's body vertical and pitch angular accelerations are significantly reduced by the designed networked controller compared with traditional passive suspen-

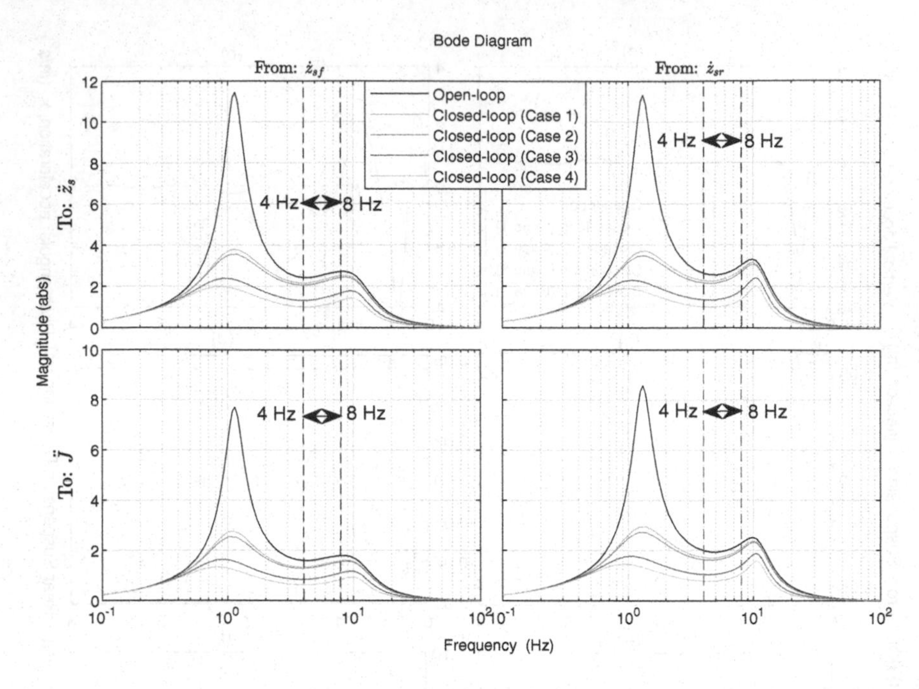

Fig. 3. Frequency responses of the performance requirements of the in-wheel motor driven electric vehicle active suspension system under the bump road disturbance and fixed masses (m_s = 621 kg, $m_{u(f,r)}$ = 39.6 kg, $m_{r(f,r)}$ = 20.9 kg), in the open-loop and closed-loop (Case 1, Case 2, Case 3 and Case 4) cases: (a) from $\dot{z}_{sf}$ to $\ddot{z}_s$, (b) from $\dot{z}_{sr}$ to $\ddot{z}_s$, (c) from $\dot{z}_{sf}$ to $\ddot{J}_s$, and (d) from $\dot{z}_{sr}$ to $\ddot{J}_s$

sion system. Furthermore, the various responses of the HVAS for different transmission delays and different sampling periods are also demonstrated in Fig. 2. A significant improvement can be noticed in the performance of the suspension system when the transmission delay and sampling period are small, e.g., in the case of $\tau_m = \bar{\tau} = 2$ ms and $\bar{h} = 3$ ms, compared with the others. The performance of the suspension system is adversely effected, as the delay and the sampling period are increased, e.g., in the case of $\tau_m = \bar{\tau} = 8$ ms and $\bar{h} = 5$ ms. On the other hand, it is clearly depicted from Fig. 2(c), (d) that both the front and rear active suspension system deflections are within the maximum allowable limit in all cases while the passive suspension deflections cross the boundary. The road holding performance is also improved by keeping the dynamic tyre load always within the bounds, as evidently seen from Fig. 2 (e), (f).

Finally, the suspension system performance analyzed in the frequency domain to further demonstrate the efficacy of the proposed control methodology. The frequency response results are shown in Fig. 3. Notice that the human body is sensitive to the vibrations in the frequency range of 4 Hz–8 Hz, represented by the vertical dashed lines in the figure. From Fig. 3, it is obvious that significant improvements are achieved when the transmission delay and sampling period are small. As the transmission delay and sampling period increase, the response of the suspension system gets worsen. However, the figure shows that the frequency responses can be significantly improved in all cases of transmission delays and sampling periods of the closed-loop system as compared to the open-loop system in the frequency range of 4Hz – 8Hz.

5 Conclusion

The networked active suspension control problem of in-wheel motor driven electric vehicles equipped with dynamic dampers has been addressed. All the masses have been considered to be uncertain. A networked state feedback fuzzy controller has been designed to tackle both the time-varying transmission delays and uncertain sampling periods in such a way that the asymptotic stability of the developed suspension system model is guaranteed and the desired suspension performance requirements are preserved. Numerical simulations in both the time and frequency domains have been provided to show the effectiveness of the proposed control approach.

References

1. Bridgestone Dynamic-Damping In-Wheel Motor Drive System. http://enginuitysystems.com/files/In-WheelMotor.pdf, October 2017
2. Guan, Y., Han, Q.-L., Yao, H., Ge, X.: Robust event-triggered H_∞ controller design for vehicle active suspension systems. Nonlinear Dyn. **94**(1), 627–638 (2018)
3. Ning, D., Sun, S., Zhang, F., Du, H., Li, W., Zhang, B.: Disturbance observer based Takagi-Sugeno fuzzy control for an active seat suspension. Mech. Syst. Signal Process. **93**(1), 515–530 (2017)
4. Li, W., Xie, Z., Zhao, J., Wong, P.K., Wang, H., Wang, X.: Static-output-feedback based robust fuzzy wheelbase preview control for uncertain active suspensions with time delay and finite frequency constraint. IEEE/CAA J. Automatica Sinica. **8**(3), 664–678 (2020)
5. Sibielak, M., Raczka, W., Konieczny, J., Kowal, J.: Optimal control based on a modified quadratic performance index for systems disturbed by sinusoidal signals. Mech. Syst. Signal Process. **64–65**, 498–519 (2015)
6. Shao, X., Naghdy, F., Du, H.: Reliable fuzzy H_∞ control for active suspension of in-wheel motor driven electric vehicles with dynamic damping. Mech. Syst. Signal Process. **87**, 365–383 (2017)
7. Wang, R., Jing, H., Yan, F., Karimi, H.R., Chen, N.: Optimization and finite-frequency H_∞ control of active suspensions in in-wheel motor driven electric ground vehicles. J. Frankline Inst. **352**(2), 468–484 (2015)
8. Qin, Y., He, C., Shao, X., Du, H., Xiang, C., Dong, M.: Vibration mitigation for in-wheel switched reluctance motor driven electric vehicle with dynamic vibration absorbing structures. J. Sound Vib. **419**, 249–267 (2018)
9. Ahmad, I., Ge, X., Han, Q.-L.: Decentralized dynamic event-triggered communication and active suspension control of in-wheel motor driven electric vehicles with dynamic damping. IEEE/CAA J. Automatica Sinica. **8**(5), 971–986 (2021)

10. Ge, X., Han, Q.-L., Zhang, X.-M., Ding, D.: Dynamic event-triggered control and estimation: a survey. Int. J. Autom. Comput., 1–30 (2021). https://doi.org/10.1007/s11633-021-1306-z
11. Ge, X., Ahmad, I., Han, Q.-L., Wang, J., Zhang, X.-M.: Dynamic event-triggered scheduling and control for vehicle active suspension over controller area network. Mech. Syst. Signal Process. **152**(1), 107–481 (2021)
12. Li, H., Jing, X., Lam, H.-K., Shi, P.: Fuzzy sampled-data control for uncertain vehicle suspension systems. IEEE Trans. Cybern. **44**(7), 1111–1126 (2014)
13. Gao, H., Sun, W., Shi, P.: Robust sampled-data H_∞ control for vehicle active suspension systems. IEEE Trans. Control Syst. Technol. **18**(1), 238–245 (2010)
14. Li, W., Xie, Z., Cao, Y., Wong, P.K., Zhao, J.: Sampled-data asynchronous fuzzy output feedback control for active suspension systems in restricted frequency domain. IEEE/CAA J. Automatica Sinica. **8**(5), 1052–1066 (2021)

Sparse Control of Decentralized Systems

Xiaoning Li, Yanpeng Guan(✉), and Junpeng Du

School of Automation and Software Engineering, Shanxi University, Taiyuan 030006, China
y.guan@sxu.edu.cn

Abstract. Two methods for sparse control of decentralized systems are used in this paper. Firstly, a decentralized event-triggered transmission scheme is performed to sparse sampled signals before transmitting them to the controller. Secondly, the number of the decentralized controller-actuators channels is constrained, and is equivalently transformed to the number of non-zero rows in the controller gain matrix. The cardinality constraint problem is solved by using a mixed integer method. In addition, H_∞ stability analysis and controller design are performed for the proposed system using Lyapunov-Krasovskii functional method. Finally, the effectiveness of the proposed method is illustrated by a simulation example.

Keywords: Sparse control · Decentralized control · Event-triggered · H_∞ Control

1 Introduction

In recent years, decentralized control systems have regained a wide interest due to the advent of wireless communication, see [1–4]. In the design of large-scale systems, there are many problems such as data packet loss and limited number of components due to economic and computational costs [5, 6]. This is particularly prominent in decentralized control systems, whose idea of decentralized control and centralized operation requires the system to process a great deal of data simultaneously. In addition, because most of the components of a decentralized system are distributed in space, using network communication. This generates numerous communication channels, which is not conducive to the post maintenance work of the system. Therefore, it is necessary to perform sparse optimization for decentralized systems. In this paper, we mainly consider the sparse control sampled signals and the number of controller-actuators channels.

A decentralized control system deploys a set of geographically distributed sensors to sample the system's state signals. This deployment generates a large amount of sampled data in the sampling channel of the control system. If the traditional periodic transmission is used, a large amount of redundant data will be generated to increase the computational burden [7]. To solve this problem, this paper applies an event-triggered transmission scheme (ETTS), where data can only be transmitted under a predefined scheme [8]. Doing so enables the sparse processing of data from the control system sampling channels to avoid too much sampled data being transmitted. It is because of the important advantages of ETTS in resource-constrained systems that they have received increasing attention in the last years, see [9–13].

K. Li et al. (Eds.): LSMS 2021/ICSEE 2021, CCIS 1468, pp. 285–294, 2021.
https://doi.org/10.1007/978-981-16-7210-1_27

It is a hot issue to consider sparse optimization of the communication channels of the system in order to simplify the system model and save the system building cost. However, as described in [14] sparse optimization problems are np-hard and have no known analytic solutions. Current approaches to solving sparse problems can be broadly classified into two categories One approach is to define the structural constraints based on some prior knowledge before the system design. It is kept constant during the system design and implementation, and then a control strategy is developed to find the optimal solution, as in [15]. Another approach is to convert the sparse problem into an optimization problem with a composite function. It consists of two components reflecting the system performance and sparsity, where the component related to sparsity is usually replaced using the ℓ_1-norm, as in [14, 16, 17]. This approach improves the sparsity of the system while maintaining its performance, offering more flexibility than the pre-defined fixed structure model approach at a relatively low computational cost.

In this paper, we limit the number of controller-actuators channels in advance. Let the system operate stably even with a small number of control channels. In addition, we equate it to the number of non-zero rows in the controller gain matrix, and using a mixed-integer approach can impose cardinality constraint on the K. This achieves the sparse optimization of the controller-actuators channels of the decentralized system.

The main contribution of this paper is to introduce a decentralized ETTS for asynchronous sampling to achieve data sparsity in the sampling channel. A controller-actuators channels sparse optimization algorithm is also proposed. Unlike the previously proposed algorithms, the method proposed in this paper does not require any initial condition and a complex iterative process. Only the gain matrix K needs to satisfy a certain condition, and this condition is reasonable in practice.

2 Problem Formulation

Consider the following class of linear time-invariant systems

$$\begin{aligned} \dot{x}(t) &= Ax(t) + Bu(t) + B_{\omega}\omega(t) \\ z(t) &= Cx(t) + Du(t), \ \mathrm{t} \geq 0 \end{aligned} \tag{1}$$

where $x(t) \in \mathbb{R}^n$, $u(t) \in \mathbb{R}^m$, $\omega(t)$ and $z(t) \in \mathbb{R}^r$ are the state signal, control input, external disturbances and the controlled output of the system, respectively.A, B, B_{ω}, C, D are constant matrices of appropriate dimensions, and the initial condition of the system (1) is $x(0) = x_0$.

As is shown in Fig. 1, we sample the full state of the system using n sensors, where the i th sensor samples the i th component $x_i(t)$ of state $x(t)$, $i = 1, 2, \ldots, n$, all with sampling period $h > 0$. All sensors have the same sampling rate, but do not have to start sampling at the same time, i.e. the whole sampling process is asynchronous.

In addition, not all sampled data will be used for feedback control in order to save limited resources. To this end, we adopt a decentralized ETTS to determine whether the i th state measurement is passed to the controller for use in control input calculations. The current sampled state value can be transmitted only when the error between the current sampled state value $x_i(k_i h)$ and the last transmitted sampled state value $x_i(t^i_{k_i}h)$ breaks the event-triggered threshold condition

$$(x_i(k_i h) - x_i(t^i_{k_i}h))^2 < \delta_i x_i^2(k_i h) \tag{2}$$

where, $\delta_i \in (0, 1)$ is the event threshold parameter.

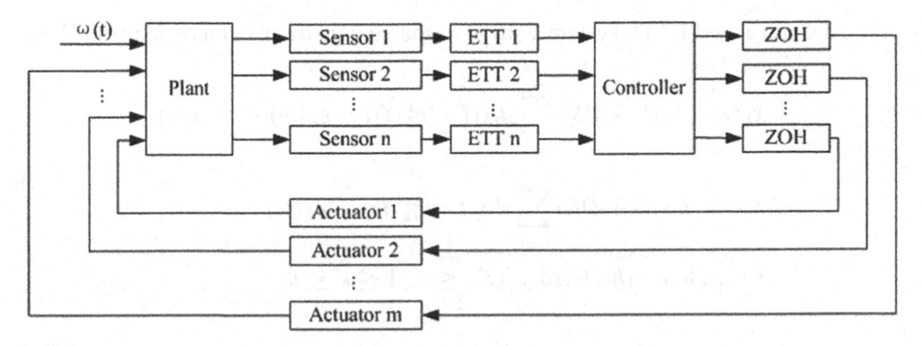

Fig. 1. A framework of decentralized control system

The state feedback control law is chosen in the following form

$$u(t) = K\left[\, x_1(t^1_{k_1}h)\ x_2(t^2_{k_2}h)\ \cdots\ x_n(t^n_{k_n}h) \,\right]^T,\ t \in \left[t_k h, t_{k+1}h\right) \tag{3}$$

where $t_k h = \max_{i=1,2,\ldots,n}\{t^i_{k_i}h\}$, $t_{k+1}h = \min_{i=1,2,\ldots,n}\{t^i_{k_i+1}h\}$, $K \in \mathbb{R}^{m\times n}$ is to be determined.

In practice, systems with sparse optimization of the communication channel have additional advantages. On the one hand, it can save the system setup cost. On the other hand, it can reduce the risk of cyber-attack. Therefore, how to use fewer communication channels to attain the desired control performance is the main purpose of the system to achieve control channel sparse optimization. To this end, we equate the number of controller-actuators channels to the number of non-zero rows of the feedback gain matrix K, and limit the number of non-zero rows of the matrix K to no more than κ.

$$\mathrm{Card}_{\mathrm{row}}(K) \le \kappa,\ 1 \le \kappa \le m \tag{4}$$

Remark 1. When the number of non-zero rows of K is κ, the system needs to select κ control channels to continue working in order to maintain the required performance. The value of κ is smaller, it means that the system needs fewer control channels to maintain a certain performance, thus saving network resources and reducing costs.

In the interval $t \in \left[t_k h, t_{k+1}h\right)$, let

$$\eta^i_k(t) = t - t_k h - \max_{l\in\mathbb{Z}^+}\{lh | t^i_{k_i}h + lh \le t\} \tag{5}$$

$$e^i_k(t) = x_i(t - \eta^i_k(t)) - x_i(t^i_{k_i}h),\quad i = 1, 2, \ldots, n. \tag{6}$$

Then from (3) we have

$$u(t) = K(\sum_{i=1}^{n} d_i x(t - \eta_k^i(t)) - e_k(t)),\ t \in [t_k h, t_{k+1} h) \tag{7}$$

where $d_i = \text{diag}\{\underbrace{0, \cdots, 1,}_{i} \cdots, 0\}$, $e_k(t) = [e_k^1(t), e_k^2(t), \cdots, e_k^n(t)]^T$.

The decentralized ETTS system with row-sparse constraints can be expressed as

$$\begin{aligned} \dot{x}(t) &= Ax(t) + BK(\sum_{i=1}^{n} d_i x(t - \eta_k^i(t)) - e_k(t)) + B_\omega \omega(t) \\ z(t) &= Cx(t) + DK(\sum_{i=1}^{n} d_i x(t - \eta_k^i(t)) - e_k(t)) \\ t &\in [t_k h, t_{k+1} h),\ \text{Card}_{\text{row}}(K) \le \kappa,\ 1 \le \kappa \le m. \end{aligned} \tag{8}$$

3 H_∞ Stability Analysis

In this section, we give a sufficient condition for H_∞ stability of a closed-loop system (8). This condition is also used to give a design approach for the controller.

Theorem 1. For given $\gamma > 0, h > 0$, the system (8) is asymptotically stable with H_∞ performance level γ if there exist real matrices $P > 0, Q > 0, \{M_i = M_i^T\}_{i=1}^n, \{N_i\}_{i=1}^n$, and the appropriate dimensions of Y_1, Y_2, K, for all $\theta_i \in \{0, 1\}, i = 1, 2, \ldots, n$, such that

$$\begin{bmatrix} P + h\sum_{i=1}^{n} \theta_i M_i & h\Gamma_1\Theta \\ * & (h\Gamma_2 - I)\Theta + I \end{bmatrix} > 0 \tag{9}$$

$$\begin{bmatrix} \Pi_{11} & \Pi_{12} & 0 & \Pi_{14} & -Y_1^T BK & Y_1^T B_W & C^T \\ * & \Pi_{22} & 0 & \Pi_{24} & -Y_2^T BK & Y_2^T B_W & 0 \\ * & * & -Q & 0 & 0 & 0 & 0 \\ * & * & * & \Pi_{44} & 0 & 0 & \Pi_{47} \\ * & * & * & * & -I & 0 & -K^T D^T \\ * & * & * & * & * & -\gamma^2 I & 0 \\ * & * & * & * & * & * & -I \end{bmatrix} < 0 \tag{10}$$

$\Gamma_1 = \begin{bmatrix} N_1 - M_1 & N_2 - M_2 & \cdots & N_n - M_n \end{bmatrix}$, $\Gamma_2 = \text{diag}\{M_1 - N_1 - N_1^T, \cdots, M_n - N_n - N_n^T\}$,

$\Theta = \text{diag}\{\theta_1 I, \theta_2 I, \ldots, \theta_n I\}$, $\Pi_{11} = A^T Y_1 + Y_1^T A + Q - \sum_{i=1}^{n} M_i$, $\Pi_{47} = [d_1, d_2, \ldots, d_n]^T K^T D^T$,

$\Pi_{12} = A^T Y_2 + P - Y_1^T + h\sum_{i=1}^{n} \theta_i M_i$, $\Pi_{14} = -\Gamma_1 + Y_1^T BK\begin{bmatrix} d_1 & d_2 & \cdots & d_n \end{bmatrix}$, $\Pi_{22} = -Y_2 - Y_2^T$,

$\Pi_{24} = h\Gamma_1\Theta + Y_2^T BK\begin{bmatrix} d_1 & d_2 & \cdots & d_n \end{bmatrix}$, $\Pi_{44} = -\Gamma_2 + \text{diag}\{\delta_1 d_1, \delta_2 d_2, \ldots, \delta_n d_n\}$.

Proof. Choose the following Lyapunov-Krasovskii functional candidate,

$$V(t,x(t)) = x^T(t)Px(t) + \int_{t-h}^{t} x^T(s)Qx(s)ds + \sum_{i=1}^{n}(h-\eta_k^i(t))\left[x^T(t)\; x^T(t-\eta_k^i(t))\right] \times \begin{bmatrix} M_i & N_i - M_i \\ * & M_i - N_i - N_i^T \end{bmatrix} \begin{bmatrix} x(t) \\ x(t-\eta_k^i(t)) \end{bmatrix}. \tag{11}$$

Taking the right derivative of $V(t,x(t))$ with respect to t along the trajectory of (8) yields

$$\begin{aligned} \dot{V}(t,x(t)) &= 2x^T(t)P\dot{x}(t) - x^T(t-h)Qx(t-h) + x^T(t)Qx(t) \\ &+2\sum_{i=1}^{n}(h-\eta_k^i(t))(x^T(t)M_i\dot{x}(t) + x^T(t-\eta_k^i(t))(N_i^T - M_i)\dot{x}(t)) \\ &-\sum_{i=1}^{n}\left[x^T(t)\;\; x^T(t-\eta_k^i(t))\right] \times \begin{bmatrix} M_i & N_i - M_i \\ * & M_i - N_i - N_i^T \end{bmatrix} \begin{bmatrix} x(t) \\ x(t-\eta_k^i(t)) \end{bmatrix}. \end{aligned} \tag{12}$$

There exists a real non-singular matrix Y_1, Y_2 such that

$$2(x^T(t)Y_1^T + \dot{x}^T(t)Y_2^T)(Ax(t) + BK(\sum_{i=1}^{n} d_i x(t-\eta_k^i(t)) - e_k(t)) + B_\omega \omega(t) - \dot{x}(t)) = 0. \tag{13}$$

According to the event-triggered threshold condition (2), we have

$$e_k^T(t)e_k(t) < \sum_{i=1}^{n} \delta_i x^T(t-\eta_k^i(t))d_i x(t-\eta_k^i(t)). \tag{14}$$

By replacing $h - \eta_i(t)$ with $\theta_i h$, we know that

$$\dot{V}(t,x(t)) + z^T(t)z(t) - \gamma^2\omega^T(t)\omega(t) \le \varphi^T(t)(\Pi(\eta(t)) + M^TM)\varphi(t) \tag{15}$$

where, $M = [C\ 0\ 0\ m\ {-DK}\ 0]$, $m = DK[d_1, d_2, \ldots, d_n]$,

$$\varphi(t) = \left[x^T(t)\ \dot{x}^T(t)\ x^T(t-h)\ x^T(t-\eta_k^1(t))\ x^T(t-\eta_k^2(t))\ \cdots\ x^T(t-\eta_k^n(t))\ e_k^T(t)\ \omega^T(t)\right]^T$$

$$\Pi(\eta(t)) = \begin{bmatrix} \Pi_{11} & \Pi_{12} & 0 & \Pi_{14} & -Y_1^TBK & Y_1^TB_W \\ * & \Pi_{22} & 0 & \Pi_{24} & -Y_2^TBK & Y_2^TB_W \\ * & * & -Q & 0 & 0 & 0 \\ * & * & * & \Pi_{44} & 0 & 0 \\ * & * & * & * & -I & 0 \\ * & * & * & * & * & -\gamma^2 I \end{bmatrix}$$

Then using Schur complement lemma, we obtain (9)-(10). If (9)-(10) hold, it is shown that (16) holds.

$$\dot{V}(t,x(t)) + z^T(t)z(t) - \gamma^2\omega^T(t)\omega(t) < 0 \tag{16}$$

From inequation (16) it is clear that when the system is under zero initial conditions, $\|\omega(t)\|_2 \neq 0$, $\|z(t)\|_2 \leq \gamma\|\omega(t)\|_2 < \infty$ it holds for any constant. When $\omega(t) = 0$, inequation (16) becomes $\dot{V}(t,x(t)) < -z^T(t)z(t) < 0$. Therefore, the system (8) is asymptotically stable with H_∞ performance level γ.

4 Design of a H_∞ Controller with Control Channel Sparsity

In this section, the results of H_∞ stability analysis and the mixed integer method are used to design a controller with sparse constraints on the controller-actuators channels.

Lemmas 1. For a matrix $K \in R^{m\times n}$, suppose that the maximum of the absolute values of the elements of the matrix K does not exceed M, i.e. the matrix satisfies the condition $\|K\|_{\max} \leq M$, M is a positive number, and if the $\mathrm{Card}_{\mathrm{row}}(K) \leq \kappa$ holds, then there exists a binary sequence $b_i \in \{0, 1\}$, $i = 1, 2 \ldots, m$, such that.

$$|K(i,j)| \leq M(1-b_i),\ \sum_{i=1}^{m} b_i \geq m-\kappa. \tag{17}$$

Theorem 2. For given scalars $\gamma > 0, h > 0$, and an integer κ satisfying $1 \leq \kappa \leq m$, system (8) is asymptotically stable with H_∞ performance level γ and gain matrix $K = \sum_{i=1}^{n} F_i\tilde{Y}^{-1} = F\tilde{Y}^{-1}$ satisfies the condition $\mathrm{Card}_{\mathrm{row}}(K) \leq \kappa$ if there exists real matrices $\tilde{P} > 0, \tilde{Q} > 0$, $\{\tilde{M}_i = \tilde{M}_i^T\}_{i=1}^n$,$\{\tilde{N}_i\}_{i=1}^n$, $\{F_i\}_{i=1}^n$, $\tilde{Y}$, U, real constants $\{\tilde{\delta}_i > 0\}_{i=1}^n$, ν,$M > 0$ and a binary sequence $b_i \in \{0, 1\}$, $i = 1, 2, \ldots, m$, such that

$$|F(i,j)| \leq M(1-b_i),\ \sum_{i=1}^{m} b_i \geq m-\kappa \tag{18}$$

$$\begin{bmatrix} \tilde{P} + h\sum_{i=1}^{n}\theta_i\tilde{M}_i & h\tilde{\Gamma}_1\Theta \\ * & h\tilde{\Gamma}_2\Theta + h\widetilde{\mathrm{P}}(I-\Theta) \end{bmatrix} > 0 \tag{19}$$

where, $\tilde{\Gamma}_1 = \left[\tilde{N}_1 - \tilde{M}_1\ \tilde{N}_2 - \tilde{M}_2 \cdots \tilde{N}_n - \tilde{M}_n\right]$, $\Theta = \mathrm{diag}\{\theta_1 I, \theta_2 I, \ldots, \theta_n I\}$, $\tilde{\Gamma}_2 = \mathrm{diag}\left\{\tilde{M}_1 - \tilde{N}_1 - \tilde{N}_1^T, \cdots, \tilde{M}_n - \tilde{N}_n - \tilde{N}_n^T\right\}$, $\widetilde{\mathrm{P}} = \mathrm{diag}\{\tilde{P}, \tilde{P}, \ldots, \tilde{P}\}$

$$\begin{bmatrix} \tilde{\Pi}_{11} & \tilde{\Pi}_{12} & 0 & \tilde{\Pi}_{14} & -U^T B \sum_{i=1}^{n} F_i & U^T B_\omega & \tilde{Y}^T C^T & 0 \\ * & -\tilde{Y} - \tilde{Y}^T & 0 & \tilde{\Pi}_{24} & -B \sum_{i=1}^{n} F_i & B_\omega & 0 & 0 \\ * & * & -\tilde{Q} & 0 & 0 & 0 & 0 & 0 \\ * & * & * & -\tilde{\Gamma}_2 & 0 & 0 & \tilde{\Pi}_{47} & \tilde{\Pi}_{48} \\ * & * & * & * & \nu^2 I - \nu\tilde{Y} - \nu\tilde{Y}^T & 0 & -\sum_{i=1}^{n} F_i^T D^T & 0 \\ * & * & * & * & * & -\gamma^2 I & 0 & 0 \\ * & * & * & * & * & * & -I & 0 \\ * & * & * & * & * & * & * & \tilde{\Pi}_{88} \end{bmatrix} < 0 \quad (20)$$

where, $\tilde{\Pi}_{11} = \tilde{Q} + U^T A\tilde{Y} + \tilde{Y}^T A^T U - \sum_{i=1}^{n} \tilde{M}_i$, $\tilde{\Pi}_{12} = \tilde{P} + \tilde{Y}^T A^T - U^T \tilde{Y} + h\sum_{i=1}^{n} \theta_i \tilde{M}_i$,

$\tilde{\Pi}_{14} = -\tilde{\Gamma}_1 + U^T B\begin{bmatrix} F_1 & F_2 & \cdots & F_n \end{bmatrix}$, $\tilde{\Pi}_{24} = h\tilde{\Gamma}_1\Theta + B\begin{bmatrix} F_1 & F_2 & \cdots & F_n \end{bmatrix}$,

$\tilde{\Pi}_{47} = \begin{bmatrix} DF_1 & DF_2 & \cdots & DF_n \end{bmatrix}^T$, $\tilde{\Pi}_{48} = \mathrm{diag}\{\tilde{Y}^T E_1, \tilde{Y}^T E_2, \ldots, \tilde{Y}^T E_n\}$,

$\tilde{\Pi}_{88} = \mathrm{diag}\{-\tilde{\delta}_1, -\tilde{\delta}_2, \ldots, -\tilde{\delta}_n\}$.

Proof. Assuming that Theorem 1 holds, let $\tilde{Y} = Y_2^{-1}$, $U = Y_1 Y_2^{-1}$. By some matrix congruence transformations, let $\tilde{Q} = \tilde{Y}^T Q\tilde{Y}, \tilde{P} = \tilde{Y}^T P\tilde{Y}, \tilde{M}_i = \tilde{Y}^T M_i \tilde{Y}, \tilde{N}_i = \tilde{Y}^T N_i \tilde{Y}, E_i = d_i\begin{bmatrix} 1 & 1 & \cdots & 1 \end{bmatrix}^T$, $\tilde{\delta}_i = 1/\delta_i$, $i = 1, 2, \ldots, n$. Denote Y and K respectively as $\tilde{Y} = [\,\tilde{Y}_1^T \; \tilde{Y}_2^T \; \cdots \; \tilde{Y}_n^T\,]^T, K = [\,K_1 \; K_2 \; \cdots \; K_n\,]^T$, and let $F_i = K_i\tilde{Y}_i, i = 1, 2, \ldots, n$. It is obtained that $\tilde{Y}^T d_i \tilde{Y} = \tilde{Y}^T E_i E_i^T \tilde{Y}$, $Kd_i\tilde{Y} = F_i, K\tilde{Y} = \sum_{i=1}^{n} F_i = F$. Finally, by some linearization techniques, we can obtain the inequalities (19) and (20).

From Lemma 1, if (18) holds then the matrix F satisfies the sparsity constraint $\mathrm{Card}_{\mathrm{row}}(F) \leq \kappa$. And since $K = F\tilde{Y}^{-1}$, Y is a non-singular matrix, so the number of non-zero rows of the matrix K does not exceed κ.

5 Simulation Examples

In this section, we will illustrate by a simulation example that the proposed method can effectively sparse the sampling data and the number of controller-actuators channels.

Taking the lateral dynamics model of the Boeing 747 aircraft as a simulation example, its linearized model at 4000 feet level flight and a forward speed of 774 feet per second can be described as

$$\dot{x}(t) = \begin{bmatrix} -0.0558 & -0.9968 & 0.0802 & 0.0415 \\ 0.598 & -0.115 & -0.0318 & 0 \\ -3.05 & 0.388 & -0.465 & 0 \\ 0 & 0.0805 & 1 & 0 \end{bmatrix} x(t)$$

$$+ \begin{bmatrix} 0.00729 & 0.01 & 0.05 \\ -0.475 & -0.5 & -0.3 \\ 0.153 & 0.2 & 0.1 \\ 0 & 0 & 0 \end{bmatrix} u(t) + \begin{bmatrix} 0.1 \\ 0 \\ 0.1 \\ 0 \end{bmatrix} w(t) \tag{21}$$

$$z(t) = \begin{bmatrix} 0 \; 1 \; 0 \; 0 \end{bmatrix} x(t) \, x(t) = \begin{bmatrix} \beta \; y_r \; p \; \phi \end{bmatrix}^T$$

where β represents the sideslip angle, y_r represents the yaw rate, p is the roll rate, ϕ is the roll angle. We choose $\gamma = 0.9$, $U = 0.3I$, the sampling period $h = 0.1s$, external disturbances $w(t) = e^{-t}\cos(t)$. In the case that $\kappa \leq 2$, controller parameter matrix K and the decentralized ETTS parameters are obtained as follows.

$$K = \begin{bmatrix} 8.5216 & 10.9145 & 0.2538 & 0.2631 \\ -8.2776 & -7.0039 & -0.2595 & -0.2419 \\ 0 & 0 & 0 & 0 \end{bmatrix}$$

$$\delta_1 = 0.056, \; \delta_2 = 0.075, \; \delta_3 = 0.022, \; \delta_4 = 0.177$$

Accordingly, the event-triggered release interval of the four sensors of the system in this case is shown in Fig. 2, and their state responses are shown in Fig. 3.

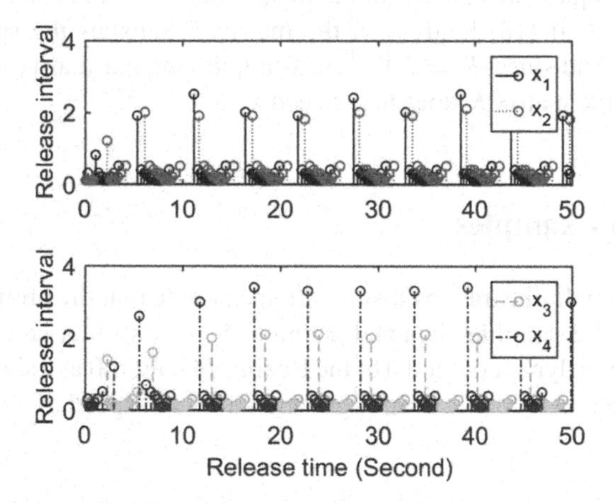

Fig. 2. Release time intervals of the system with control channel sparsity

From the above simulation results, we can see that $\text{Card}_{\text{row}}(K) = 2$, which means that two rows of the controller gain matrix K of the system are non-zero. According to Fig. 3, the system can remain stable with one control channel discarded, thus indicating

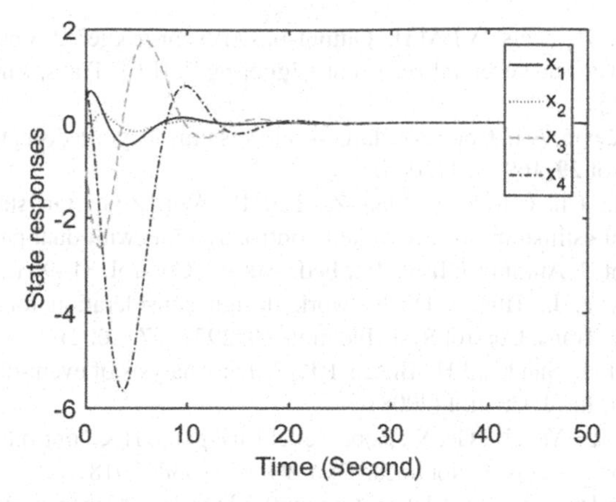

Fig. 3. State response of the system with control channel sparsity

that the system only needs to select two controller-actuators channels to maintain the required performance of the system. It shows that the controller-actuators channels of the decentralized system have been sparsely optimized.

In the case of the system simulation process of 50 s, while the numbers of transmitted sampled signals are respectively 103, 158, 210,103 in four sensors. The amount of data transmission is significantly reduced compared with the number of samples, indicating that the system using ETTS can sparse the data of the sampled channels to effectively save network resources and thus achieve the goal of cost reduction.

6 Conclusion

In this paper, the proposed decentralized control system is sparsely optimized using two approaches. First, a decentralized ETTS with asynchronous sampling is applied to sparse the data of the sampled channels of thc controlled object. Then a mixed integer optimization algorithm is applied to constrain the specific number of controller communication coupling channels. The stability analysis and controller design of the proposed system are carried out using Lyapunov-Krasovskii functional. Finally, the effectiveness of the proposed method is demonstrated by a concrete example.

References

1. Ahmad, I., Ge, X., Han, Q.L.: Decentralized dynamic event-triggered communication and active suspension control of in-wheel motor driven electric vehicles with dynamic damping. J. IEEE/CAA J. Autom. Sinica. **8**, 971–986 (2021)
2. Yang, D., Ren, W., Liu, X., Chen, W.: Decentralized event-triggered consensus for linear multi-agent systems under general directed graphs. J. Automatica. **69**, 242–249 (2016)
3. Yang, L., Liu, T., Hill, D.J.: decentralized event-triggered frequency regulation for multi-area power systems. J. Automatica. **126**, 109479 (2021)

4. Donkers, M.C.F., Heemels, W.P.M.H.: Output-based event-triggered control with guaranteed -gain and improved and decentralized event-triggering. J. IEEE Trans. Automatic Control **57**, 1362–1376 (2012)
5. Šiljak, D.D., Zečević, A.I.: Control of large-scale systems: beyond decentralized feedback. J. Ann. Rev. Control **29**, 169–179 (2005)
6. Lin, H., Su, H., Chen, M.Z.Q., Shu, Z., Lu, R., Wu, Z.G.: On stability and convergence of optimal estimation for networked control systems with dual packet losses without acknowledgment. J. Autom.: J. IFAC Int. Fed. Autom. Control, 81–90 (2018)
7. Lian, F.L., Moyne, J., Tilbury, D.: Network design consideration for distributed control systems. J. IEEE Trans. Control Syst. Technol. **10**, 297–307 (2002)
8. Heemels, W.P.M.H., Sandee, J.H., Bosch, P.P.J.V.D.: Analysis of event-driven controllers for linear systems. J. Int. J. Control (2008)
9. Guan, Y., Han, Q.L., Yao, H., Ge, X.: Robust event-triggered H∞ controller design for vehicle active suspension systems. J. Nonlinear Dyn. **94**, 627–638 (2018)
10. Yue, D., Tian, E., Han, Q.L.: A delay system method for designing event-triggered controllers of networked control systems. J. IEEE Trans. Autom. Control **58**, 475–481 (2013)
11. Postoyan, R., Anta, A., Heemels, W.P.M.H., Tabuada, P., Nešić, D.: Periodic event-triggered control for nonlinear systems. In: 52nd IEEE Conference on Decision and Control, pp. 7397–7402 (2013)
12. Zhang, X.M., Han, Q.L., Zhang, B.L.: An overview and deep investigation on sampled-data-based event-triggered control and filtering for networked systems. J. IEEE Trans. Ind. Inf. **13**, 4–16 (2017)
13. Ge, X., Yang, F., Han, Q.L.: Distributed networked control systems: a brief overview. J. Inf. Sci. **380**, 117–131 (2017)
14. Babazadeh, M., Nobakhti, A.: Sparsity promotion in state feedback controller design. J. IEEE Trans. Automat. Contr. **62**, 4066–4072 (2017)
15. Fattahi, S., Fazelnia, G., Lavaei, J.: Transformation of optimal centralized controllers into near-global static distributed controllers. In: 2015 54th IEEE Conference on Decision and Control (CDC), pp. 4915–4922 (2015)
16. Lin, F., Fardad, M., Jovanović, M.R.: Design of optimal sparse feedback gains via the alternating direction method of multipliers. J. IEEE Trans. Automat. Contr. **58**, 2426–2431 (2013)
17. Automated model construction for combined sewer overflow prediction based on efficient LASSO algorithm. IEEE J. Mag. IEEE, Xplore

Intelligent Control Methods in Energy Infrastructure Development and Distributed Power Generation Systems

Design of Gas Flow Standard Device by the Master Meter Method Based on Synchronous Pulse Counting Method

Bin Li, Sheng Qiang(✉), Jie Chen, and Wangcheng Zhao

School of Mechatronic Engineering and Automation, Shanghai University, Shanghai 200444, China

Abstract. This paper introduces the hardware, software composition and working principle of field service gas flow standard device by the master meter method. The synchronous pulse counting method is proposed and implemented, and the key points in software design are described in detail. It is proved that the field service gas flow standard device by the master meter method can be used for the verification of various types of gas flowmeter and high accuracy verification service.

Keywords: Gas flow standard device · Master meter method · Synchronous pulse counting method · Incremental digital PID

1 Introduction

In recent years, the rapid development of economy has led to the deterioration of the environment, the concentration of air pollutants is gradually increasing, the Urban Haze is more and more serious, and the coverage is more and more wide. Facing the situation of frequent haze weather in recent years, and actively coping with China's climate change, the State Council reformed from the supply side and made use of the characteristics of natural gas, such as no pollution, high calorific value and large reserves. A new way of clean energy focusing on natural gas has been opened.

Natural gas needs to be measured in the process from production to sales and then to customers. There are millions of flowmeters used in the measurement. In addition to natural gas, in the measurement of compressed air, oxygen, nitrogen and similar types of energy, such as liquefied petroleum gas to natural gas, coal to gas, a large number of gas flow meters are also needed, and the flow meters used to measure these gases are increasing year by year. If it is used in trade settlement, its measurement accuracy and reliability must be high, and some of them can be connected with the international market. The buyer also hopes that the expensive flowmeter can get value for money, which puts forward higher requirements for maintaining the accuracy of the flowmeter [1].

K. Li et al. (Eds.): LSMS 2021/ICSEE 2021, CCIS 1468, pp. 297–305, 2021.
https://doi.org/10.1007/978-981-16-7210-1_28

For the detection of gas flowmeter, sonic nozzle gas flow device, master meter method gas flow calibration device and bell type gas flow calibration device are generally used. These devices are generally large in volume and very heavy in weight [2]. When the gas flowmeters need annual inspection, it have to be sent to the designated testing unit for offline detection. The whole process will take at least one week, which greatly increases the shutdown time of the enterprise and reduces the production efficiency. How to save time for users, improve production efficiency, and accurately test has become a difficult problem in the field of inspection.

In this paper, we design a mobile gas flow standard device by the master meter method to solve the above problem. It has the advantages of simple structure, small volume, high verification efficiency and high calibration accuracy by using high-precision flowmeter as master meter.

2 Device Overview and Overall Design

The overall design idea of the device is to reduce the occupied space, move to the site, improve the calibration efficiency, reduce energy consumption and provide high accuracy calibration service. Figure 1 shows the overall appearance of the device. It can be seen that the structure of the device is very compact and occupies very little space. The gas flowmeter with diameter below DN80 can be verified on a desk sized device.

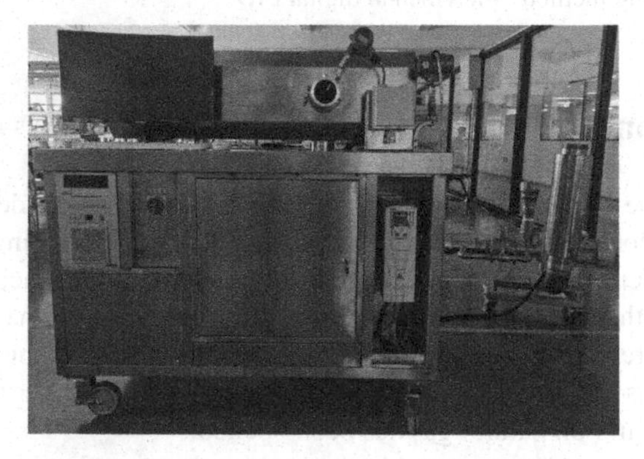

Fig. 1. Overall appearance of the device.

In terms of hardware, two Elster gas waist wheel flowmeters (DN50, DN100) are selected as the master meter of the device. This kind of flowmeter has excellent repeatability and long-term stability. The device is equipped with Rosemount temperature and pressure transmitter for temperature and pressure compensation. The flow regulation is realized through the combination of ABB Inverter and Elektror fan, which can quickly and accurately adjust to the required flow within the range and greatly improve the calibration efficiency. The controller unit is composed of Advantech's IPC-900 industrial computer, PCI series board and ADAM data acquisition module, and its control structure is shown in Fig. 2. ADAM-4117 module realizes the data acquisition function of the temperature and pressure transmitter. PCI-1753 digital I/O board detects the pulse of master meter as external interrupt source to start and stop the pulse counting function. PCI-1780 counter/timer card can count the pulse output of master meter and tested meter, and open and shut the ball valve.

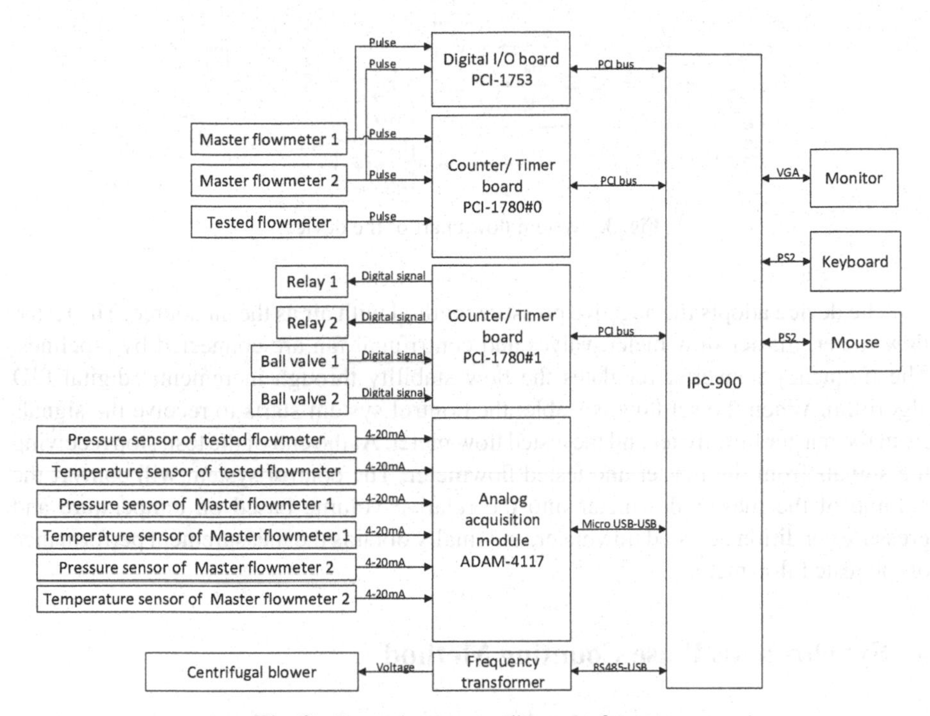

Fig. 2. Control structure diagram of system.

In terms of software, with the development of detection technology, virtual instrument technology plays an increasingly important role in the field of automatic detection. LabWindows/CVI is an interactive C language development software launched by NI company, which combines powerful and flexible C language platform with professional measurement and control tools for data acquisition, analysis and display [3]. The software system of the device provides a simple and easy-to-use program for calibration based on LabWindows/CVI and MySQL.

3 Device Principle

The system flow of the whole device is shown in Fig. 3. The designed flow range is 0.5–270 m^3/h, which can be used to verify DN8–DN80 gas flowmeter.

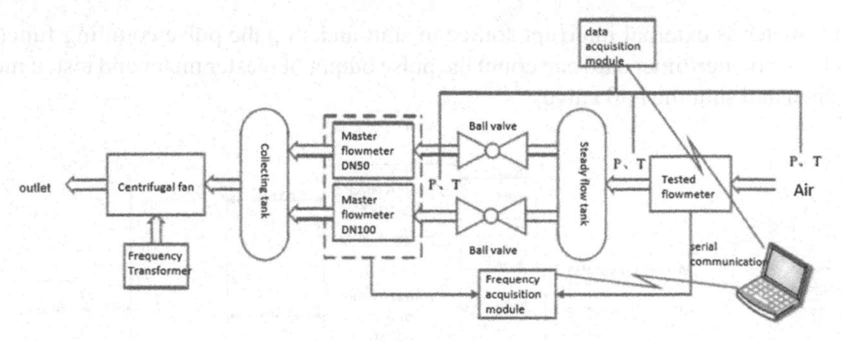

Fig. 3. System flow chart of the device.

The device adopts the negative pressure method with air as the air source. The tested flow meter, master flow meter, valves and centrifugal fan are connected by pipelines. The frequency converter regulates the flow stability through incremental digital PID algorithm. When the set flow is stable, the control system starts to receive the signals from the master flow meter and the tested flow meter. At the end of the test, stop receiving the signals from the master and tested flowmeter. The control system will convert the volume of the master flowmeter into the relative volume based on temperature and pressure condition in tested flowmeter, and finally obtains the measurement performance of the tested flowmeter.

4 Synchronous Pulse Counting Method

Because the commonly used counter chip counts at the rising or falling edge of the pulse signal, in the computer control system, the start and stop signals of the accumulated pulse of the counter are all sent by the computer, as shown in Fig. 4. Taking the following falling edge acquisition pulse as an example, it may have lost nearly a cycle time from the time when the computer sends the "start" command to the time when the falling edge of the next pulse is counted. The same is true for the "end" command. Because the frequency of the output signal of the standard flowmeter is sometimes very low, the non synchronization of the pulse and time will cause errors in the measurement results. For example, the frequency of the DN50 master gas waist wheel flowmeter in this flow standard device is 127 Hz at the maximum flow rate. If one pulse is lost in the total number of pulses after 10 s, the measurement error will be 0.08%. This obviously can not meet the requirements of establishing high accuracy flow standard device. In order to ensure its measurement accuracy, the master flowmeter needs a certain number of digits, so it needs a longer detection time T, which will lengthen the verification time, increase the energy consumption and reduce the efficiency of verification. In order to overcome this shortcoming, the author puts forward a fast and high-precision counting method--synchronous pulse counting method.

The principle of synchronous pulse counting method is shown in Fig. 4, which requires recording the complete pulses number N_M of the tested flowmeter between the computer start and stop commands in the verification process, and the actual time T_2 taken by these pulses. At the same time, record the complete pulse number N_N of the master flowmeter in this process, and actual time T_1. At the end of verification, what we really want to get is the corresponding pulse number N_S of the master flowmeter in the time T_2 when the tested flowmeter sends out pulse N_M. Considering that the flow standard device has the characteristics of stable flow in the verification of flowmeter, and the frequency f of the gas waist wheel flowmeter is certain under a certain flow rate, the pulse interpolation can be used to correct:

$$f = \frac{N_S}{T_2} = \frac{N_N}{T_1}. \tag{1}$$

$$N_S = N_N \frac{T_2}{T_1}. \tag{2}$$

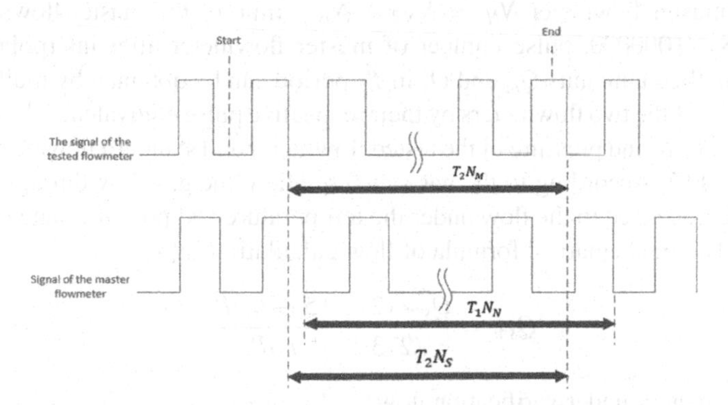

Fig. 4. Principle of synchronous pulse counting method.

As can be seen from Fig. 4, the implementation of this method can ensure that after the computer sends the start/stop signal, the time of collecting the master flowmeter or the tested flowmeter is completely synchronized with the number of pulses, so that the error of collecting frequency signal only depends on the timer error of timing T_1 and T_2.

Theoretically speaking, the use of synchronous pulse method can shorten the verification time of standard meter method to short enough. If the timer with timing accuracy of 1 μs is used, the verification time is set as 10 s, and the relative error of signal pulse count is only 10–5%, which can be ignored.

4.1 Realization of Synchronous Pulse Counting Method

Refer to Fig. 3. When the flow reaches the set point and is stable, make PCI-1780 send 1 MHz high frequency pulse, and start four pulse counters at the same time to record the pulse number of the tested flowmeter, high frequency pulse, master flowmeter 1 and master flowmeter 2. Port 0 of PCI-1780 is used as the interrupt control of the tested flowmeter pulse. When the calibration function is started, it does not count immediately. When the tested flowmeter pulse of port 0 reaches the rising edge of the next pulse, it triggers the interrupt and records the pulse number N_{M1} of the tested flowmeter at this time. At the same time, it records the high frequency pulse number HFC1 at this time. According to the set flow, judge and start the interrupt of master flowmeter on PCI-1753, and record the pulse number N_{N1} of master flowmeter 1 or 2 and the high frequency pulse number HFS1 at this time.

When the set calibration time is reached, the meter will not count immediately. When the tested flowmeter pulse of port 0 reaches the rising edge of the next pulse, it triggers the interrupt and records the pulse number N_{M2} of the tested flowmeter at this time. At the same time, it records the high frequency pulse number HFC2 at this time. According to the set flow, judge and start the interrupt of master flowmeter on PCI-1753, and record the pulse number N_{N2} of master flowmeter 1 or 2 and the high frequency pulse number HFS2 at this time.

The relevant data can be obtained as follows: pulse number of the tested flowmeter $N_M = N_{M2} - N_{M1}$, time of tested flowmeter $T_2 = (HFC2 - HFC1)/1000000$, pulse number of master flowmeter $N_N = N_{N2} - N_{N1}$, time of the master flowmeter $T_1 =$ (HFS2-HFS1)/1000000, pulse number of master flowmeter after interpolation $N_S = N_N \frac{T_2}{T_1}$. Then, the cumulants Q_m and Q_s in T_2 period can be obtained by multiplying the pulse number of the two flowmeters by their respective pulse equivalent. Then according to the temperature and pressure of the tested flowmeter and standard flowmeter collected by ADAM-4117. According to the gas state equation, the gas flow through the master flowmeter is corrected to the flow under the temperature and pressure state of the tested flowmeter. The mathematical formula of flow calculation is [4]:

$$Q_{sm} = \frac{Q_s * (273.15 + t_m)P_s}{(273.15 + t_s)P_m}. \tag{3}$$

Indication error under verification flow:

$$E = \frac{Q_m - Q_{sm}}{Q_{sm}}. \tag{4}$$

Where: Q_s is the cumulative flow of master flowmeter, t_m is the temperature at the tested flowmeter, t_s is the temperature at the master flowmeter, P_m is the pressure at the tested flowmeter, P_s is the pressure at the master flowmeter. Q_m is the cumulative volume flow of the tested flowmeter. Q_{sm} is the cumulative volume flow of the master flowmeter after temperature and pressure compensation.

The use of synchronous pulse counting method not only improves the accuracy of pulse counting, but also improves the verification efficiency of standard meter method.

5 Software Design

Software design includes the following processes: interface design, data structure design and function module design. The overall design of the software is shown in Fig. 5.

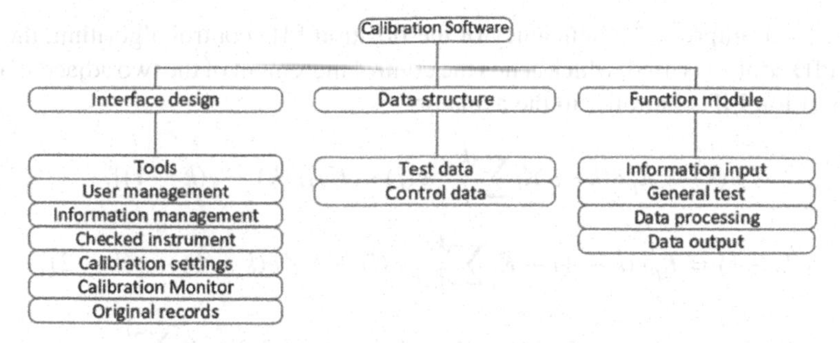

Fig. 5. Overall design of the software.

5.1 Interface Design

Because CVI provides abundant callback functions and graphical controls, the easy-to-use software interface can be designed. The software includes two main interfaces, login interface and main interface. The login interface allocates the permissions and functions that the current login personnel can use according to the entered account information. The main interface consists of seven sub interfaces: tool interface, user management interface, information management interface, checked instrument interface, calibration setting interface, calibration monitor interface and original records interface. The tool interface is used to input and modify the hardware parameters of the device. The user management interface is used to add, modify and delete account information. The information management interface is used for user to input the current commission order and sample information. The checked instrument interface is used to select the current sample. The calibration monitor interface mainly realizes the verification function of the tested flowmeter, and the original record interface is used to export the test information.

5.2 Data Structure Design

The data of calibration monitoring process can be saved in different tables simply and effectively through MySQL database management. Based on the powerful functions of MySQL database, the software can accurately realize data operation.

5.3 Function Module Design

The function modules mainly include information input, calibration, data processing and data output. The calibration function includes sending hardware control instructions, data acquisition and real-time display of hardware information. The data processing function realizes the analysis, processing and storage of verification data. The data output module outputs the verification data to the specified format of word file, users can easily generate certificates and original records.

In the calibration function, the PID program in the software is used to control the frequency converter to make the fan run stably at the set flow point.

In order to improve the deficiency of the position PID control algorithm, the incremental PID control is used, which sends the control increment of the two adjacent outputs of the position PID controller to the actuator.

$$u(k) = K_p e(k) + K_i \sum\nolimits_{j=0}^{k} e(j) + K_d[e(k) - e(k-1)]. \tag{5}$$

$$u(k-1) = K_p e(k-1) + K_i \sum\nolimits_{j=0}^{k} e(j) + K_d[e(k-1) - e(k-2)]. \tag{6}$$

$$\Delta u(k) = u(k) - u(k-1) = K_p[e(k) - e(k-1)] + K_i \sum\nolimits_{j=0}^{k} e(j) + K_d[e(k) - 2e(k-1) + e(k-2)]. \tag{7}$$

Compared with positional PID, incremental PID outputs increment. It reduces the influence of misoperation of the control system, and avoids the system to produce large vibration.

6 Verification of Measurement Results

The combined uncertainty of the whole set of device have been verified by Shanghai Institute of Measurement and Testing Technology (SIMT) based on verification regulation JJG 643-2003 [5]. Expanded uncertainty $U_{rel} = 0.59\%$ (k = 2).

The transfer comparison method is used to verify the measurement results [6]. A DN25 gas waist wheel flowmeter is selected as the measured object, the measurement range is 1–35 m^3/h, and the accuracy level is 1.5. This device, the measurement range is 0.5–270 m^3/h, the uncertainty is $U_{lab} = 0.59\%$ (k = 2); The sonic nozzle gas flow standard device from Shanghai Anjun Intelligent Technology Co., Ltd., whose measurement range of 0.5–3700 m^3/h, and the uncertainty is $U_{ref} = 0.26\%$ (k = 2). The results are shown in Table 1. The tested flow point followed verification regulation JJG 633-2005 for gas displacement meters.

Table 1. Comparison results.

Flow point (m^3/h)	Average error of master meter $y_{lab}(\%)$	Average error of sonic nozzle $y_{ref}(\%)$	$\lvert y_{lab} - y_{ref} \rvert$ (%)
1	−1.083	−1.164	0.081
7	1.019	1.164	0.145
35	0.570	0.785	0.215

It can be seen from Table 1 that $\lvert y_{lab} - y_{ref} \rvert \leq \sqrt{U_{lab}^2 + U_{ref}^2} = 0.645$. It is proved that the device is qualified. At the same time, the accuracy and repeatability of the standard meter device developed by the proposed method can be improved to almost the same as that of the sonic nozzle method. Calibration time of master meter (synchronous pulse counting method) was 30 s, and calibration time of sonic nozzle is 60 s. So the verification efficiency can be increased by 50%. Therefore, the device can be used for high accuracy verification service.

7 Conclusion

The device has the advantages of small volume, light weight, easy to move, good repeatability, high efficiency and low cost. The biggest innovation is to solve the problem of field calibration. The whole set of device and all equipped meters have been verified by Shanghai Institute of Measurement and Testing Technology (SIMT), and the results meet the design requirements. It can be used for the verification of various types of gas flowmeter. The field calibration service provided by the device can save the standby time of the production line when the flowmeter is sent for calibration, improve the use efficiency of enterprise equipment, and bring significantly improvement in economic benefits.

References

1. Ji, G.: Application Skills of Flow Measurement Instrument. Chemical Industry Press, Beijing (2003)
2. Xu, Y.H.: Development of gas flow standard device. Mod. Meas. Lab. Manag. **12**(2), 7–8 (2004)
3. Chen, Y.J., Huang, R.: Design and implementation of software based on LabWindows/CVI. Inf. Comput. (Theor. Ed.) **16**, 55–58 (2019)
4. Yuan, M.: Improvement of gas flowmeter verification ideas. Ind. Metrol. **21**(4), 48–49+61 (2011)
5. General Administration of Quality Supervision, Inspection and Quarantine of the People's Republic of China. J. JJG 643-2003 Verification Regulation of Flow Standard Facilities by Master Meter Method (2003)
6. General Administration of Quality Supervision, Inspection and Quarantine of the People's Republic of China. J. JJF 1033-2016 Rule for the Examination of Measurement Standards (2016)

Distributed Photovoltaic Grid-Connected Evaluation Method Based on Analytic Hierarchy Process and Radar Chart

Jianyu Chen[1], Jing Li[1](✉), Weimin Chen[1], Qian Guo[1], Jianming Chen[2], and Ruijin Dai[2]

[1] College of Mechanical and Electrical Engineering, China Jiliang University, Hangzhou, China

[2] Zhejiang Huayun Information Technology Co. Ltd., Hangzhou, China

Abstract. The large-scale use of distributed photovoltaic brings challenges to the safe and reliable operation of power grid. Therefore, it is important to explore the impact of distributed photovoltaic on power grid. Firstly, this paper introduces five distributed photovoltaic grid-connected reliability evaluation indexes, including photovoltaic absorption rate, voltage fluctuation rate, voltage over-limit rate, harmonic distortion rate, and inverse power flow. Secondly, the research sets the weight of above five indexes by analytic hierarchy process and obtains the experimental data after accessing to photovoltaic power supply by MATLAB simulation. Finally, this method is used to analyze real users in a certain area. The results show that the proposed distributed photovoltaic grid-connected evaluation method has practical value, which can provide reference for power companies in installing photovoltaic power scientifically, reducing system planning cost and improving grid operation reliability.

Keywords: Distributed photovoltaic · Evaluation index · Analytic hierarchy process · Radar chart

1 Introduction

The rapid development of the world economy has brought massive energy consumption and serious environmental pollution. In order to avoid such impact and realize the sustainable development strategy, it is indispensable to use sustainable energy [1, 2]. Different kinds of clean and renewable energy have been explored continuously. Because of the strong support of the country, distributed photovoltaic power has been well developed.

With the continuous investment of distributed photovoltaic power, more and more issues are appearing. For example, voltage fluctuation and flicker lead to the change of light intensity [3, 4], excessive power of photovoltaic power brings about voltage limit and so on. These problems will not only damage the electrical equipment, but also affect the safety and stability of distribution network [5]. Therefore, it is necessary to evaluate the impact of distributed photovoltaic access to the grid, so as to formulate the best access scheme and monitor the impact of the formal grid connection of distributed photovoltaic in real time.

K. Li et al. (Eds.): LSMS 2021/ICSEE 2021, CCIS 1468, pp. 306–314, 2021.
https://doi.org/10.1007/978-981-16-7210-1_29

Currently, scholars have a lot of research on the evaluation method of distributed photovoltaic. For instance, PCNN model is adopted in Literature [6] to evaluate the power quality of power grid. Nevertheless, the classification is not detailed and the result is not intuitive. Literature [7] analyzes the impact of distributed photovoltaic access area through three dimensions involving system power balance, power quality and grid stability. However, there is no detailed analysis of the indicators, and the numbers are too few to be accountable. Reference [8] evaluates the impact of distributed photovoltaic on the distribution network from five directions including reliability, power quality, relay protection, environmental protection level and efficiency change, but the definition of indicators was very rough. In reference [9], the analytic hierarchy process and entropy weight method are used to analyze each power index. Though the weight matching results are good, the use of a variety of methods makes the amount of calculation larger. Moreover, the results are not intuitive enough.

This paper shows the influence of distributed photovoltaic on distribution network through five evaluation indexes: photovoltaic accommodation rate, voltage fluctuation rate, voltage out of limit rate, harmonic distortion rate and counter current flow rate. In the meantime, analytic hierarchy process is applied to quantitatively evaluate the impact of distributed photovoltaic on distribution network. Finally, the research uses radar chart to present the results. Quantitative results make the results reliable and improve the credibility, and the use of radar chart makes the results more intuitive and convenient for the analysis of the results.

2 Design of Evaluation Index

There are five evaluation indexes adopted in this paper: photovoltaic absorption rate, voltage fluctuation rate, voltage out of limit rate, harmonic distortion rate and counter current flow rate. The specific definitions of the indicators are as follows:

(1) Photovoltaic Accommodation Rate

Photovoltaic accommodation rate refers to whether the users of a node can reasonably use the output power of photovoltaic power or not after photovoltaic is connected to the grid. Photovoltaic absorptivity A is defined as:

$$A = \frac{\sum_i^x A_i}{x}. \tag{1}$$

$$A_i = \begin{cases} 1, & P^{DG} < P^{av} \\ \frac{P_i^{av}}{P_i^{DG}}, & P^{DG} \geq P^{av}. \end{cases} \tag{2}$$

In these formulas, P_i^{av} is the daily average load of installation node, P_i^{DG} is the nominal power of distributed generation, and x is the number of installed photovoltaic.

(2) Voltage Fluctuation Rate

Voltage fluctuation rate is defined as the actual voltage of household electricity changes rapidly due to some reasons. Voltage fluctuation rate F is defined as:

$$F = \frac{\sum_i^n F_i}{n}. \tag{3}$$

$$F_i = \begin{cases} 1 - 4 * \Delta U_i & \Delta U_i < 0.1 \\ 0.6 - 4 * \Delta U_i & \Delta U_i > 0.1 \end{cases}. \tag{4}$$

$$\Delta U_i = \frac{U_i^{\max} - U_i^{\min}}{U_i^{\min}}. \tag{5}$$

In these formulas, ΔU_i is the gradient of each node voltage after photovoltaic access, U_{i} is the voltage of the ith user on the line, n is the number of users on the line, $U_i^{\max}$ is the maximum node voltage, $U_i^{\min}$ is the minimum node voltage.

(3) Out-of-Limit Voltage Rate

Voltage overrun rate refers to the ratio that the voltage of each node exceeds the rated voltage after the distributed photovoltaic is connected to the distribution network. According to the electrical national standard [9], the maximum household voltage should not exceed 7% of the rated voltage, So, voltage over limit O is defined as:

$$\frac{\sum_i^n O_{i.}}{n}. \tag{6}$$

$$O_i = \begin{cases} 1 - k1 * \left(220 - U_i^{\max}\right) \ldots U_i^{\max} < 220 \\ 1 - k2 * \left(U_i^{\max} - 220\right) \ldots 220 < U_i^{\max} < 235.4. \\ 0.6 - k3 * \left(U_i^{\max} - 235.4\right) \ldots U_i^{\max} > 235.4 \end{cases} \tag{7}$$

In these formulas, $k1$, $k2$, $k3$ are the coefficients of voltage in different ranges. To ensure that the evaluation score is 0.6 when the limit is properly reached, this paper defines $k1 = 0.018$, $k2 = 0.026$, $k3 = 0.03$.

(4) Inverse Tidal Current

Inverse tidal current indicates that when the output of distributed photovoltaic power is excessively large, current back to the transformer will leads to the damage of the transformer. The inverse current flow R is defined as:

$$R = \begin{cases} 1, & p_{\mathrm{DG}}^{\max} \le p_{\mathrm{L}}^{\min} \\ 1 - \frac{p_{\mathrm{DG}}^{\max} - p_{\mathrm{L}}^{\min}}{p_{\mathrm{DG}}^{\max}}, & p_{\mathrm{DG}}^{\max} \le p_{\mathrm{L}}^{\min} \end{cases}. \tag{8}$$

In the formula, $p_{\mathrm{DG}}^{\max}$ is the maximum output power for photovoltaic, $p_{\mathrm{L}}^{\min}$ is the total load for station area.

(1) Harmonic Distortion Rate

Harmonic distortion rate suggests that different amount of harmonics will be generated when different amount of photovoltaic is connected. Assuming that the number of installed photovoltaic power sources is x, the harmonic distortion rate [10] H is defined as:

$$H = 1 - \left(0.0617x^3 - 0.6875 * x^2 + 3.013 * x + 0.01\right)/100. \tag{9}$$

3 Algorithm Principle

3.1 Analytical Hierarchy Process

Analytic Hierarchy Process (AHP) is used to deal with complex decision-making when some indicators are difficult to quantify or interrelated. The specific steps of AHP are as follows:

1) Establish hierarchical model structure. The evaluation object is set to be distribution network area system, the judgment layer is the score result and the criterion layer is the five indicators.

2) Build the judgment matrix. Comparing five indicators with each other according to the scaling method. The importance of the ith and jth indicators is expressed as c_{ij}, then the judgment matrix is as follows:

$$C = \begin{bmatrix} c_{11} & c_{12} & \cdots & c_{1m} \\ c_{21} & c_{22} & \cdots & c_{2m} \\ \vdots & \vdots & \ddots & \vdots \\ c_{n1} & c_{n2} & \cdots & c_{nm} \end{bmatrix}. \tag{10}$$

3) Calculate the relative weights of indicators are based on matrix C, as follows:

$$f(\lambda) = |\lambda E - C| = 0. \tag{11}$$

In the formula, E is the identity matrix of $n * m$.

Obtain the latent root of $\lambda_1, \lambda_2, \ldots, \lambda_{\mathrm{n}}$, choose the maximum latent root $\lambda_{\max}$. let

$$|\lambda_{\max} E - C| = 0. \tag{12}$$

Solve and get the vector of the maximum latent root $w = [w_1, w_2, \ldots, w_{\mathrm{n}}]$.

4) Consistency testing. Inset the λ_{max} into the consistency index CI, if CI is zero, the λ_{max} have the full consistency; The closer to zero of CI, the consistency is better; On the other hand, the consistency is worse. Consistency ratio (*CR*) is used to measure the

consistency, if CR is less than 0.1, the matrix will have the satisfactory consistency. The calculation formulas of CI and CR are as follows:

$$CI = \frac{\lambda_{\max} - n}{n - 1}. \tag{13}$$

$$CR = \frac{CI}{RI}. \tag{14}$$

In formulas, n is the number of index, the value of RI can be referenced in one-time indicator RI table [12].

3.2 Rate Chart Method

As a multi-variable graphical and visual analysis method, radar chart method is beneficial to intuitively reflect the evaluation results.

The evaluation process of this method is as follows:

1) Confirm the weight of every index. Every index has the different influence extent. This paper defines the weight w_i by analytical hierarchy process.

2) Taking a point as the center point, draw a ray of unit length for this point. Taking this ray as the benchmark, rotate the ray by a certain degree θ_i according to the w_i, and get the secondary. Similarly, plot the remaining ray, and get a unit circle at last. The calculation formula of the angle θ_i corresponding to the weight w_i is as follows:

$$\theta_i = w_i * 360^{\circ}. \tag{15}$$

3) Starting from the center of the circle, do the angle dividing line of each region, and mark at a distance of d_i (d_i is the corresponding index calculation result) from the dot on each corner line.

4) Connect the labeled points in 3) with the two sides of the arc of the corresponding sector area to form a new closed graph. The formulas of the area and gold are as follows:

$$S = \sum\nolimits_{i}^{n} sin\frac{1}{2}\theta_i d_i. \tag{16}$$

$$G = \frac{S}{S_{\max}}. \tag{17}$$

In formulas, n is the number of the index, $S_{\max}$ is the area of full score, its value is 2.92 cm^2.

4 Example Analysis

4.1 Weight Calculation

Determine the relative importance of indicators through expert surveys. It is defined as $c_{12} = 0.6$, $c_{23} = 1$, $c_{34} = 1.5$, $c_{45} = 0.6$. the judgment matrix C as follows:

$$C = \begin{bmatrix} 1 & 0.6 & 0.6 & 1.5 & 0.6 \\ 5/3 & 1 & 1 & 2.5 & 1 \\ 5/3 & 1 & 1 & 2.5 & 1 \\ 2/3 & 2/5 & 2/5 & 1 & 2/5 \\ 5/3 & 1 & 1 & 5/2 & 1 \end{bmatrix}.$$

According to the formula (10) and (11), calculate the weight of every index:

$$w = [w_1, w_2, w_3, w_4, w_5] = [0.15, 0.26, 0.25, 0.1, 0.24].$$

4.2 Multifactorial Evaluation

Use the algorithm model in this paper to analyze the electricity consumption data of a residential area in June 2020. The topology of the residential area is shown in Fig. 1, Use MATLAB to build a photovoltaic grid-connected simulation circuit to obtain the simulation data after accessing to photovoltaic power supply.

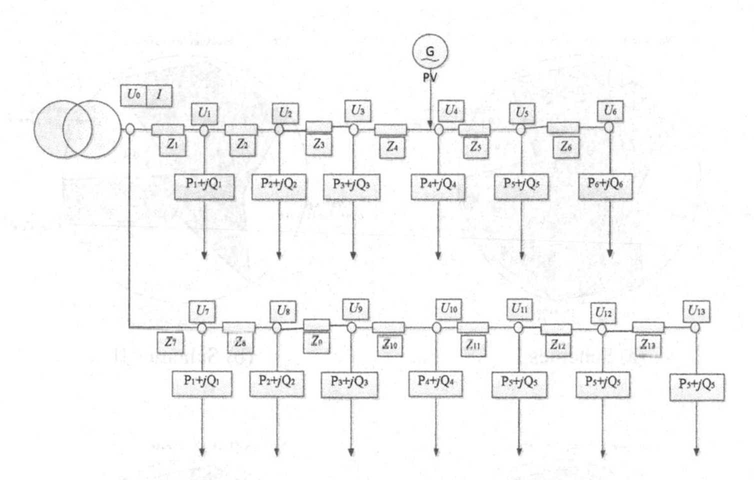

Fig. 1. Distribution line topology of a residential quarter.

A day was selected randomly from June 2020 to get the load data of the day. In this paper, according to the rated power, installation location and installation number of distributed photovoltaic power supply, four kinds of grid-connected schemes are designed respectively. Scheme I, connecting a photovoltaic node with rated output power of 1,000 W and maximum outputting power of 1, 500 W to the first branch 4; Scheme II, connecting a photovoltaic with rated output power of 5000 W and maximum outputting power of 10,000 W is to Node 4 of the first branch; Scheme III, connecting two PVs with rated output power of 500 W and maximum outputting power of 750 W to node 4 and node 6 of the first branch respectively; In Scheme IV, connecting a photovoltaic with rated output power of 1,000 W and maximum outputting power of 1,500 W to Node 8 of the second branch.

According to formula (1)–(8), the evaluation indexes and scores of four schemes are calculated respectively, as shown in Table 1.

Draw radar charts for each scenario as shown in Fig. 2.

As can be observed from Fig. 2(a), the scores of PV consumption rate and voltage over-limit rate in scheme I are not high. indicating that the output power of PV cannot be consumed. In addition, the index score of voltage overrun rate is not high, which indicates that the voltage in the station area is on the high side.

Table 1. Evaluation results of four schemes.

Scheme	Photovoltaic absorption rate	Voltage fluctuation rate	Voltage over-limit rate	Inverse power flow	Harmonic distortion rate	Gold
Scheme I	0.395	0.972	0.496	1.000	0.976	0.76
Scheme II	0.079	0.967	0.486	0.305	0.976	0.52
Scheme III	0.896	0.972	0.496	1.000	0.962	0.84
Scheme IV	1	0.973	0.497	1.000	0.976	0.86

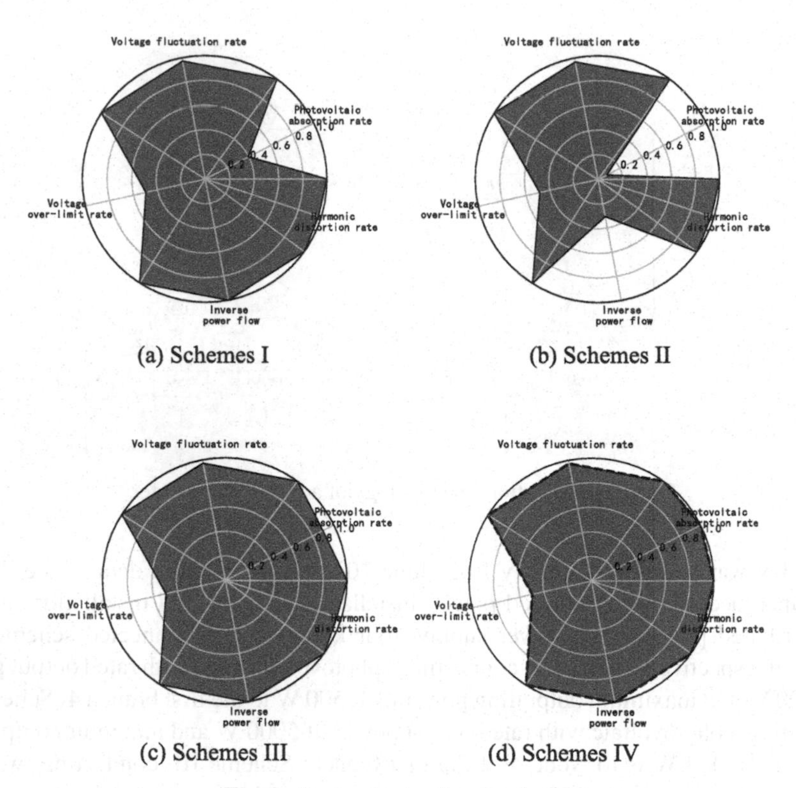

Fig. 2. Radar charts of each scheme.

It can be concluded from Fig. 2(b) that the scores of the three indexes of the consumption rate, the voltage exceeding limit rate and the inverse power flow rate in Scheme II are considerably low. The node can not consume the output power of photovoltaic, and high-power photovoltaic makes the voltage higher.

It can be seen from Fig. 2(c) that two small capacity photovoltaic devices are installed at different nodes of the same branch, which makes the area of photovoltaic absorption rate index increase significantly, while the area of other indexes basically remains unchanged, and the overall score increases.

It can be observed from Fig. 2(d) that after changing the access location of distributed PV in scheme 4, the score of PV absorption rate index also increases obviously, and the score of voltage fluctuation rate and voltage out of limit rate increases slightly.

Through the evaluation and analysis of the four schemes, the suggestions and conclusions for the installation of photovoltaic in the station area are as follows:

(1) The fraction of low-power photovoltaic access voltage out of limit rate is very low, which indicates that the voltage of the station area is too high, so it is not recommended to install distributed photovoltaic in the station area. Moreover, the consumption of the daily power of multiple users in the station area is low. The power of distributed photovoltaic output cannot be completely consumed.

(2) If it is ineluctable to install distributed photovoltaic power in the station area, it would be necessary to connect one or more small power distributed photovoltaic power to nodes with high power consumption, such as node 2 and node 7 in the station area.

5 Conclusion

In this paper, the evaluation index system is established for the impact of distributed photovoltaic power access to distribution network. The analytic hierarchy process is introduced to obtain the weight of each index. Besides, the evaluation results are intuitively displayed by radar chart, which provides an effective basis for Assessment of the impact of distributed photovoltaic access area.

In the future. This may have more impact on the grid. In the future, the current evaluation method can be optimized to make it more widely applicable, and more influencing factors can be considered to increase the number of indicators. This will *make the evaluation results more persuasive.

Acknowledgments. This research is supported by the Zhejiang Provincial public welfare Foundation (LGG20E070003), China.

References

1. Wang, X., Liu, G., Yuan, Z., et al.: Analysis on the effects of Volt/Var control method considering distributed generation. Power Syst. Prot. Control **42**(1), 47–53 (2014)
2. Chen, Y.: The Impact Study of Distribution Generation Access to Distribution Grid and Active Management Technology. North China Electric Power University (2014)
3. Xiao, Q., Fang, Z., Sun, H., et al.: Study on load prediction of smart grid with distributed power supply. Foreign Electron. Measure. Technol. **39**(2), 77–82 (2020)
4. Raghavendra, P., Gaonkar, D.N.: Online voltage estimation and control for smart distribution networks. J. Mod. Power Syst. Clean Energy **4**(1), 40–46 (2016). https://doi.org/10.1007/s40565-016-0187-6
5. Ma, L., Yang, T., Dou, C.: Research on intelligent evaluation and method of distributed power grid- connected power quality. Electron. Measure. Technol. **43**(11), 74–78 (2020)
6. Qin, Y., Lu, Z., Liu, J., et al.: The power quality evaluation method of microgrid based on PCNN model. J. Jiangsu Inst. Technol. **22**(6), 35–40 (2016)
7. Dong, L., Qiaoyu, K., Peng, W., et al.: Multidimensional comprehensive impact analysis of distributed photovoltaic access on power grid. China Equip. Eng. **04**, 186–187 (2021)

8. Hu, S., Shi, Y., Han, J., et al.: The evaluation method of the impact of distributed generation on distribution network. Rural Electrification **01**, 10–14 (2019)
9. Li, F., Sun, B., Wang, X., et al.: Power quality assessment for rural rooftop photovoltaic access system based on analytic hierarchy process and entropy weight method. Trans. Chin. Soc. Agric. Eng. (Trans. CSAE) **35**(11), 159–166 (2019)
10. GB / T 12326, power quality voltage fluctuation and flicker (2008)
11. Gao, J., Zheng, Y., Ni, J., et al.: Analysis of the causes of liquefied gas leakage accidents in tank truck area based on fault tree-analytic hierarchy process. Oil Gas Field Ground Eng. **40**(03), 44–49 (2021)

Coordinated Setting Method of PSS and PSVR Parameters in Generator

Jiatong Yu, Kewen Wang(✉), Mengru Chen, Songfang Jiang, and Jian Chen

School of Electrical Engineering, Zhengzhou University, Zhengzhou 450001, China
kwwang@zzu.edu.cn

Abstract. Power system stabilizer (PSS) is widely used to damp power angle oscillation in power systems, and power system voltage regulator (PSVR) can improve the voltage characteristics. Both devices can be installed in synchronous generators as additional controllers, while it is suggested that PSS and PSVR should be installed simultaneously on the generator to improve the angle and voltage oscillation modes of the system. In this paper, an approach for setting PSS and PSVR parameters in generators is proposed. The time constants and gain coefficients of PSS and PSVR are set based on the results of analyzing phase-frequency characteristics and root locus. The interaction between PSS and PSVR is considered in the proposed method. It is verified in the eight-machine system that after setting PSS and PSVR parameters using this method, both the power angle and voltage oscillation modes of the system achieve high damping levels.

Keywords: Power system stabilizer · Power system voltage regulator · Small-signal stability

1 Introduction

Applying PSS to improve system damping and suppress low-frequency oscillations is a simple, economic and effective measure, which is widely used worldwide [1, 2]. When the tie line breaks down, the problem of dynamic voltage stability is particularly prominent. Compared with the conventional terminal voltage control, PSVR adopts the high voltage side voltage control to ensure the voltage stability of the remote high voltage bus to improve the voltage stability of the power system [3]. The different structures and different control strategies of generator high voltage side voltage controllers such as PSVR are proposed [4, 5]. The method of PSS parameters setting is relatively mature, while the related research of PSVR is in the initial stage. Most of the existing literature adopts the time-domain simulation method to set PSVR parameters in a single infinite machine system and analyze the influence of PSVR on the stability of transient voltage [6, 7]. The method of eigenvalue analysis of the multi-machine system is seldom used to analyze the dynamic voltage characteristics of the system and to analyze the PSVR parameters setting method.

In this paper, the method of setting PSS and PSVR parameters in the multi-machine system is proposed, which provides a reference for setting PSVR parameters. The calculation method of dynamic angle characteristics [8] is extended to the analysis of dynamic

K. Li et al. (Eds.): LSMS 2021/ICSEE 2021, CCIS 1468, pp. 315–324, 2021.
https://doi.org/10.1007/978-981-16-7210-1_30

voltage, and the amplitude-frequency and phase-frequency characteristics of voltage are calculated and the oscillation mode is selected. PSVR parameters are set according to the setting method of PSS parameters, that is, the time constants of PSVR are set according to the voltage phase-frequency characteristics, and the gain coefficients is are set according to the root locus of voltage mode.

2 PSS and PSVR

PSS can effectively improve the damping of the power system and suppress low-frequency oscillations. At present, PSS is widely used in the world and with mature technology. The basic principle is based on the torque decomposition method, compensating the phase lag characteristics of the generator excitation system to maximize the damping component of the electromagnetic torque provided by the excitation system, and finally, achieve the goal of improving the system damping [8].

PSVR can improve the voltage characteristics of the system. In this paper, PSVR adopts 8-PSS in the Power System Analysis Comprehensive Program (PSASP). PSVR control is introduced based on AVR control of generator terminal voltage by conventional excitation system. The input signal of PSVR is the voltage variation of the high voltage side of the main transformer and it can maintain the stability of the high-voltage bus voltage. The generator excitation voltage control with PSVR is shown in Fig. 1. The control block diagram of PSVR is shown in Fig. 2. The main parameters include voltage control gain coefficient K_{qv}, transmitter time constant T_{qv}, phase shift link-time constant T_q and T'_q, and limiter link parameters.

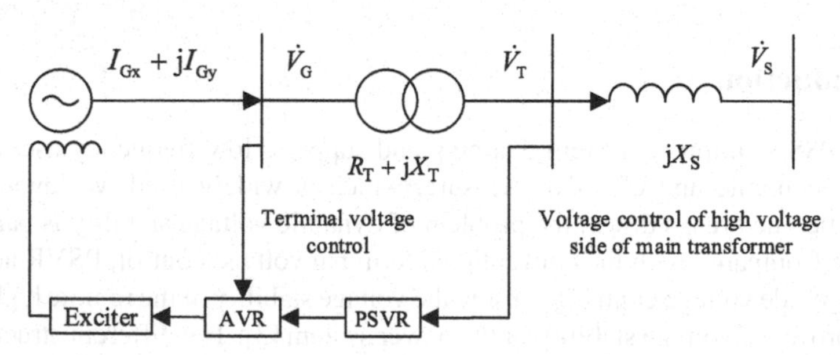

Fig. 1. Excitation control of the generator.

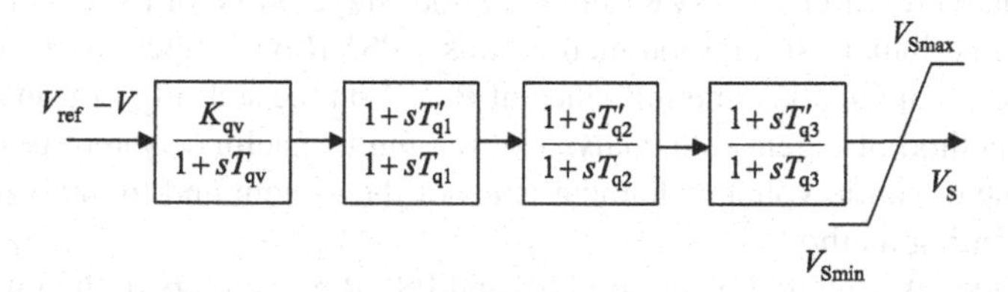

Fig. 2. Control block diagram of PSVR.

3 Power Angle Oscillation Modes and Voltage Oscillation Modes

3.1 Eigenvalue Analysis

In the analysis of small-signal stability of power system, the linearized system model can be expressed as:

$$\begin{bmatrix} \Delta\dot{X} \\ \mathbf{0} \end{bmatrix} = \begin{bmatrix} A' & B' \\ C' & D' \end{bmatrix} \begin{bmatrix} \Delta X \\ \Delta Y \end{bmatrix}. \tag{1}$$

Where $\Delta\boldsymbol{X}$ is the state variable of the system, $\Delta\boldsymbol{Y}$ is the non-state variable. $\boldsymbol{A}'$, $\boldsymbol{B}'$, $\boldsymbol{C}'$, and $\boldsymbol{D}'$ are the correlation submatrices.

The non-state variables are eliminated to obtain the state space equation:

$$\Delta\dot{X} = A\Delta X. \tag{2}$$

Where $\boldsymbol{A} = \boldsymbol{A}' - \boldsymbol{B}'\boldsymbol{D}'^{-1}\boldsymbol{C}'$, and it is the system state matrix, and the small-signal stability of the system can be analyzed according to its eigenvalue ($\lambda_i = \alpha_i + j\beta_i, i = 1, 2, \cdots$). The real part α_i describes the attenuation degree of the oscillation, the imaginary part β_i corresponds to the oscillation frequency, and the corresponding damping ratio ($\xi_i = -\alpha_i \Big/ \sqrt{\alpha_i^2 + \beta_i^2}$) describes the dynamic quality of the system.

Among the state variables in Eq. 1, some variables are closely related to the power angle, such as power angle $\Delta\delta$ and angular velocity $\Delta\omega$ of the generator. Some are greatly affected by voltage, such as generator transient potential $\Delta E'_{\mathrm{d}}$ and $\Delta E'_{\mathrm{q}}$, and sub-transient potential $\Delta E''_{\mathrm{q}}$. In this paper, two types of oscillation modes are considered and analyzed respectively with corresponding eigenvalues which are power angle modes and voltage modes.

3.2 Power Angle Oscillation Modes of Multi-machine System

In this paper, the participation factor is firstly calculated, and then the angle oscillation modes are selected by using the electromechanical correlation ratio index.

Firstly, the participation factors P_{ki} of $\Delta\delta$ and $\Delta\omega$ in the state variables are calculated. For the eigenvalue λ_k (corresponding to the angle oscillation mode), the participation degree of the state variable ΔX_i in the corresponding mode can be conveniently calculated by using its left and right eigenvectors:

$$P_{ki} = W_{ki} U_{ki}. \tag{3}$$

W_{ki} and U_{ki} respectively represent the left and right eigenvector elements corresponding to the eigenvalue λ_k.

P_{ki} can be used to calculate the electromechanical correlation ratio $\rho_{\delta,k}$:

$$\rho_{\delta,k} = \sum_{\Delta X_i \in C_{P\delta}}^{i} P_{ki} \Big/ \sum_{\Delta X_j \notin C_{P\delta}}^{j} P_{kj}. \tag{4}$$

$C_{P\delta}$ represents the set of state variables, which is related to the power angle. $\rho_{\delta,k}$ describe the proportion of the state variable $\Delta\delta$ and $\Delta\omega$ in the corresponding angle oscillation mode. The angle oscillation modes of the system can be selected according to the value of $\rho_{\delta,k}$ and the frequency range of angle oscillation. The participation factor is used to determine the participation of each generator in each mode. There are $N-1$ angle oscillation modes in the N-machine system.

3.3 Voltage Oscillation Modes of Multi-machine System

In this paper, the selection of voltage oscillation modes is carried out according to the processing method of angle oscillation modes. The state variables $\Delta E_{\mathrm{d}}'$, $\Delta E_{\mathrm{q}}'$ and $\Delta E_{\mathrm{q}}''$ of the generator are used to calculate the participating factor. k is the sequence number of the eigenvalue (corresponding to the voltage oscillation mode), and i is the sequence number of the state variable. The voltage correlation ratio is calculated according to Eq. 5, and then the voltage oscillation mode is selected. C_{QV} represents the set of voltage state variables.

$$\rho_{\mathrm{V},k} = \sum_{\Delta X_i \in C_{\mathrm{QV}}}^{i} P_{ki} \Big/ \sum_{\Delta X_j \notin C_{\mathrm{QV}}}^{j} P_{kj}. \tag{5}$$

In the eigenvalue analysis model, all state variables are based on $\Delta\delta = 0$, so N voltage oscillation modes need to be considered in the N-generator system.

4 Parameters Setting Method of PSS and PSVR

4.1 Parameters Setting Method of PSS

PSS can improve the damping of the power system, and PSS parameters should be adjusted according to the power angle stability characteristics of the generator.

Firstly, the phase-frequency characteristics of PSS (φ_{PSS}) are determined according to the hysteresis characteristics of the excitation system (φ_{EXC}) and the input signal of PSS. Then the parameters of the phase-shifting link are set. In this paper, the electromagnetic power variation ΔP_{e} is selected as the input signal of PSS, that is, according to $\varphi_{\mathrm{EXC}} + \varphi_{\mathrm{PSS}}(\Delta P_{\mathrm{e}}) = 0°$, the time constant of PSS phase-shifting link is set.

According to the root locus method, the gain coefficient of PSS is determined. The root locus of the angle oscillation modes of the system can be obtained by using the eigenvalues of the multi-machine system, and the critical gain of the PSS can be obtained according to the root locus of the corresponding angle oscillation mode. According to the input signal, 1/8–1/2 of the critical gain can be selected.

Referring to the above PSS parameters setting method, setting the time constant of PSVR according to the voltage characteristics of the generator, and setting the gain coefficient of PSVR according to the root path of the voltage oscillation mode. In this paper, the calculation method of dynamic power angle characteristics of the system is extended to the analysis of dynamic voltage, and the amplitude-frequency and phase-frequency characteristics of reactive voltage are calculated. Referring to the method of calculating the dynamic characteristic of power angle, this paper uses the sixth-order model of the generator to derive the analytical expression of voltage characteristics.

4.2 Dynamic Voltage Characteristic of the Generator

Generator voltage characteristics can represent the changes in the bus voltage on the high voltage side (V_T) and its ability to maintain stability when the infinite bus voltage (V_S) is disturbed.

In this paper, the generator voltage characteristics are calculated for the single machine system as shown in Fig. 1. The sixth-order model is adopted for the generator, and the related equations of excitation system, speed regulating system, the transfer functions of PSS and PSVR are added into the relevant expressions to form the differential and algebraic equations for the single machine infinite system. After linearization and calculation, the analytical expression for calculating voltage characteristics can be obtained. The solution idea is as follows:

(1) Calculate the relationship between generator terminal voltage (ΔV_G) and current (ΔI).
(2) The relationship between ΔV_T and ΔV_G is shown in Eq. 6. Substituting the calculation results in step (1), ΔV_{Gx} and ΔV_{Gy} in Eq. 6 are eliminated, and then the expression of ΔV_T on the current is obtained.

$$\begin{cases} \Delta V_{Tx} = \Delta V_{Gx} - R_T \Delta I_x + X_T \Delta I_y \\ \Delta V_{Ty} = \Delta V_{Gy} - R_T \Delta I_y - X_T \Delta I_x \end{cases}. \tag{6}$$

(3) The relation between ΔV_T and ΔV_S is shown in Eq. 7.

$$\Delta \dot{V}_T = \Delta V_S + jX_S \Delta \dot{I}. \tag{7}$$

Substitute the expression obtained in step (2) into Eq. 7 to obtain the analytical expression of the generator voltage characteristics:

$$\Delta \dot{V}_T = F_{TS}(p, V_{Gx}, V_{Gy}, I_x, I_y, G_{EXC}, G_{GOV}, G_{PSS}, G_{PSVR}) \Delta V_S. \tag{8}$$

F_{TS} is the functional expression of abbreviation and p is the differential operator. V_{Gx}, V_{Gy}, I_x and I_y are the components of terminal voltage and current in the static coordinate system. G_{EXC}, G_{GOV}, G_{PSS} and G_{PSVR} are the transfer functions of excitation system, speed regulation system, PSS, PSVR.

Due to the existence of the differential operator p in the calculation process, although ΔV_S only changes its amplitude, both $\Delta \dot{V}_T$-amplitude and $\Delta \dot{V}_T$-angle will change. The change of $\Delta \dot{V}_T$-angle is caused by the phase shift in the control unit. Therefore, the voltage amplitude-frequency characteristics reflect the influence of amplitude, which belongs to the voltage reactive power relationship. The voltage phase-frequency characteristics reflect the influence on the $\Delta \dot{V}_T$-angle and then affect the active power and angle characteristics of the generator.

4.3 Parameters Setting Method of PSVR

From the point of view of decoupling, PSVR mainly works on reactive voltage. However, if the reactive power load of the system is heavy or the gain coefficient of PSVR is too large, the power angle characteristics of the system will be affected obviously. To deal with this interaction, the following points are considered when setting PSVR parameters in this paper.

(1) Setting time constants of PSVR: Time constants of PSVR can be set according to the voltage phase-frequency characteristics. By adjusting the PSVR time constants, the voltage phase-frequency characteristics in the considered frequency band can be improved, and the voltage angle can be reduced. Thus, the influence of PSVR on active power and angle is reduced, and the effect on voltage characteristics is enhanced.
(2) Setting gain coefficient of PSVR: The gain coefficient of PSVR can be determined by the root locus method. Calculating the eigenvalues of the multi-machine system and select the voltage oscillation modes according to the method described in Sect. 3.3. By changing the gain coefficient of PSVR and eigenvalue calculation, the root locus of the corresponding voltage oscillation mode is formed. The critical gain of PSVR can be determined from the root locus, and 1/8–1/2 of the critical gain can be selected.
(3) In addition, when determining the initial gain coefficient of PSVR, it is necessary to check whether there is excessive influence on the angle oscillation modes. For the power angle oscillation mode in which the PSVR is located, the damping ratio should not be less than the given damping ratio, such as 0.03.

5 Results and Analysis

In this section, the eight-machine system shown in Fig. 3 is used to verify the influence of the PSS gain coefficient on the dynamic voltage stability of the system and the improvement of the PSVR on the dynamic voltage characteristics of the system.

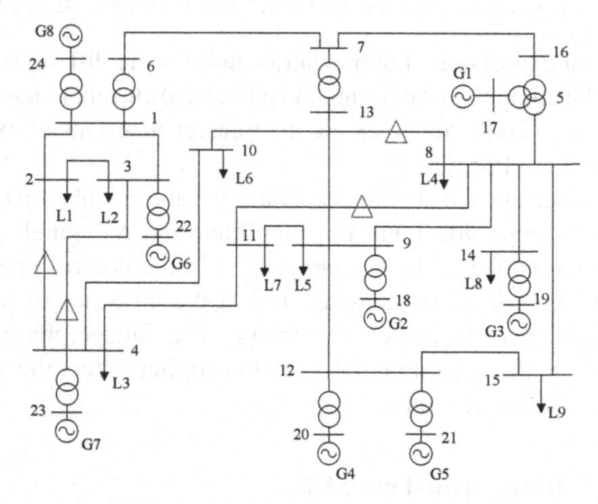

Fig. 3. The eight-machine system.

To facilitate the analysis of the influence of PSVR on the voltage oscillation modes, disconnect the lines 2–4, 3–4, 8–10 and 8–11. Power angle oscillation modes without PSS or PSVR are shown in Table 1. The damping of mode 3 is relatively small, and the damping ratio of other modes can be further improved.

Table 1. Power angle oscillation modes without PSS or PSVR.

No.	α	β	ξ	Main participating generator
1	−0.2503	8.0154	0.0312	G6
2	−0.2331	5.5304	0.0421	G7 ↔ (G5, G4, G3)
3	−0.0010	3.3050	0.0003	G7 ↔ (G8, G5, G3, G4)
4	−0.5287	7.3181	0.0721	G3 ↔ (G4, G5)
5	−0.7803	10.7812	0.0722	G2
6	−1.7537	15.6048	0.1117	G1
7	−0.6528	7.9685	0.0817	G4 ↔ G5

The specific situation of each generator participating in the oscillation modes is analyzed. Type 1 PSS in PSASP is added in G2, G3, G5, G6 and G7, and the transfer function of PSS is shown in Eq. 9.

$$F_{\text{PSS}}(s) = K_{\text{PSS}} \frac{sT_{\text{w}}}{1 + sT_{\text{w}}} \cdot \frac{1 + sT_1}{1 + sT_2} \cdot \frac{1 + sT_3}{1 + sT_4}. \tag{9}$$

K_{PSS} is the gain coefficient. T_{w} is the time constant of the separation link. T_1, T_2, T_3 and T_4 are the time constants of the phase shift link.

For all PSSs, $T_{\text{w}} = 5$ s. The initial time constant of the phase-shifting link is set according to the angle phase-frequency characteristics in the considered frequency band, and the initial gain coefficient is determined according to the root locus of the angle oscillation modes calculated by the eigenvalues of the multi-machine system. The initial parameters of PSS are listed in Table 2.

Table 2. Parameters of PSSs.

	PSS2	PSS3	PSS5	PSS6	PSS7
K_{PSS}	0.50	−0.30	0.30	0.30	−0.35
T_{W}	5.0	5.0	5.0	5.0	5.0
T_1	0.43	0.90	0.43	0.32	0.33
T_2	0.40	0.85	0.40	0.30	0.30
T_3	0.043	0.080	0.42	0.31	0.31
T_4	0.040	0.070	0.40	0.30	0.30

After installing PSSs, eight voltage oscillation modes are investigated (not given). As the PSS gain coefficient in G6 increases, it will cause instability in the first voltage mode, shown in Table 3. The other 7 voltage modes are stable.

Table 3. The influence of PSS6 gain coefficient on voltage mode.

K_{PSS6}	No.	α	β	ξ	First participant generator
0.1	1	−0.0898	1.6680	0.0538	G6
0.2	1	−0.0510	1.6483	0.0310	G6
0.3	1	−0.0121	1.6292	0.0074	G6

To improve the unstable voltage mode, PSVR can be installed on the first participating generator G6. According to the PSVR control block diagram in Fig. 2, its transfer function is shown in Eq. 10.

$$F_{PSVR}(s) = \frac{K_{qv}}{1 + sT_{qv}} \cdot \frac{1 + sT'_{q1}}{1 + sT_{q1}} \cdot \frac{1 + sT'_{q2}}{1 + sT_{q2}} \cdot \frac{1 + sT'_{q3}}{1 + sT_{q3}}. \tag{10}$$

According to the generator voltage phase-frequency characteristic of G6, the time constant of PSVR is adjusted to minimize the influence on the power angle in the considered frequency band, and the time constant is obtained as follows: $T_{qv} = 0.1$ s, $T'_{q1} = 0.5$ s, $T_{q1} = 0.2$ s, $T'_{q2} = 0.1$ s, $T_{q2} = 0.8$ s, $T'_{q3} = 0.1$ s, $T_{q3} = 0.8$ s. The root locus of the voltage oscillation modes is calculated according to the eigenvalues of the multi-machine system, and then the limit value of the gain coefficient of PSVR can be obtained by the root locus method, which was about 4.5. Taking into account the influence on the power angle modes, the gain value of PSVR is 0.5. The power angle oscillation modes and voltage oscillation modes are listed in Table 4 and Table 5. The results show that PSS can improve the power angle modes of the system. By adding PSVR and parameters setting of the proposed method, the voltage oscillation modes of the system can be effectively improved on the premise of reducing the influence on the power angle characteristics.

Table 4. Power angle oscillation modes with PSS and PSVR.

No.	α	β	ξ
1	−0.5967	7.0745	0.0841
2	−0.9627	5.8445	0.1625
3	−1.1163	3.8463	0.2787
4	−1.2086	8.7446	0.1369
5	−1.3385	10.3141	0.1287
6	−1.7651	15.6240	0.1123
7	−0.8928	8.1059	0.1095

Table 5. Voltage oscillation modes with PSS and PSVR.

No.	α	β	ξ
1	−0.0775	1.6292	0.0475
2	−0.6573	0.7677	0.6504
3	−1.4700	0.9365	0.8434
4	−17.0831	3.6080	0.9784
5	−0.5966	0.9355	0.5377
6	−0.6504	0.7028	0.6792
7	−0.4903	0.7264	0.5595
8	−0.6504	0.7028	0.6792

6 Conclusions

In this paper, the function and working principles of PSS and PSVR are briefly introduced. The calculation method of dynamic power angle characteristic is extended to the analysis of dynamic reactive power and voltage, and the method of calculating voltage phase-frequency characteristic and voltage oscillation modes is carried out. Referring to the parameters setting method of PSS, a parameters setting method of PSVR is given. Finally, it is verified in the eight-machine system. The results show that PSS can improve the system damping, but it may also affect the voltage stability, and the PSVR may be able to compensate for the influence of PSS. In other words, PSVR can improve the system voltage stability to a certain extent and expand the adjustment range of PSS gain coefficient. The parameters setting method is feasible, which can improve the dynamic power angle and dynamic voltage stability of the system.

References

1. Sreedivya, K.M., Jeyanthy, P.A., Devaraj, D.: An effective AVR-PSS design for electromechanical oscillations damping in power system. In: 2019 IEEE International Conference on Clean Energy and Energy Efficient Electronics Circuit for Sustainable Development, pp. 1–5 (2019)
2. Agrawal, V., Rathor, B., Bhadu, M., Bishnoi, S.K.: Discrete time mode PSS controller techniques to improve stability of AC microgrid. In: 2018 8th India International Conference on Power Electronics, pp. 1–5 (2018)
3. Srun, C., Samman, F.A., Sadjad, R.S.: A high voltage gain DC-DC converter design based on charge pump circuit configuration with a voltage controller. In: 2018 2nd International Conference on Applied Electromagnetic Technology, pp. 79–84 (2018)
4. Dong, Y., Xie, X., Zhou, B., Shi, W., Jiang, Q.: An integrated high side var-voltage control strategy to improve short-term voltage stability of receiving-end power systems. IEEE Trans. Power Syst. **31**, 2105–2115 (2016)
5. Noguchi, S., Shimomura, M., Paserba, J.: Improvement to an advanced high side voltage control. IEEE Trans. Power Syst. **21**, 683–692 (2006)

6. Kim, C.K., Kine, J.Y., Lee, S.D., et al.: Stability enhancement in HVDC system with STATCOM. Engineering **03**, 1072–1081 (2011)
7. Maslennikov, V.A., Ustinov, S.M.: Method and software for coordinated tuning of power system regulators. IEEE Trans. Power Syst. **12**, 1419–1424 (1997)
8. Chung, C.Y., Wang, K.W., Tse, C.T., Bian, X.Y., David, A.K.: Probabilistic eigenvalue sensitivity analysis and PSS design in multimachine systems. IEEE Trans. Power Syst. **18**, 1439–1445 (2003)

Planning of Rail Connected PV Operation and Maintenance Robot Based on NSGA-II Algorithm

Yidi Fan(✉), Jian Sun, and Hongwei Xu

College of Mechanical and Electrical Engineering, China Jiliang University, Hangzhou 310018, Zhejiang, China

Abstract. In order to reconcile the contradiction between the economy and efficiency of photovoltaic module operation and maintenance in photovoltaic power plant and improve the investment efficiency, a new planning method of intelligent operation and maintenance robots based on the nondominated sorting genetic algorithm II (NSGA-II) is proposed. Based on the basic pattern of photovoltaic power plant, a track that connects the photovoltaic operation and maintenance system is designed. The total cost and time of operation with multi robots are taken as the planning objective. Considering the number and path of robots, a Pareto optimal solution set is obtained by the NSGA-II algorithm. The simulation results prove the effectiveness of the method, which can provide a variety of decision optimization schemes to meet the preferences of different decision makers.

Keywords: Multi-objective optimization · Nondominated sorting genetic algorithm II · Pareto front

1 Introduction

The upper surface of photovoltaic modules in photovoltaic power station is easy to accumulate dust, forming hot spot effect, which leads to the reduction of power generation efficiency of the whole system. Therefore, the detection and cleaning of photovoltaic modules is an important part of operation and maintenance of photovoltaic power station [1]. The layout of photovoltaic array in different scale power stations is different, and the requirements for input cost and operation and maintenance time are also different [2]. Decision makers need to optimize operation and maintenance strategy and plan the most suitable number and path of operation and maintenance robots, so as to reduce power generation cost and improve efficiency.

In recent years, the number of photovoltaic power plants in China has increased rapidly, and the intelligent operation and maintenance technology has also developed accordingly [3–5]. For the multi-robot multi-objective optimization problem, there are some studies at present. Yi-fei Wan [6] and others proposed cooperative non dominated sorting genetic algorithm to solve the multi-objective multi robot path planning problem; Jiang-bo Fan [7] and others used firefly algorithm to introduce Pareto dominance relation

K. Li et al. (Eds.): LSMS 2021/ICSEE 2021, CCIS 1468, pp. 325–333, 2021.
https://doi.org/10.1007/978-981-16-7210-1_31

to search and optimize the motion path. Jing Li [8] and others proposed a distributed cooperative multi-robot multi-task traversal path planning strategy; Zi-xuan Luo [9] and others constructed a multi-objective optimization model of path planning, and then solved it by ant colony algorithm.

Through the above situation, the multi robot multi-objective optimization problem has not been applied in the field of photovoltaic intelligent operation and maintenance. To this end, this paper is based on the straddle array track connected photovoltaic operation and maintenance system. In view of the planning problem of photovoltaic intelligent operation and maintenance robot, a multi-objective optimization model with minimum input cost and shortest operation and maintenance time is constructed. Under the constraint conditions, the multi-objective model is solved by the nondominated sorting genetic algorithm II (NSGA-II) with elite strategy, so as to obtain the Pareto front, and the number and path scheme of operation and maintenance robots corresponding to each scheme. Finally, the effectiveness of the method is verified by simulation experiments.

2 Optimization Model of Intelligent Operation and Maintenance of Photovoltaic Panel

2.1 Problem Description

In this system, longitudinal transmission tracks are set on both sides of each row of photovoltaic modules, and each row of longitudinal transmission track is connected to the transverse transmission track. A control center is set on the transverse transmission track. All operation and maintenance robots are stored in the control center. The operation and maintenance robots can slide on the transmission track and the surface of photovoltaic modules and carry out operation and maintenance operations [1]. The schematic diagram of the rail-connected photovoltaic operation and maintenance system is shown in Fig. 1. This system is applicable in general photovoltaic power station.

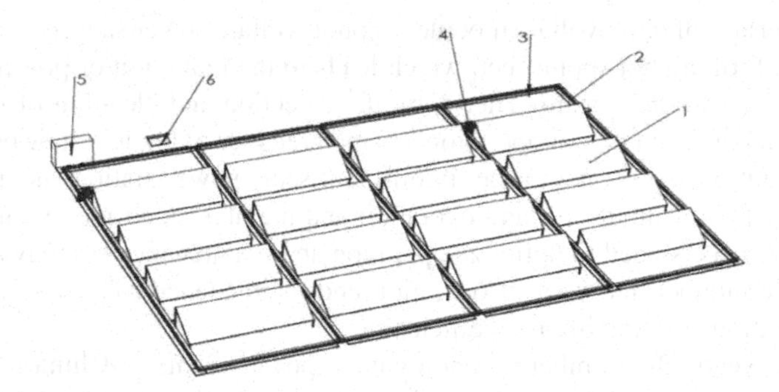

1 - Photovoltaic array ; 2 - longitudinal transport orbit ; 3 - Transverse transport orbits ; 4 - operation and maintenance robot ; 5 - Control center ; 6 - Control terminal

Fig. 1. Schematic diagram of the rail-connected photovoltaic operation and maintenance system.

The research problem can be described as: there is one operation and maintenance control center and multiple operation and maintenance points of photovoltaic array in the photovoltaic power station. The control center allocates operation and maintenance robots to operate and maintain each PV array point. One robot can operate and maintain multiple PV array points and determine the number of robots. Multiple robots are dispatched from the control center for operation and maintenance. After the task is completed, the robot must return to the control center.

2.2 Problem Assumptions

Combined with the actual situation, this paper makes the following assumptions.

(1) The positions of all photovoltaic arrays are known and the grid point coordinates and numbers are assigned to them.
(2) The starting point of operation and maintenance is the control center, and there is only one control center, that is, all operation and maintenance robots can only start from the control center.
(3) The operation and maintenance efficiency of the operation and maintenance robot is the same, and they all travel on the track at a uniform speed.

2.3 Symbol Definition

M is the set of all photovoltaic array points $(i = 0, 1, 2, \cdots, m)$, $i = 0$ represents the control center, and the remaining points represent photovoltaic array points; N is the set of operations robot k, $(k = 0, 1, 2 \cdots, n)$; t Represents the time required to maintain a photovoltaic array; d_{ij} represents the distance from the i PV array to the j PV array in orbit $(i, j \in M)$; T_k represents the total time required for the kth operation and maintenance robot to complete its tasks; c_1 represents the cost of an operation and maintenance robot; c_2 represents the cost per meter of the transmission track; l represents the total length of the transmission track required for a photovoltaic power station; a represents the loss cost coefficient of the operation and maintenance robot per meter; T represents the total time; C Represents total cost.

Decision variables:

(1) $x_{ij} \in \{0, 1\}$, x_{ij} takes 1,indicating that the operation and maintenance robot travels from the i PV array to the j PV array, otherwise $x_{ij} = 0$;
(2) When $x_{ijk} \in \{0, 1\}$, x_{ijk} takes 1, it means that the kth operation and maintenance robot runs from the i PV array to the j PV array, otherwise $x_{ijk} = 0$;
(3) When $y_{ik} \in \{0, 1\}$, y_{ik} takes 1, it means that the kth operation and maintenance robot completes the i PV array operation and maintenance task, otherwise $y_{ik} = 0$.

2.4 Establishment of Multi-objective Models

Considering various influencing factors, this paper takes the minimum operation and maintenance time and minimum input cost as the objective functions to establish the following model:

(1) The shortest operation and maintenance time

$$T_k = \sum_{i \in M} \sum_{k \in N} ty_{ik} + \left(\sum_{i \in M} \sum_{j \in M} \sum_{k \in N} d_{ij} x_{ijk} \right) / v. \tag{1}$$

$$minT = max\{T_1, T_2, \cdots T_n\}. \tag{2}$$

The driving time includes two parts, namely, the time required for the operation and maintenance of photovoltaic arrays and the time required for driving on the transmission track. All operation and maintenance robots start to perform operation and maintenance tasks at the same time, then the total time required for a scheme is the time spent by the robot that returns to the latest.

(2) Minimum input costs

$$minC = c_1 n + c_2 l + a \sum_{i \in M} \sum_{j \in M} \sum_{k \in N} d_{ij} x_{ijk}. \tag{3}$$

The cost includes three parts, namely, the fixed cost of the operation and maintenance robot, the cost of the transmission track of the operation and maintenance system, and the operation loss cost of the operation and maintenance robot.

The constraint conditions of the problem are as follows:

$$\sum_{k \in \mathrm{N}} y_{ik} = \begin{cases} 1, & i = 1, 2, \cdots m \\ n, & i = 0 \end{cases}. \tag{4}$$

$$\sum_{i \in \mathrm{M}} x_{ijk} = y_{jk}, \ \forall j \in \mathrm{M}, \ k \in \mathrm{N}. \tag{5}$$

$$\sum_{j \in \mathrm{M}} x_{ijk} = y_{ik}, \ \forall i \in \mathrm{M}, \ k \in \mathrm{N}. \tag{6}$$

$$\sum_{j \in \mathrm{M}} \sum_{i \in M} x_{ij} \leq m + 1. \tag{7}$$

The Eq. (4) denotes that each PV module is maintained only once and that all operations are performed by n robots; Eq. (5) indicates that each robot can only access one photovoltaic array point after starting from the previous array point; Eq. (6) indicates that each robot can only operate and maintain one photovoltaic array point at most once when it arrives; Eq. (7) indicates that each operation and maintenance robot starts from the control center and returns to the control center after operation and maintenance.

3 NSGA-II Algorithm Solving Model

NSGA-II is a relatively good algorithm in multi-objective genetic algorithm [10]. NSGA-II has the characteristics of fast running speed, high precision of optimization results and good convergence of solution set. The NSGA-II algorithm first randomly generates the initial population with the size of N, and defines the iteration number Gen and the maximum iteration number max_ Gen and other parameters. Then fast non dominated sorting and crowding degree calculation are carried out to generate new species groups according to Pareto level and individual crowding degree until the number of iterations reaches the maximum value [11].

The flow chart of NSGA-II algorithm to solve the model is shown in Fig. 2.

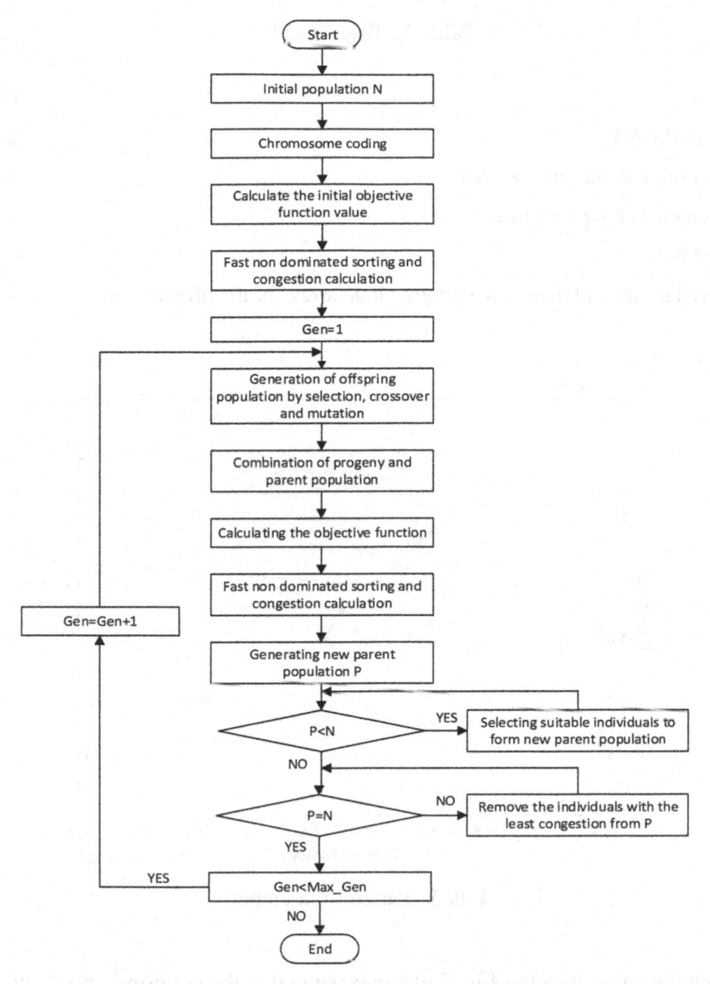

Fig.2. Flow chart of NSGA-II algorithm.

4 Example Analysis

This paper takes the photovoltaic project of Hangzhou tooth mountain north plant of Zhengtai new energy as an example to verify the practical operation and feasibility of the algorithm. The photovoltaic array of one of the projects is arranged in 10 rows and 5 columns. During initialization, some basic data are set as shown in Table 1, and the maximum number of iterations is set as 250, the size of subpopulation is 150, the crossover probability is 0.9, the mutation probability is 0.1, and the size of mutation pool is 1/2 of the size of subpopulation N. In the simulation, it is assumed that the control center point is (0, 0), and the driving loss cost coefficient of the operation and maintenance robot is 0.4. The robot is running at uniform speed and the driving speed of each robot is the same. The obtained Pareto optimal frontier is shown in Fig. 3.

Table 1. Basic data list.

Project	Data
Each robot cost/CNY	2000
Rail connection price per meter/CNY	10
The track length of the project/m	850
Robot speed m/s	0.6
Time required to run and maintain a photovoltaic array for the project/min	33

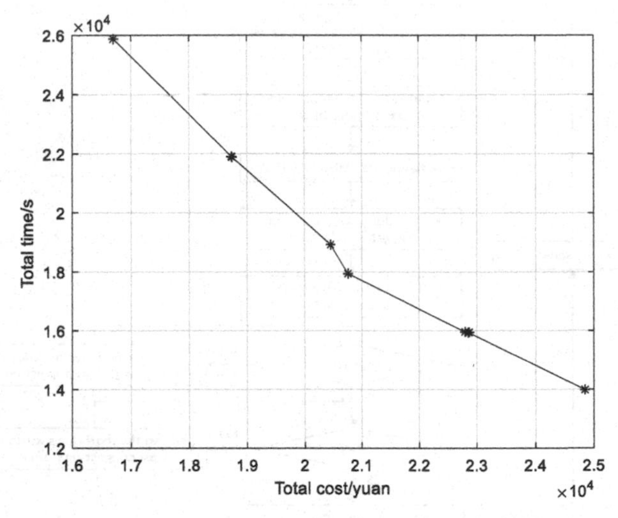

Fig. 3. Pareto front graph.

Through the curve trend of Fig. 3, it can be seen that the economic goal and efficiency goal of operation and maintenance robot planning are negatively correlated. If you want to have higher efficiency and shorter time, you need to invest more cost. The Pareto front points in Fig. 3 are the non-inferior combination of input cost and operation and maintenance time. If more attention is paid to economic objectives and less cost is invested, it can be selected in the first half of the curve. If you pay more attention to efficiency objectives and reduce operation and maintenance time, you can choose in the latter half of the curve. According to the actual situation of photovoltaic power station, decision makers can choose one of the combinations to invest in the corresponding operation and maintenance robot. The above Pareto optimal solution set and the corresponding number of operation and maintenance robots are shown in Table 2.

Each scheme has its corresponding number of robots and operation and maintenance path schemes. Randomly select the path diagram of one scheme as shown in Fig. 4 (a) and (b) are the solution paths of each robot in scheme 8. Each linetype represents the operation and maintenance path of a robot, and different graphic markers in the path

Table 2. Pareto solution set.

Number	Total cost/CNY	Total time/second	Number of operation and maintenance robots /one
1	16700.4	25875	4
2	18731.6	21906	5
3	18731.6	21906	5
4	18743.6	21888	5
5	20458.0	18926	6
6	20762.8	17940	6
7	22796.4	15960	7
8	22796.4	15960	7
9	22870.8	15945	7
10	22854.0	15939	7
11	24846.0	13986	8

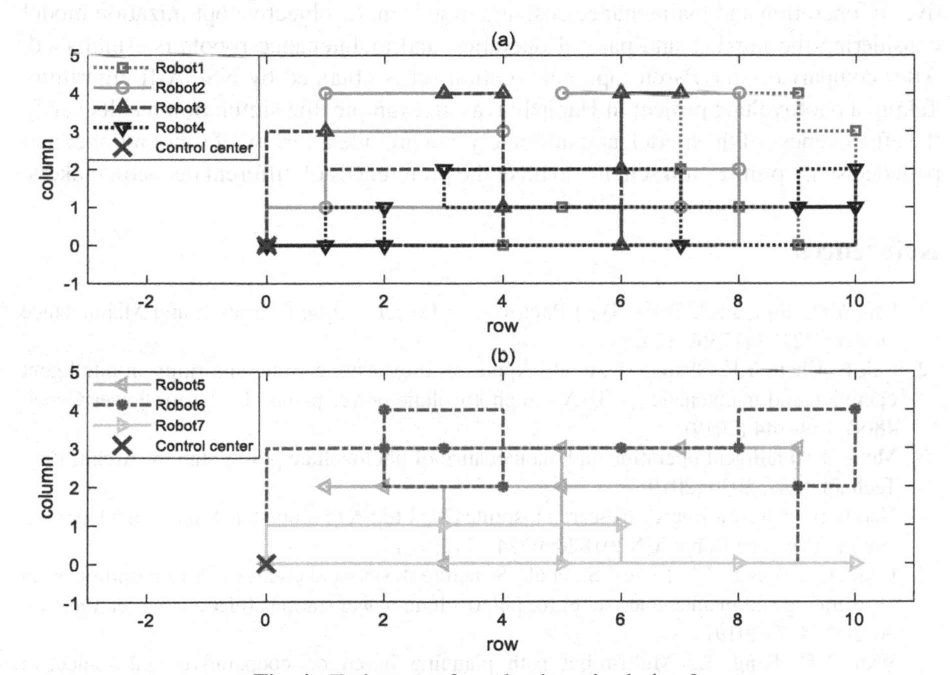

Fig. 4. Trajectory of nondominated solution 8.

represent the operation and maintenance photovoltaic array points of the robot (0, 0) is the control center, from which the robots start and return to the zero action orbit.

In order to show more clearly the array points to be cleaned for each robot in scheme 8, Table 3 is listed as follows.

Table 3. Paths of non-dominated solution 8.

Robot number	Operation and maintenance array point path
Robot1	(0,0)->{8,1}->{5,1}->{9,0}->{10,3}->{9,4}->{7,2}->{4,0}->(0,0)
Robot2	(0,0)->{8,4}->{5,4}->{8,2}->{7,1}->{1,4}->{4,3}->{1,1}->(0,0)
Robot3	(0,0)->{4,4}->{3,4}->{4,1}->{6,0}->{6,4}->{7,4}->{6,2}->{1,3}->(0,0)
Robot4	(0,0)->{10,2}->{9,1}->{10,1}->{3,2}->{7,0}->{2,1}->{1,0}->{2,0}->(0,0)
Robot5	(0,0)->{5,3}->{5,2}->{1,2}->{9,3}->{7,3}->>{2,2}->{3,0}- > (0,0)
Robot6	(0,0)->{9,2}->{10,4}->{8,3}->{6,3}->{3,3}->{2,4}->{4,2}->{2,3}->(0,0)
Robot7	(0,0)->{10,0}->{8,0}->{5,0}->{6,1}->{3,1}->(0,0)

5 Conclusions

Based on the basic pattern of photovoltaic power station, a track connected photovoltaic operation and maintenance system is designed. Aiming at the two conflicting objectives of operation and maintenance cost and time, a multi-objective optimization model considering the number and path of operation and maintenance robots is established. After comparison, the Pareto optimal solution set is obtained by NSGA-II algorithm. Taking a photovoltaic project in Hangzhou as an example, the simulation results verify the effectiveness of the model, and ultimately can provide a variety of robot number and path decision optimization scheme to meet the preferences of different decision makers.

References

1. Fan, Y.D., Sun, J.: A Track Type Photovoltaic Power Station Operation and Maintenance Device. N212341736U (2021)
2. Lai, F., Chen, Y.P., Shan, Z.T., et al.: Application of deep learning algorithm in intelligent operation and maintenance of UAVs in photovoltaic power plants. J. Therm. Power Gener. **48**(9), 139–144 (2019)
3. Meng, J.: Intelligent operation and maintenance of photovoltaic power station. Archit. Eng. Technol. Des., 2030 (2019)
4. Nanjing Green New Energy Research Institute Co., Ltd.: A Photovoltaic Intelligent Operation and Maintenance Robot. CN201820906747.7 (2019)
5. Duan, C.Y., Feng, Z.J., Lian, J.S., et al.: Structure design and control function optimization of operating and maintenance robot for photovoltaic power station. Telecom Power Technol. **36**(2), 1–4, 7 (2019)
6. Wan, Y.F., Peng, L.: Multi-robot path planning based on cooperative multi-objective algorithm. Inf. Control **49**(2), 139–146 (2020)
7. Fan, J.B., Wang, Y., Zheng, K.: Multi-objective path planning for mobile robot based on improved firefly algorithm. J. Univ. Jinan Sci. Technol. **34**(5), 459–469 (2020)
8. Li, J., Yang, F.: Distributed cooperative multi-robot traversing multi-task target path planning. J. Tianjin Polytech. Univ. **39**(6), 68–75 (2020)
9. Luo, Z.X., Liu, X.W.: Research on logistics distribution path optimization based on ant swarm algorithm. J. Chongqing Technol. Bus. Univ. (Nat. Sci. Edn.) **37**(4), 89–94 (2020)

10. Konak, A., Coit, D.W., Smith, A.E.: Multi-objective optimization using genetic algorithms: a tutorial. Reliab. Eng. Syst. Saf. **91**(9), 992–1007 (2006)
11. Zheng, Y.F., Fang, R.C., Zhou, Y.J., et al.: Planning of distribution network reliability optimization based on NSGA2 algorithm. Electr. Autom. **42**(6), 7–72+79 (2020)

A Defect Detection Method for Optical Fiber Preform Based on Machine Vision

Xinzhen Ren[1], Wenju Zhou[1(✉)], Xiaogang Gu[1], and Qiang Liu[2(✉)]

[1] School of Mechatronic Engineering and Automation, Shanghai University, Shanghai 200444, China
[2] Department of Psychiatry, University of Oxford, Oxford OX3 7JX, UK
qiang.liu@psych.ox.ac.uk

Abstract. In order to ensure reliable product quality, fiber optics need to be detected for defects during the manufacturing process. Most domestic manufacturers use manual visual inspection and comparison with tool cards for defect detection. However, manual inspection is slow and subjective, and problems such as missed inspections and false inspections are extremely easy to occur, which seriously affects the quality of products. This paper proposes a defect detection method for optical fiber preforms based on machine vision. First, rely on the optical fiber preform defect detection experimental platform to obtain the full-angle image of the optical fiber preform. Then, the boundary of the optical fiber preform in the full-angle image is determined by the proposed algorithm and the full-angle image is preprocessed. Further, the defects are tracked through the full-angle image and the tracking information is recorded. Finally, the location, size and category of the defect are calculated by tracking information. It is verified by experiments that this algorithm can realize the detection of defects in the optical fiber preform.

Keywords: Optical fiber preform · Defect detection · Machine vision

1 Introduction

The optical fiber is drawn from the optical fiber preform, which is an important intermediate product [1]. Internationally, the manufacturing methods of optical fiber preforms can be roughly divided into modified chemical vapor deposition, outside vapor deposition, vapor axial deposition, advance plasma vapor deposition, and plasma chemical vapor deposition [2]. However, due to the limitation of the manufacturing principle, there will inevitably be small bubbles, air lines and other defects inside the optical fiber preform no matter which method is adopted. The slight mechanical friction during transportation can also cause scratches, cracks on the surface. These defects will affect the drawing effect of

Supported by Natural Science Foundation of China (61877065) and 111 Project (No. D18003).

K. Li et al. (Eds.): LSMS 2021/ICSEE 2021, CCIS 1468, pp. 334–343, 2021.
https://doi.org/10.1007/978-981-16-7210-1_32

the optical fiber preform, resulting in poor optical fiber communication effect, and even breakage in severe cases.

In order to ensure reliable product quality, manufacturers need to inspect the optical fiber preforms and classify the optical fiber preforms according to the number and size of the defects. Most domestic manufacturers use manual visual inspection and comparison with tool cards for inspection, but manual inspection has many shortcomings: (1) The detection process is complicated. First, the inspector needs to put the optical fiber preform into the inspection room that is isolated from external light sources. Then, the inspector irradiates the optical fiber preform with a high-intensity light source to find the defect location, and compares it with the tool card to record the defect information. Finally, inspector judges the quality of the optical fiber preform subjectively. (2) The detection speed is slow. It takes nearly 5 min to detect an optical fiber preform with a length of one meter, which is very incompatible with the speed of the optical fiber preform production line. (3) Manual judgment has subjective factors. Due to the strong subjectivity of manual detection, different inspectors will have different detection results, which reduces the objective accuracy of the detection results. (4) Employees are prone to fatigue. Inspectors are prone to physical discomfort in the test room that is isolated from external light sources for a long time, which not only does cause great physical harm to employees, but also is prone to problems such as missed inspections and false inspections.

Fig. 1. Manual inspection of optical fiber preform

Fig. 2. Standard comparison card

Fig. 3. Optical fiber preform with a lot of defects

Machine vision technology is an effective way to solve the above problems. Paper [3] proposes a method that combines industrial cameras, mobile platforms, beam expanders, and laser light sources. An algorithm for detecting internal defects in optical fiber preforms using scattered light generated by lasers at bubbles or impurities is proposed. However, this method needs to place the optical fiber preform in the matching liquid to prevent stray light, and cannot detect overlapping defects. Paper [4] proposes a method to detect the defects of the optical fiber preform from the top view. This method uses the industrial camera matrix, laser light source and moving device to detect the defects of the optical fiber preform from the top view. This method also has the problem of not being able to detect overlapping defects. Paper [5] introduces a cognex optical fiber preform visual inspection system based on Siemens S7-1200 control system.

With the help of high-precision Siemens control system, the system can detect optical fiber preform size information and bubble defects, but it is costly.

This paper proposes a defect detection method for optical fiber preforms based on machine vision. First, Obtain the full-angle image of the optical fiber preform through the built-up optical fiber preform defect detection experimental platform. Then, find the boundary of the optical fiber preform in the full-angle image and prepossess the full-angle image by the algorithm proposed in this paper. Further, the defects are tracked through the full-angle image and the tracking information is recorded. Finally, the location, size and category of the defect are calculated by tracking information. And we verified the feasibility of this algorithm in Sect. 4.2

2 Types of Fiber Preform Defects

The defects of the optical fiber preform are mostly gas defects which are transparent, so the characteristics of gaseous defects are not obvious under natural light. The Fig. 3 shows a fiber preform with a large number of defects. We can clearly see the larger-sized defects, but it also contains a large number of small defects, which is extremely difficult to distinguish with the naked eye.

The defects of the optical fiber preform are mainly divided into two categories: internal defects and external defects. The internal defects mainly contain bubbles, air lines, blue spots and opaque crystal impurities. External defects mainly involve scratches and bumps on the surface. The specific defects are classified as follows. The defects are shown in Fig. 4. (1) Bubble defects and airline defects: Bubble defects and airline defects are difficult to capture by the camera under sunlight. But gas-based defects mainly show bright features and their shapes are relatively regular illuminated by special light source. (2) Blue spots: Blue spot defects show a dull blue-white characteristic illuminated by special light source. (3) Internal impurity defects: The internal impurity defects show irregular highlights and its texture is messy illuminated by special light source. (4) Scratches and bumps on the external surface: The surface scratches and bumps show the characteristics of continuous large-area linear highlighting illuminated by special light source.

(a) Bubbles and air line defects

(b) Blue spots

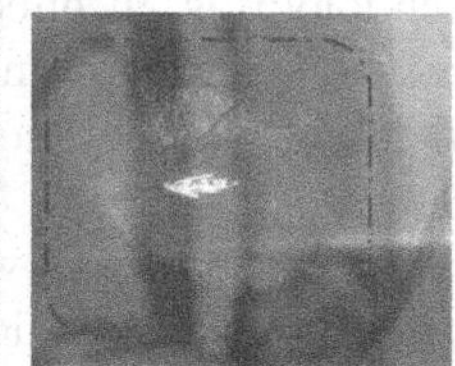
(c) Internal impurities

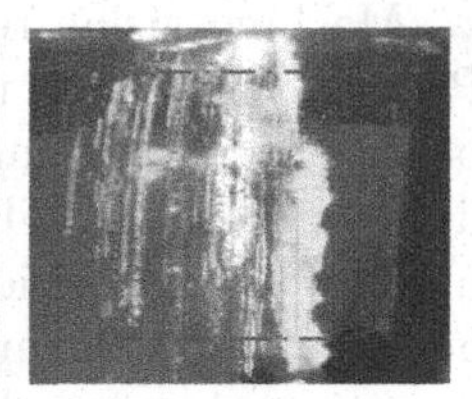
(d) Surface scratches

Fig. 4. Types of fiber preform defects

3 Defect Detection Algorithm of Optical Fiber Preform Based on Full-Angle Image

In practical applications, the location information, size information, and type of defects determine the quality classification of the optical fiber preform. Therefore, it is necessary to accurately obtain the position information and size information of various defects to judge the quality. In this paper, a defect detection algorithm for optical fiber preforms based on the full-angle image is adopted. The algorithm is mainly divided into five steps: 1. Regional positioning. Determine the position of the optical fiber preform in the image to increase the processing speed; 2. Image preprocessing. Deal with the whitening problem of the optical fiber preform caused by the light source problem to improve the accuracy of defect detection; 3. Preliminary screening of defects. The optical fiber preform image at an angle of 0 angles is used as the preliminary screening image to detect defects in the optical fiber preform and open up storage space to record the number and height information H of the defect and the distance between the defect and the boundary M; 4. Track the defects and use the tracking information to calculate the defect information. Use the full-angle image to track the defects, record the initial movement direction of the defect (Move right or Move left) and the angle θ at which the movement direction changes, and use the tracking information and The calculation formula finally obtains the location information of the defect; 5. The size information and type of defects are obtained by the theory of Medical MRI through the full-angle image.

We will introduce in detail in the next few subsections. We establish a coordinate system as shown in Fig. 5 to facilitate the elaboration of the algorithm.

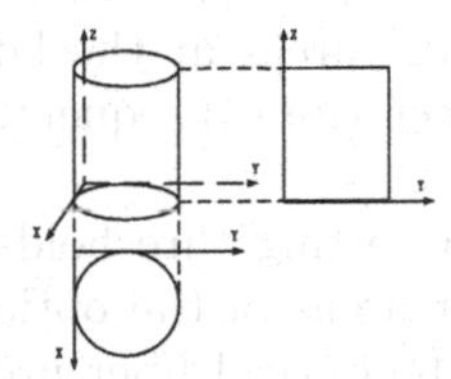

Fig. 5. Coordinate system

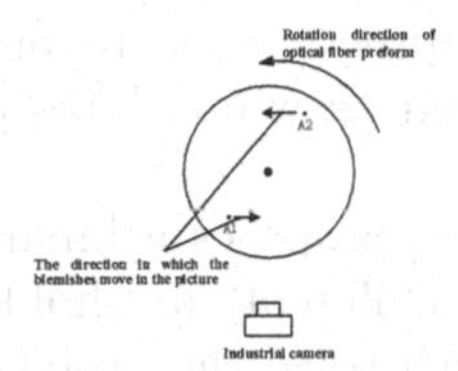

Fig. 6. Sectional view of optical fiber preform

3.1 Positioning the Boundary of Optical Fiber Preform

The volume of the optical fiber preform is very large, but the defects are small. It is necessary to ensure that a sufficiently large field of view and accuracy can meet the requirements at the same time. This makes the pixels of the photographed pictures continue to increase. From Fig. 13(a), we can see that the optical fiber preform cannot fill the entire image. If the image processing is performed on the entire image, the speed of the algorithm and the detection effect will be reduced. So the boundary of the optical fiber preform needs to be positioned.

There is a big difference between the borders of the optical fiber preform and the background. However, the size of the image of the optical fiber preform and the number of full-angle images are large. In order to further accelerate the detection speed, this paper adopts a method of two-way detection combined with probability to determine the boundary. The specific steps are: 1. Preset pixel mutation threshold δ; 2. Starting from the first row of the image. Perform horizontal sobel calculations at the same time from the left and right sides of the image; 3. When the first pixel that exceeds the threshold δ is currently detected, stop the calculation of this row and record the ordinate of the pixel; 4. Repeat steps 2 and 3 until the last row of the image; 5. Calculate the average value of the vertical coordinates of the pixels found in step 3 and eliminate pixels with excessive deviations.

This method finds the first abrupt pixel point on the left and right sides of a row as the temporary boundary point, and no longer detects the inside of the optical fiber preform, which greatly accelerates the calculation speed. By counting the ordinates of the temporary boundary points of each row, the misjudgment of the boundary points caused by noise is eliminated, and the boundary of the optical fiber preform is finally located. The effect will be shown in Sect. 4.2.

3.2 Image Preprocessing

The image after positioning the boundary of the optical fiber preform cannot be directly detected for defects. It is not difficult to see from Fig. 13(a) that both sides of the Optical fiber preform are white. If the otsu algorithm of the global threshold is directly used to binarize the image, many areas at the boundary will be defined as white. This phenomenon will affect the subsequent defects tracking.

This paper proposes a binarization algorithm for setting thresholds in different areas. Different thresholds are set for different areas of the optical fiber preform, which solve the problem that the edge of the optical fiber preform is whitened in the image. The specific steps are: 1. Preset the binarization threshold value α of the optical fiber preform, and preset the binarization threshold value β of the whitening area of the optical fiber preform; 2. Binarize the original image I with α as the threshold to obtain the image I_1; 3. Perform an open operation on the image I_1 to remove small external noises; 4. Determine the whitening area by calculating the area of the connected domain; 5. Binarize the whitening area of the original image I with β as the threshold to obtain the binary image I_2 of the whitening area; 6. Replace the white area in the image I_1 with the image I_2.

In theory, $\alpha < \beta$. In step 2, the white area of the optical fiber preform will become the foreground, and its area will be much larger than defects. Finding this area and binarizing this area with a new threshold can find out the blemishes hidden in the white area. The results will be shown in Sect. 4.2.

3.3 Track Defects and Calculate the Location of Defects

The pre-processed 0-angle image is used as the preliminary screening image to find the defects in the image. The defects are numbered uniformly, and storage spaces are opened to store the defect number, height information H, and the distance M between the defect and the left edge of the optical fiber preform. Reserve storage space for subsequent storage of the plane information, size and type of the defect.

Through the initial screening results of the defects of the 0-angle image, we can know the height information H, and the distance M between the defect and the left edge of the optical fiber preform. But the type, size and plane coordinate information of the defects cannot be obtained. In this paper, the full-angle image is used to track the defects, and the information of defects is calculated by the tracking information.

As the optical fiber preform rotates, defects will follow the movement. The movement of defects in the full-angle image is horizontal movement. When the optical fiber preform rotates for a certain angle, the moving direction of the defects in the full-angle image will change. Record the initial movement direction of each defect in the full-angle image and the angle at which the movement direction changes to form defect tracking information.

According to the initial movement direction of the defect, we can conclude that the defect is located in the front or the back of the optical fiber preform. As shown in Fig. 6, this picture is the XY section of the optical fiber preform. $A1$ and $A2$ are defects. Assuming that the optical fiber preform rotates counterclockwise, the defect $A1$ moves to the right and the defect $A2$ move to the left in the full-angle image. It can be judged that $A1$ is at the front of the optical fiber preform and $A2$ is at the back. The movement of defects in the optical fiber preform is shown in Fig. 7.

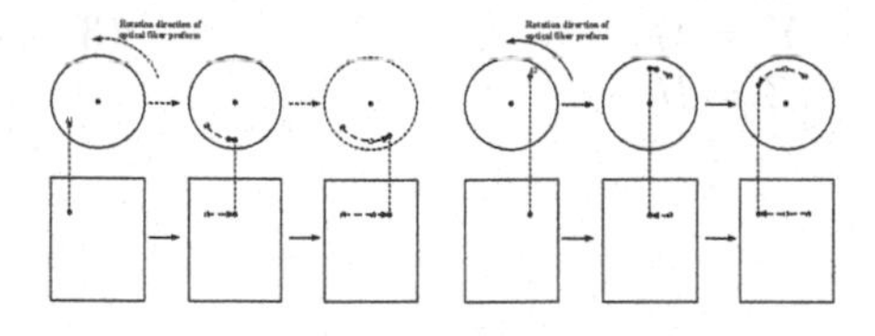

Fig. 7. Schematic diagram of the movement of defects on the front and back of the optical fiber preform

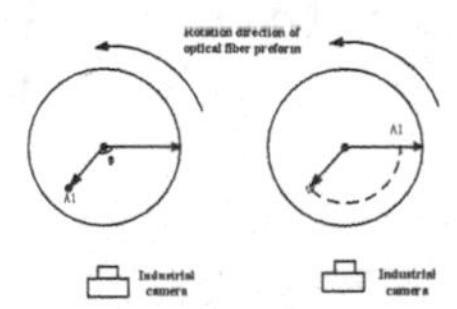

Fig. 8. Sectional view of the optical fiber preform at 0 angles and rotated by θ angles

It is now stipulated that the optical fiber preform moves counterclockwise along the axis at a constant speed. The initial movement direction of the defect A in the image is divided into the right (Move right) and the left (Move left). The formula for judging whether the initial position of defect A is at the front or back of the optical fiber preform is as follows:

$$\text{Position (A)} = \begin{cases} \text{Front,} & \text{Move right} \\ \text{Back,} & \text{Move left} \end{cases} \tag{1}$$

When the optical fiber preform is rotated at a certain angle, the direction of movement of the defect will change on the full-angle image. Figure 8 is a cross-sectional view of the optical fiber preform before and after rotating a certain angle. The left picture is before rotation, and the right picture is after rotation. The black point $A1$ in the figure is a defect in the front of the optical fiber preform. The position of the camera is shown in the figure, and $A1$ forms an angle θ with the positive direction of the X axis. As shown in the figure on the right, if the optical fiber preform rotates θ, the defect $A1$ is at the critical position. If the optical fiber preform continues to be rotated, the defect $A1$ will reach the second half of the optical fiber preform, and its moving direction in the full-angle image will change (moving right becomes moving to left).

The specific plane position of the defect can be calculated according to the initial movement direction of the defect and the angle at which the movement direction of the defect changes. It is now known that the radius of the optical fiber preform is R, the distance between the defect A and the left edge of the optical fiber preform in the image with an angle of 0 angles is M, the initial movement direction of the defect A is rightward, and the angle at which the moving direction of defect A changes or disappears is θ, then the plane coordinate $A : (X, Y)$ calculation formula of defect A is:

$$A : (X, Y) = \begin{cases} X = \frac{R}{2} + \tan(\pi - \theta) \times (R - M) \\ Y = M \end{cases} \tag{2}$$

With the tracking information shown in Fig. 9, we can calculate that the location of the defect is expressed in the coordinate system of Fig. 5 as:

$$A : (X, Y, Z) = \begin{cases} X = \frac{R}{2} + \tan(\pi - \theta) \times (R - M) \\ Y = M \\ Z = H \end{cases} \tag{3}$$

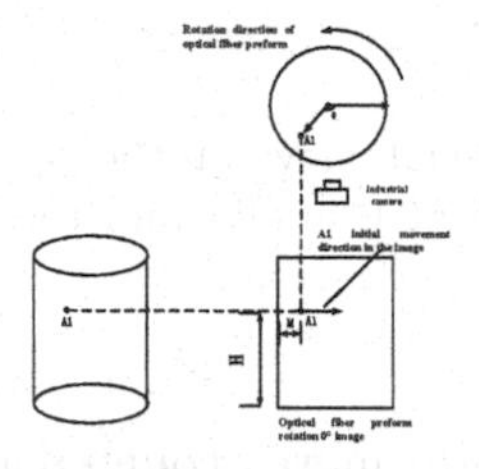

Fig. 9. Tracked information

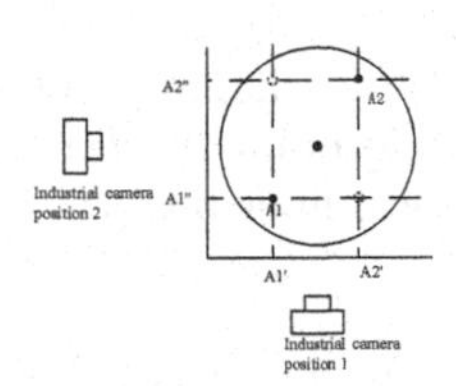

Fig. 10. Sectional view of optical fiber preform

3.4 Calculate Defect Size

In a full-angle image, the shape of the defect will change with the rotation angle, and the size and type of the defect cannot be obtained by simple calculation of the area of the connected domain. The algorithm of this paper uses a method similar to medical MRI scanning imaging, using the full-angle image to simulate the plane containing the defects, thereby obtaining the size information of the defects and judging the type of defects.

The theory of the method briefly described below. In the actual situation, the camera position is fixed, and the optical fiber preform is rotated at a constant speed, then the controller trigger the camera to continuously take pictures to obtain the full-angle image. In order to clearly explain the theory, we take the optical fiber preform as a reference, assuming that the optical fiber preform does not move, and the camera rotates around the optical fiber preform. As shown in Fig. 10, this picture is a cross-sectional view of the XY plane of the optical fiber preform with defects. $A1$ and $A2$ are the two defects in the optical fiber preform. The initial position of the camera is at the front of the optical fiber preform. In the 0-angle image obtained at this time, $A1$ and $A2$ will be highlighted at the corresponding height and position of the image ($A1'$ and $A2'$). In the same way, when the optical fiber preform is rotated 90 angles counterclockwise, the camera reaches position 2. At this time, the defects $A1$ and $A2$ are highlighted in the corresponding height and position in the 90 angles image($A1''$ and $A2''$). We use this theory to reverse the coordinates of $A1$ and $A2$. That is to say, when passing the information of $A1'$ and $A2'$ at 0 angles, the information of $A1''$ and $A2''$ at 90 angles is drawn as a vertical line in the XY plane, and the intersection point may be the coordinate of the defect. But there is a problem that needs to be paid attention to. The result of the reverse inference by this method will have 4 intersections ($A1$, $A2$ and two dashed circles in Fig. 10), of which only two intersections are the correct positions of the defects. It is not possible to get the location of the defect only from two-angle images, but use the full-angle image to calculate the point of intersection. Theoretically, the vertical line of each angle will intersect at the defect position. The size information of the defect is obtained by calculating the area of the connected domain, and the type of defect is judged by the shape of the connected domain.

4 Experimental Platform and Experimental Analysis

4.1 Experimental Platform

In order to verify the feasibility of the method proposed in this paper, we built an experimental platform for defect detection of optical fiber preforms. The system composition of the experimental platform is shown in the Fig. 11.

The optical fiber preform is placed vertically in the center of the tray. The industrial camera is placed in front of the optical fiber preform. The arc light source is placed on one side of the optical fiber preform. The controller is connected to industrial cameras, trays and arc-shaped light sources. The computer is

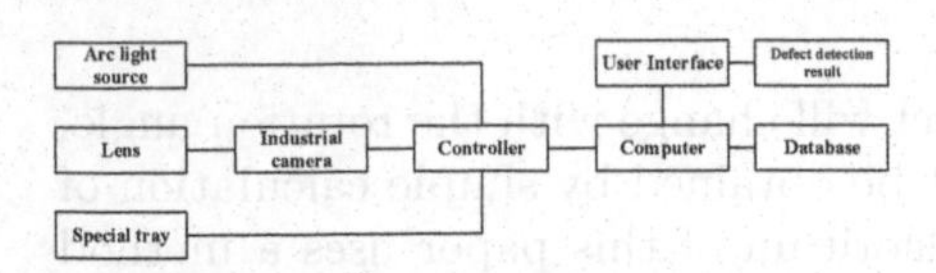

Fig. 11. System composition

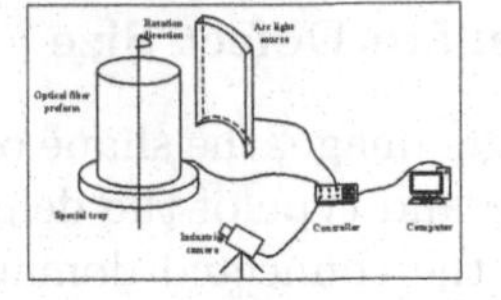

Fig. 12. Experimental platform

connected to the controller. The controller turns on the arc-shaped light source, controls the tray to drive the optical fiber preform to rotate at a uniform speed, and triggers the industrial camera to continuously collect images of the optical fiber preform. The computer obtains the rotation angle of the optical fiber preform and the image of the optical fiber preform taken by the industrial camera at this angle from the controller. The computer matches the rotation angle with the image to generate a full-angle image of the optical fiber preform. The computer uses the method proposed in the paper to detect the defects. The experimental platform is shown in Fig. 12. The full-angle image is shown in Fig. 13(a).

4.2 Experimental Analysis

The Fig. 13(b) shows the result of positioning the boundary of the optical fiber preform. From the figure, we can see that the border of the optical fiber preform is marked by a green line, which separates the optical fiber preform from the background.

The Fig. 13(c) and 13(d) shows the result of preprocessing the image. The Fig. 13(c) is the result obtained by the global threshold OTSU algorithm, and the Fig. 13(d) is the result obtained by the algorithm proposed in this paper. We can see that the whitening area on both sides is almost eliminated in Fig. 13(d), only a small amount of whitening is contained at the boundary of the optical fiber preform, which has no influence on the overall defect detection.

After preliminary screening of defects, tracking of defects, and calculating the location of defects based on tracking information. The results are shown in Fig. 13(e) and 13(f). The red box in the figure indicates the internal defects of the optical fiber preform, and the yellow box indicates the surface defects of the optical fiber preform.

The Fig. 13(g) and 13(h)show the result of calculating the defect area of the optical fiber preform. The Fig. 13(g) is the initial image obtained by the above algorithm, because the channel value of the defect area must be 255, and the result obtained by filtering is shown in Fig. 13(h). The white area in the center is a bubble defect, and the upper left corner is the scratch on the surface of the optical fiber preform.

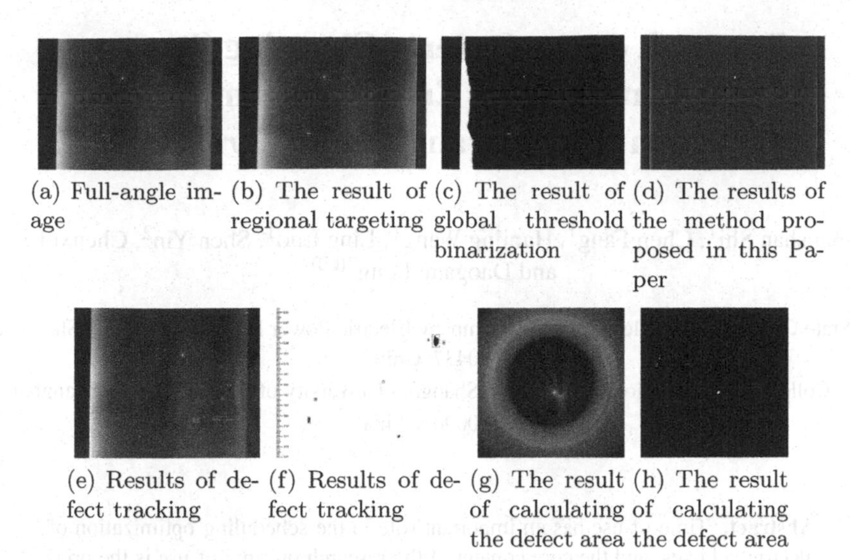

(a) Full-angle image (b) The result of regional targeting (c) The result of global threshold binarization (d) The results of the method proposed in this Paper

(e) Results of defect tracking (f) Results of defect tracking (g) The result of calculating the defect area (h) The result of calculating the defect area

Fig. 13. Experimental result

5 Conclusion

This paper introduces a defect detection method for optical fiber preform based on machine vision, which realizes the defect detection of optical fiber preform. First, establish an experimental platform for defect detection of optical fiber preforms. The computer fuses the image and rotation angle information into the full-angle image of the optical fiber preform. Then, find the boundary of the optical fiber preform in the full-angle image and perform preprocessing operations on the full-angle image. Further, the defect is tracked according to the full-angle image, and the tracking information is recorded. Finally, the location, size and type of the defect arc obtained based on the tracking information. The effectiveness of this method is fully demonstrated by the experimental results. This method helps to improve the quality of the optical fiber and reduce unnecessary waste.

References

1. Zhao, H.: Research and practice on intelligent manufacturing technology of optical fiber preform. Smart Manufacturing **293**(03), 36–40 (2020)
2. Tang, R.: The latest development of optical fiber preform technology. Optical Commun. Res. (05), 54–58 (2000)
3. Wang, F.: Machine Vision Based Internal Defects Inspection for Optical Fiber Preform. D. Zhejiang Normal University (2014)
4. Li, H.: Research and Implementation of Preform Inspection Detection System Based on Machine Vision. Wuhan University of Technology (2018)
5. Li, J.: Optical fiber preform visual inspection project based on Siemens control system. Mechatronics Technol. Home Abroad **136**(04), 49–52 (2018)

Research on the Orderly Charging Strategy of EV in Regional Energy System Based on *NSGA-II* and *Monte Carlo*

Shanshan Shi[1], Chen Fang[1], Haojing Wang[1], Ling Luo[1], Shen Yin[2], Chenxi Li[2], and Daogang Peng[2](✉)

[1] State Grid Shanghai Electric Power Company Electric Power Research Institute, Shanghai 200437, China

[2] College of Automation Engineering, Shanghai University of Electric Power, Shanghai 200090, China

Abstract. Time-of-use has an important role in the scheduling optimization of electric vehicles, and the core content of the research on time-of-use is the pricing of tariff and the division of time periods. In this paper, firstly, the *K-means* algorithm is used to divide the peak and valley periods, and the tariff of different time periods is determined according to the actual situation; then an orderly charging and discharging model of EV based on time-of-use is established, and two different methods are used to solve the model, the Pareto optimal boundary of multi-objective function is found by *NSGA-II* algorithm, through analysis and calculation, it can be found that the EV orderly charging strategy based on time-of-use not only reduces the charging cost of users, but also reduces the average value of net load in the system and improves the stability of system operation.

Keywords: Electric vehicles · Time-of-use · *K-means* clustering · *Monte Carlo* simulation

1 Introduction

With the promotion of "Carbon neutral, carbon peaking" [1], the development of electric vehicle (Electric Vehicle, EV) in China has been further stimulated, and the ownership of EV will continue to show a rapid increase [2]. The energy Internet is a new form of energy industry development, which will break the traditional top-down centralized decision-making control of resource allocation model and form a centralized and distributed system structure [3].

Large-scale EV entering the grid can have a large impact on the system and threaten the safe and stable operation of the system, so the strategy to guide EV orderly charging is particularly important [4]; the most effective and commonly used method today is the electricity price guidance strategy [5], real-time electricity price and time-of-use (time-of-use, TOU) are two typical demand response mechanisms based on electricity prices, which change the electricity price by changing electricity prices at different times to change customers' electricity consumption behavior [6]. Time-of-use can effectively

K. Li et al. (Eds.): LSMS 2021/ICSEE 2021, CCIS 1468, pp. 344–356, 2021.
https://doi.org/10.1007/978-981-16-7210-1_33

balance electricity price fluctuations and demand response, and are more acceptable and operable for users. More research has been conducted on demand response and time-of-use pricing. The literature [7] firstly analyzes the load distribution and peak and valley hours using fuzzy half-gradient affiliation function, and then constructs a time-of-day pricing model with minimized peak-to-valley difference of daily load curve. In [8], a clustering method based on dimensionality reduction of characteristic indicators is firstly used to identify customer load characteristics, and then a customer selection behavior model is constructed based on multinomial *Logit* model, and finally an optimization model for tariff package design is proposed. The literature [9] proposed a method to identify the cost coefficients of load transfer satisfaction of residential customers for the problem of time-of-use pricing, considered the probability distribution of the cost coefficients of load transfer satisfaction, and established a non-complete information *Stackelberg* game model between electricity sales companies and residential customers. The above research scenarios and methods are different, but the general idea of pricing is similar: firstly, users are subdivided according to their characteristics to form user groups, and then tariffs are set based on the optimization model.

This paper argues that time-of-use need to further consider customer response characteristics (including EV customers and general electricity customers). The current study focuses on clustering users based on the characteristics of the electricity load. However, for users with the same electricity load characteristics and different degrees of demand response, even if the same time-of-use is implemented, it will produce different effects. Demand response characteristics are factors that cannot be ignored in user clustering and time-of-use pricing, and the time-of-use is continuously and dynamically adjusted on top of the initial relatively optimal electricity value so that the degree of response of users can reach a better level.

Therefore, based on the park, based on the demand response mechanism of time-of-use, this paper proposes a time-of-use pricing method for the park in order to realize the local consumption of distributed power, reduce the overall energy cost and improve the overall economy of the park, and verify its feasibility through the analysis of calculation cases.

2 Orderly Charging Strategy for EV Based on Time-of-Use

In order to make the whole charging process more process-oriented, the specific steps for optimizing the time-of-use are roughly divided into three steps.

Firstly, based on the historical load data of the distribution network, the load is divided into four different quarterly average daily load curves, and then the corresponding average daily load curve is selected as the object of study according to the quarter in which it is currently located.

1) *K-means* algorithm was used to divide the load in the daily average load curve into three categories: peak, flat and valley, and different levels of tariffs were applied in different load segments, thus generating a time-of-use curve, which is only for the load side, and the subsidy cost for the reverse energy delivery from EV to the system through the function is set in subsequent sections.

2) To establish the EV charging and discharging model in the system, simulate to obtain the daily load curve of the system on top of the time-of-use, relative to the

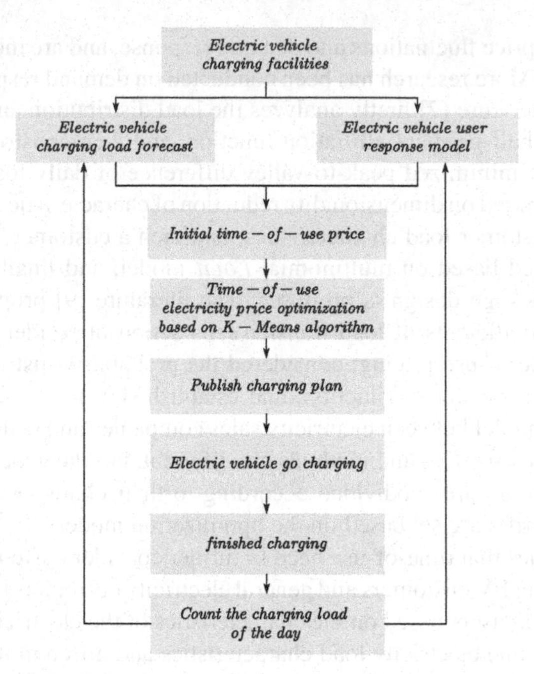

Fig. 1. Optimization flow chart of electric vehicle time-of-use

changes generated by the daily load curve in the system when the EV is in the disordered charging state, and analyze the obtained results. Specifically, the basic flow of time-of-day optimization in the park is shown in Figure.

Specifically, the basic flow of time-of-day optimization in the park is shown in Fig. 1.

3 EV Charging Response Model

From the perspective of the regional energy system (source side), guiding EV to shift peaks and fill valleys through a reasonable dispatch optimization strategy will significantly improve the stability and reliability of the system operation, and at the same time, if EV can form a complementary with RES, the operation cost of the source side will also be reduced; and for the EV user side, vehicle owners can make good use of the time-of-use mechanism in the system, or participate in the cooperative dispatch of the system optimization to reduce their charging cost, which is not a win-win approach, and for this reason, a multi-objective optimization model for EV charging and discharging based on time-of-use tariff is developed in this section:

3.1 Objective Function

1) Charging Cost for EV Users

Since the study is about the effectiveness of time-of-use, one of the most important indicators studied is the cost of charging for the customer, which is calculated as follows:

$$f_1 = \sum_{l=1}^{N}\sum_{t=1}^{T} c(t) \cdot P_{c,l}(t). \tag{1}$$

In the formula, C(t) is a time-of-use tariff within the regional energy system, $P_{c,l}(t)$ is the total charge of the electric vehicle l at sampling point t.

2) Load Fluctuation Level

For a regional energy system, measuring the level of load fluctuation within the system cannot only consider the maximum and minimum values of the daily load curve unilaterally, but also need to consider the fluctuation degree of the overall load curve, so the magnitude of the mean squared deviation of the load is chosen here to measure the fluctuation level of the daily load, and the expression of the objective function is shown as follows.

$$f_2 = \sqrt{\frac{1}{T}\sum_{t=1}^{T}\left(P(t) - \overline{P}\right)^2}. \tag{2}$$

$$P(t) = L_B(t) + \sum_{l=1}^{N} P_{c,j}(t) \tag{3}$$

$$\overline{P} = \frac{1}{T}\sum_{t=1}^{T} P(t). \tag{4}$$

In the above equation, P(t) is the sum of charging load and base load of EV in time period t, $\overline{P}$ is the average value of the system load during a day.

Since the above objective functions do not have the same magnitudes, they need to be normalized in magnitudes before converting them into a single objective function. It can be dimensionless normalized by a linear method with an evaluation matrix of, $D = (d_{lj})_{n\times m}$ among them:

$$d_{lj} = 1 - \frac{\left|b_{lj} - b_j^*\right|}{\xi_j} \tag{5}$$

$$\xi_j = \max_{l=1,2,\ldots,n}\left\{\left|b_{lj} - b_j^*\right|\right\} \tag{6}$$

where d_{lj} is the jth value of the dimensionless vehicle l, b_j^* is the extreme or minimal value of the jth value, and ξ_j is the absolute value of the extreme value of the difference between b_{lj} and b_j^*.

3.2 Constraint Conditions

1) The state of the battery charge after the end of EV charging should be between the desired value and the maximum value 1:

$$S_{E,L} \leq S_{out,l} \leq 1. \tag{7}$$

2) Relationship between charging state and charging time of electric vehicles:

$$S_{out,l} - S_{in,l} = \frac{P_l \cdot (T_{out,,l} - T_{in,l})}{Q_l}. \tag{8}$$

Where Sin,l is the charging state of the battery before the charging of the lth electric vehicle; SE,l is the expected value of the charging state of the battery after the end of the charging of the lth electric vehicle user; Sout,l is the charging state of the battery after the completion of the charging of the lth EV; Tout,l is the time at the end of the charging of the EV; Tin,l is the time at the beginning of the charging of the EV; Pl is the charging power of the lth electric vehicle; Ql is the battery capacity of the lth electric vehicle.

3) EV charging constraints: In a regional energy system, the number of EV (charging and discharging, parking, etc.) in the campus should not be greater than the total number of EV in the campus at any given time.

$$a(t) \leq N, \; t = 1, 2, \ldots, T \tag{9}$$

In the equation, $\alpha(t)$ denotes the number of EV at charging moment t.

4) To ensure the effectiveness of the time-of-use guidance, the cost of charging for vehicle owners under the time-of-use should be lower than the cost when the time-of-use is implemented:

$$F_0 = q_0 \cdot p_0. \tag{10}$$

$$F_1 = \sum_{t=1}^{T} (q_i + \Delta q_i) \cdot p_i. \tag{11}$$

$$q_i = \sum q_{in}. \tag{12}$$

In the equation, where F_0 is the total cost of EV in the original regional energy system when they are charged and discharged in an unordered manner; F_1 is the tariff cost of EV in the regional energy system when the time-of-use scheme is implemented; q_o represents the total load in the system when EV are charged in an unordered manner, and p_o represents the one-day tariff in the original system.

3.3 Solving Method

The multi-objective optimization model of EV charging based on time-of-use is solved using *NSGA-II* algorithm, which is a fast non-dominated multi-objective optimization algorithm with elite retention strategy and also a multi-objective optimization algorithm based on Pareto optimal solution, and the basic steps of the algorithm are shown in Fig. 2.

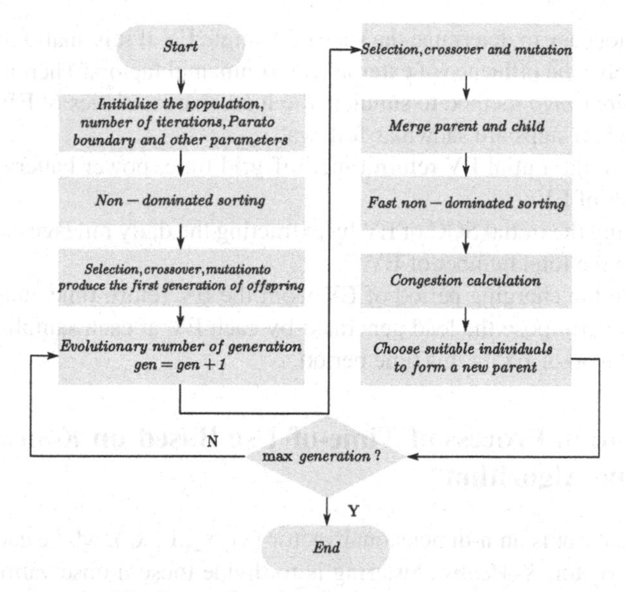

Fig. 2. Basic steps of NSGA-II algorithm

4 Charging Load Prediction Based on Monte Carlo

EV users' charging and discharging behaviors are uncertain not only in time but also in space, for example, factors such as origin, destination and driving path can have an impact on user behavior. According to the research needs, this section analyzes the EV load from the perspective of dual spatial and temporal scales, so as to accurately describe the travel process of EV users in a day.

The main idea of the *Monte Carlo* simulation method is the process of constructing a suitable mathematical probability model based on the data derived from the statistics of a physical phenomenon that will involve some random variables and characterize the probability by frequency approximation.

The *Monte Carlo* simulation method is based on the theoretical foundations of "*Theorem of Large Numbers*" and "*Central Limit Theorem*", which establish the reference for error evaluation in the *Monte Carlo* simulation method. The steps of the *Monte Carlo* simulation method are as follows [10]:

1) obtaining a probability distribution model of the actual physical process and representing it mathematically, and then using statistics to find the mean, variance and other characteristic quantities of this type of mathematical model;

2) generating random numbers that fit this probability distribution model based on the characteristic quantities found in 1), and it is known from *Monte Carlo*'s principles (*Theorem of Large Numbers* and *Central Limit Theorem*) that a large sample of random numbers must be taken in the simulation;

3) Sampling of the random variables to be studied;

4) Simulation of the sampled samples and analysis of the results of n times random sampling, taking their average, a relatively correct value.

Since it is not easy to determine the state of a single EV if it is analyzed as an object, it is easy to receive the influence of external environmental factors. Therefore, this paper adopts the *Monte Carlo* method to simulate the load characteristics of EV clusters, and the specific method steps are shown as follows.

1) Determine the initial EV return time, off-grid time, power battery capacity and the total number of EV;

2) Calculating the initial SOC of EV by extracting the daily mileage and arrival time of EV based on the total number of EV.

3) Calculate the charging period of EV from the EV return time and daily driving mileage, and superimpose the load generated by each EV at each sampling moment to obtain the total load of EV in this time period.

5 Optimization Process of Time-of-Use Based on *K-means* Clustering Algorithm

The available dataset is an n-dimensional vector (x_1, x_2, L, x_n), where each dataset is a d-dimensional vector. *K-Means* clustering is to divide these n observations into k sets ($k \leq n$) that the sum of squares of the distances from each point within this cluster to the center of mass is minimized. That is, the goal of this algorithm is to find a cluster S_i such that it satisfies the following expression.

$$\arg_S \min \sum_{i=1}^{k} \sum_{x \in S_i} ||x - u_i||^2. \tag{13}$$

In the equation, S_i is the mean value of all points.

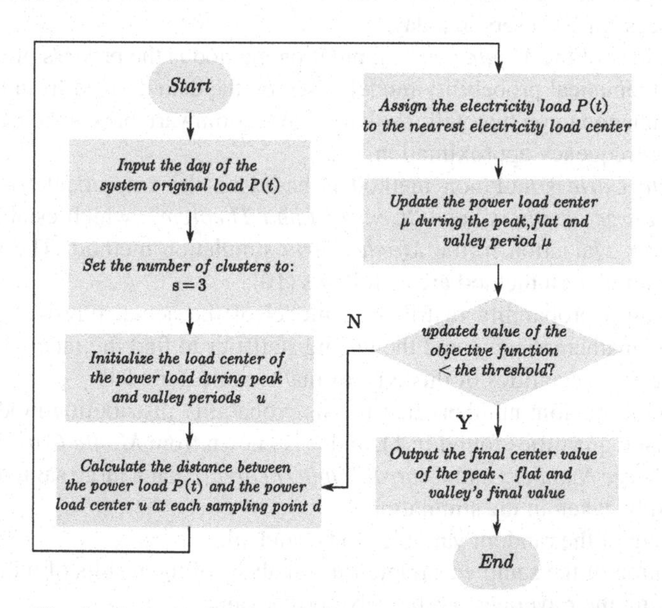

Fig. 3. Optimization flow of *K-means* algorithm in the system

The specific flow chart for determining the time-sharing periods based on the *K-Means* clustering algorithm is shown in Fig. 3, 4, 5, 6, and in subsequent sections we will apply the *K-Means*-algorithm to make a division of the daily load curve.

6 Algorithm Validation

6.1 Scenario Simulation

A specific regional energy system is simulated, and the daily load curve of the system is shown in Fig. 4. There are 90 EVs with a capacity of 42 kWh and a rated charging power of 7 kW, and 10 cabs with a capacity of 60 kWh, and the charging mode of the cabs are all fast charging, and their rated charging power is 50 kW.

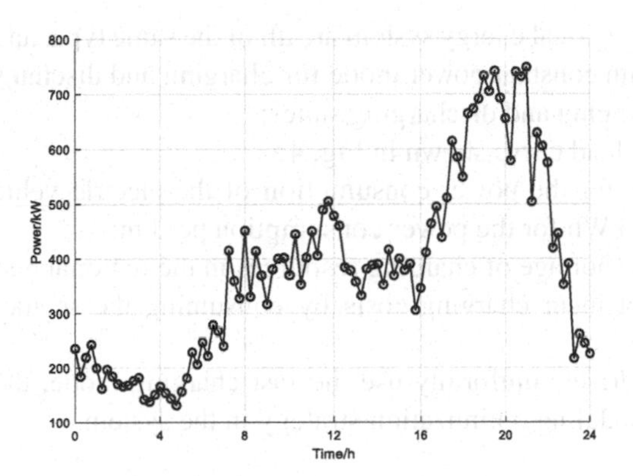

Fig. 4. Daily load curve of the cell

K-Means clustering algorithm can classify the load curve power shown in Fig. 3 into three categories, and the results of the classification are shown in Table 1.

Table 1. Results of daily load curve classification

Load classification	First category	Second category	Third category
Power/kW	109.14	223.78	367.70

The visualization results are shown in Fig. 5, from top to bottom, green, orange and blue represent the load power values of the peak and trough periods at each sampling moment, and it can be seen that the load in the normal period is the majority.

Then, according to the idea of equal division, the peak load can be set above 295 kW, the load in the weekday period can be set between 166 and 295 kW, and the load value in the low time period can be set below 166 kW.

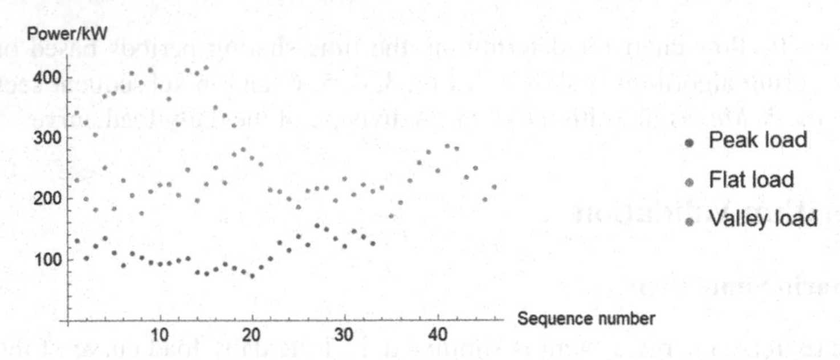

Fig. 5. Classification results of peak and trough times (Color figure online)

6.2 Model Assumptions

(1) The EVs in the regional energy system are all of the same type, and the day is divided into 96 periods, with constant power mode for charging and discharging in each period and continuous charging and discharging states;

(2) $L_B(t)$ is the load curve shown in Fig. 4;

(3) 0.172 kWh for the power consumption of the electric vehicle per 100 km of driving, i.e., 0.172 kWh for the power consumption per km;

(4) There is no shortage of charging resources in the regional energy system;

(5) EVs do not incur charging costs by consuming the scenic power output for charging;

(6) Since cab drivers uniformly use the fast charging mode, their behavior is not affected by the scheduling optimization strategy in the system.

6.3 Time-of-Use Scheme

According to reference [11], the customer's response to the tariff strategy can reach a relatively optimal level when the time-of-use is as shown in Table 2.

Table 2. Optimal time-of-use tariff table

Time period	Time	Electricity price (yuan/kWh)
Peak hours	17:00–22:00 22:00–23:00	1.79
Usual hours	7:00–17:00	1.47
Valley hours	23:00–7:00 (next day)	1.18

The whole day 24 h is divided into 96 time periods, and the corresponding values of electricity prices for peak and flat periods are 1.18 (yuan/kWh) for the valley period,

1.47 (yuan/kWh) for the flat period, and 1.79 (yuan/kWh) for the peak period.

$$C_l(t) = \begin{cases} 1.18, \ L_B(t) \leq 166 \\ 1.47, \ 166 < L_B(t) \leq 295 \\ 1.79, \ L_B(t) > 295 \end{cases} \tag{14}$$

In the equation, $L_B(t)$ denotes the real-time value of the original load in the regional energy system.

The specific time-of-use is shown in Fig. 6.

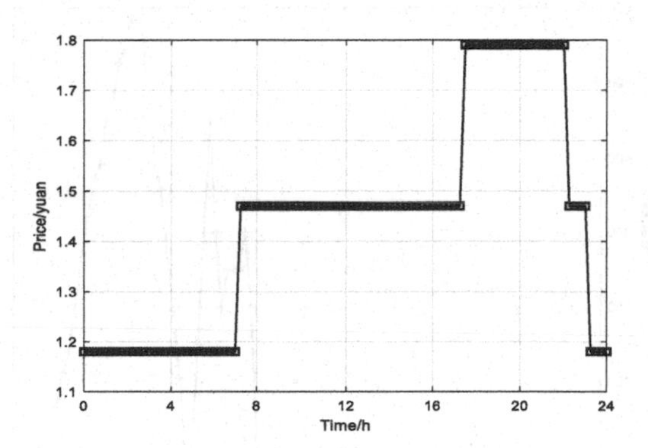

Fig. 6. The determined time-of-use

6.4 Solving for EV Charging Load Based on Time-of-Use

The time-of-use tariff-based EV charging model in section is solved by the *NSGA-II* algorithm, setting the initial population size to 300, the number of generations of iteration to 100, and the lower bound of the decision variable to 0 and the upper bound to 700, and the obtained Pareto optimal bound is shown in Fig. 7.

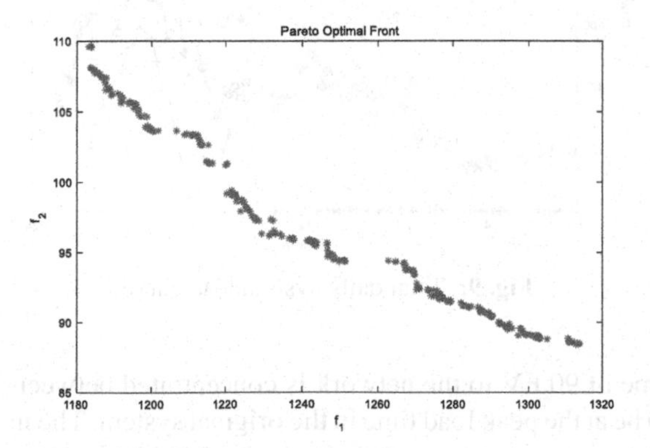

Fig. 7. Pareto optimal bound

The horizontal coordinate is the first optimization objective f_1, which indicates the minimum charging cost for EV users, and the vertical coordinate is the second optimization objective f_2, which is the minimum net load mean in the whole regional energy system.

The charging loads of EV before and after the time-of-use are implemented separately for *Monte Carlo* simulation, and the charging load curves of EV before the time-of-use are obtained as shown in Fig. 8. The load curve of the system at this point is shown in Fig. 9.

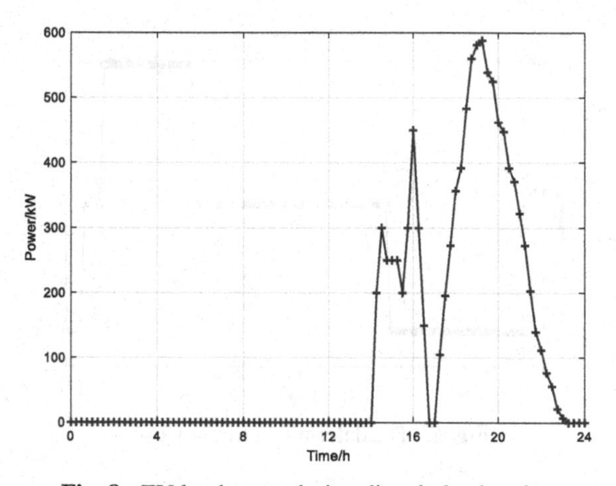

Fig. 8. EV load curve during disorderly charging

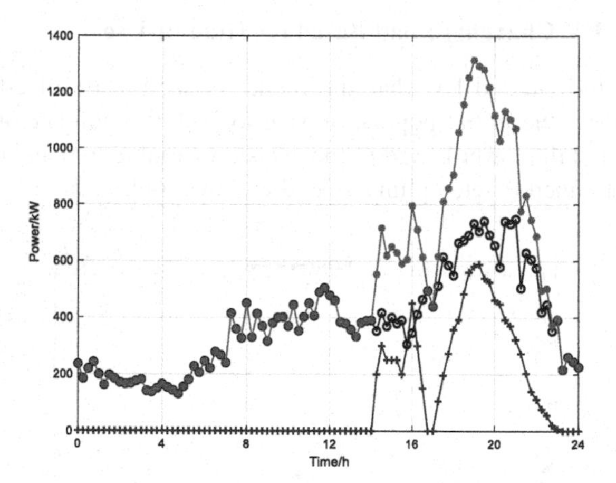

Fig. 9. Total daily system load curve

The return time of 90 EV to the network is concentrated between 17:30 and 21:00, which happens to be at the peak load time in the original system. The maximum charging load occurs at 19:15, with a maximum load value of 588 kW and 84 charging vehicles, which is a peak on top of the load level. The maximum load of the system is 1314 kW, which is nearly double the original maximum value of 748.6 kW and poses a serious threat to the stable operation of the system; the charging time of cabs in the park is concentrated in 14:00–16:00, and according to the hypothesis 6), the cabs uniformly use the charging mode of fast charging, the maximum charging load appears at 14:00, the maximum load value is 450 kW and the number of charging vehicles is 9, the charging time is located in the off-peak period.

The EV charging load curve after the time-of-use is shown in Fig. 10. The daily load curve in the system at this point is shown as the red curve in Fig. 11.

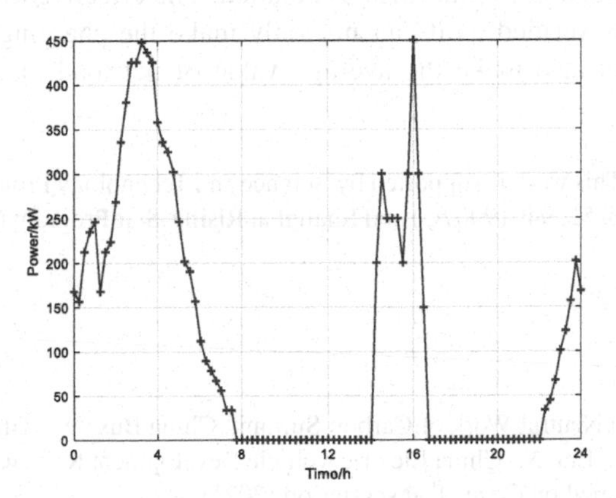

Fig. 10. EV orderly charging load curve after time-of-use

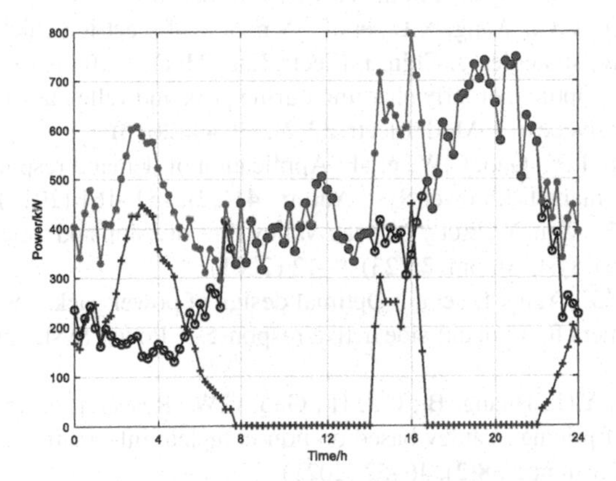

Fig. 11. System load after time-of-use

From the figure, it can be seen that 90 ordinary EV users will orderly charge under the guidance of time-of-use, which makes the average net load value change from 319.2007 kW to 126.3822 kW, a decrease of 60.5%, and the cost of EV charging is reduced from RMB 4301.25 to RMB 3431.75 when charging is unordered, and the feasibility of EV orderly charging strategy based on time-of-use is verified.

In summary, the way of guiding users to orderly charging based on time-of-use can achieve the effect of reducing users' charging cost and lowering the net load average.

7 Summary

The K-Means algorithm is used to divide the peak and valley periods, and then the tariffs for different periods are determined according to the actual situation. At the same time, an EV orderly charging model based on time-of-use is established, and two different methods are used to solve the model. The Pareto-optimal bound of multi-objective function is found using *NSGA-II* algorithm, and an Pareto-optimal solution set of EV charging load is obtained, and then the EV charging load after time-of-use is obtained by the method of *MonteCarlo* simulation. The effectiveness of the sequential charging strategy is verified as it can not only make the charging cost of EV users greatly reduced, but also make the average value of net load of regional integrated energy decrease.

Acknowledgment. This work is supported by Science and Technology Project of State Grid Corporation of China (No. 52094019007G) and Shanghai Rising-Star Program (No. 20QB1400300).

References

1. Chen, Y.: Carbon Neutral Work of Carbon Summit. China Business Times (2021)
2. Zhu, J., Xiong, Y., Liu, X.: China Electric Vehicle Development Review and Future Outlook. International Council on Clean Transportation (2021)
3. Liu, D., Tang, T.Q., Yang, J.H., et al.: Design of microbalance scheduling transactions for energy internet. J. Power Syst. Autom. **41**(10), 1–8 (2017)
4. Wang, X.F., Shao, C.C., Wang, X.L., et al.: A review of electric vehicle charging load and scheduling control strategies. J. Chin. J. Electr. Eng. **33**(1), 1–10 (2013)
5. Wang, B., Ai, X.: Optimal orderly charging during peak and valley tariff periods considering V2G user responsiveness. J. Mod. Electr. **33**(2), 39–44 (2016)
6. Shen, Y., Curtain, L.Y., Gao, C.W., et al.: Application of demand response in the electricity auxiliary service market. J. Power Syst. Autom. **41**(22), 151–161 (2017)
7. Ding, N., Wu, J.K., Zou, Y.: Study on peak-valley time division and time-sharing tariff based on DSM. J. Power Syst. Autom. **25**(23), 9–12 (2001)
8. Hou, J.X., Lin, Z.Z., Yang, L., et al.: Optimal design of power packages for commercial and industrial customers for demand-side active response. J. Power Syst. Autom. **42**(24), 11–19 (2018)
9. Wang, G., Song, Y.H., Huang, B., Cai, H., Gao, C.W.: Research on residential customers' time-of-use tariff pricing strategy based on non-complete information Stackelberg game. J. Electr. Supply Consumpt. **38**(2), 46–52 (2021)
10. Xue, Y.X.: Orderly charging strategy for electric vehicles based on time-sharing tariff. D. Xi'an University of Technology (2019)
11. Liu, D.N., Xu, E.F., Liu, M.G., et al.: A time-of-use tariff pricing method for local consumption of distributed power sources in parks. J. Power Syst. Autom. **44**(20), 19–28 (2020)

Intelligent Modeling, Simulation and Control of Power Electronics and Power Networks

Interface Modeling and Analysis of Interphase-Delay Induced Charge Imbalance in Neural Stimulator Design

Juzhe Li[1], Kaishan Zheng[2], Xu Liu[1](✉), Tao Chen[3], Peiyuan Wan[1], and Zhijie Chen[1]

[1] College of Microelectronics, Beijing University of Technology, Ping Le Yuan 100, Beijing, China
liuxu16@bjut.edu.cn

[2] Institute of Microelectronics, Chinese Academy of Sciences, Bei Tu Cheng 3, Beijing, China

[3] Advanced Photonics Institute, Beijing University of Technology, Ping Le Yuan 100, Beijing, China

Abstract. Neural stimulators design should take safety as the top priority and many methods have been developed in order to achieve biphasic charge balancing in biphasic stimulation with interphase. Based on a widely-used neural stimulation configuration and two electrode-electrolyte interface models, this work gives the quantitative analysis of the voltage change on the electrode-electrolyte interface in each stimulation phase and explores the causes of charge imbalance in the whole biphasic stimulation process especially with an interphase delay. The parameters analysis in this paper is beneficial to the designers for improving the safety of neural stimulators.

Keywords: Neural stimulator · Biphasic current · Charge imbalance · Interphase delay

1 Introduction

Neural stimulators and other implanted medical devices have been proven to be effective in the treatment of neurological diseases such as Parkinson, epilepsy, visual impairment and so on [1]. Instead of medication, neural stimulation has persistent effectiveness and less side effect [2]. A safe long-term stimulation requires the stimulator to realize a charge balanced stimulation with minimum direct current injection because a direct current leakage of 100 nA will cause permanent damage to the tissue [3, 4].

Previous work has shown that the biphasic current-mode stimulation with interphase can achieve both safety and effectiveness but less charge balancing [5]. The stimulation waveform is showed in Fig. 1(a). It usually contains a cathodic phase, an anodic phase, an interphase, and a recovery phase. The cathodic phase is used for depolarization and bringing for muscle or nerve response. The anodic phase is used for eliminating charge accumulation. The interphase with a delay between cathodic phase and anodic phase can be used to let the action potential spread. In the recovery phase, the unbalanced charge

K. Li et al. (Eds.): LSMS 2021/ICSEE 2021, CCIS 1468, pp. 359–368, 2021.
https://doi.org/10.1007/978-981-16-7210-1_34

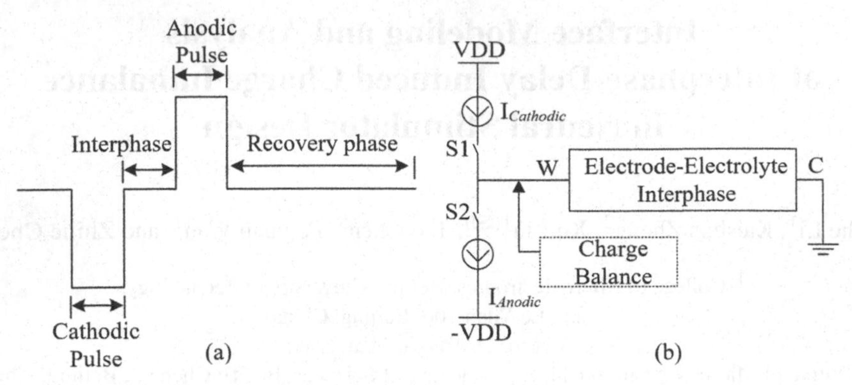

Fig. 1. **(a)** Waveform of the biphasic stimulation with an interphase delay. **(b)** Configuration of the biphasic current stimulation with the stimulator generating bidirectional currents to the electrode-electrolyte.

flows out through a certain path and charge accumulation on the electrode-electrolyte interface is avoided [6]. In order to output such waveforms in Fig. 1(a), a biphasic current stimulation configuration shown in Fig. 1(b) is widely used [7]. It usually has two paths to generate current stimulus. In cathodic phase, S1 turns on and S2 turns off, and the stimulation current flows from W to C. During interphase delay, S1 and S2 turn off and the evoked action potential spreads. In the anodic phase, S2 turns on and the stimulation current flows from C to W. In the recovery phase, S1 and S2 both turn off again and an extra circuit would be used for eliminating charge accumulation on the electrode.

The charge balancing techniques include active ways and passive ways. A lot of work has been done in such technology and can nearly achieve zero charge accumulation [4, 8, 9]. Those approaches can solve the problem of charge accumulation in the practical way. However, there are still considerable work for charge balancing as the researchers have not theoretically explained the reason why the charge imbalance occurs and yet given precise mathematical expressions to illustrate the charge imbalance with its influence factors. In this work we aim at analyzing the voltage change on the electrode-electrolyte interface during the whole stimulation process, and firstly analyze the interphase delay induced charge imbalance in a theoretical way with quantitative formula derivation. Then, a discussion of the relationship between the charge imbalance and the stimulation interphase delay is given. Finally, the simulation is done to verify our derived mathematical expressions. So, the parameters of the stimulators could be further optimized based on our quantitative analysis.

2 Electrode-Electrolyte Interface Model

Establishing accurate electrode-electrolyte interface model is the foundation of the analysis of biphasic stimulation. The essential process during an electrical stimulation includes charge transfer and redistribution across the electrode and electrolyte interface. It contains two primary mechanisms, Faradic charge transfer and non-Faradaic charge redistribution [10, 11]. Modelling the electrode-electrolyte interface is a rather complicated task. To simplifying the analysis process, two widely used electrode-electrolyte interface models are used in this paper, shown in Fig. 2(a) and Fig. 2(b). The electrode-electrolyte interface model shown in Fig. 2(a) is also called one-order model. Here, C_{dl} represents the Helmholtz (double layer) capacitor, while R_F represent the Faradaic resistors. R_S is the solution spreading resistance, which is determined by the fluid resistance and the electrode geometry. The electrode-electrolyte interface model shown in Fig. 2(b) is named two-order model in this paper. C_W and C_C represent the charge accumulation effect on the electrode-electrolyte or cell membrane. R_W and R_C represent the current leakage effect caused by the electrochemical reaction. R_T means tissue impedance. These models have been widely used in neural interface studies, and been proved to be effective in estimating the electrode characteristics during neural recording and stimulation [12, 13].

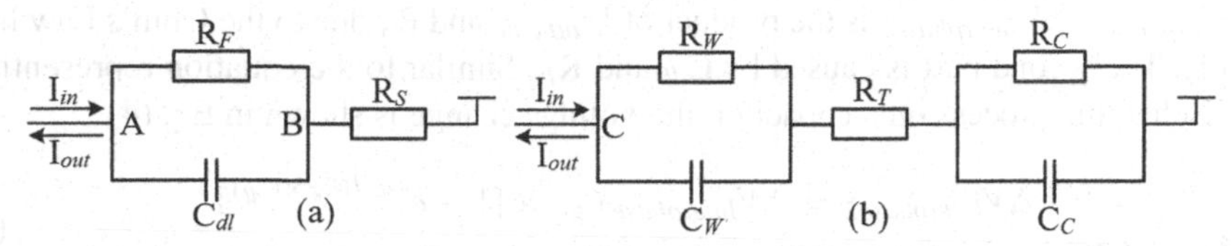

Fig. 2. (a) One-order electrode-electrolyte interface model. (b) Two-order electrode-electrolyte interface model.

3 Quantitative Analysis Based on Electrode-Electrolyte Interface Model

3.1 Voltage Analysis on the Electrode-Electrolyte Interface Model

Modelling of one-order electrode-electrolyte interface can be divided into 4 phases: the cathodic phase, interphase, anodic phase and the recovery phase.

In the cathodic phase, current I_{in} flows from A to B shown in Fig. 2(a). The voltage change at point A can be divided into two parts by superposition theorem. The first part is caused by R_S, and the voltage at point A increases as shown in Eq. (1).

$$\Delta V_{Cathodic1} = I_{Cathodic} \times R_S. \tag{1}$$

$\Delta V_{Cathodic1}$ represents the voltage change constituting the first part during the cathodic phase. $I_{Cathodic}$ represents the amplitude of the cathodic current. The second part is caused by C_{dl} and R_F. Similar to the process of charging a capacitor, the voltage change constituting the second part can be written as Eq. (2).

$$\Delta V_{Cathodic2} = I_{Cathodic} \times R_F \times [1 - e^{-t_1/(R_F \times C_{dl})}] \ (0 < t_1 \leq T_{Cathodic}). \tag{2}$$

$\Delta V_{Cathodic2}$ represents the increased voltage at point A constituting the second part when analyzing the voltage change in the cathodic phase, with a time variable t_1. $T_{Cathodic}$ represents the cathodic pulse width. Due to the superposition theorem, in the second part,

R_S does not influence $\Delta V_{Cathodic2}$, so it can be considered as shorted. Thus, combining the first part and the second part, the voltage change at point A during the cathodic phase can be expressed as Eq. 3.

$$\begin{aligned}\Delta V_{Cathodic} &= \Delta V_{Cathodic1} + \Delta V_{Cathodic2}\\ &= I_{Cathodic} \times R_S + I_{Cathodic} \times R_F \times [1 - e^{-t_1/(R_F \times C_{dl})}] \;\; (0 < t_1 \leq T_{Cathodic}).\end{aligned} \tag{3}$$

$\Delta V_{Cathodic}$ represents the whole voltage change at point A in the cathodic phase.

In the interphase, all switches are turned off and the current stops flowing from outside into the model. The voltage change at point A can also be divided into two parts using superposition theorem. The first part is caused by R_S. As the stimulation current is turned off during the interphase delay time, the voltage change at point A decreases as $\Delta V_{Interphase1}$. $\Delta V_{Interphase1}$ is the product of $I_{Cathodic}$ and R_S due to the Ohm's Law like Eq. (1). The second part is caused by C_{dl} and R_F. Similar to the equation representing the discharging process on a capacitor, the voltage change is shown in Eq. (4).

$$\begin{aligned}\Delta V_{Interphase2} &= \Delta V_{Interphase2(t_0)} \times [1 - e^{-t_2/(R_F \times C_{dl})}]\\ \Delta V_{Interphase2(t_0)} &= \Delta V_{Cathodic} - \Delta V_{Interphase1} \; (0 < t_2 \leq T_{Interphase}).\end{aligned} \tag{4}$$

$\Delta V_{Interphase2}$ represents the voltage change at point A in the interphase, with a time variable t_2, which constitutes the second part when calculating the voltage. $T_{Interphase}$ represents the time of the interphase delay. $T_{Interphase(t0)}$ means the initial voltage at point A and it depends on both the previous voltage value in the cathodic phase and the value in the first part of the expression representing the voltage change in the interphase. Therefore, combining the first and the second parts, the whole voltage decrease at point A during the interphase can be described in Eq. (5).

$$\begin{aligned}\Delta V_{Interphase} &= \Delta V_{Interphase1} + \Delta V_{Interphase2}\\ &= I_{Cathodic} \times R_S + V_{Interphase2(t0)} \times [1 - e^{-t_2/(R_F \times C_{dl})}] \; (0 < t_2 \leq T_{Interphase}).\end{aligned} \tag{5}$$

$\Delta V_{Interphase}$ represents the voltage change during the interphase and its value is related to the voltage in the previous stimulation phase.

In the anodic phase, the current I_{out} flows from B to A, as shown in Fig. 2(a). The derivation process of the voltage change is the same to the cathodic phase and the voltage change can also be divided into two parts to be calculated separately. The first part is caused by R_S, and the second part is caused by C_{dl} and R_F. In the first part, the voltage change at point A, $\Delta V_{Anodic1}$, is the product of I_{Anodic} and R_S due to the Ohm's Law. In the second part, the voltage change shows an exponential feature due to the charging of the capacitor. So, the voltage change can be written as Eq. (6).

$$\Delta V_{Anodic2} = I_{Anodic} \times R_F \times [1 - e^{-t_3/(R_F \times C_{dl})}] \;\; (0 < t_3 \leq T_{Anodic}). \tag{6}$$

$\Delta V_{Anodic2}$ represents the voltage change in the anodic phase, which constitutes the second part when analyzing the voltage change, with a time variable t_3. T_{Anodic} represents the pulse width of the anodic stimulation current. Therefore, combining the first and the second parts, the voltage decrease at point A in the anodic phase can be expressed in Eq. (7).

$$\begin{aligned}\Delta V_{Anodic} &= \Delta V_{Anodic1} + \Delta V_{Anodic2} \\ &= I_{Anodic} \times R_S + I_{Anodic} \times R_F \times [1 - e^{-t_3/(R_F \times C_{dl})}] \ (0 < t_3 \leq T_{Anodic}).\end{aligned} \tag{7}$$

ΔV_{Anodic} represents the whole voltage change in the anodic phase.

In the recovery phase, all switches are cut off and the stimulation current stops flowing from outside into the model. To calculate the voltage change in the recovery phase, according to superposition theorem, the first part is caused by current disappearing and the second part is caused by capacitance discharge. $\Delta V_{Recovery1}$ is the product of I_{Anodic} and R_S due to the Ohm's Law, which represents the voltage change constituting the first part when analyzing the voltage in this recovery phase. While, the voltage change ($\Delta V_{Recorvery2}$) constituting the second part can be written as Eq. (8).

$$\begin{gathered}\Delta V_{Recovery2} = \Delta V_{Recovery2(t0)} \times [1 - e^{-t_4/(R_F \times C_{dl})}] \\ \Delta V_{Recovery2(t0)} = \Delta V_{Cathodic} - \Delta V_{Interphase} - \Delta V_{Anodic} + \Delta V_{Recovery1} \ (0 < t_4 \leq T_{Recovery}).\end{gathered} \tag{8}$$

$T_{Recovery}$ represents the duration of the recovery phase. $\Delta V_{Recorvery2(t0)}$ is the initial voltage in the second part in the recovery phase at point A and it depends on all the previous voltage values in the cathodic phase, interphase, anodic phase and the recovery phase. Therefore, combining the first part and the second part, the whole voltage increase at point A in the recovery phase can be expressed as Eq. (9).

$$\begin{aligned}&\Delta V_{Recovery} = \Delta V_{Recovery1} + \Delta V_{Recovery2} \\ &= I_{Anodic} \times R_S + V_{Recovery2(t_0)} \times [1 - e^{-t_4/(R_F \times C_{dl})}] \ (0 < t_4 \leq T_{Recovery}).\end{aligned} \tag{9}$$

$\Delta V_{Recovery}$ represents the whole voltage change in the recovery phase and it depends on not only the recovery phase itself but also the whole past phases during the neural stimulation.

The modelling of the two-order electrode-electrolyte interface is the same as one-order modelling. However, due to the differences between the two electrode-electrolyte interface models, the resistor and capacitor parameters are changed in the expression. By using superposition theorem, R_T in Fig. 2(b) is shorted in each stimulation phase and the two-order model can be simplified as a capacitor C_W with a series capacitor C_C, and a resistor R_W with a series resistor R_C. Thus, in the derivation of the voltage in the two-order electrode-electrolyte interface model, the calculations of the voltages in each phase are shown below. The voltage change in cathodic phase is expressed as Eq. (10).

$$\begin{aligned}&\Delta V_{Cathodic} = \Delta V_{Cathodic1} + \Delta V_{Cathodic2} \\ &= I_{Cathodic} \times R_T + I_{Cathodic} \times (R_W + R_C) \times [1 - e^{-t_1/[(R_W + R_C) \times (C_W // C_C)]}] \ (0 < t_1 \leq T_{Cathodic}).\end{aligned} \tag{10}$$

And the voltage change in the interphase is shown in Eq. (11).

$$\Delta V_{Interphase} = \Delta V_{Interphase1} + \Delta V_{Interphase2}$$
$$= I_{Cathodic} \times R_T + V_{Interphase2(t0)} \times [1 - e^{-t_2/[(R_W+R_C)\times(C_W//C_C)]}]\ \ 0 < t_2 \le T_{Interphase}). \tag{11}$$

The voltage change in the anodic phase can be expressed in Eq. (12).

$$\Delta V_{Anodic} = \Delta V_{Anodic1} + \Delta V_{Anodic2}$$
$$= I_{Anodic} \times R_T + I_{Anodic} \times (R_W + R_C) \times [1 - e^{-t_3/[(R_W+R_C)\times(C_W//C_C)]}]\ \ (0 < t_3 \le T_{Anodic}). \tag{12}$$

The voltage change in the recovery phase is shown in Eq. (13).

$$\Delta V_{Recovery} = \Delta V_{Recovery1} + \Delta V_{Recovery2}$$
$$= I_{Anodic} \times R_T + V_{Recovery2(t_0)} \times [1 - e^{-t_4/[(R_W+R_C)\times(C_W//C_C)]}]\ \ (0 < t_4 \le T_{Recovery}). \tag{13}$$

3.2 Analysis of Charge Imbalance in the Electrode-Electrolyte Interface Model

As mentioned above, the cathodic stimulation pulse is for action potential evoking and the anodic pulse is for charge offsetting. Thus, in order to keep charge balance, the quantity of the charge inflow and outflow should be same in the cathodic and anodic phases. However, according to the voltage analysis above, the voltage ($\Delta V_{\text{Interphase2}}$ and $\Delta V_{Recorvery2}$) is changed in the interphase and the recovery phase due to a discharging process happened at point A. Since the $\Delta V_{Interphase2}$ is not strictly equal to $\Delta V_{Recorvery2}$, the charge imbalance occurs. Therefore, due to the existence of interphase and recovery phase, nearly all biphasic neural stimulation naturally cause residual charge accumulation and tissue damage. The amount of residual charge in one stimulation cycle is equal to the subtraction of the quantities during capacitance discharge in the interphase and recovery phase. In the interphase, the charge remained on the interface is expressed in Eq. (14).

$$Q_{Interphase} = C_{dl} \times \Delta V_{Interphase2}$$
$$= C_{dl} \times I_{Cathodic} \times R_F \times [1 - e^{-T_{Cathodic}/(R_F \times C_{dl})}] \times [1 - e^{-T_{Interphase}/(R_F \times C_{dl})}]. \tag{14}$$

$Q_{Interphase}$ represents the charge variation in the interphase. In the recovery phase, the charge is calculated as shown in Eq. (15).

$$Q_{Recovery} = C_{dl} \times \Delta V_{Recovery2}$$
$$= C_{dl} \times [1 - e^{-T_{Recovery}/(R_F \times C_{dl})}] \times [\Delta V_{cathodic} - \Delta V_{Interphase} - \Delta V_{Anodic} + \Delta V_{Recovery1}]. \tag{15}$$

$Q_{Recovery}$ represents the charge variation in the recovery phase.

Combining these two equations above, the imbalanced charge during the biphasic current stimulation process accumulates as shown in Eq. (16).

$$Q_{Residue} = |Q_{\ Interphase} - Q_{Recovery}|$$

$$= C_{dl} \times I_{Cathodic} \times R_F \times [1 - e^{-T_{Cathodic}/(R_F \times C_{dl})}] \times [1 - e^{-T_{Recovery}/(R_F \times C_{dl})}] \times e^{-T_{Interphase}/(R_F \times C_{dl})}$$
$$- C_{dl} \times I_{Cathodic} \times R_F \times [1 - e^{-T_{Cathodic}/(R_F \times C_{dl})}] \times [1 - e^{-T_{Interphase}/(R_F \times C_{dl})}]$$
$$- C_{dl} \times I_{Anodic} \times R_F \times [1 - e^{-T_{Anodic}/(R_F \times C_{dl})}] \times [1 - e^{-T_{Recovery}/(R_F \times C_{dl})}]. \quad (16)$$

$Q_{Residue}$ represents the imbalanced charge during the biphasic current stimulation process. According to the expression (16), we can find that the time duration of the interphase and the recovery phase can influence the amount of imbalanced charge and the time duration has an exponential relationship with $Q_{Residue}$ as a variable.

Similar to the theoretical analysis for one-order interface model, in the two-order interface model, the imbalanced charge is caused by inequality of $\Delta V_{Interphase2}$ and $\Delta V_{Recorvery2}$. Therefore, the imbalanced charge in the two-order electrode-electrolyte interface can be expressed in Eq. (17).

$$Q_{Residue} = (C_W//C_C) \times I_{Cathodic} \times (R_W + R_C) \times [1 - e^{-T_{Cathodic}/[(R_W+R_C)\times(C_W//C_C)]}]$$
$$\times [1 - e^{-T_{Recovery}/[(R_W+R_C)\times(C_W//C_C)]}] \times e^{-T_{Interphase}/[(R_W+R_C)\times(C_W//C_C)]} - (C_W//C_C) \times I_{Cathodic}$$
$$\times (R_W + R_C) \times [1 - e^{-T_{Cathodic}/[(R_W+R_C)\times(C_W//C_C)]}] \times [1 - e^{-T_{Interphase}/[(R_W+R_C)\times(C_W//C_C)]}]$$
$$- (C_W//C_C) \times I_{Anodic} \times (R_W + R_C) \times [1 - e^{-T_{Anodic}/[(R_W+R_C)\times(C_W//C_C)]}] \times [1 - e^{-T_{Recovery}/[(R_W+R_C)\times(C_W//C_C)]}]. \quad (17)$$

4 Model Accuracy Verification and Imbalanced Charge Analysis

In order to test the accuracy of the modelling, the simulations in Cadence Virtuoso have been done with a neural stimulation circuit and the results show that the voltage waveforms on both electrode-electrolyte interfaces fit the equations in this paper correspondingly and separately. The simulation waveforms and the comparison with calculation results are shown in Fig. 3 (a)–(d) respectively. The black lines represent the voltage curve in Cadence and the red squares represent the fitting results in MATLAB, and the green triangles represent the error between the voltage curve and the theoretical result from the model. These results are obtained under the conditions where various electrode models and stimulator parameters are used. From these figures, the results show that the derived quantitative equations in this paper accord with stimulation waveforms well in different stimulation conditions. And the maximum error is within 4%. The accuracy of the theoretical analysis of the voltage change between electrode-electrolyte interfaces in this paper is verified.

As for the error, it comes from the superposition theorem used to analyze the voltages on two interface models. For one-order interface model, R_S is not strictly shorted in each phase and still causes a tiny voltage drop. For two-order interface model, due to the voltage drop between R_T, the capacitor C_W is not strictly series connected with C_C and the resistor R_W is also not strictly series connected with R_C. Although there are small errors, this analysis method proposed in this paper effectively reduces the difficulty and complexity and reaches high accuracy.

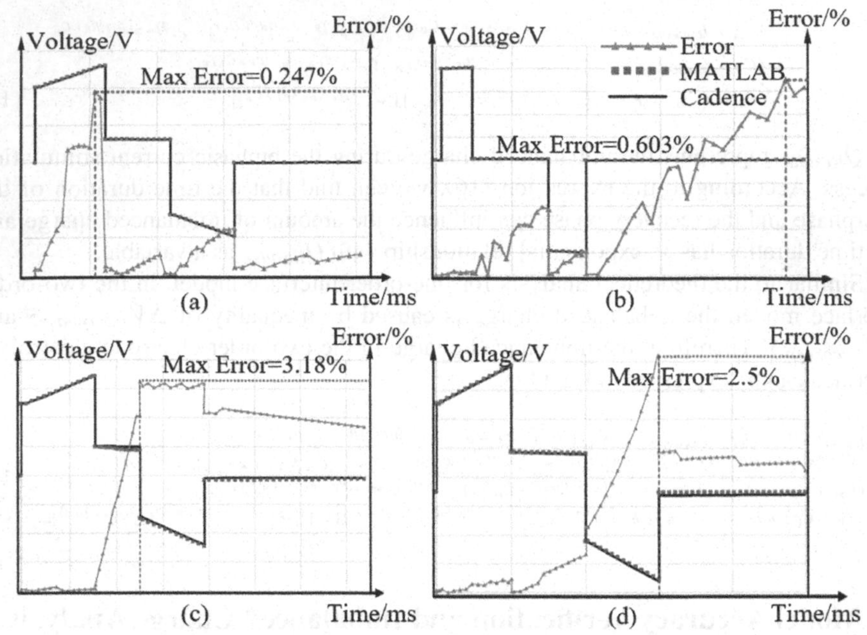

Fig. 3. One-order model: **(a)** $R_S = 20\ K\Omega$, $R_F = 500\ K\Omega$, $C_{dl} = 80$ nF, $I_{Cathodic} = 200\ \mu A$, $I_{Anodic} = 200\ \mu A$, $T_{Cathodic} = 500\ \mu s$, $T_{Anodic} = 500\ \mu s$, $T_{Interphase} = 500\ \mu s$, $T_{Recovery} = 1$ ms. **(b)** $R_S = 50\ K\Omega$, $R_F = 1.5\ K\Omega$, $C_{dl} = 700$ nF, $I_{Cathodic} = 80\ \mu A$, $I_{Anodic} = 80\ \mu A$, $T_{Cathodic} = 100\ \mu s$, $T_{Anodic} = 200\ \mu s$, $T_{Interphase} = 200\ \mu s$, $T_{Recovery} = 600\ \mu s$. Two-order model: **(c)** $R_T = 50\ K\Omega$, $R_W = 50\ K\Omega$, $C_W = 50$ nF, $C_C = 500$ nF, $I_{Cathodic} = 80\ \mu A$, $I_{Anodic} = 80\ \mu A$, $T_{Cathodic} = 300\ \mu s$, $T_{Anodic} = 300\ \mu s$, $T_{Interphase} = 200\ \mu s$, $T_{Recovery} = 700\ \mu s$. **(d)** $R_T = 18\ K\Omega$, $R_W = 50\ K\Omega$, $C_W = 350$ nF, $C_C = 40$ nF, $I_{Cathodic} = 200\ \mu A$, $I_{Anodic} = 200\ \mu A$, $T_{Cathodic} = 500\ \mu s$, $T_{Anodic} = 500\ \mu s$, $T_{Interphase} = 500\ \mu s$, $T_{Recovery} = 1$ ms.

This work further analyzes the relationship between $Q_{Residue}$ and $T_{Interphase}$, $T_{Recovery}$ respectively with the time variable changing ±50% (take 500 μs as the middle value) and other variables controlled by two models showed in Fig. 4 (a)–(d). The results show that, even the quantity of the injected charge in cathodic and anodic phases is equal, the residual charge still exists due to the charging effect in the interphase and recovery phases. Moreover, the results show that the shorter interphase and the longer recovery phase will lead to less residual charge. Figure 5 gives the 3D graphs showing the relationship of $Q_{Residue}$ and $T_{Interphase}$, $T_{Recovery}$ in two models. The results imply that the interphase delay should be minimized and the recovery phase between each stimulation cycle must be considered in the stimulator design. Moreover, all parameters in the stimulator design are highly relevant with the interface model and should be carefully optimized using our derived expressions of voltage and imbalanced charge.

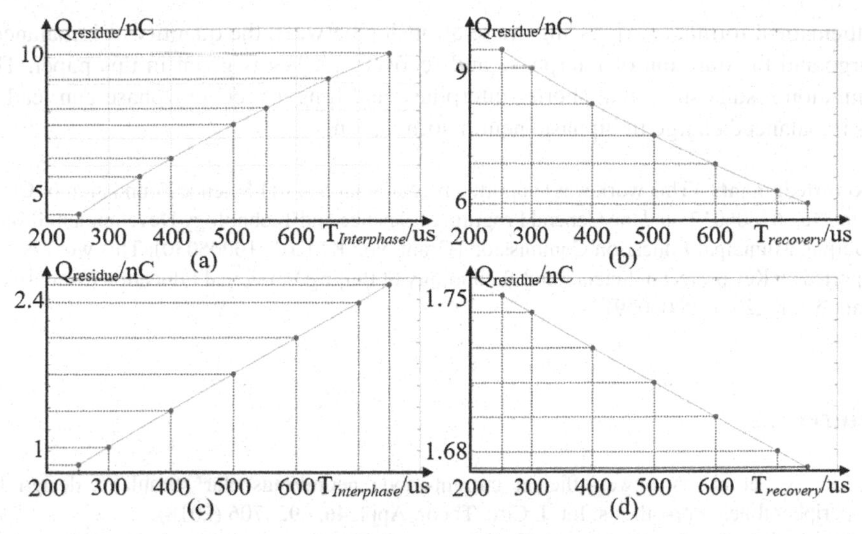

Fig. 4. In one-order model (R_S = 20 KΩ, R_F = 500 KΩ, C_{dl} = 80 nF, $I_{Cathodic}$ = 200 μA, I_{Anodic} = 200 μA, $T_{Cathodic}$ = 500 μs, T_{Anodic} = 500 μs) **(a)** $Q_{Residue}$ vs $T_{Interphase}$, $T_{Recovery}$ = 1 ms. **(b)** $Q_{Residue}$ vs $T_{Recovery}$, $T_{Interphase}$ = 500 μs. In two-order model (R_T = 18 KΩ, R_W = 50 KΩ, C_W = 350 nF, R_C = 250 KΩ, C_C = 40 nF, $I_{Cathodic}$ = 200 μA, I_{Anodic} = 200 μA, $T_{Cathodic}$ = 500 μs, T_{Anodic} = 500 μs) **(c)** $Q_{Residue}$ vs $T_{Interphase}$, $T_{Recovery}$ = 1 ms. **(d)** $Q_{Residue}$ vs $T_{Recovery}$, $T_{Interphase}$ = 500 μs.

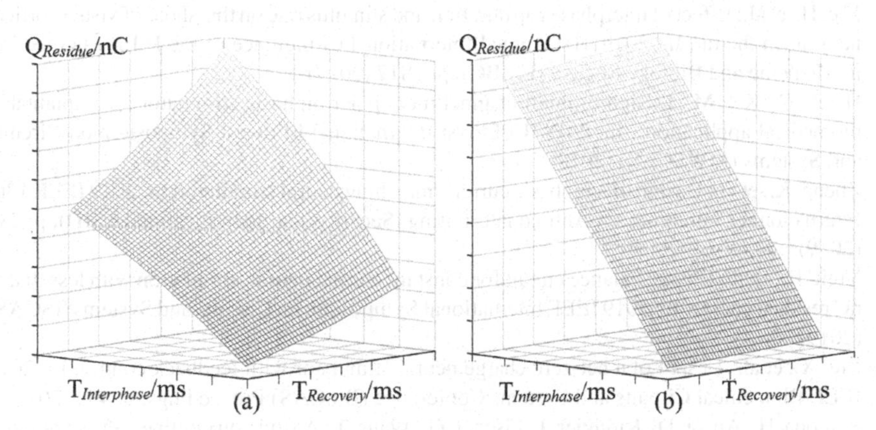

Fig. 5. Relationship of $Q_{Residue}$ and $T_{Interphase}$, $T_{Recovery}$ in 3D graphs using (a) one-order model and (b) two-order model.

5 Conclusion

This is the first work quantitatively analyzing the voltage on the electrode-electrolyte interface and the interphase delay induced charge imbalance. For calculation and verification, two widely used interface models are used with a biphasic current-mode stimulation configuration. This work proves that the interphase and the recovery phase can naturally result in charge imbalance during biphasic current stimulation. Moreover, exact mathematical formulas expressing the relationship between the quantity of imbalanced charge and the duration of interphase and recovery phases is given in this paper. The simulation results show that shorter interphase and longer recovery phase can lead to less imbalanced charge in biphasic neural stimulation.

Acknowledgments. This work was supported by National Natural Science Foundation of China (Grant No.: 61806012) and the General Program of Science and Technology Development Project of Beijing Municipal Education Commission (Grant No.: KM201910005030). This work is also supported by Key project of Science and Technology of Beijing Municipal Education Commission (Grant No. KZ201910005009).

References

1. Liu, X., et al.: A power-efficient current-mode neural/muscular stimulator design for peripheral nerve prosthesis. Int. J. Circ. Theor. Appl. **46**, 692–706 (2018)
2. Goetz, C.G., Poewe, W., Rascol, O., Sampaio, C.: Evidence-based medical review update: pharmacological and surgical treatments of Parkinson's disease: 2001 to 2004. Mov. Disord. **20**(5), 523–539 (2005)
3. Gak, J., et al.: Integrated ultra-low power precision rectifiers for implantable medical devices. In: 2019 Argentine Conference on Electronics (CAE), p. 27 (2019)
4. Faizollah, M., et al.: A low-power, generic biostimulator with arbitrary pulse shape, based on a central control core. IEICE Electron. Exp. **10**, 20120868 (2013)
5. Xie, H., et al.: Effect of interphase gap duration and stimulus rate on threshold of visual cortical neurons in the rat. In: 2019 41st Annual International Conference of the IEEE Engineering in Medicine and Biology Society (EMBC), p. 1817 (2019)
6. Hsieh, C., Ker, M.: Design of multi-channel monopolar biphasic stimulator for implantable biomedical applications. In: 2018 IEEE 61st International Midwest Symposium on Circuits and Systems (MWSCAS) (2018)
7. Zheng, K., et al.: Design of a biphasic current-matching neural stimulator. In: 2019 IEEE 13th International Conference on Anti-counterfeiting, Security, and Identification (ASID), p. 286 (2019)
8. Yiğit, H., et al.: Charge balance circuit for constant current neural stimulation with less than 8 nC residual charge. In: 2019 IEEE International Symposium on Circuits and Systems (ISCAS) (2019)
9. Liu, X., et al.: Design of a net-zero charge neural stimulator with feedback control. In: 2014 IEEE Biomedical Circuits and Systems Conference (BioCAS) Proceedings, p. 492 (2014)
10. Shinoda, H., Anzai, D., Kirchner, J., Fischer, G., Wang, J.: A study on nonlinear effect of modulated low-frequency electromagnetic waves on stimulus response. IEICE Trans. Commun. **E102.B**(6), 1097–1103 (2019)
11. Baharin, R.H.M., et al.: Effects of the permittivity and conductivity of human body for normal-mode helical antenna performance. IEICE Electron. Exp. **16**(16), 20190395 (2019)
12. Yu, W., Jiang, D., Liu, X., Bayford, R., Demosthenous, A.: A human–machine interface using electrical impedance tomography for hand prosthesis control. IEEE Trans. Biomed. Circ. Syst. **12**(6), 1322–1333 (2018). https://doi.org/10.1109/TBCAS.2018.2878395
13. Nasrollaholhosseini, S., et al.: Electrode–electrolyte interface modeling and impedance characterizing of tripolar concentric ring electrode. J. IEEE Trans. Biomed. Eng. **66**, 2897 (2019)

Non-fragile Sliding Mode Control for Enterprise Knowledge Workers System with Time Delay

Yang-Shun Ma[1], Bao-Lin Zhang[2(✉)], and Yu-Xiang Yang[1]

[1] College of Economics and Management, China Jiliang University, Hangzhou 310018, China
[2] College of Automation and Electronic Engineering, Qingdao University of Science and Technology, Qingdao 266061, China

Abstract. This paper deals with the dynamic modeling and non-fragile sliding mode control of an enterprise knowledge workers system. First, from the perspective of the change in the number of knowledge workers, an uncertain system with time delay is established. Second, in order to reduce the influence of uncertainties on the system, an integral sliding mode surface and a non-fragile sliding mode control law are designed, and sufficient conditions for the stability of the system are derived based on the Lyapunov-Krasovskii functional method. Simulation results show that the non-fragile sliding mode control law designed in this paper is effective in maintaining the asymptotic stability of the knowledge workers system. Finally, the effects of time delay on the performance of the system are analyzed. It is found that the smaller the time delay of job-transfer, the better the system performance.

Keywords: Knowledge workers · Dynamic stability · Non-fragile · Sliding mode control · Time delay

1 Introduction

In the era of knowledge economy, knowledge plays an important role in the development of the economy and enterprises, and human resources are the important strategic resources for enterprises [1]. As important carriers of knowledge, knowledge workers are the vitality of organizational innovation and the significant components of core competence, which have received more and more attention [2, 3]. In the complex and changeable market environment, knowledge workers can create huge productivity with the ability of continuous learning and innovation [4]. In order to achieve economic benefits and development goals, the demand for knowledge workers will be increased. It means that enterprises should emphasize the importance of knowledge and strengthen human resources management, especially the management of knowledge workers.

How to effectively manage knowledge workers has aroused widespread concern in academia. For example, in [5], the integration of game mechanisms into the work of knowledge workers has been researched, which has shown that the use of game mechanisms can create a good working atmosphere and improve employees' work capabilities.

K. Li et al. (Eds.): LSMS 2021/ICSEE 2021, CCIS 1468, pp. 369–378, 2021.
https://doi.org/10.1007/978-981-16-7210-1_35

In [6], relying on experts and acquiring new knowledge has been considered as an effective way for knowledge workers to cope with the challenges brought by the technological turbulence in the external environment. From the view of dynamic perspective, it will be found that uncertain factors will affect the dynamic change in the number of knowledge workers, even cause instability of the knowledge workers system. Therefore, considering the dynamic stability of the number of knowledge workers can promote the healthy development of knowledge workers.

The knowledge worker system is inevitably affected by uncertain factors. Fortunately, sliding mode control can provide ways to deal with the uncertainties in knowledge workers system. The sliding mode controller can put the state trajectory into the designed sliding mode surface, on which the system is asymptotically stable and insensitive to uncertainties [7]. At present, sliding mode control has been widely used in supply chain system [8], power system [9], UAV [10] and other fields. Moreover, in the control of knowledge workers system, small changes in the controller parameters may affect the control effect, even destroy the stability of the system [11, 12]. Consequently, it is important to design a non-fragile controller that is not sensitive to the parameters perturbation.

Based on the above analysis, to reduce the influence of uncertainties and controller gain perturbation on knowledge workers system, a design method of non-fragile sliding mode controller (recruitment strategies) is proposed. Based on the Lyapunov-Krasovskii functional method, the sufficient conditions for the robust stability of the system are derived. Finally, simulation results show the effectiveness of the proposed method, and the effects of job-transfer delay on system stability are analyzed.

2 Problem Formulation

In order to carry out the human resources management smoothly, it is necessary to know the changing trend in the number of employees. From the perspective of the dynamic change in the number of employees, this paper takes the n $(n \geq 1)$ types of knowledge workers in the enterprise as the research object to model. The number of the type i knowledge workers at time t can be expressed by $x_i(t)$, and the job-transfer rate or natural resignation rate of different types of knowledge workers at time t can be expressed by $\theta_{ij}(t)$, where $0 \leq \theta_{ij}(t) \leq 1$, $i = 0, 1, 2, \cdots, n$ and $j = 1, 2, \cdots, n$. In particular, when $i \neq j$ and $i, j = 1, 2, \cdots, n$, $\theta_{ij}(t)$ represents the time-varying job-transfer rate of the knowledge workers from type j to type i; when $i = 1, 2, \cdots, n$, $\theta_{ii}(t)$ and $\theta_{0i}(t)$ represents the time-varying rate of natural resignation and voluntary resignation of type i knowledge workers, respectively. The promotion rate of the type i knowledge workers at time t can be expressed by $p_i(t)$. Variable $u_i(t)$ represents the number of type i knowledge workers recruited or dismissed at time t. If $u_i(t) \geq 0$, the enterprise should recruit to fill the gap in the number of employees. Otherwise, the redundant employees should be dismissed. In fact, due to the different time required for different knowledge workers to adapt to the new position, some knowledge workers may not be competent for the new position immediately after the job-transfer. In other words, there is a time delay in job-transfer, which may affect the stability of knowledge workers system.

To sum up, the change in the number of knowledge workers can be described by

$$\dot{x}_i(t) = \sum_{j=1}^{n} b_{ij}(t)x_j(t) + \sum_{j=1, j\neq i}^{n} b_{ij}^d(t-d)x_j(t-d) + u_i(t),\ i = 1, 2, \cdots, n. \quad (1)$$

where, for $i = j$, $b_{ij}(t) = p_i(t) - \sum_{m=0}^{n} \theta_{mi}(t)$, $b_{ij}^d(t-d) = 0$; for $i \neq j$, $b_{ij}(t) = \theta_{ij}(t)$, $b_{ij}^d(t-d) = \theta_{ij}^d(t-d)$. In addition, $d \geq 0$ is the time delay of job-transfer.

Let $\theta_{ij}(t) = c_{ij} + \tilde{\theta}_{ij}(t)$, $\theta_{ij}^d(t-d) = c_{ij}^d + \tilde{\theta}_{ij}^d(t-d)$, $p_i(t) = p_i + \tilde{p}_i(t)$. If $i \neq j$ and $i, j = 1, 2, \cdots, n$, c_{ij} and $\tilde{\theta}_{ij}(t)$ are the nominal value and time-varying perturbation value of job-transfer rate without time delay, respectively; c_{ij}^d and $\tilde{\theta}_{ij}^d(t-d)$ are the nominal value and time-varying perturbation value of job-transfer rate with time delay, respectively; if $i = 1, 2, \cdots, n$, c_{0i} and $\tilde{\theta}_{0i}(t)$ are the nominal value and time-varying perturbation value of voluntary resignation rate, respectively; c_{ii} and $\tilde{\theta}_{ii}(t)$ are the nominal value and time-varying perturbation value of natural resignation rate, respectively; p_i and $\tilde{p}_i(t)$ are the nominal value and time-varying perturbation value of promotion rate, respectively. For the above parameters, set $0 \leq c_{ij}, c_{ij}^d \leq 1$, $\left|\tilde{\theta}_{ij}^d(t-d)\right| < 1$, $\left|\tilde{\theta}_{ij}(t)\right| < 1$ and $|\tilde{p}_i(t)| < 1$.

Then, one has

$$b_{ij}(t) = \begin{cases} p_i - \sum_{m=0}^{n} c_{mi} + \tilde{p}_i(t) - \sum_{m=0}^{n} \tilde{\theta}_{mi}(t), & i = j \\ c_{ij} + \tilde{\theta}_{ij}(t), & i \neq j \end{cases}, \quad b_{ij}^d(t-d) = \begin{cases} 0, & i = j \\ c_{ij}^d + \tilde{\theta}_{ij}^d(t-d), & i \neq j \end{cases}. \quad (2)$$

and the dynamic model (1) can be rewritten as

$$\dot{x}_i(t) = \sum_{j=1}^{n} \left[a_{ij} + \tilde{a}_{ij}(t)\right]x_j(t) + \sum_{j=1}^{n} \left[a_{ij}^d + \tilde{a}_{ij}^d(t-d)\right]x_j(t-d) + u_i(t),\ i = 1, 2, \cdots, n. \quad (3)$$

where

$$\begin{cases} a_{ij}^d = c_{ij}^d \\ \tilde{a}_{ij}^d(t-d) = \tilde{\theta}_{ij}^d(t-d) \end{cases}, \quad a_{ij} = \begin{cases} p_i - \sum_{i=0}^{n} c_{mi}, & i = j \\ c_{ij}, & i \neq j \end{cases}, \quad \tilde{a}_{ij}(t) = \begin{cases} \tilde{p}_i(t) - \sum_{m=0}^{n} \tilde{\theta}_{mi}(t), & i = j \\ \tilde{\theta}_{ij}(t), & i \neq j \end{cases}. \quad (4)$$

Let $x(t) = \left[x_1(t)\ x_2(t) \cdots x_n(t)\right]^T$, $x(t-d) = \left[x_1(t-d)\ x_2(t-d) \cdots x_n(t-d)\right]^T$, $u(t) = \left[u_1(t)\ u_2(t) \cdots u_n(t)\right]^T$. Then, the system can be written as

$$\begin{cases} \dot{x}(t) = (A + \tilde{A}(t))x(t) + (A_d + \tilde{A}_d(t-d))x(t-d) + u(t) \\ x(\varphi) = x_0, \varphi \in [-d, 0] \end{cases}. \quad (5)$$

Where x_0 is the initial vector, and

$$A = \begin{bmatrix} p_1 - \sum_{m=0}^{n} c_{m1} & c_{12} & \cdots & c_{1n} \\ c_{21} & p_2 - \sum_{m=0}^{n} c_{m2} & \cdots & c_{2n} \\ \vdots & \vdots & \ddots & \vdots \\ c_{n1} & c_{n2} & \cdots & p_n - \sum_{m=0}^{n} c_{mn} \end{bmatrix},$$

$$\tilde{A}(t) = \begin{bmatrix} \tilde{p}_1(t) - \sum_{m=0}^{n} \tilde{\theta}_{m1}(t) & \tilde{\theta}_{12}(t) & \cdots & \tilde{\theta}_{1n}(t) \\ \tilde{\theta}_{21}(t) & \tilde{p}_2(t) - \sum_{m=0}^{n} \tilde{\theta}_{m2}(t) & \cdots & \tilde{\theta}_{2n}(t) \\ \vdots & \vdots & \ddots & \vdots \\ \tilde{\theta}_{n1}(t) & \tilde{\theta}_{n2}(t) & \cdots & \tilde{p}_n(t) - \sum_{m=0}^{n} \tilde{\theta}_{mn}(t) \end{bmatrix},$$

$$A_d = \begin{bmatrix} 0 & c_{12}^d & \cdots & c_{1n}^d \\ c_{21}^d & 0 & \cdots & c_{2n}^d \\ \vdots & \vdots & \ddots & \vdots \\ c_{n1}^d & c_{n2}^d & \cdots & 0 \end{bmatrix}, \tilde{A}_d(t-d) = \begin{bmatrix} 0 & \tilde{\theta}_{12}^d(t-d) & \cdots & \tilde{\theta}_{1n}^d(t-d) \\ \tilde{\theta}_{21}^d(t-d) & 0 & \cdots & \tilde{\theta}_{2n}^d(t-d) \\ \vdots & \vdots & \ddots & \vdots \\ \tilde{\theta}_{n1}^d(t-d) & \tilde{\theta}_{n2}^d(t-d) & \cdots & 0 \end{bmatrix}. \quad (6)$$

In (5), Set $f(t, x_t, x_\tau) = \tilde{A}(t)x(t) + \tilde{A}_d(t-d)x(t-d)$, which is bounded and satisfies

$$\|f(t, x_t, x_\tau)\| \le \alpha_1 \|x(t)\| + \alpha_2 \|x(t-d)\|. \quad (7)$$

where $x_t = x(t)$, $x_\tau = x(t-d)$, α_1 and α_2 are positive scalars.

Then, system (5) can be rewritten as

$$\begin{cases} \dot{x}(t) = Ax(t) + A_d x(t-d) + u(t) + f(x, x_\tau, t) \\ x(\varphi) = x_0, \varphi \in [-d, 0] \end{cases}. \quad (8)$$

3 Design of Non-fragile Integral Sliding Mode Control

3.1 Design of Integral Sliding Surface

The implementation of recruitment strategies (including recruitment and dismissal) may be affected by other uncertain factors and cause deviation of actual control effect. The small deviation may reduce the performance of system, and even destroy the stability of the system. Thus, the integral sliding surface can be designed as

$$s(t) = Gx(t) - \int_0^t G(A - K(s))x(s)ds - \int_0^t GA_d x(s-d)ds. \quad (9)$$

where G is real matrix with proper dimensions which is nonsingular, and $K(t) = K + EF(t)H$, where K is the controller gain matrix to be determined, E and H are constant matrix with appropriate dimensions, $F(t)$ is an unknown matrix satisfying $F^T(t)F(t) \leq I$.

As the trajectory of system enters the sliding surface, we have $s(t) = 0$, $\dot{s}(t) = 0$. Then, from (8) and (9), the equivalent control law can be obtained as

$$u_{eq} = -K(t)x(t) - f(t, x_t, x_\tau). \tag{10}$$

Therefore, under the equivalent control law, the sliding motion can be written as

$$\dot{x}(t) = (A - K(t))x(t) + A_d x(t-d). \tag{11}$$

3.2 Stability Analysis of Sliding Motion

Proposition 1. For a given scalar $d > 0$, the sliding motion (11) is asymptotically stable if there exist 4×4 real symmetric matrices $N_1 > 0$, $N_2 > 0$, $P > 0$, $Q > 0$, $R > 0$, a 4×4 matrix K, and a scalar $\varepsilon > 0$ such that

$$\begin{bmatrix} \varphi_{11} & R + N_1 A_d & \varphi_{13} & -N_1 E & H^T \\ * & -Q - R & A_d^T N_2 & 0 & 0 \\ * & * & \varphi_{33} & -N_2 E & 0 \\ * & * & * & -\varepsilon^{-1} I & 0 \\ * & * & * & * & -\varepsilon I \end{bmatrix} < 0. \tag{12}$$

where

$$\begin{cases} \varphi_{11} = Q - R + N_1 A + A^T N_1 - N_1 K - K^T N_1 \\ \varphi_{13} = P - N_1 + A^T N_2 - K^T N_2, \ \varphi_{33} = d^2 R - 2N_2 \end{cases}. \tag{13}$$

Proof. Choose the following Lyapunov-Krasovskii functional candidate as

$$V_1(t) = x^T(t)Px(t) + \int_{t-d}^{t} x^T(s)Qx(s)ds + d\int_{-d}^{0} ds \int_{t+s}^{t} \dot{x}^T(r)R\dot{x}(r)dr. \tag{14}$$

where P, Q and R are real symmetric positive definite matrices.

Calculating the time derivative of $V_1(t)$ along with the trajectory of (11), one yields

$$\begin{aligned} \dot{V}_1(t) &= 2x^T(t)P\dot{x}(t) + x^T(t)Qx(t) - x^T(t-d)Qx(t-d) \\ &\quad + d^2\dot{x}^T(t)R\dot{x}(t) - d\int_{t-d}^{t} \dot{x}^T(r)R\dot{x}(r)dr. \end{aligned} \tag{15}$$

Notice from (11), for real symmetric matrices $N_1 > 0$ and $N_2 > 0$, one has

$$2\left[x^T(t)N_1 + \dot{x}^T(t)N_2\right][-\dot{x}(t) + (A - K(t))x(t) + A_d x(t-d)] = 0. \tag{16}$$

Let $\eta^T(t) = \left[x^T(t)\, x^T(t-d)\, \dot{x}^T(t)\right]$. From (15), (16) and Lemma 2 in [13], one yields

$$\dot{V}_1(t) \leq \eta^T(t)\Omega\eta(t)\,. \tag{17}$$

where

$$\Omega = \begin{bmatrix} \phi_{11} & R+N_1A_d & \phi_{13} \\ * & -Q-R & A_d^TN_2 \\ * & * & \varphi_{33} \end{bmatrix}, \begin{cases} \phi_{11} = Q-R+N_1A+A^TN_1-N_1K(t)-K(t)^TN_1 \\ \phi_{13} = P-N_1+A^TN_2-K(t)^TN_2 \\ \varphi_{33} = d^2R-2N_2 \end{cases}. \tag{18}$$

Let $\Lambda = \begin{bmatrix} \varepsilon I & 0 \\ * & \varepsilon^{-1}I \end{bmatrix}$ and $\Gamma = \left[\Theta\ \Psi^T\right]$. Then by Lemma 1 in [14], one obtains

$$\Omega = \tilde{\Omega} + \varepsilon\Theta\Theta^T + \varepsilon^{-1}\Psi^T\Psi = \tilde{\Omega} + \Gamma\Lambda\Gamma^T\,. \tag{19}$$

where

$$\tilde{\Omega} = \begin{bmatrix} \varphi_{11} & R+N_1A_d & \varphi_{13} \\ * & -Q-R & A_d^TN_2 \\ * & * & \varphi_{33} \end{bmatrix}, \Theta = \begin{bmatrix} -N_1E \\ 0 \\ -N_2E \end{bmatrix}, \Psi = \begin{bmatrix} H \\ 0 \\ 0 \end{bmatrix}. \tag{20}$$

According to Lyapunov-Krasovskii stability theory, if $\Omega < 0$, from (17), the sliding motion (11) is asymptotically stable. By the Schur complement, we can deduce that the inequality (12) holds if and only if (19) holds. This completes the proof.

Notice that the inequality (12) is nonlinear, to solve the controller gain K, set $N_2 = \lambda N_1$ in (12), then pre- and post-multiply the left-hand side the (12) by $diag\left\{N_1^{-1}, N_1^{-1}, N_1^{-1}, \varepsilon I, I\right\}$ and its transpose, respectively, and set $\overline{N}_1 = N_1^{-1}$, $\overline{P} = N_1^{-1}PN_1^{-1}$, $\overline{Q} = N_1^{-1}QN_1^{-1}$ and $\overline{K} = KN_1^{-1}$, then Proposition 2 can be obtained.

Proposition 2. For a given scalar $d > 0$, the sliding motion (11) is asymptotically stable if there exist 4×4 real symmetric matrices $\overline{N}_1 > 0$, $\overline{P} > 0$, $\overline{Q} > 0$, $\overline{R} > 0$, a 4×4 matrix $\overline{K}$, and a scalar $\varepsilon > 0$ such that

$$\begin{bmatrix} \overline{\varphi}_{11} & \overline{R}+A_d\overline{N}_1 & \overline{\varphi}_{13} & -\varepsilon E & \overline{N}_1H^T \\ * & -\overline{Q}-\overline{R} & \lambda\overline{N}_1A_d^T & 0 & 0 \\ * & * & \overline{\varphi}_{33} & -\lambda\varepsilon E & 0 \\ * & * & * & -\varepsilon I & 0 \\ * & * & * & * & -\varepsilon I \end{bmatrix} < 0\,. \tag{21}$$

where

$$\begin{cases} \overline{\varphi}_{11} = \overline{Q} - \overline{R} + A\overline{N}_1 + \overline{N}_1A^T - \overline{K} - \overline{K}^T \\ \overline{\varphi}_{13} = \overline{P} - \overline{N}_1 + \lambda\overline{N}_1A^T - \lambda\overline{K}^T,\ \overline{\varphi}_{33} = d^2\overline{R} - 2\lambda\overline{N}_1 \end{cases}. \tag{22}$$

In addition, the controller gain matrix K can be obtain by $K = \overline{KN_1^{-1}}$.

3.3 Design of Non-fragile Sliding Mode Control Law

To ensure the reachability of the sliding surface, a non-fragile sliding mode control law with the following form is designed as

$$u(t) = -K(t)x(t) - G^{-1}[\|G\|(\alpha_1\|x(t)\| + \alpha_2\|x(t-d)\|) + \eta]\mathrm{sgn}(s(t)). \quad (23)$$

where $\mathrm{sgn}(\cdot)$ is the sign function and η is a positive scalar.

Proposition 3. Under the non-fragile sliding mode control law (23), the state trajectory of knowledge workers system (8) can reach the sliding mode surface $s(t) = 0$ in a finite time and maintain on it thereafter.

Proof. The Lyapunov function can be selected as $V_2(t) = \frac{1}{2}s^T(t)s(t)$. From (8), (9) and (23), one has

$$\dot{s}(t) = -[\|G\|(\alpha_1\|x(t)\| + \alpha_2\|x(t-d)\|) + \eta]\mathrm{sgn}(s(t)) + Gf(t, x_t, x_\tau). \quad (24)$$

Taking the derivate of $V_2(t)$, from (7) and (24), we can obtain

$$\dot{V}_2(t) \le -\eta\|s(t)\|_1 < 0, \ \forall\|s(t)\|_1 \ne 0. \quad (25)$$

Therefore, this completes the proof.

4 Simulation Results

In order to illustrate the effectiveness of the non-fragile sliding mode control law (recruitment strategies including recruitment and dismissal), the knowledge workers dynamic system (8) with $n = 4$ is taken into consideration. To simplify, the four types of knowledge workers will be denoted by KW-1, KW-2, KW-3 and KW-4. The relevant parameters of four types of knowledge workers are given in the following.

$$A = \begin{bmatrix} -0.0946 & 0.0546 & 0 & 0 \\ 0 & -0.1255 & 0.0525 & 0 \\ 0.0532 & 0.0541 & -0.246 & 0 \\ 0 & 0 & 0.0727 & -0.1228 \end{bmatrix}, A_d = \begin{bmatrix} 0 & 0.0375 & 0 & 0 \\ 0 & 0 & 0.0437 & 0 \\ 0.0322 & 0.0491 & 0 & 0 \\ 0 & 0 & 0.0324 & 0 \end{bmatrix}. \quad (26)$$

From the optimal number and the current number of the four types of knowledge workers, the initial state of the system (8) can be given as follows $x(0) = \left[-20\ 16\ -17\ 19\right]^T$. Moreover, set $G = diag\{100,\ 50,\ 100,\ 50\}$, $\alpha_1 = 0.1$, $\alpha_2 = 0.1$, $\eta = 0.5$, $\lambda = 0.01$, $F(t) = \sin t$ and

$$E = \begin{bmatrix} 0.1 & 0.2 & 0.1 & 0.1 \\ 0.1 & 0.12 & 0.1 & 0.08 \\ 0.15 & 0.12 & 0.1 & 0.1 \\ 0.1 & 0.08 & 0.2 & 0.1 \end{bmatrix}, H = \begin{bmatrix} 0.1 & 0.12 & 0.2 & 0.1 \\ 0.08 & 0.1 & 0.12 & 0.08 \\ 0.1 & 0.15 & 0.1 & 0.1 \\ 0.1 & 0.2 & 0.12 & 0.1 \end{bmatrix}. \quad (27)$$

In order to analyze the influence of job-transfer delay on the stability of the system, the values of time delay d can be chosen as 0.8, 1.6, 2.2 and 2.6, respectively. By solving the matrix inequality (21), the gain matrices with different time delays can be obtained, which are denoted as K_1, K_2, K_3 and K_4, respectively

$$K_1 = \begin{bmatrix} 0.6547 & 0.1654 & 0.2130 & 0.1033 \\ 0.1635 & 0.7084 & 0.2362 & 0.1074 \\ 0.1051 & 0.2287 & 0.5711 & 0.1744 \\ 0.0950 & 0.1481 & 0.1484 & 0.6143 \end{bmatrix}, K_2 = \begin{bmatrix} 0.7552 & 0.1366 & 0.1858 & 0.0887 \\ 0.1459 & 0.7903 & 0.1870 & 0.0905 \\ 0.0860 & 0.1821 & 0.6595 & 0.1569 \\ 0.0855 & 0.1303 & 0.1249 & 0.7200 \end{bmatrix}. \quad (28)$$

$$K_3 = \begin{bmatrix} 0.8051 & 0.1226 & 0.1760 & 0.0862 \\ 0.1644 & 0.8463 & 0.2014 & 0.1076 \\ 0.1006 & 0.1866 & 0.7160 & 0.1710 \\ 0.0805 & 0.1111 & 0.1124 & 0.7696 \end{bmatrix}, K_4 = \begin{bmatrix} 0.8486 & 0.1180 & 0.1725 & 0.0850 \\ 0.1681 & 0.8870 & 0.2056 & 0.1094 \\ 0.1032 & 0.1878 & 0.7569 & 0.1737 \\ 0.0793 & 0.1049 & 0.1093 & 0.8136 \end{bmatrix}. \quad (29)$$

Then, from (23), (28) and (29), four kinds of non-fragile sliding mode control laws are obtained. Under the non-fragile sliding mode control laws, the simulation results of knowledge workers dynamic system (8) are given. The dynamic curves of the number of knowledge workers with different time delays are shown in Figs. 1, 2, 3 and 4.

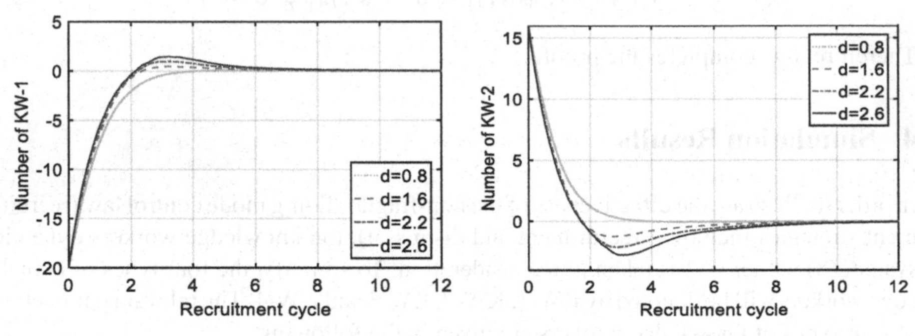

Fig. 1. The number of KW-1

Fig. 2. The number of KW-2

From Figs. 1, 2, 3 and 4, the non-fragile sliding mode control laws designed in this paper can make the knowledge workers system asymptotically stable. In other words, for the influence of natural resignation, voluntary resignation, promotion, job-transfer and its delay, the number of knowledge workers can be kept dynamic stable under the control laws (i.e. recruitment strategies including recruitment and dismissal). However, With the increase of d, the time for the number of same type knowledge workers being stable gets longer, and the number fluctuation becomes bigger. Therefore, the job-transfer delay affects the dynamic performance of the knowledge workers system, which is not conducive to the asymptotic stability of the system.

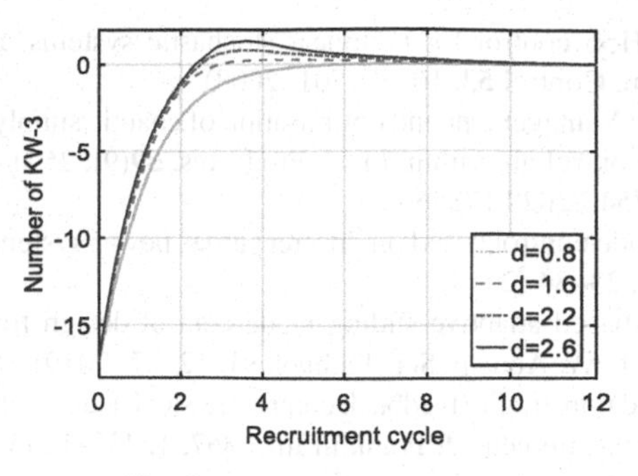

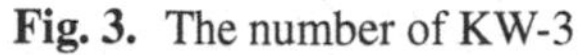
Fig. 3. The number of KW-3

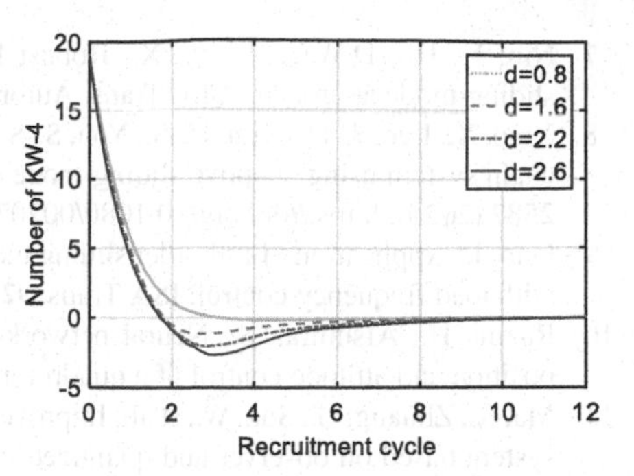

Fig. 4. The number of KW-4

5 Conclusion

This paper deals with the modeling, controller design and stability of a time delay model of knowledge workers affected by time-varying factors such as natural resignation, voluntary resignation, promotion and job transfer. In the modeling, the above time-varying factors are divided into the deterministic part and the uncertain part, and the effects of job-transfer delay, recruitment and dismissal are also considered. To reduce the influence of uncertainties and controller gain perturbation on the system, based on the Lyapunov-Krasovskii functional method, the design method of non-fragile sliding mode control law and the sufficient conditions for the asymptotic stability of the system are given. The simulation results show that the non-fragile sliding mode control law designed in this paper can make the knowledge worker system asymptotically stable. In addition, through the comparison, it is found that the time delay of job-transfer has an impact on the stability of the system. The smaller the time delay of job-transfer, the better the system performance. This study has reference values in the management practice for enterprises.

References

1. Todericiu, R., Şerban, A., Dumitraşcu, O.: Particularities of knowledge worker's motivation strategies in Romanian organizations. Procedia Econ. Fin. **6**, 405–413 (2013)
2. Tzortzaki, A.M., Mihiotis, A.: A review of knowledge management theory and future directions. Knowl. Process Manage. **21**, 29–41 (2014)
3. Lu, J., Lv, Y.: Research of electric power enterprise knowledge workers' incentive based on fuzzy model. Procedia Eng. **16**, 695–701 (2011)
4. Iazzolino, G., Laise, D., Gabriele, R., et al.: Knowledge-based strategies and sustainability: a framework and a case study application. Meas. Bus. Excellence **21**, 152–174 (2017)
5. Spanellis, A., Dörfler, V., MacBryde, J.: Investigating the potential for using gamification to empower knowledge workers. Exp. Syst. Appl. **160**, 113694 (2020)
6. Chen, T., Li, F., Chen, X.-P., Zhanying, O.: Innovate or die: how should knowledge-worker teams respond to technological turbulence? Organ. Behav. HuM. Decis. Process. **149**, 1–16 (2018). https://doi.org/10.1016/j.obhdp.2018.08.008

7. Niu, Y., Ho, D.W.C., Wang, X.: Robust H∞ control for nonlinear stochastic systems: a sliding-mode approach. IEEE Trans. Autom. Control **53**, 1695–1701 (2008)
8. Xiao, X., Lee, S.-D., Kim, H.-S., You, S.-S.: Management and optimisation of chaotic supply chain system using adaptive sliding mode control algorithm. Int. J. Prod. Res. **59**(9), 2571–2587 (2020). https://doi.org/10.1080/00207543.2020.1735662
9. Guo, J.: Application of full order sliding mode control based on different areas power system with load frequency control. ISA Trans. **92**, 23–34 (2019)
10. Razmi, H., Afshinfar, S.: Neural network-based adaptive sliding mode control design for position and attitude control of a quadrotor UAV. Aerosp. Sci. Technol. **91**, 12–27 (2019)
11. Ma, K., Zhuang, G., Sun, W., et al.: Improved non-fragile feedback control for stochastic jump system based on observer and quantized measurement. J. Franklin Inst. **357**, 12433–12453 (2020)
12. Wang, D., Shi, P., Wang, W., Karimi, H.R.: Non-fragile $H\infty$ control for switched stochastic delay systems with application to water quality process. Int. J. Robust Nonlinear Control **24**(11), 1677–1693 (2014). https://doi.org/10.1002/rnc.2956
13. Han, Q.-L.: Absolute stability of time-delay systems with sector-bounded nonlinearity. Automatica **41**, 2171–2176 (2005)
14. Wo, S., Yun, Z., Chen, Q., et al.: Non-fragile controller design for discrete descriptor systems. J. Franklin Inst. **346**, 914–922 (2009)

Adaptive Load Frequency Control in Multi-area Power Systems with Distributed Traffic Attacks Based on Deterministic Network Calculus

Yajian Zhang[1], Chen Peng[1(✉)], Yuqing Niu[2], and Min Zheng[1]

[1] Department of Automation, School of Mechatronic Engineering and Automation, Shanghai University, Shanghai, China
{zhang_ya_jian,c.peng,zhengmin203}@shu.edu.cn

[2] Key Laboratory of Smart Grid of Ministry of Education, School of Electrical Automation and Information Engineering, Tianjin University, Tianjin, China
niuyuqing8051@tju.edu.cn

Abstract. This paper investigates an adaptive load frequency control (LFC) strategy in multi-area power systems with malicious distributed traffic attacks. Based on the deterministic network calculus theory, the analytical relationship between transmission delay and key network parameters is established under different attack levels. Secondly, constraints of controller gain under each attack level are derived by constructing a Lyapunov function. Finally, the most matching controller gain is switched into the controller loop by taking the actual packet transmission delay as the decision variable. Simulation results show that the proposed method can effectively resist the malicious traffic attacks while enhance the reliability of physical-cyber coupling power systems.

Keywords: Load frequency control · Distributed traffic attack · Deterministic network calculus · Adaptive control

1 Introduction

Load frequency control (LFC) is a key method to maintain frequency and tie-line power exchanges at the rated values in power systems [1]. With the scale expansions of power systems, exchanges of operation status and control instructions gradually rely on the communication network [2]. However, the introduction of network unavoidably suffer from the malicious network attacks [3–6]. Specifically, the traffic attack, as a common attack approach, may result in significant delay increase of the normal data flows [7]. The basic principle is to preempt the communication bandwidth by injecting abnormal data packets through loopholes or security defects.

For LFC systems, the transmission performance deterioration will result in a failure of the control law designed according to the normal network environment. In most

This work was supported by the National Key Research and Development Program of China under Grant 2020YFB1708202.

K. Li et al. (Eds.): LSMS 2021/ICSEE 2021, CCIS 1468, pp. 379–389, 2021.
https://doi.org/10.1007/978-981-16-7210-1_36

literatures, a single and fixed controller is designed to stabilize the LFC system under normal and attack scenarios. In [8, 9], the event-triggered schemes with DoS attacks are designed. The operation status transmission can be triggered only when the frequency deviation exceeds a certain threshold. In [10], the delay attack process is described as a Bernoulli model. To ensure all the closed-loop poles are in the stable region, the controller constrains are obtained by pole placement technique. In [11], a gradient descent-based delay estimator is designed to obtain actual delays. However, since using one controller to stabilize normal and attack scenarios, it is difficult to ensure a satisfactory performance under normal and low-level attack scenarios.

Different from fixed gain schemes, some researchers have proposed the adaptive schemes to improve the control performance under different attack levels. The basic idea is to pre-design a series of controllers based on estimating the transmission delay ranges under different attack levels. The most matching controller is switched into the control loop by designing an appropriate selection mechanism. For example, in [10], for the LFC with TDS attacks, the most matching controller is selected by using the actual transmission delay as the decisions. The controller is designed with pole placement technique. However, the positions of poles have a great influence on the dynamic performance. The relationship between pole positions and dynamic performance was not discussed. Besides, the current researches rarely analyze the transmission characteristics of normal data flows under different traffic attack levels.

Hence, an adaptive LFC scheme based on deterministic network calculus (DNC) is proposed in this paper. Main contributions are as follows: 1) Based on DNC, the upper transmission delay bound of the normal data flows under traffic attacks is deduced. Furtherly, the attack levels are determined according to the theoretical delay bounds; 2) By deducing the controller gain constraints to guarantee asymptotical stability at each attack level, an adaptive LFC scheme is designed aiming at improving the dynamic performance under normal and low-level attack scenarios.

2 Upper Transmission Delay Bound Calculation Under Attack

In DNC theory, the amounts of data flows arrived at forwarding nodes, and the forwarding capability provided for data flows can be described as the arrival curve (denoted as $\alpha(t)$) and service curve (denoted as $\beta(t)$), respectively [13].

$$\alpha(t) = \sigma + \rho t. \tag{1}$$

$$\beta(t) = r(t - T)^{+} = \begin{cases} 0, & 0 < t \leq T \\ r(t - T), & \text{others} \end{cases} \tag{2}$$

where σ and ρ are the size and transmission rate of data packet; r and T are the forwarding rate and inherent cache delay of forwarding node. Let $R(t)$ and $R^{*}(t)$ denote the cumulative amounts of input and output data flows in a forwarding node, the following inequalities hold during the time interval $0 \leq s \leq t$:

$$R(t) - R(s) \leq \alpha(t - s). \tag{3}$$

$$R^*(t) \geq (R \otimes \beta)(t) \triangleq \inf_{0 \leq s \leq t} \{R(t-s) + \beta(t)\}. \tag{4}$$

Theorem 1 (Equivalent service curve): For a given data flow which is transmitted by the forwarding nodes with service curves as $\beta_1(t), \beta_2(t), \ldots, \beta_{\mathbb{K}}(t)$, the equivalent service curve (denoted as $\beta(t)$) during multi-hop transmission process is given by

$$\beta(t) = \beta_1 \otimes \beta_2(t) \ldots \otimes \beta_{\mathbb{K}}(t) \tag{5}$$

Proof: Take $\mathbb{K} = 2$ as an example. The cumulative function of output data flow from the first forwarding node (i.e., $R_1^*(t)$) is also the cumulative function of input data flow arrived at the second forwarding node (i.e., $R_2(t)$). Hence, the cumulative function of output data flow from the second forwarding node (i.e., $R_2^*(t)$) satisfies

$$R_2^*(t) \geq R_1^* \otimes \beta_2(t) \geq (R_1 \otimes \beta_1(t)) \otimes \beta_2(t) = R_1 \otimes (\beta_1(t) \otimes \beta_2(t)). \tag{6}$$

Corollary 1: For a data flow transferred by forwarding nodes with service curves $\{\beta^k = r^k(t - T^k)^+, k = 1, 2, \ldots, \mathbb{K}\}$, the equivalent service curve $\beta_0(t)$ is given by.

$$\beta_0 = \beta^1 \otimes \beta^2 \cdots \otimes \beta^{\mathbb{K}}(t) = \min\{r^k\}(t - \sum_{k=1}^{\mathbb{K}} T^k)^+. \tag{7}$$

Proof: Without losing of generality, take $\mathbb{K} = 2$ as an example. For the situation as $0 \leq s \leq t \leq T^2$, the following equation holds according to Eq. (4).

$$\beta^1 \otimes \beta^2(t) = \inf_{0 \leq s \leq t} \{r^1(t-s-T^1)^+ + r^2(s-T^2)^+\} = 0. \tag{8}$$

For another situation as $t > T^2$, we have

$$\begin{aligned}
\beta^1 \otimes \beta^2(t) &= \min \left\{ \begin{array}{l} \inf\limits_{0 \leq s \leq T^2} \{r^1(t-s-T^1)^+ + r^2(s-T^2)^+\}, \\ \inf\limits_{T^2 < s < t} \{r^1(t-s-T^1)^+ + r^2(s-T^2)^+\}, \\ \inf\limits_{s=t} \{r^1(t-s-T^1)^+ + r^2(s-T^2)^+\} \end{array} \right\} \\
&= \min\left\{r^1(t-T^2-T^1)^+, r^2(t-T^1-T^2)^+, r^2(t-T^2)^+\right\}. \\
&= \min\{r^1, r^2\}(t-T^2-T^1)^+
\end{aligned} \tag{9}$$

Consider a normal data flow with arrival curve $\alpha_0(t) = \sigma_0 + \rho_0 t$. Other data flows transmitted by the k-th forwarding node contain: 1) normal data flows from other PMUs; 2) abnormal data flows injected by attackers. The arrival curves are given by

$$\alpha_p^k(t) = \sigma_p^k + \rho_p^k t,\ p = 1, 2, \ldots, P \tag{10a}$$

$$\overline{\alpha}_q^k(t) = \overline{\sigma}_q^k + \overline{\rho}_q^k t,\ q = 1, 2, \ldots, Q \tag{10b}$$

where $\alpha_q^k(t)$ and $\overline{\alpha}_q^k(t)$ are the arrival curves of normal and abnormal data flows, respectively. The service curve provided by the k-th forwarding node for the investigated normal data flow (denoted as $\beta_0^k(t)$) is given by [13].

$$\beta_0^k(t) = (\beta^k - \sum_{p=1}^{P} \alpha_p^k - \sum_{q=1}^{Q} \overline{\alpha}_q^k)^+(t) = r_0^k(t - T_0^k)^+. \tag{11}$$

$$r_0^k = r^k - \sum_{p=1}^{P} \rho_p^k - \sum_{q=1}^{Q} \overline{\rho}_q^k. \tag{12}$$

$$T_0^k = \frac{r^k T^k + \sum_{p=1}^{P} \sigma_p^k + \sum_{q=1}^{Q} \overline{\sigma}_q^k}{r_0^k}. \tag{13}$$

Moreover, the upper queuing delay bound with $\mathbb{K}$ hops is given by

$$d_{\max} = \frac{\sigma_0}{\min\{r_0^k\}} + \sum_{k=1}^{\mathbb{K}} T_0^k. \tag{14}$$

Finally, according to the analysis process presented in [13], the upper transmission delay bound containing queuing delay, serial delay and propagation delay is given by

$$T_{\text{delay\_max}} = d_{\max} + \sum_{k=1}^{\mathbb{K}} \frac{\sigma_0}{r_0^k} + \frac{L_\Sigma}{v}. \tag{15}$$

where L_Σ is the total length of links; v is the propagation velocity of optical fiber.

3 Controller Design

Figure 1 shows the LFC diagram where n is the number of generators; N is the number of interconnected areas; M_i and D_i are inertia and damping coefficient; Δf_i and Δf_j are the frequency deviations; $\Delta P_{\text{tie\_}i}$ is the tie-line power fluctuation; L_{ij} is the tie-line synchronization coefficient; $\Delta P_{\text{m}pi}$ is the output mechanical power increment; $\Delta P_{\text{v}pi}$ is the position deviation of turbine; $\Delta P_{\text{d}i}$ is the load fluctuation; T_{tpi} and T_{gpi} are time constants of turbine and governor; $\alpha_{\text{p}i}$ is the participation factor; U_i is the control input; R_{pi} is the droop coefficient. Hence, dynamics of LFC is given by [12, 13].

$$\begin{cases} X_i(k+1) = \overline{A}_i X_i(k) + \overline{B}_i U_i(k) + \overline{H}_i \omega_i(k) \\ Y_i(k) = C_i X_i(k) \end{cases}. \tag{16}$$

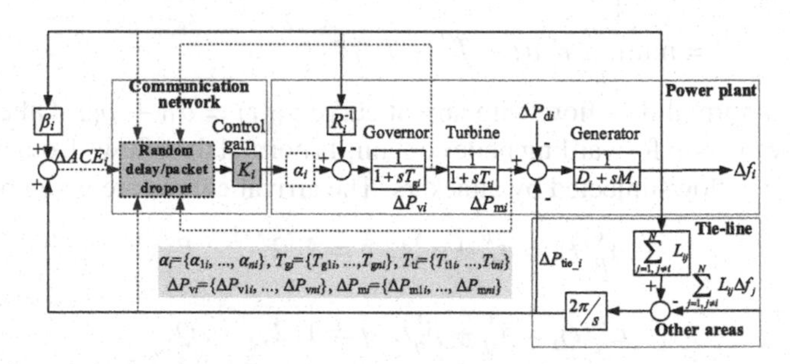

Fig. 1. Block diagram of LFC system in Area i.

Since the upper transmission delay bound of data flows can be calculated, the LFC system can be divided into a series of operating scenarios. Furtherly, a series of controllers can be pre-designed. The following assumptions are given as:

1) The upper transmission delay bounds with different attack levels satisfy

$$\left\{T^{\mathrm{i}}_{\mathrm{delay\_max0}}, T^{\mathrm{i}}_{\mathrm{delay\_max1}}, \ldots, T^{\mathrm{i}}_{\mathrm{delay\_}maxN}\right\}. \tag{17}$$

where $0 < T^{\mathrm{i}}_{\mathrm{delay\_max0}} < T^{\mathrm{i}}_{\mathrm{delay\_max1}} < \ldots < T^{\mathrm{i}}_{\mathrm{delay\_}maxN}$. Besides, for the arbitrary $T^{\mathrm{i}}_{\mathrm{delay\_max}\,n}$ ($n = 0, 1, \ldots, N$), $T^{\mathrm{i}}_{\mathrm{delay\_max}\,n} \le N^i_n T_{\mathrm{s}}$, $N^i_n \in \mathbf{Z}^+$ holds. Hence, the attack levels can be determined as $[0, N^i_0]$, $[N^i_0, N^i_1]$, …, $[N^i_{N-1}, N^i_N]$ where $N^i_{-1} = 0$ denote the ideal scenario without networks.

2) Static output feedback control struct is used to generate the control input U_i. Denote $\overline{A}_{\mathrm{d}i} = \overline{B}_i K_i C_i$, the dynamics of closed-loop LFC system is given by

$$X_i(k+1) = \overline{A}_i X_i(k) + \overline{A}_{\mathrm{d}i} X_i(k - N^i_n) + \overline{H}_i \omega_i(k). \tag{18}$$

Let $X^*_i(k)$ denote the equilibrium point. The following equation holds

$$0 = \overline{A}_i X^*_i(k) + \overline{A}_{\mathrm{d}i} X^*_i(k - N^i_n) + \overline{H}_i \omega_i(k). \tag{19}$$

Define $X_{\mathrm{s}i}(k) = X_i(k) - X^*_i(k)$. Subtracting Eq. (18) from Eq. (19), we have

$$X_{\mathrm{s}i}(k+1) = \overline{A}_i X_{\mathrm{s}i}(k) + \overline{A}_{\mathrm{d}i} X_{\mathrm{s}i}(k - N^i_n). \tag{20}$$

Then the stability constraints can be transferred into analyzing the stability of the closed-loop LFC system at the equilibrium point with bounded delay $[N^i_{n-1}T_{\mathrm{s}}, N^i_n T_{\mathrm{s}}]$.

Let $\chi_i(k) = [X^{\mathrm{T}}_{si}(k), X^{\mathrm{T}}_{si}(k - N^i_{n-1}), X^{\mathrm{T}}_{si}(k - N^i_n)]^T$ be the augmented vector where

$$X_{si}(k) = [I\,0\,0]\chi_i(k) = \varXi_1 \chi_i(k). \tag{21a}$$

$$X_{si}(k - N^i_{n-1}) = [0\,I\,0]\chi_i(k) = \varXi_2 \chi_i(k). \tag{21b}$$

$$X_{si}(k - N^i_n) = [0\,0\,I]\chi_i(k) = \varXi_3 \chi_i(k). \tag{21c}$$

$$X_{si}(k+1) = [\overline{A}_i\, \overline{A}_{\mathrm{d}i}\, 0]\chi_i(k) = \varXi_4 \chi_i(k). \tag{21d}$$

where I is the identity matrix. A Lyapunov function is chose as

$$V(k) = V_1(k) + V_2(k) + V_3(k) + V_4(k) + V_5(k). \tag{22a}$$

$$V_1(k) = X^{\mathrm{T}}_{\mathrm{s}i}(k)\varPi_1 X_{\mathrm{s}i}(k),\ \varPi_1 > 0. \tag{22b}$$

$$V_2(k) = \sum_{\vartheta = k - N^i_{n-1}}^{k} X^{\mathrm{T}}_{\mathrm{s}i}(\vartheta)\varPi_2 X_{\mathrm{s}i}(\vartheta),\ \varPi_2 > 0. \tag{22c}$$

$$V_3(k) = \sum_{\vartheta = k - N^i_n}^{k} X^{\mathrm{T}}_{\mathrm{s}i}(\vartheta)\varPi_3 X_{\mathrm{s}i}(\vartheta),\ \varPi_3 > 0. \tag{22d}$$

$$V_4(k) = N_{n-1}^i \sum_{\vartheta=-N_{n-1}^i}^{0} \sum_{\vartheta'=k+\vartheta}^{k} X_{si}^{\mathrm{T}}(\vartheta')\Pi_4 X_{si}(\vartheta'),\ \Pi_4 > 0. \tag{22e}$$

$$V_5(k) = \left(N_n^i - N_{n-1}^i\right) \sum_{\vartheta=-N_n^i}^{-N_{n-1}^i} \sum_{\vartheta'=k+\vartheta}^{k} X_{si}^{\mathrm{T}}(\vartheta')\Pi_5 X_{si}(\vartheta'),\ \Pi_5 > 0. \tag{22f}$$

With Jensen inequality [14], the increment of Eq. (22 a, b, c, d, e and f) can be calculated as follows.

$$\begin{aligned}\Delta V_1(k) &= X_{si}^{\mathrm{T}}(k+1)\Pi_1 X_{si}(k+1) - X_{si}^{\mathrm{T}}(k)\Pi_1 X_{si}(k) \\ &= \chi_i^{\mathrm{T}}(k)\left(\Xi_4^{\mathrm{T}}\Pi_1\Xi_4 - \Xi_1^{\mathrm{T}}\Pi_1\Xi_1\right)\chi_i(k).\end{aligned} \tag{23}$$

$$\begin{aligned}\Delta V_2(k) &= X_{si}^{\mathrm{T}}(k+1)\Pi_2 X_{si}(k+1) - X_{si}^{\mathrm{T}}(k-N_{n-1}^i)\Pi_2 X_{si}(k-N_{n-1}^i) \\ &= \chi_i^{\mathrm{T}}(k)\left(\Xi_4^{\mathrm{T}}\Pi_2\Xi_4 - \Xi_2^{\mathrm{T}}\Pi_2\Xi_2\right)\chi_i(k).\end{aligned} \tag{24}$$

$$\begin{aligned}\Delta V_3(k) &= X_{si}^{\mathrm{T}}(k+1)\Pi_2 X_{si}(k+1) - X_{si}^{\mathrm{T}}(k-N_n^i)\Pi_2 X_{si}(k-N_n^i) \\ &= \chi_i^{\mathrm{T}}(k)\left(\Xi_4^{\mathrm{T}}\Pi_2\Xi_4 - \Xi_3^{\mathrm{T}}\Pi_2\Xi_3\right)\chi_i(k).\end{aligned} \tag{25}$$

$$\begin{aligned}\Delta V_4(k) &= N_{n-1}^i\left(N_{n-1}^i X_{si}^{\mathrm{T}}(k+1)\Pi_4 X_{si}(k+1) - \sum_{\vartheta'=k-N_{n-1}^i}^{k} X_{si}^{\mathrm{T}}(\vartheta')\Pi_4 X_{si}(\vartheta')\right) \\ &\le \left(N_{n-1}^i\right)^2 X_{si}^{\mathrm{T}}(k+1)\Pi_4 X_{si}(k+1) - \left(\sum_{\vartheta'=k-N_{n-1}^i}^{k} X_{si}(\vartheta')\right)^{\mathrm{T}} \Pi_4 \sum_{\vartheta'=k-N_{n-1}^i}^{k} X_{si}(\vartheta').\end{aligned} \tag{26}$$

$$\begin{aligned}\Delta V_5(k) &= (N_n^i - N_{n-1}^i)^2 X_{si}^{\mathrm{T}}(k+1)\Pi_5 X_{si}(k+1) + (N_n^i - N_{n-1}^i)\left(\begin{array}{c}\sum_{\vartheta'=k-N_n^i+1}^{k-N_{n-1}^i-1} X_{si}^{\mathrm{T}}(\vartheta')\Pi_5 X_{si}(\vartheta') \\ +X_{si}^{\mathrm{T}}(k+1)\Pi_5 X_{si}(k+1)\end{array}\right) \\ &\le \left[(N_n^i - N_{n-1}^i)^2 + (N_n^i - N_{n-1}^i)\right] X_{si}^{\mathrm{T}}(k+1)\Pi_5 X_{si}(k+1) \\ &\quad + (N_n^i - N_{n-1}^i)\left(\sum_{\vartheta'=k-N_n^i+1}^{k-N_{n-1}^i-1} X_{si}(\vartheta')\right)^{\mathrm{T}} \Pi_5 \sum_{\vartheta'=k-N_n^i+1}^{k-N_{n-1}^i-1} X_{si}(\vartheta').\end{aligned} \tag{27}$$

For Eq. (26), we have

$$\begin{aligned}\sum_{\vartheta'=k-N_{n-1}^i}^{k} X_{si}(\vartheta') &= \left(\sum_{h=1}^{N_{n-1}^i} (\overline{A}_i)^h\right) X_{si}(k-N_{n-1}^i) + \left(I + \sum_{h'}^{N_{n-1}^i-1} \sum_{h=0}^{h'} (\overline{A}_i)^h \overline{A}_{\mathrm{d}i}\right) X_{si}(k-N_n^i) \\ &= \sum_{h=1}^{N_{n-1}^i} (\overline{A}_i)^h \Xi_2\chi_i(k) + \left(I + \sum_{h'}^{N_{n-1}^i-1} \sum_{h=0}^{h'} (\overline{A}_i)^h \overline{A}_{\mathrm{d}i}\right) \Xi_3\chi_i(k) \\ &= \Upsilon_1\Xi_2\chi_i(k) + \Upsilon_2\Xi_3\chi_i(k).\end{aligned} \tag{28}$$

Then the following inequality holds.

$$\Delta V_4(k) \leq \chi_i^{\mathrm{T}}(k)\Big((N_{n-1}^i)^2 \Xi_4^{\mathrm{T}} \Pi_4 \Xi_4 - \Xi_2^{\mathrm{T}} \Upsilon_1^{\mathrm{T}} \Pi_4 \Upsilon_1 \Xi_2 - \Xi_3^{\mathrm{T}} \Upsilon_2^{\mathrm{T}} \Pi_4 \Upsilon_2 \Xi_3\Big)\chi_i(k). \tag{29}$$

Similarly, for Eq. (27), the follow inequality holds.

$$\begin{aligned}\Delta V_5(k) \leq & \Big[(N_n^i - N_{n-1}^i)^2 + (N_n^i - N_{n-1}^i)\Big]\chi_i^{\mathrm{T}}(k)\Xi_4^{\mathrm{T}} \Pi_5 \Xi_4 \chi_i(k) \\ & + (N_n^i - N_{n-1}^i)\chi_i^{\mathrm{T}}(k)\Xi_3^{\mathrm{T}} \Upsilon_3^{\mathrm{T}} \Pi_5 \Upsilon_3 \Xi_3 \chi_i(k).\end{aligned} \tag{30}$$

Furtherly, we have

$$\Delta V(k) \leq \chi_i^{\mathrm{T}}(k)\left\{\begin{array}{l} \Xi_4^{\mathrm{T}} \Pi_1 \Xi_4 - \Xi_1^{\mathrm{T}} \Pi_1 \Xi_1 + 2\Xi_4^{\mathrm{T}} \Pi_2 \Xi_4 - \Xi_2^{\mathrm{T}} \Pi_2 \Xi_2 - \Xi_3^{\mathrm{T}} \Pi_2 \Xi_3 \\ + (N_{n-1}^i)^2 \Xi_4^{\mathrm{T}} \Pi_4 \Xi_4 - \Xi_2^{\mathrm{T}} \Upsilon_1^{\mathrm{T}} \Pi_4 \Upsilon_1 \Xi_2 - \Xi_3^{\mathrm{T}} \Upsilon_2^{\mathrm{T}} \Pi_4 \Upsilon_2 \Xi_3 \\ + \Big[(N_n^i - N_{n-1}^i)^2 + (N_n^i - N_{n-1}^i)\Big]\Xi_4^{\mathrm{T}} \Pi_5 \Xi_4 \\ + (N_n^i - N_{n-1}^i)\Xi_3^{\mathrm{T}} \Upsilon_3^{\mathrm{T}} \Pi_5 \Upsilon_3 \Xi_3 \end{array}\right\}\chi_i(k). \tag{31}$$

Therefore, if Eq. (32) holds, the LFC system is asymptotically stable.

$$\left\{\begin{array}{l} \left\{\begin{array}{l} \Xi_4^{\mathrm{T}} \Pi_1 \Xi_4 - \Xi_1^{\mathrm{T}} \Pi_1 \Xi_1 + 2\Xi_4^{\mathrm{T}} \Pi_2 \Xi_4 - \Xi_2^{\mathrm{T}} \Pi_2 \Xi_2 - \Xi_3^{\mathrm{T}} \Pi_2 \Xi_3 \\ + (N_{n-1}^i)^2 \Xi_4^{\mathrm{T}} \Pi_4 \Xi_4 - \Xi_2^{\mathrm{T}} \Upsilon_1^{\mathrm{T}} \Pi_4 \Upsilon_1 \Xi_2 - \Xi_3^{\mathrm{T}} \Upsilon_2^{\mathrm{T}} \Pi_4 \Upsilon_2 \Xi_3 \\ + \big[(N_n^i - N_{n-1}^i)^2 + (N_n^i - N_{n-1}^i)\big]\Xi_4^{\mathrm{T}} \Pi_5 \Xi_4 + (N_n^i - N_{n-1}^i)\Xi_3^{\mathrm{T}} \Upsilon_3^{\mathrm{T}} \Pi_5 \Upsilon_3 \Xi_3 \end{array}\right\} < 0. \\ \Pi_1, \Pi_2, \Pi_3, \Pi_4, \Pi_5 > 0 \end{array}\right. \tag{32}$$

4 Simulation Results and Discussions

4.1 Transmission Performance Analysis Under Distributed Traffic Attack

To verify the effectiveness of our proposed method, a three-area LFC system is simulated [16]. Assume that the number of other normal PMU data flows is 10 and the arrival curves of abnormal data flows arrived at each forwarding node are same. The transmission performance of communication network under distributed traffic attacks is analyzed in OPNET environment. Five scenarios are simulated and discussed: 1) $(\overline{\sigma}_q, \overline{\rho}_q, Q, \mathbb{K}) = (0, 0, 0, 5)$; 2) $(\overline{\sigma}_q, \overline{\rho}_q, Q, \mathbb{K}) =$ (9.36 kbits, 0.936 Mbps, 10, 5); 3) $(\overline{\sigma}_q, \overline{\rho}_q, Q, \mathbb{K}) =$ (18.72 kbits, 1.872 Mbps, 15, 10); 4) $(\overline{\sigma}_q, \overline{\rho}_q, Q, \mathbb{K}) =$ (28.08 kbits, 2.808 Mbps, 20, 15); 5) $(\overline{\sigma}_q, \overline{\rho}_q, Q, \mathbb{K}) =$ (56.16 kbits, 5.616 Mbps, 25, 20).

Figure 2 illstruates the theorical and simulation results. It can be seen that the errors compared theorical results with the simulation results in the above five scenarios are 15.625%, 14.329%, 3.240%, 1.229%, and 1.717%, respectively. It should be noted that although the error in scenario 1) is relatively larger compared with scenarios 2)–5), the actual difference between theorical and simulation results is within 0.94 ms. The reason may be the simplifications during simulation processes. For example, adopting the average value as the size of PMU data packet may cause the actual packet size in simulation to be less than this value. However, in all scenarios, because the upper simulated delay bounds are less than the theorical values, the theorical upper delay bounds calculated by DNC theory can cover the possible attack scenarios.

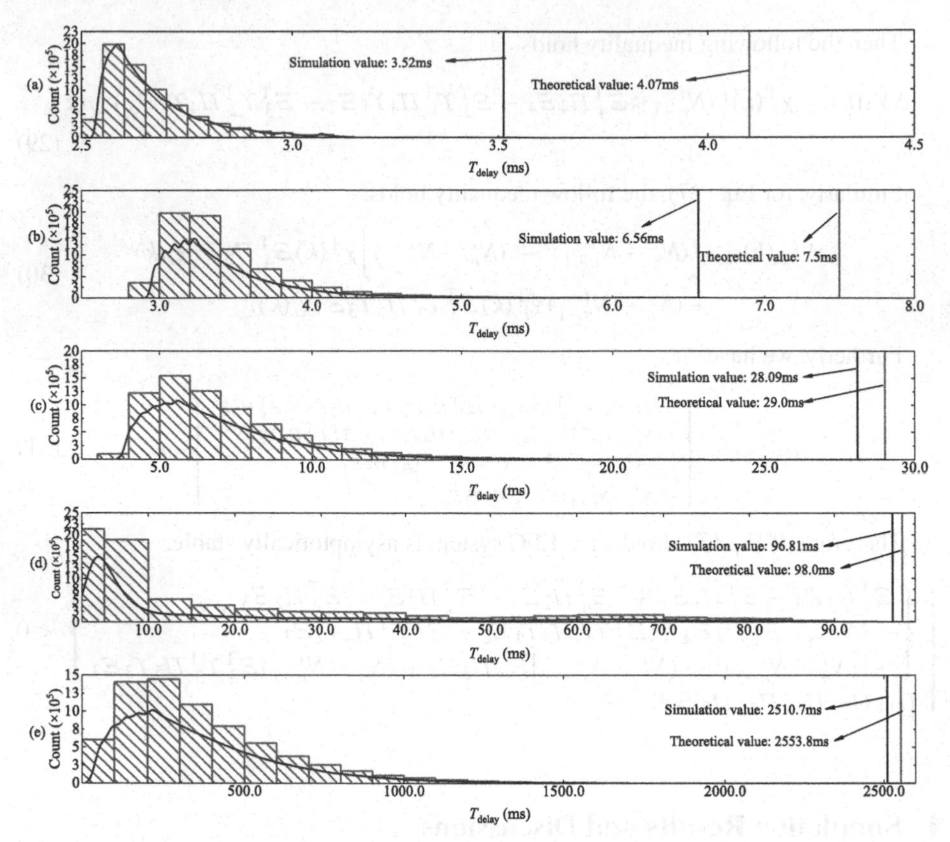

Fig. 2. Theorical and simulation results in different attack levels: (a)–(b) scenarios 1)–5).

4.2 Control Performance Analysis

Furtherly, four operating scenarios are simulated and discussed: 1) Normal scenario with delay (0, 100 ms]; 2) Low-level attack scenario with delay (100 ms, 500 ms]; 3) Medium-level attack scenario with delay (500 ms, 2500 ms]; 4) High-level attack scenario with delay (2500 ms, 7500 ms]. The following three methods are selected as the comparisons: 1) Event-triggering based method proposed in [8]; 2) Delay estimator-based method proposed in [11]; 3) Pole placement-based method proposed in [10]. The controller gains in [8] and [11] remain unchanged in both normal and attack scenarios and are designed according to the maximum delay under attacks. Similar to our proposed method, the method in [10] is with an adaptive struct. Figures 3 (a) and (b) illustrate the transmission delays and external power fluctuations. Figure 4 illustrates the frequency deviations, tie-line power fluctuations and ACEs.

Table 1 shows the sums of IAEs of the above three physical quantities. Although the four control strategies can stabilize the LFC system under different attack levels. However, the dynamic performances with four strategies show obvious differences. Compared with traditional fixed-gain methods (i.e., event-triggered and delay-estimator methods), the IAE of frequency deviations with our proposed method can be reduced by

Table 1. Sums of IAEs with different control strategies.

Variables	Event-triggered method [8]	Delay-estimator method [11]	Pole placement method [10]	Proposed method
Δf_{Σ}(Hz)	0.173	0.197	0.183	0.169
$\Delta P_{\text{tie}\_\Sigma}$(pu)	4.617	5.278	3.331	3.781
ACE_{Σ}(pu)	8.109	10.421	9.694	7.967

2.31% and 14.21% while IAE of tie-line power fluctuations can be reduced by 1.75% and 23.55%. This is because that a controller set corresponding to the specific attack levels are designed in this paper. Therefore, the resilience and reliability of LFC system in presence of traffic attacks can be enhanced.

Moreover, the main drawback of zero-pole placement method is that the pole positions have great influences on the dynamic characteristics. Besides, how to determine the optimal closed-loop pole positions has not been discussed in [10]. Compared with pole placement method, by using our proposed method, IAE of frequency deviations is reduced by 7.65% while IAE of tie-line power fluctuations is increased by 13.51%. However, due to the optimization objective in this paper is to minimize the total IAE of ACEs, the final performance in ACE with our proposed method is improved by 17.82% compared with the pole placement method.

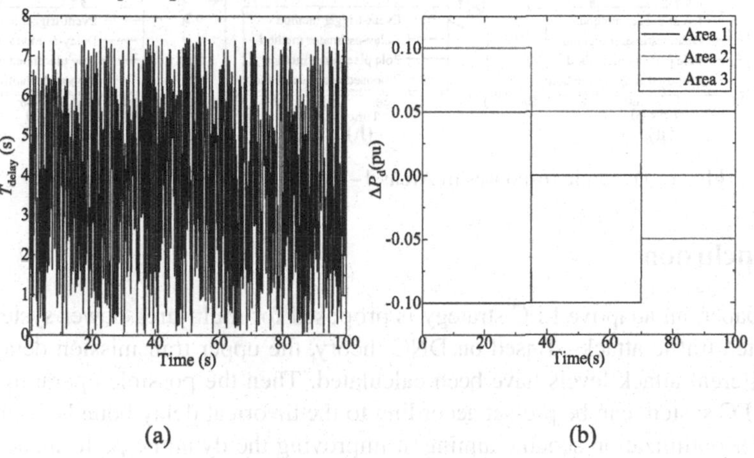

Fig. 3. (a) Transmission delays of data packets; (b) power fluctuations in Areas 1–3.

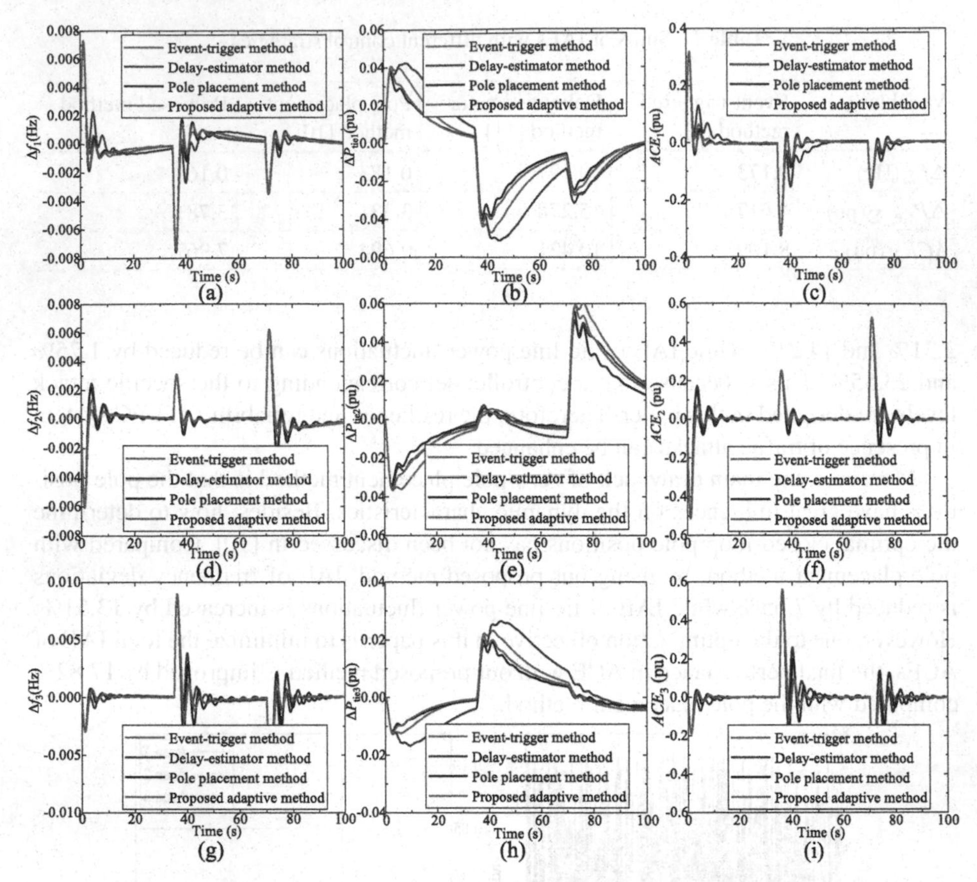

Fig. 4. Dynamic responses in Areas 1–3 with different control strategies.

5 Conclusion

In this paper, an adaptive LFC strategy is proposed for multi-area power systems with distributed traffic attacks. Based on DNC theory, the upper transmission delay bound with different attack levels have been calculated. Then the possible operating scenarios of LFC system can be pre-set according to the theorical delay bounds. Moreover, a controller optimization scheme aiming at improving the dynamic performance is proposed. The most matching controller is switched into control loop according to the actual transmission delays. The three-area simulation results show that, our proposed adaptive method have better the resilience and dynamic performance in the presence of distributed traffic attacks with different levels.

References

1. Jood, P., Aggarwal, S.K., Chopra, V.: Performance assessment of a neuro-fuzzy load frequency controller in the presence of system non-linearities and renewable penetration. J. Comput. Electr. Eng. **74**, 362–378 (2019)

2. Jin, L., Zhang, C.K., He, Y., et al.: Delay-dependent stability analysis of multi-area load frequency control with enhanced accuracy and computation efficiency. J. IEEE. T. Power. Syst. **34**(5), 3687–3696 (2019)
3. Li, Y., Huang, R., Ma, L.: False data injection attack and defense method on load frequency control. J. IEEE Internet Things **8**(4), 2910–2919 (2021)
4. Xu, Y., Fang, M., Wu, Z., et al.: Input-based event-triggering consensus of multiagent systems under denial-of-service attacks. J. IEEE T Syst. Man Cybern. **50**(4), 1455–1464 (2018)
5. Wang, X., Lin, C., Zhong, Z., et al.: Estimation and compensation for discrete-time IT2 fuzzy systems against time delay switch attacks. In: IEEE 9th Annual International Conference on Cyber Technology in Automation, Control, and Intelligent Systems (CYBER), Suzhou, China, pp. 1585–1590. IEEE (2019)
6. Shafique, M., Iqbal, N.: Load frequency resilient control of power system against delayed input cyber attack. In: 2015 Symposium on Recent Advances in Electrical Engineering (RAEE), Islamabad, pp. 1–6. IEEE (2015)
7. Cheng, Z., Yue, D., Hu, S., et al.: Detection-based weighted H_∞ LFC for multiarea power systems under DoS attacks. J. IET Control Theor. A **13**(12), 1909–1919 (2019)
8. Peng, C., Li, J., Fei, M.: Resilient event-triggering load frequency control for multi-area power systems with energy-limited DoS attacks. J. IEEE. T. Power. Syst. **32**(5), 4110–4118 (2017)
9. Liu, J., Gu, Y., Zha, L., et al.: Event-triggered H∞ load frequency control for multiarea power systems under hybrid cyber attacks. J. IEEE T Syst Man Cybern. **49**(8), 1665–1678 (2019)
10. Sargolzaei, A., Yen, K., Abdelghani, M., et al.: Resilient design of networked control systems under time delay switch attacks, application in smart grid. IEEE Access **5**(1), 15901–15912 (2017)
11. Sargolzaei, A., Yen, K., Abdelghani, M.: Preventing time-delay switch attack on load frequency control in distributed power systems. J. IEEE T Smart Grid. **7**(2), 1176–1185 (2016)
12. Zhang, Y.J., Yang, T.: Decentralized switching control strategy for load frequency control in multi-area power systems with time delay and packet lose. IEEE Access **8**, 15838–15850 (2020)
13. Yang, T., Zhang, Y.J., Li, W., Zomaya, A.: Decentralized networked load frequency control in interconnected power systems based on stochastic jump system theory. IEEE T Smart Grid. **11**(5), 4427–4439 (2020)
14. Peng, C., Zhang, J., Yan, H.: Adaptive event-triggering H∞ load frequency control for network-based power systems. IEEE Trans. Ind. Electron. **65**(2), 1685–1694 (2018)
15. Chen, L., Alwi, H., Edwards, C.: Integrated LPV controller/estimator scheme using sliding modes. In: 54th IEEE Conference on Decision and Control (CDC), Osaka, pp. 5124–5129 (2015)
16. Zhang, Y.J., Yang, T., Tang, Z.H.: Active fault-tolerant control for load frequency control in multi-area power systems with physical faults and cyber attacks. J. Int. Trans. Electr. Energy Syst. **31**(7), e12906 (2021). https://doi.org/10.1002/2050-7038.12906

Simulation of Electrofusion Parameters of PE Saddle Pipe Based on Finite Element Analysis of Temperature Field

Wenchao Yu[1], Qihua Yang[1(✉)], Wei Zhang[1], Yuan Chen[1], and Jie Sun[2]

[1] School of Mechanical and Electrical Engineering, China Jiliang University, Hangzhou 310018, China

[2] Ningbo Heng Yuan Precision Pipe Valve Technology Co., Ltd., Ningbo 315000, China

Abstract. This paper compares the electrofusion welding of PE saddle pipe with that of conventional straight pipe, and many differences are identified under the two structure types. Then, a one-dimensional axisymmetric transient heat transfer model and a three-dimensional model of saddle pipe during electrofusion welding are established. Finite element (FE) software is used to analyze the relationship between the distribution of resistance wire, the material of resistance wire, the ambient temperature, and the formation and evolution of electrofusion temperature field. Based on FE analysis results, this paper points out possible problems of welding quality due to the distribution of resistance wire and the selection of wire materials, which can serve as reference for the formulation and the optimization of actual process parameters.

Keywords: Polyethylene saddle pipe · Electrofusion welding · Finite element method · Temperature field

1 Introduction

The products of pipe and valve made of high-density polyethylene (HDPE), have been widely used in medium and low pressure gas pipelines due to their advantages of corrosion resistance, aging resistance, long service life, light weight and flexibility [1–3]. In the network of PE gas pipe, the connection of pipe/valve assembly has two ways: hot fusion and electrofusion. The electrofusion is realized by heating the pre-embedded resistance wire on the inner wall of PE electrofusion pipe. Saddle electrofusion fittings are generally used for the newly added branch pipe of main pipe, with the outlet branch pipe section.

Compared with hot fusion, electrofusion has better convenience in PE pipe/valve welding construction. The control of welding quality is closely related to multiple factors such as the structural parameters and welding process parameters of electrofusion fittings. The products of different manufacturers are not completely unified in such aspects as buried depth of resistance wire and the control of power density of electrofusion pipe fittings, so there may be obvious differences in power-on parameters. The

K. Li et al. (Eds.): LSMS 2021/ICSEE 2021, CCIS 1468, pp. 390–399, 2021.
https://doi.org/10.1007/978-981-16-7210-1_37

welding quality is easily affected by the connection and environment of different pipe fittings on site. It is of great significance to study the distribution and variation of temperature field in welding process to optimize welding parameters and improve welding quality [4, 5].

During welding, the temperature field of PE electrofusion pipe fittings is a ring column centered on a spiral wound rang by resistance wire. Many scholars have established the heat transfer model, analyzed the steady and unsteady temperature field [6–8], and studied the relationship between temperature field and electrofusion time [9]. However, the related researches are still insufficient in the influence of material and structural parameters on the temperature field, as well as the interaction between temperature field and environment in the evolution process [10]. The structure of PE saddle pipe is different from that of straight through type. The multi loop resistance wire is arranged inside the curved surface of saddle pipe in a way of encircling into a hollow circle. This kind of special boundary structure may bring more influence factors, such as phase transition of materials, release of internal stress, movement of fluid and transfer of ambient temperature. Relying on simple experiments to determine the electrofusion process parameters of the PE saddle pipe, may bring some risks to the quality assurance of welding.

2 Analysis Model of Temperature Field

2.1 Heat Conduction Model

Heat conduction [11] and heat convection are two main heat transfer modes in the process of electrofusion welding. For a continuous medium with volume V and surface area ψ, its heat conduction law can be expressed as:

$$-\frac{\partial}{\partial x_i}\left(k_{ij}\frac{\partial T}{\partial x_j}\right)+Q_v-\rho_a C_a\frac{\partial T}{\delta t}=0. \tag{1}$$

Where T is the temperature; q_s is the normal unit vector; k_{ij} is the heat conductivity coefficient; ρ_a and C_a are the density and specific heat capacity of PE saddle pipe; Q_V is the heat flow per unit volume per unit time; t is time.

Because of the thermal convection between the surface of saddle pipe and the external environment, the convection boundary conditions are:

$$q_s = h_{du}(T_{sur}-T_t). \tag{2}$$

Where q_s is convection heat flux; h_{du} is convection coefficient; T_{sur} is the surface temperature of saddle pipe; T_t is the surface contact temperature of saddle pipe.

According to the relevant knowledge of engineering thermodynamics, the model can be transformed into a one-dimensional axisymmetric unsteady heat transfer model, and the differential equation for controlling welding process can be expressed as follows:

$$\frac{\partial T}{\partial t}=\frac{k_a}{\rho_a C_a}\left(\frac{\partial^2 T}{\partial r_a}+\frac{1}{r_a}\cdot\frac{\partial T}{\partial r_a}\right), r_{in}<r_a<r_{out}. \tag{3}$$

Where k_a is the thermal conductivity of saddle pipe; r_{in} and r_{out} represent the inner diameter and outer diameter of saddle pipe.

In combination with formula (2), the following boundary conditions are obtained:

$$k_a \frac{\partial T}{\partial r_a}\bigg|_{r_a=r_{in}} = h_1(T - T_e). \quad (4)$$

$$k_a \frac{\partial T}{\partial r_a}\bigg|_{r_a=r_{out}} = -h_2(T - T_e). \quad (5)$$

Where *h1* and *h2* are the convection coefficients of the inner wall of pipe and the outer wall of the sleeve; *Te* is the ambient temperature.

The heat generated by the resistance wire diffuses to the inner wall of pipe and the outer wall of the sleeve, and the diffusion heat can be expressed as:

$$q_i = -k_a \frac{\partial T}{\partial r_i}(i = 1, 2), r_1 = R_{din}, r_2 = R_{dout}. \quad (6)$$

Where q1 and q2 are the heat fluxes flowing from the resistance wire into the inner wall of the pipe and the outer wall of the sleeve respectively; R_{din} and R_{dout} are the distances from the resistance wire to the inner wall of the pipe and the outer wall of the sleeve respectively.

In unit time, the heat generated by the resistance wire passing through the unit cross section of the object is:

$$q_d = \frac{U^2}{4\pi R_d l_d[1 + a(T - T')]R'}. \quad (7)$$

Where U is the constant voltage (39.5 V) output by the welding machine; R_d and l_d are the wiring radius of resistance wire and the length of single side zone of resistance wire; α is the temperature coefficient of resistance; R' is the resistance at the calibrated temperature T'.

According to the second law of thermodynamics, all the heat generated by the resistance wire is used to heat the saddle pipe:

$$q_d = h_d(T_d - T_t). \quad (8)$$

Where h_d is the convection coefficient of the saddle pipe; T_d is the temperature of resistance wire; T_t is the temperature of polyethylene in contact with it.

Combining formulas (7) with (8), the relationship between input heat flux and temperature is:

$$q_d = \frac{U^2}{4\pi R_d l_d\left[1 + a\left(\frac{q_d}{h_d} + T_t - T'\right)\right]R'}. \quad (9)$$

2.2 Simulation Geometry Model

To simplify the calculation, the following assumptions were made: the heat transfer inside the saddle pipe was even; the material parameters of saddle pipe only change with temperature; simulate the welding process by heating and melting with resistance wire.

Geometric model of the saddle type electrofusion fitting shown in Fig. 1(a), ignoring bypass section existing on saddle pipe (with little effect on welding results). Because the model had the symmetry of spatial structure, only 1/4 part was selected for analysis, as shown in Fig. 1(b).

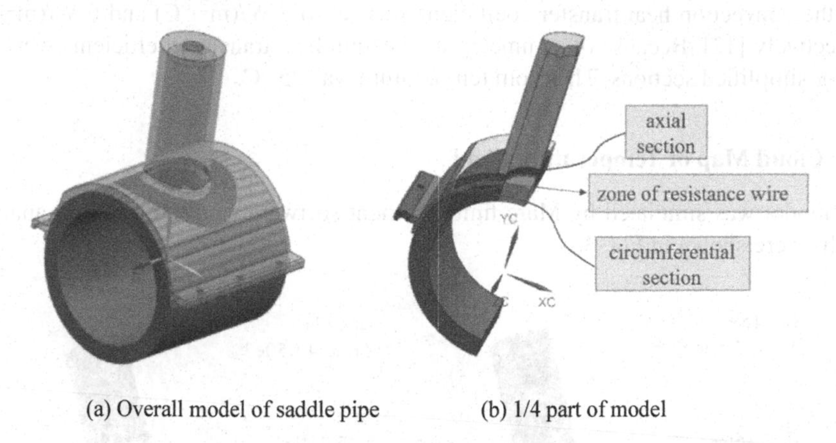

(a) Overall model of saddle pipe (b) 1/4 part of model

Fig. 1. Physical model and 3D model of the saddle pipe.

In order to be as close to the fact as possible, the actual structure of the resistance wire embedded in the inner wall of the pipe fitting was reserved, and the resistance wire was the only heat source of the joint. In order to strengthen the simulation effect and speed up the calculation speed, the mesh partition near the resistance wire was relatively dense, other parts were relatively sparse, and the length of edge was set to 0.3–5 mm. As the saddle pipe was of irregular structure, the grid near the weld would be distorted by sweeping type division, resulting in non-convergence of calculation. Therefore, the grid around the weld would be divided into pentahedron and the rest would be hexahedron. The results were shown in Fig. 2. In this paper, DN250 $\times$ 90 saddle pipe was selected as the research object. The number of simulated grids was 476411, among which the number of resistive wire grids was 13448 and the number of nodes was 357596.

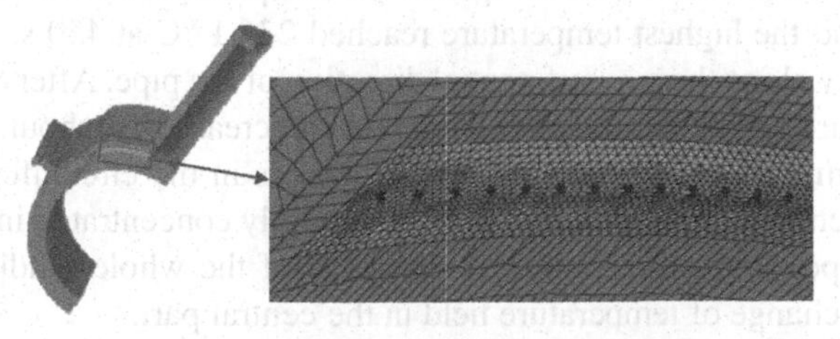

Fig. 2. Saddle pipe grid division diagram.

3 Analysis on Temperature Field in Electrofusion Joint

3.1 Analysis Condition Setting

The outer wall of saddle and the inner wall of pipe fittings were natural heat dissipation, and the convection heat transfer coefficient was set to 8 W/(m^2·°C) and 6 W/(m^2·°C) respectively [12]. Because of symmetry, convection heat transfer coefficients were not set for simplified sections. The room temperature was 25 °C.

3.2 Cloud Map of Temperature Field

The model was simulated by Marc finite element software, and the transient analysis results were shown in Fig. 3.

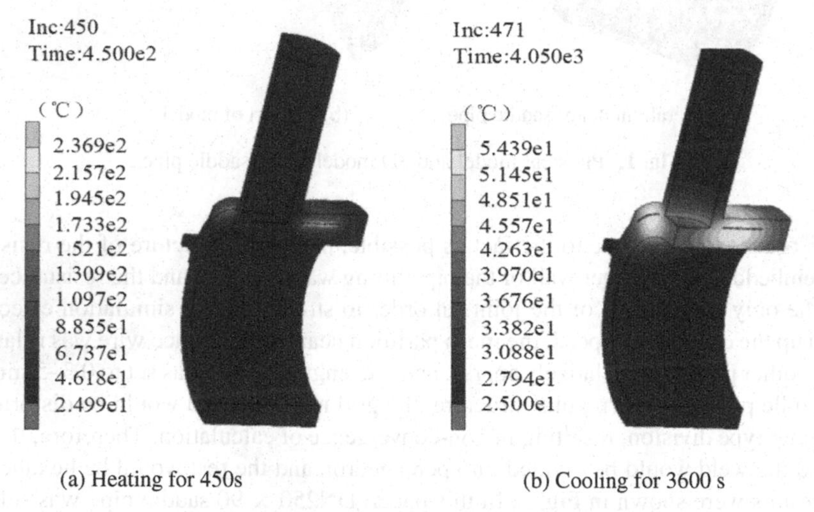

Fig. 3. Temperature field after heating and cooling.

When the resistance wire was energized, the temperature of central part increased continuously, and the highest temperature reached 236.1 °C at 450 s; the temperature decreased rapidly along the circumferential direction of the pipe. After cooling for 3600 s, the temperature of the central resistance wire decreased to about 50 °C, and the difference of temperature between different distances in the circumferential direction was small. The change of temperature field was mainly concentrated in the central part of the saddle pipe, so the temperature distribution of the whole saddle pipe could be reflected by the change of temperature field in the central part.

3.3 Analysis of Temperature Field of Each Section

In order to compare the relationship between the resistances of different materials and the evolution of temperature field. At room temperature (25 °C), the temperature-time

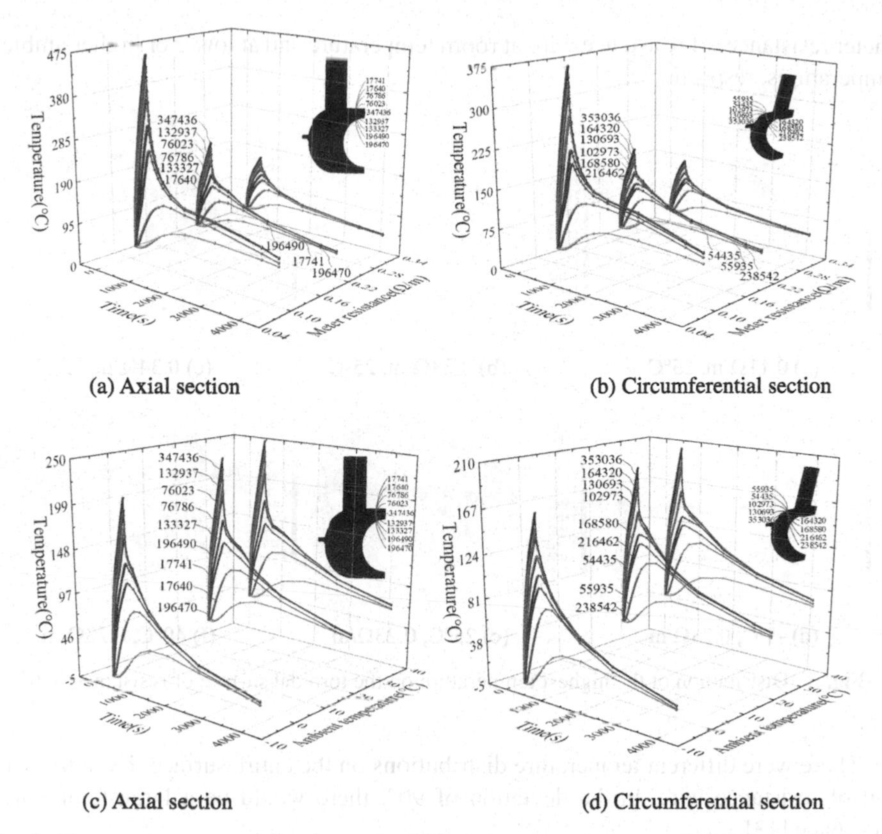

(a) Axial section (b) Circumferential section

(c) Axial section (d) Circumferential section

Fig. 4. Temperature curves of nodes under different meter resistance and ambient temperature.

characteristics of each node in two sections were analyzed with resistance wires of 0.11 Ω/m, 0.23 Ω/m and 0.34 Ω/m respectively. Three sets of curves were obtained by simulation (Fig. 4).

It was obvious from the relationship between the temperature of each node and the meter resistance: the smaller the meter resistance was, the greater the temperature rose; node temperature was positively correlated with ambient temperature. According to the 39.5 V constant output condition of electrofusion welding machine, the maximum temperature of electric heating could be limited to the allowable value by selecting the appropriate meter resistance value.

From the perspective of different sections, the temperature of nodes on the axial section was higher than that on the circumferential section. The reason was that the special structure of saddle pipe led to different distribution density of resistance wire in different sections. The distribution density reached maximum and minimum in axial and circumferential direction respectively. At the same heating time, the polye-thylene near the latter absorbed more heat and the temperature will be higher than the former.

Figure 5 showed the cloud map of temperature on the ring surface of resistance wire when heated for 450 s. It showed the highest temperature of the three materials (meter resistance) of resistance wire at room temperature and at lower or higher ambient temperatures, respectively.

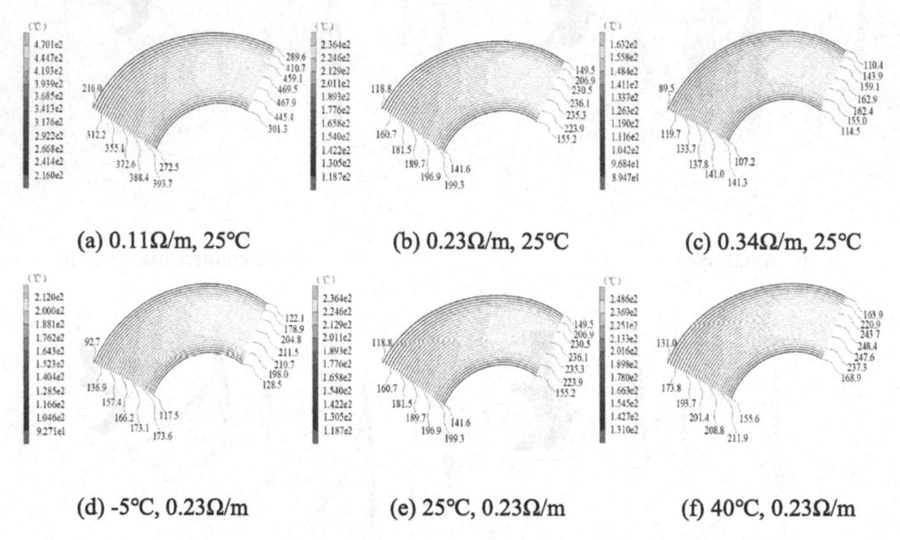

(a) 0.11Ω/m, 25°C (b) 0.23Ω/m, 25°C (c) 0.34Ω/m, 25°C

(d) -5°C, 0.23Ω/m (e) 25°C, 0.23Ω/m (f) 40°C, 0.23Ω/m

Fig. 5. Distribution of the highest temperature on the toroidal surface of resistance wire.

There were different temperature distributions on the entire surface. Even the same coil of resistance wire, in the deviation of 90°, there would be a large temperature difference [13].

Large temperature difference and asymmetric distribution, on the one hand, it was easy to cause the distortion of heat during electric melting; on the other hand, it would make the welding of saddle pipe joint uneven, and some areas might be over-melt or under-melt.

According to the data obtained from simulation, the difference of temperature between the axial and circumferential nodes tended to be consistent under different ambient temperatures. This meant that the influence of ambient temperature on the inhomogeneity of the torus temperature of resistance wire could be neglected. The key to realize the temperature uniformity of fusion welding surface lay in the reasonable matching between the volume power and heat dissipation of the wiring zone.

An easy to think of improvement design for the wiring zone is to adjust the original wiring of circular equidistant to the wring of elliptical variable distance, in which, the maximum compression area of the filament space shall be in the circumferential section. According to the simulation data, the maximum volume power of the target saddle pipe can be adjusted to 25%–30%. The theoretical volume power from axial to circumferential direction increases gradually, that is, the distance between resistances decreases gradually. At this time, it is necessary to evaluate whether the narrowing of the width of fusion zone in this direction conforms to the requirements of the overall control size of fusion zone. It may be necessary to increase the number of wire coils to compensate for the lack of up-size of the welding zone in the circumferential direction.

3.4 Analysis of Highest Temperature of Different Nodes

In order to explore the variation of the highest temperature of different nodes in the fusion zone of PE joint with different materials and ambient temperature. Some nodes were selected on the two sections for analysis. The selection principle was as follows: Considering that the heat was concentrated in the center of the wiring area. Selecting nodes at both ends, center and any two positions in the routing area of circumferential and axial section. In addition, selecting two points outside the wiring area but inside the fusion zone. When heated for 450 s, the variation curves of the highest temperature of nodes with different materials (meter resistance) and ambient temperature were shown in Fig. 6.

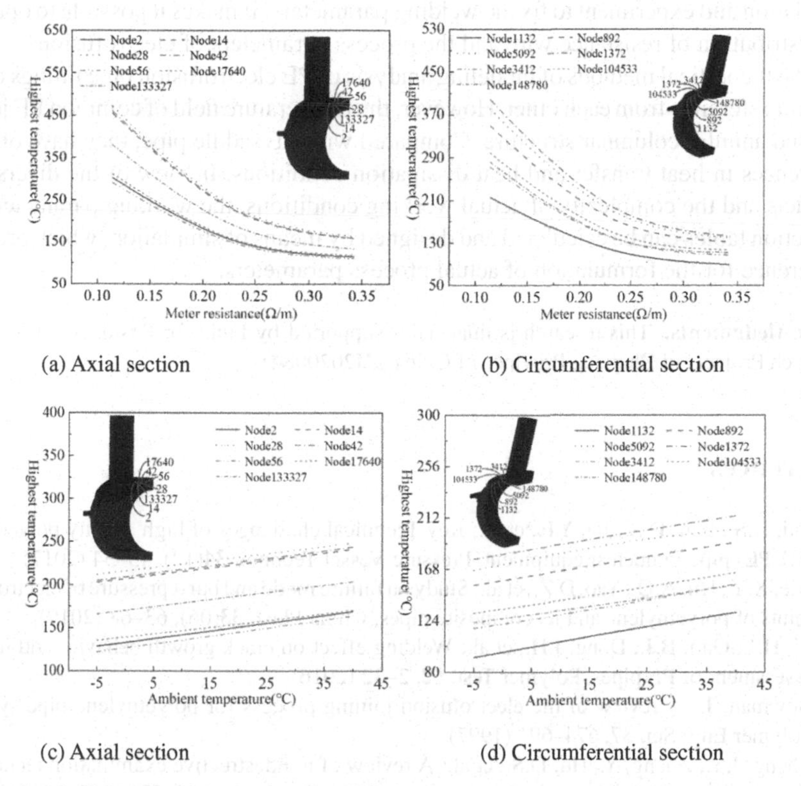

Fig. 6. Maximum temperature under different meter resistance and ambient temperature.

For high-density polyethylene (HDPE), the optimal fusing temperature was 180–230 °C. Comparing the highest temperature of each node with different parameters, the variation trended of the highest temperature of each node was consistent with the increase of meter resistance, the temperature difference between the highest temperature of different nodes decreased gradually, and the change rates of temperature presented as curves, which gradually decreased from large to small; the highest temperature increased linearly with the increase of ambient temperature, and the temperature difference of each node was not affected by it; when the meter resistance was 0.23–0.27 Ω/m and the ambient temperature was 25 °C, the suitable fusing effect could be achieved.

4 Conclusion

This paper studied the influence of welding parameters such as meter resistance and ambient temperature on the welding performance of PE saddle pipe. The result showed that the temperature of each node was inversely proportional to meter resistance and directly proportional to ambient temperature. When the meter resistance was between 0.23–0.27 Ω/m and the ambient temperature was about 25 °C, good welding effect could be achieved. This paper simulated and analyzed the methods, which utilizing traditional calculation and experiment to fix the welding parameters. It makes it possible to optimize the distribution of resistance wire and the process parameters of electrofusion.

The theoretical methods of modeling analysis of PE electrofusion pipe fittings can be used for reference from each other. However, the temperature field of common PE joint is a closed annular columnar structure. Compared with PE saddle pipe, they have obvious differences in heat transfer and heat dissipation conditions. In view of the diversity of products and the complexity of actual working conditions, the working parameters and correction tables can be calculated and designed by means of simulation, which provides a reference for the formulation of actual process parameters.

Acknowledgments. This research is financially supported by Fund for Postdoctoral Scientific Research Projects of Zhejiang Province of China (ZJ2020084).

References

1. Lai, H.S., Fan, D.S., Jia, Y.F., et al.: Key Technical challenges of high density polyethylene (HDPE) pipe in nuclear equipment. Pressure Vessel Technol. **34**(12), 45–54 (2017)
2. Nie, X.Y., Hu, A.Q., Yao, D.Z., et al.: Study on failure mode and burst pressure of electrofusion joints of polyethylene and its composite pipes. China Plast. **33**(06), 63–68 (2019)
3. Li, H.J., Gao, B.J., Dong, J.H., et al.: Welding effect on crack growth behavior and lifetime assessment of PE pipes. Polymer Test. **52**, 2–32 (2016)
4. Bowman, J.: A review of the electrofusion joining process for polyethylene pipe systems. Polymer Eng. Sci. **37**, 674–691 (1997)
5. Zheng, J.Y., Zhang, Y., Hu, D.S., et al.: A review of nondestructive examination technology for polyethylene pipe in nuclear power plant. Front. Mech. Eng. **13**, 535–545 (2018)
6. Zheng, J.Y., Shi, J.F., Guo, W.C., et al.: Analysis on temperature field in electrofusion joint for polyethylene pipe. Trans. China Weld. Inst. **30**, 5–9 (2009)
7. Egerton, J.S., Lowe, M.J.S., Huthwaite, P., et al.: Ultrasonic attenuation and phase velocity of high-density polyethylene pipe material. J. Acoust. Soc. Am. **141**(3), 1535 (2017)
8. Chebbo, Z., Mich, V., Adil, B., et al.: Numerical and experimental study of the electrofusion welding process of polyethylene pipes. Polymer Eng. Sci. **55**(1), 123–131 (2015)
9. Guo, W.C., Shi, J.F., Tang, P., et al.: A new method for determining the welding time of polyethylene electrofusion. Weld. Technol. **48**(06), 37–42 (2019)
10. Andreyev, B.I., Guerasimov, A.I., Starostin, N.P.: Determination of the parameters of coupling welding at low temperatures. Weld. Int. **29**, 826–828 (2015)

11. Shi, C.S., Li, J.J., Wang, Y.L., et al.: Modeling of temperature field in electrofusion joint for polyethylene pipe. Trans. China Weld. Inst. **21**, 36–39+3 (2000)
12. Zhang, R.Y., Yang, Q.H., Chen, J.Q.: Transient-thermal FEM simulation analysis and verification of PE electrofusion fitting welding process. Weld. Technol. **47**(01), 18–21+5 (2018)
13. Hu, A.Q., Nie, X.Y., Yao, D.Z., et al.: Stress analysis of electrofusion joint of polyethylene gas pipeline. J. Pressure Vessel Technol. **36**(05), 21–28+20 (2019)

Research on Controlling Method of Welding Temperature of PE Electrofusion Pipe Fittings Based on Simulation of Temperature Field

Wei Zhang[1], Qihua Yang[1](✉), Bingqing Wang[1], Wenchao Yu[1], and Jinjin Tong[2]

[1] China Jiliang University, Hangzhou 310018, China
[2] Ningbo Hengyuan Precision Tube Valve Technology Co., Ltd., Ningbo 315414, China

Abstract. With the focus on the welding of PE electrofusion pipe fittings, this paper proposes a concept of optimal control of temperature characteristics during melt welding. The finite element software is used to establish the simulation of the temperature field of PE electrofusion pipe fittings. Through simulation, the relationship between different supply voltages, the rate at which the temperature rises, and the extent of the melting zone under the condition of fixed duration is analyzed. Aiming at rapidly increasing temperature and maintaining reasonable melting temperature threshold, this paper proposes the strategy for controlling supply voltage output which can achieve optimal temperature-related characteristics. In addition, this paper develops a practical method for implementing temperature control of field welding.

Keywords: Simulation of temperature field · Electrofusion pipe fittings · Characteristics of control for welding temperature

1 Introduction

Pipe, pipe fittings and valves made of polyethylene (PE) are widely used in urban pipeline network and distributed in densely populated areas due to their advantages of good resistance to corrosion and creep property, small resistance to flow, convenient connection, long life and high reliability [1], so the security problem of pipeline systems is crucial [2, 3]. In the process of on-site pipeline assembly, the quality of pipe fitting joints is difficult to be guaranteed, making the electric fusion joints the weakest link in the pipeline system [4]. The welding process of PE electrofusion pipe fittings mainly includes three processes: electric heating to melt the pipe fittings, maintaining temperature and applying pressure to entangle the molecular chain of the pipe fittings and natural cooling to solidify the pipe fittings.

The control of the input of heat and the maximum temperature that PE material can withstand during welding will directly affect the welding quality of PE pipe. In order to control the input of heat and ensure that the polymer chain at the interface has enough entanglement activity, the relative ideal temperature of the phase melting zone should be reached in a relatively short time.

K. Li et al. (Eds.): LSMS 2021/ICSEE 2021, CCIS 1468, pp. 400–407, 2021.
https://doi.org/10.1007/978-981-16-7210-1_38

The traditional welding of PE electofusion pipe fittings mostly relies on the simple process of electric heating for a specific time under a constant voltage (39.5 V), constant current and constant power supply, it is difficult to flexibly control the welding parameters such as the input of heat, the rate of temperature rise and the upper limit of temperature in a certain melting zone. At present, there are two main types of electric welding machine: one can be used for welding with different parameters in multiple periods; the other can be used for welding with a single parameter in a single period, and has a function of smaller temperature difference compensation [5], the PID algorithm is used to achieve a control of a specific welding temperature [6], and it does not consider the impact of the thickness of the molten layer and the optimal welding temperature of the molten layer on the welding quality of PE welding. This paper presents the concept of optimal temperature control characteristics: (1) The connected area of the fittings rapidly increases the temperature to the ideal melting temperature range (180–230 °C); (2) Defines the depth of the best welding melting region to optimize the length of the heating time, controls the expansion of the melt pressure in the melt area and the diffusion and entanglement of molecular chains to achieve the optimal welding effect; (3) Natural cooling.

2 Simulation Model of the Temperature Field

The end of the conventional PE electric fusion pipe fitting is embedded with an annular resistance wire in the shape of spiral tube, and then the tubes are electrically heated to form the annular 3D welding temperature field. Physical diagram of docking model as shown in Fig. 1 on the left and simulation model on the right, the main body of the model used for simulation consisted of an inset tube, an electrofusion pipe fitting and an embedded electric heating wire. Considering the symmetry of the two ends of the electrofusion pipe fitting and the symmetry of the annular structure of the temperature field, in order to simplify the model of simulation and reduce the amount of calculation, the model of simulation of the temperature field was constructed by using the radial 2D section of the butt zone of the fused pipe fitting at one end. There was no air gap layer between the butt pipe and the electrofusion pipe fitting.

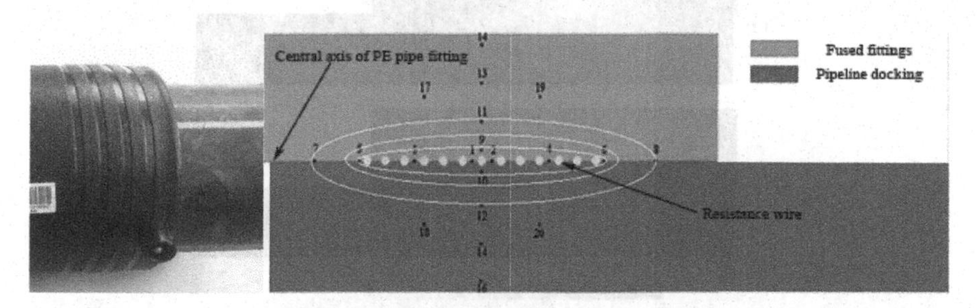

Fig. 1. Geometric model of pipe butt joint

The mesh in Fig. 1 required for simulation can be divided by software such as HyperMesh. In order to ensure the accuracy of the calculation results and minimize the consumption of time, the grid density of the structure near the resistance wire was strengthened, and the free grid was used to divide the area of pipe, sleeve and resistance wire. Taking DN400 pipe fitting as an example, the length of the smallest edge of the subdivision could be set to 0.2 mm, the length of the maximum size of the outside was 2 mm; the number of simulated grids was 27808 and the number of nodes was 14116. In order to facilitate the analysis of the law of change of temperature of the fused area in the welding process, 20 key points at which the temperature is measured were marked in Fig. 1.

3 Analysis of Simulation Data of Temperature Field

The main structure and material parameters of DN400 pipe fittings related to simulation [7]: pipe material was PE100, specific heat capacity was 0.35 J/kg · K, the inner diameter and outer diameter of the connecting pipe were 162/200 mm respectively; The outer diameter and length were 241 mm and 170 mm respectively; Resistance wire ring middle diameter were 201 mm, pitch were 5 mm, single side 13 turns; Resistance wire were brass of H65, diameter were 2 mm, resistance value were 0.83 Ω, meter resistance value were 0.21 Ω/m, specific heat capacity were 385 J/kg · K; The coefficient of convective heat transfer on the inside/outside of the sleeve 15/25 $W/m^2 \cdot K$. Welding data input for electrofusion: Voltage of 39.5 V, the power lasted 2040 s, Cooling time was 840 s; Ambient temperature was 25 °C.

Figure 2(a) and (b) respectively show the temperature field of electric melt joint after welding finishes and cooling 840 s. At the end of heating, the highest temperature of the fused joint near the electric heating wire is about 258 °C. The region of the highest temperature is approximately elliptical. The temperature in this region gradually decreases from the inside to the outside and presents a layered distribution with a large temperature gradient; after cooling for 840 s, the range of high temperature region is enlarged and the temperature and its gradient decrease under the action of heat conduction.

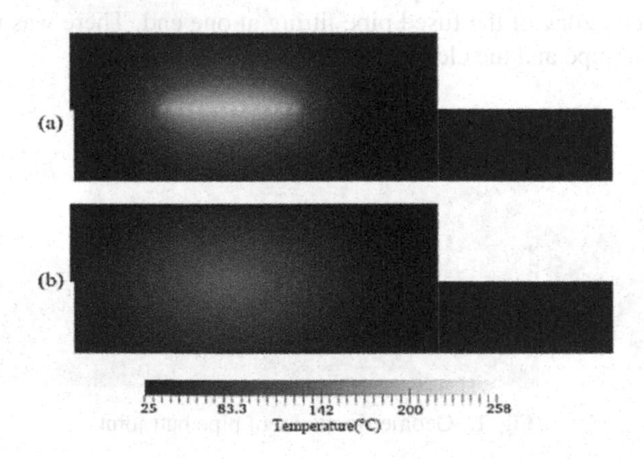

(a) At the end of heating (heating to 2040 s) **(b)** When cooling to 840 s

Fig. 2. Temperature field of constant voltage 39.5 V for 2040 s.

Figure 3 shows the characteristics of temperature of several points marked in Fig. 1 for being measured temperature. At point 3# and 4#, the material reaches the ideal melting range of temperature (≥180 °C) after electrifying for about 800 s, and the melting duration is about 1200 s. It can be seen from the figure that the temperature of the measured temperature point at any time decreases gradually from the inside to the outside, which corresponds to the law of temperature distribution in Fig. 2, and at 2040 s, the temperature of each point reaches the maximum value of the temperature-time curve, namely, a limited total time of electrification can controll the maximum temperature in the melting zone, and the temperature drops sharply after stopping the energization. The results indirectly verify the rationality of the parameters of welding process.

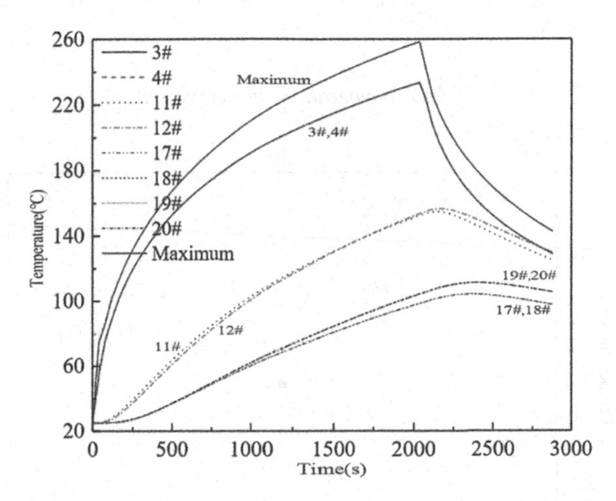

Fig. 3. Characteristic curve of *temperature versus time* corresponding to the observed point.

According to the above concept of optimal characteristics of temperature, the heating process can be accelerated by increasing the voltage. Figure 4 shows the simulation of characteristics of temperature under different voltages (loading duration: 2040 s), which indicates that increasing the voltage can significantly improve the rise rate of temperature. Takeing a 54.5 V voltage loaded into a resistance wire as an example, After about 500 s, PE material enters the ideal melting region of temperature, and the speed is about twice as fast as that of 44.5 V loading. However, after 54.5 V lasts for 850 s, the temperature will exceed the limit (>240 °C), that is the melting duration is less than 400 s, which is obviously shorter than the current actual welding duration of 1200 s. If the thickness of the melting layer and pressure expansion and molecular chain diffusion entanglement time also meet the standard, the power on time can be reduced.

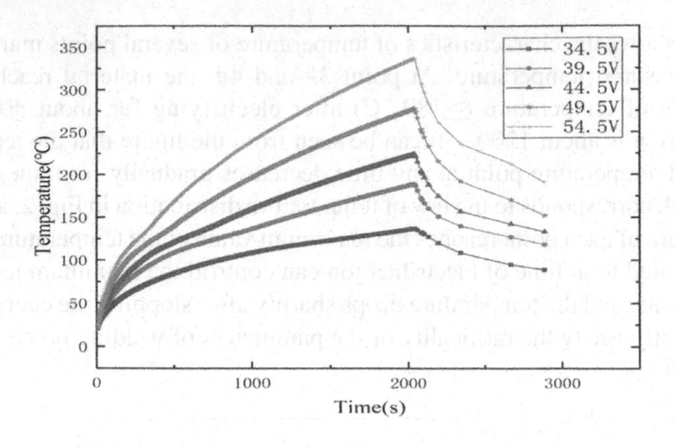

(a) *Temperature-time* curve

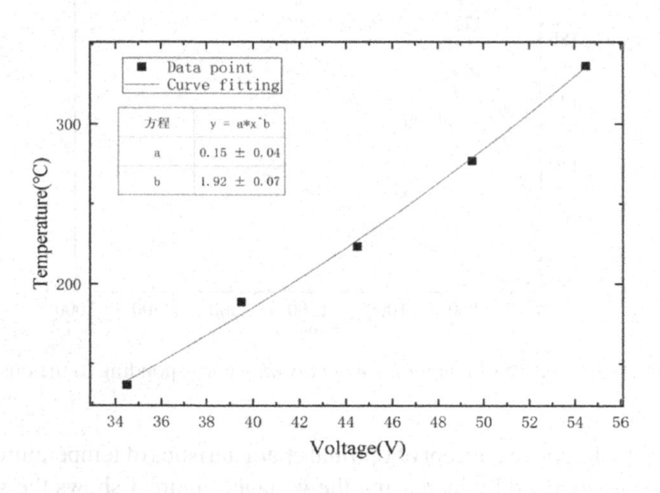

(b) *Different voltage-maximum temperature* curves

Fig. 4. Characteristics of *maximum temperature* at *different voltages* and their relationship with supply voltage.

4 Optimized Method of Variable Voltage Control Temperature

The melting area in Fig. 2 is an elliptical section. Based on the mechanism of forming the strength of the welding interface [8, 9], If a certain width of the axial direction is required to have the standard depth d of melting area (Generally in 1.1–2.6 mm [10]), The time of diffusion entanglement of polymer chain at the welding interface forming joint strength is directly related to the depth of the molten layer in the tube. As long as the depth of the molten layer in the tube reaches a certain value, the basic welding strength of the interface can be guaranteed [11]. Therefore, the range of points ① to ④ used for observating temperature in Fig. 5 should be the area where the melting temperature is controlled to the standard. For the convenience of data collection, after measuring the distance between the nodes of the fixed grid, the depth of melting zone was selected as 1. 98 mm.

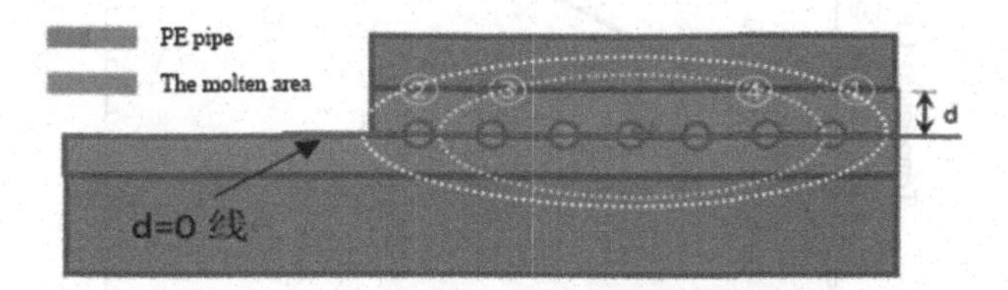

Fig. 5. *Melting zone* and the observed points on *d = 1. 98 mm line.*

In Fig. 5, the d = 0 line can be considered as the central layer of the melting zone, with the highest temperature. The temperature of the central layer needs to be controlled below 230 °C.

Table 1. Time of reaching 230 °C (the observed point on d = 0 line).

Voltage (V)	39.5	44.5	49.5	54.5	64.5	74.5	84.5
Time (s)	1566	1116	734	506	266	154	96

The data of simulation (Table 1) show that the expression related to increasing the loading voltage and shortening the welding time can be fitted for a specific fused object within a certain voltage range:

$$t = -0.02665u^3 + 6.023u^2 - 458.8u + 11940. \tag{1}$$

In order to obtain the temperature of the observed point on the line d = 1.98 along the edge of the melting zone, temperature-time characteristics under a specific supply voltage could be obtained by fitting according to the data of simulation. For example, when the supply voltage is 39.5 V, the curve fitting diagram at point ② is shown in Fig. 6.

The unified expression of temperature change at observed points ①–④ can be summarized as follows:

$$T_i(t)|_{u_i} = p_1 \cdot t^4 + p_2 \cdot t^3 + p_3 \cdot t^2 + p_4 \cdot t + p_5. \; (i = 1{\sim}4). \tag{2}$$

According to the data of simulation, the coefficient of correlation is obtained in Table 2.

According to the concept of optimal control of temperature described above, the temperature is firstly controlled by supplying power to the resistance wire at the higher (or maximum) output of voltage u_1 of the electric melting power supply, and the time of power supply is calculated by referring to the time t_1 when the observed point (such as point 3) in the central area (d = 1.98 mm line) reaches the preset temperature (such as 150 °C); The second stage of supplying power after reducing voltage starts from $T_i(t_1)|_{u1}$, calculating the temperature of the observed point in the center area under voltage u_2, and supplying power is limited within the permitted period, so that the

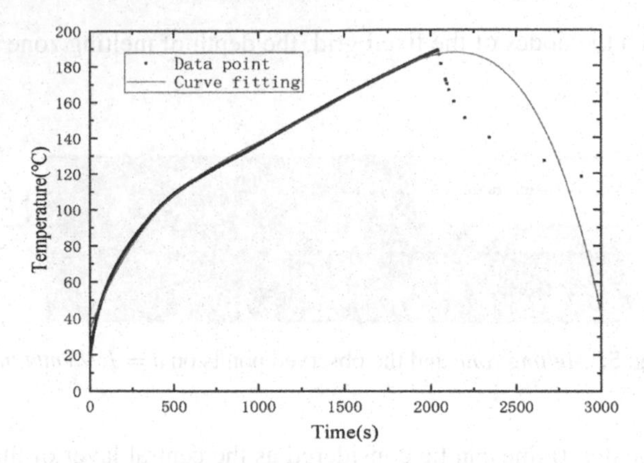

Fig. 6. Curve of *temperature over time* at point ② at voltage 39.5 V.

Table 2. Temperature-time equation coefficients at point ②.

Voltage u_i/V	p_1	p_2	p_3	p_4	p_5
39.5	-2.515×10^{-11}	1.285×10^{-7}	-2.372×10^{-4}	0.2372	34.15
44.5	-2.515×10^{-11}	1.285×10^{-7}	-2.372×10^{-4}	0.2372	34.15
49.5	-2.728×10^{-11}	1.299×10^{-7}	-2.385×10^{-4}	0.2883	43.79
54.5	-3.376×10^{-11}	1.582×10^{-7}	-2.912×10^{-4}	0.3595	45.1
64.5	-4.645×10^{-11}	2.237×10^{-7}	-4.306×10^{-4}	0.5427	47.11
74.5	-6.812×10^{-11}	3.321×10^{-7}	-6.406×10^{-4}	0.7786	49.59
84.5	-9.406×10^{-11}	4.589×10^{-7}	-8.81×10^{-4}	1.045	54.82

temperature of observed point is close to 180 °C; The third stage of supplying power after reducing voltage starts from $T_i(t_2)|_{u2}$, calculating such as before. The total melting time is the sum of each section: $\sum_{i=1,2,...} t_i$.

5 Conclusions

The data of simulation of the measured temperature point of the analysis model of temperature field has correspondence with the actual measurement, and it is universal for PE electrofusion pipe fittings. In the concept of optimal control of temperature, the premise of controlling the thickness of the welding layer is the rapid rise of the resistance wire's temperature, which can be achieved by the high output of voltage of the electric fused power supply; Combined with the heat conduction between the pipe and the environment and the heating process of reducing the output voltage during electric melting, it creates a good condition for the annular region to reach a certain thickness of the molten layer. According to the calculation data, controlling the time of supplying power and the temperature of the area under the specific voltage can make the control of the temperature and thickness of the welding layer more accurate. The simulation in this paper does not consider the influence of environmental temperature, convective conduction of heat and other factors on the data of output, in practice, should be in advance for all kinds of PE electrofusion pipe fittings of simulation of temperature field, after considering the influence of environment and correcting, all relevant data are imported into the electric melting power supply to realize the optimal control of temperature in the process of electric melting.

Acknowledgment. This research is financially supported by Fund for Postdoctoral Scientific Research Projects of Zhejiang Province of China (ZJ2020084).

References

1. Tu, X., Li, M.D., Zhang, S.J., Zhang, S.K., Wu, W.: Research status of fusion butt welding process of polyethylene pipes for the supply of gaseous fuels. Guangdong Chem. Ind. **44**(09), 160–161 (2017)
2. Xin, M.L., Li, M.D., Zhang, S.K., Yang, B., Lin, J.M.: Research progress in failure mode analysis of polyethylene gas pipelines. China Plast. **29**(03), 16–20 (2015)
3. Meng, X.L., Guo, Y.B., Meng, T., Wang, D.G., Liu, K., Chen, Q.X.: First exploration on class of safely condition for gas PE pipeline. Pressure Vessel Technol. **34**(05), 52–57 (2017)
4. Lai, H.S., Fan, D.S., Jia, Y.F., Liu, K.L., Xu, S.F., Tu, S.D.: Key technical challenges of high density polyethylene (HDPE) pipe in nuclear equipment. Pressure Vessel Technol. **34**(12), 45–54 (2017)
5. Yi, W.M.: Talking about the electrofusion welding quality control of polyethylene (PE) pipes in South Pacific area. Constr. Des. Eng. **24**, 175–177 (2018)
6. Zhang, H.: Application of intelligent PID algorithm in automatic welding machine control system. Tech. Autom. Appl. **38**(10), 37–41 (2019)
7. Zheng, J.Y., Shi, J.F., Guo, W.C., Din, S.B., Shi, P., Wang, H.: Analysis of temperature field during welding of polyethylene pipe fusion joint. Trans. China Weld. Inst. **30**(03), 5–9+113 (2009)
8. Schleger, P., Farago, B., Lartigue, C., et al.: Clear evidence of reptation in polyethylene from neutron spin-echo spectroscopy. Phys. Rev. Lett. **81**(1), 124 (1998)
9. Wool, R.P., Long, M.J.: Fractal structure of polymer interfaces. Macromolecules. **26**(19), 5227–5239 (2002)
10. Guo, W.C., Hu, Y.R., Miao, C.J., Shi, J.F.: Relationship between welding performance and depth of fusion zone for electrofusion joint of polyethylene pipes. China Plast. **33**(12), 58–62 (2019)
11. Guo, W.C., Shi, J.F., Tang, P., Zheng, M.L.: A New method for determining the welding time of polyethylene electrofusion. Weld. Technol. **48**(06), 37–42 (2019)

Wireless Network Simulation Platform Based on OPNET and Simulink

Dong Chen[1], Yang Song[1,2](✉), and Taicheng Yang[3]

[1] Department of Automation, School of Mechatronic and Automation, Shanghai University, Shanghai 200444, China
{chendong19,y_song}@shu.edu.cn

[2] Shanghai Key Laboratory of Power Station Automation Technology, Shanghai University, Shanghai 200444, China

[3] Department of Engineering and Design, University of Sussex, Brighton, UK
taiyang@sussex.ac.uk

Abstract. In this paper, we presented a WNCS co-simulation platform based on OPNET/Simulink to verify the validity of the theory and method proposed in the study of wireless network control system (WNCS). The platform combines OPNET with Simulink, and produces the mathematical model of the control system by Simulink. OPNET simulates the network part, forming a co-simulation platform based on OPNET and Simulink. In order to simulate the real network environment more accurately, this paper added the Mobile Ad Hoc Network Routing Protocol (MANET) to the platform and carried on the secondary development of it, so as to realize the co-simulation of the control system and network environment under the framework of network protocol. Finally, through the simulation of the single-stage inverted pendulum control system in the wireless network environment, the effectiveness and good scalability of the platform were verified.

Keywords: WNCS · Co-simulation · OPNET · Simulink · MANET

1 Introduction

OPNET Modeler is a powerful network simulation software that accurately analyzes the performance and behavior of complex networks. By using OPNET and MATLAB/Simulink to build a co-simulation platform, using OPNET to simulate the dynamic characteristics of the wireless network simulation environment and MATLAB to build a control system, can achieve full digital and hardware-in-the-loop co-simulation. The co-simulation platform of wireless network control system can overcome the shortcoming of single simulation platform and the dynamic changes of co-simulation control system and communication network really reflect the behavior of the network and the whole wireless network control system. Through the simulation platform, the controller algorithm can be optimized and the parameters of the control system and network can be simulated comprehensively, which provides a simulational platform for the theoretical research of the stability and real-time performance of the networked control system. Co-simulation research on control system based on wireless network can provide technical

K. Li et al. (Eds.): LSMS 2021/ICSEE 2021, CCIS 1468, pp. 408–417, 2021.
https://doi.org/10.1007/978-981-16-7210-1_39

support and simulation platform for the application of wireless network control, and further research on wireless network has important theoretical significance and practical application value [1–5]. This paper mainly designs a WNCS co-simulation platform based on OPNET/Simulink. This platform not only combines the network simulation software OPNET with the control system simulation software MATLAB/Simulink, but also adds the routing protocol MANET and carries on the secondary development, so as to realize the co-simulation of the control system and the wireless network under the framework of the network protocol. Finally, the influence of packet loss rate on the performance of the wireless network control system is analyzed through the simulation of the single-stage inverted pendulum, and the effectiveness of the platform is verified [6–8].

2 Design Principle

2.1 OPNET/Simulink Co-simulation Platform

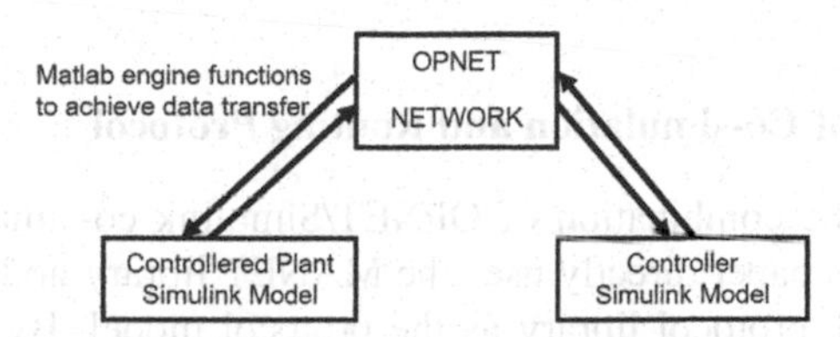

Fig. 1. Co-simulation schematic diagram

OPNET and MATLAB/Simulink are two simulation software that run independently of each other. In order to realize the combination of network simulation software OPNET and control system simulation software MATLAB/Simulink, it is necessary to add an interface module between the two software to complete the data transfer and exchange between the two simulation software. Also will be defined in OPNET simulation commands passed to the MATLAB workspace to perform Simulink simulation, OPNET to ensure the dominance and control of Simulink. Because the development process of using Matlab engine is simpler and can effectively reduce the development time on the software interface, this paper chooses the engine program provided by Matlab as the interface between software. In the networked control system, the controlled object and the controller are separated, and they are connected by the network [9–11]. According to this characteristic, the idea of modularization is adopted in the co-simulation. The co-simulation platform designed in this paper is composed of three modules: controlled object, OPNET wireless network and controller. The simulation of control system is carried out in two Simulink models of Plant and Controller modules, while OPNET is mainly used to simulate the network environment and transfer the data of control system. It connects the controlled object module and controller module to form a real-time closed-loop control system [12–15]. The simulation principle is shown in Fig. 1.

Users can realize the simulation of different types of NCS by changing the controlled object module, controller module, or changing the Settings of OPNET network module.

For example, replace the OPNET network module to analyze the impact of changes in the network environment on the control system, or replace the controller module to verify the impact of various control algorithms on the entire NCS. This modular design makes the co-simulation platform widely applicable and can be used for any NCS simulation.

2.2 MANET Routing Protocol

Mobile Ad Hoc Network (MANET) is a wireless mobile communication network composed of a group of mobile nodes with wireless transceiver devices. MANET belongs to distributed control network, without the traditional control node in the network center, and uses is the organization to build the network, constructing the network does not need to provide additional network infrastructure, so that we can be used in accordance with the requirements for users to quickly set up network, and a MANET network has very strong anti-destroying ability and robustness [16, 17]. These advantages make MANET attracted the attention of many scholars at home and abroad, and it has been applied in many fields such as emergency rescue, wireless local area network, personal area network and military communication.

2.3 Combination of Co-simulation and Routing Protocol

In order to realize the combination of OPNET/Simulink co-simulation and MANET routing protocol, this paper directly uses the MANET library node manet_station_adv provided by OPNET protocol library as the protocol model. By developing the data link layer of the IP protocol stack in the protocol model and modifying the data source module in the application layer, the co-simulation is carried out. Manet_station_adv is one of the most important wireless node models in MANET network model. It can simulate and implement an IP packet traffic generator. In this model, the data source module sends the generated data packet to the ip_encap layer. After receiving the data packet, the ip_encap layer wraps the packet into IP form and sends it to the IP layer. The IP layer then unpacks the packet according to the protocol requirements and sends the packet to the MANET network for communication [18, 19].

The main contents of the modification include: secondary development of the IP protocol stack data link layer process model; On the premise of ensuring that the data source module can communicate normally with other layers of the protocol, the definition and call of Matlab engine are added to start the co-simulation, and various Simulink simulation control commands and custom call functions are added to realize the creation, sending and receiving of the co-simulation data packets. At the same time, the routing information required by routing protocol is defined to meet the protocol requirements of IP layer. At the same time, in order to ensure that the modified node can communicate with other MANET nodes normally, the declaration of attributes, operation of various statistics and other parts should be reserved [20].

3 Model Design

3.1 Network Model

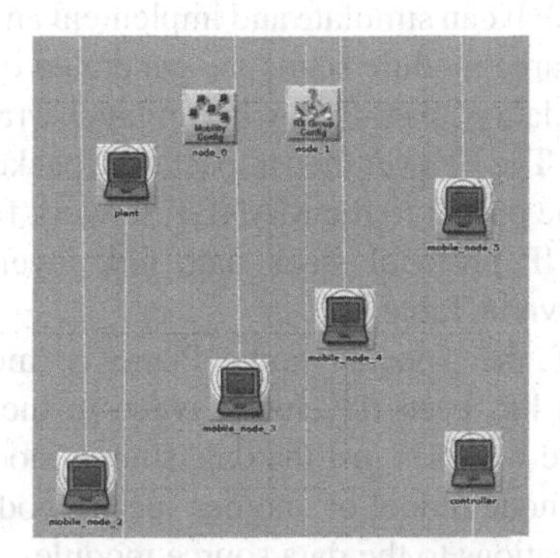

Fig. 2. Wireless network model

The network model uses subnet, node and link to describe the network topology. The wireless network simulation platform designed in this paper has a total of 6 mobile nodes for the secondary development of the IP protocol stack data link layer process model, among which two special nodes, Plant and Controller, also need to be redeveloped for the data source module. They are distributed in the area of 1000 m × 1000 m and each node moves randomly in the area. The wireless network model is shown in Fig. 2.

3.2 Node Model

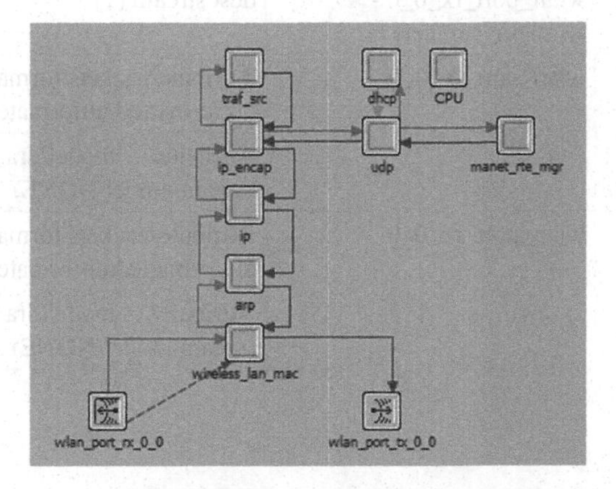

Fig. 3. Intermediate node model

The node model reflects the characteristics of the device. It is composed of multiple process modules, which are connected by packet streamline or statistical line. The process modules in the node model include three types: processor class, data streamline class and receiving/sending machine class. The 6 node modules are all modified by the IP protocol stack data link layer module of the library node manet_station_adv, as shown in Fig. 3. Manet_station_adv is one of the most important wireless node models in MANET network model. It can simulate and implement an IP packet traffic generator. In this model, the data source module sends the generated data packet to the ip_encap layer. After receiving the data packet, the ip_encap layer wraps the packet into IP form and sends it to the IP layer. The IP layer then unpacks the packet according to the protocol requirements and sends the packet to the MANET network for communication.

The modification of IP protocol stack data link layer module of library node manet_station_adv is shown in Table 1.

The node model of the special node Plant is modified on the basis of manet_station_adv which has been developed twice in the data link layer. It is different from the former one which is just the data source module, and the other aspects are exactly the same.The node model of another special node, Controller, is similar to that of Plant, with modifications to the data source module.

Table 1. Manet_station_adv's IP protocol stack data link layer modification table

	Position	Modification
Node	wireless_lan_mac	Attribute/process model/mac_csma
Statistic wire	wlan_port_tx_0_0 -> wireless_lan_mac	delete
Statistic wire	wlan_port_rx_0_0 -> wireless_lan_mac	Attribute/src stst/radio receiver.busy
strm_4110	arp -> wireless_lan_mac	dest stream [0]
rx	wlan_port_rx_0_0 -> wireless_lan_mac	dest stream [1]
Node	wlan_port_tx_0_0	Attribute/packets formats/ all formatted,unformated
		Attribute/… model/dra_… (tagain model/NONE)
Node	wlan_port_rx_0_0	Attribute/packets formats/ all formatted,unformated
		Attribute/… model/dra_… (ragain model/NONE)

3.3 Process Model

The process model consists of finite-state machine, OPNET core functions and standard C and C++ languages. Figure 4 shows the process model of the object node

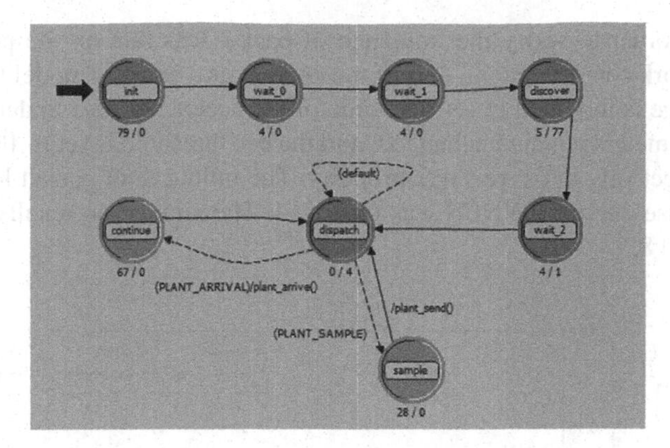

Fig. 4. Process model of the object node data source module

data source module. The first five state machines mainly complete the initialization of modules and statistics registration, and open the Matlab engine to call the Simulink model of the controlled object. It then enters the idle state waiting for interrupts, and the process model responds to two kinds of interrupts, one is a self-interrupt (PLANT_SAMPLE) with scheduled data sampling, and the other is a packet ARRIVAL interrupt (PLANT_ARRIVAL). Sample module mainly completes the reading of Simulink model data of the controlled object. When the simulation time of OPNET reaches the sampling time, it will respond to the self-interrupt, read the data of the Simulink model of the controlled object from the Matlab workspace, and then call the Plant_ Send function to put the current state into the data packet. The IP addresses of both the source node (Plant) and destination node (Contoller) are defined and written into the ICI of the IP_ENCAP model. The packet is then sent to the IP_ENCAP module, so that the routing information in the ICI is also sent along with the packet. The Continue module receives packets returned from the controller and restarts Simulink simulation. Once the packet arrival interrupt is received, the plant_arrive function is called to receive the data from the controller and the controlled object model is instructed to run with the previous input parameters to compensate for the measured delay time. Then the updated control information data is written into the MATLAB workspace to update the relevant parameters of the Simulink model of the controlled object. Finally, the simulation of Simulink is restarted until the next sampling moment.

4 Simulation Results

In order to prove that the improved joint simulation platform simulation ability for WNCS theory research, this section of single-stage inverted pendulum control system, will be added to the system of wireless network, make it a complete WNCS, then to the simulation of the WNCS, finally by observing different packet loss rate affect the performance of the control system, to verify the effectiveness of the platform.

The dynamic source routing protocol (DSR) was adopted for all nodes in the wireless network model, and the sampling period was set as 0.01 s, and the simulation time was set as 5 s. In order to verify the influence of packet loss rate on the performance of wireless network control system, part of the code in dra_ecc.ps.c model file in the error correction stage is modified to set the value of the accept variable under the condition that other parameters remain unchanged, and the ecc threshold is set to 10%, 20%, 30% and 40% respectively in the receiver property. The influence of packet loss rate on the system response curve of WNCS was observed. The simulation results are shown in Fig. 5, 6, 7 and 8.

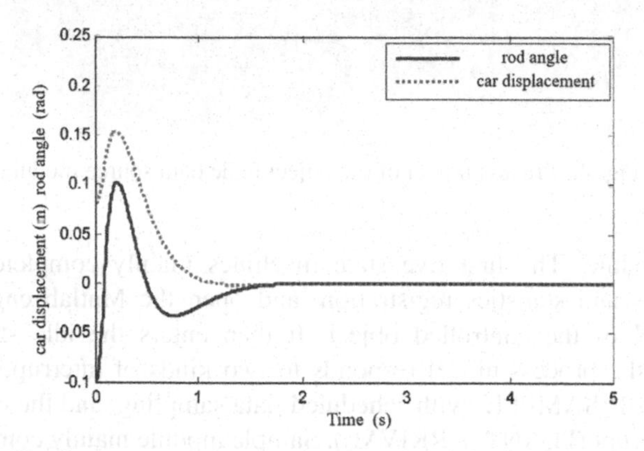

Fig. 5. System response curve when packet loss rate is 10%

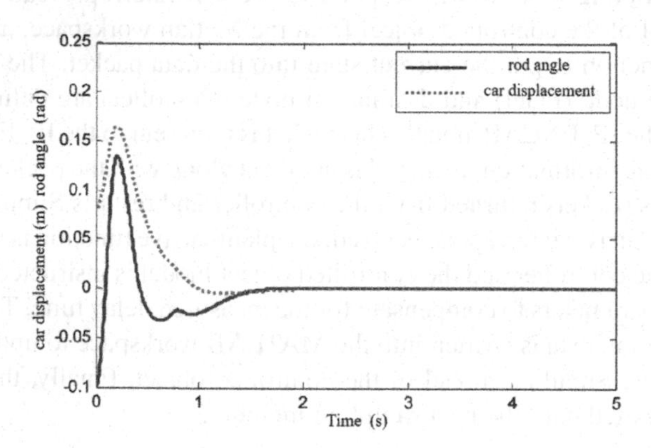

Fig. 6. System response curve when packet loss rate is 20%

As can be seen from the above figure, when the packet loss rate in the wireless network is around 10%, the system response curve does not change significantly, indicating that 10% packet loss rate has little impact on the performance of WNCS. However, with the increase of packet loss rate, the response curve began to change and the oscillation became more and more violent. When the packet loss rate is about 40%, the stability of the system will be seriously damaged and the system will be unstable. This is basically consistent with the theoretical results, which proves the effectiveness of the co-simulation platform.

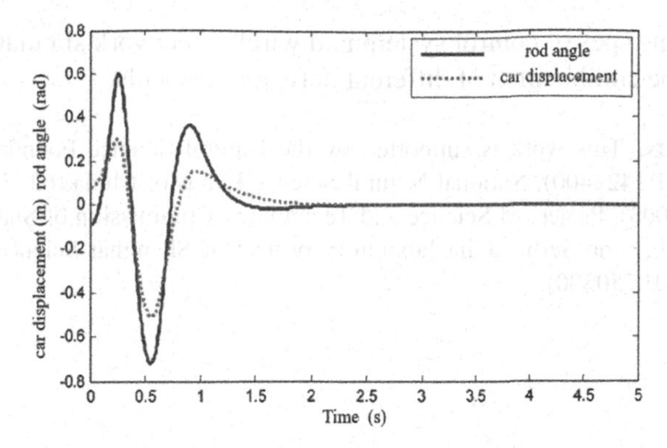

Fig. 7. System response curve when packet loss rate is 30%

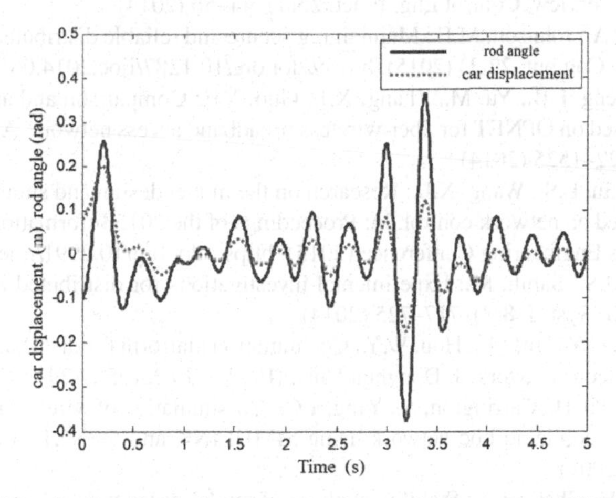

Fig. 8. System response curve when packet loss rate is 40%.

5 Conclusion

Based on the secondary development of MANET routing protocol, this paper designs a wireless network control system co-simulation platform based on OPNET/Simulink, which enables the control system to transmit data in the wireless network according to MANET routing protocol, and realizes the co-simulation of control system and network environment under the framework of network protocol.

Through the simulation of single-stage inverted pendulum control system in wireless network environment, the influence of packet loss rate on the control system performance is observed and analyzed, and the effectiveness of the platform is verified.

In this paper, the development of wireless network control system simulation platform can not only truly simulate the characteristics of the wireless network control system for the wireless network control system of the theoretical results provide powerful verification platform, but also has good expansibility: use of this platform can not only for different types of control system and wireless network simulation, also can be achieved with the combination of different network protocols.

Acknowledgments. This work is supported by the Natural Science Foundation of Shanghai Municipality(21ZR1423400), National Natural Science Funds of China (62173217, 61633016), 111 Project (D18003), Project of Science and Technology Commission of Shanghai Municipality (19500712300), Sino-Serbia Joint laboratory project of Shanghai Science and Technology Commission (19510750300).

References

1. Li, W.L., Zhang, X.B., Li, H.M.: Co-simulation platforms for codesign of networked control systems: an overview. Control Eng. Pract. **23**(1), 44–56 (2014)
2. Sleptchenko, A., Johnson, M.E.: Maintaining secure and reliable distributed control systems. INFORMS J. Comput. **27**(1) (2015). https://doi.org/10.1287/ijoc.2014.0613
3. Xu, C.X., Meng, F.B., Yu, M., Zhang, X.J., Guo, Y.F.: Comparison and analysis of routing protocols based on OPNET for fiber-wireless broadband access network. Appl. Mech. Mater. **635–637**, 1522–1525 (2014)
4. Wang, F.F., Liu, L.S., Wang, X.L.: Research on the model design and simulation for training platform based on network control. In: Proceedings of the 2015 Information Technology and Mechatronics Engineering Conference (2015). https://doi.org/10.2991/itoec-15.2015.33
5. Mahmoud, M.S., Sabih, M.: Experimental investigations for distributed networked control systems. IEEE Syst. J. **8**(3), 717–725 (2014)
6. Song, Y., Hu, C.Y., Hu, J.L., Hou, W.Y.: Cosimulation platform for distributed control system via heterogeneous network. J. Donghua Univ. (Engl. Ed.) **33**(05), 729–733 (2016)
7. Hasan, M.S., Yu, H., Carrington, A., Yang, T.C.: Co-simulation of wireless networked control systems over mobile ad hoc network using SIMULINK and OPNET. IET Commun. **3**(8), 1297–1310 (2009)
8. Kazempour, F., Ghaisari, J.: Stability analysis of model-based networked distributed control systems. J. Process Control **23**(3), 444–452 (2013)
9. Harding, C., Griffiths, A., Yu, H.: An interface between MATLAB and OPNET to allow simulation of WNCS with MANETs. In: Proceeding of IEEE International Conference on Networking, Sensing and Control, London, UK, pp. 711–716 (2007)
10. Sadi, M.A.H., Ali, M.H., Dasgupta, D., et al.: Opnet/Simulink based testbed for disturbance detection in the smart grid. In: Proceedings of the 10th Annual Cyber and Information Security Research Conference, Oak Ridge, Tennessee, USA (2015). https://doi.org/10.1145/2746266.2746283
11. Cheng, Q., Dong, H.R.: Simulation of wireless networked control systems. In: The 44th IEEE Conference on Decision and Control, Seville, Spain, pp. 476–481 (2005)
12. Qu, Z.G., Yuan, T.: The radar carrier landing simulation system design based on HLA. Sci. Technol. Inf. **25**, 57–58 (2012)
13. Garrido, P.P., Malumbres, M.P., Calafate, C.T.: NS-2 vs. OPNET: a comparative study of the IEEE 802.11e technology on MANET environments. In: Proceedings of the 1st International Conference on Simulation Tools and Techniques for Communications, Networks and Systems & Workshops, Marseille, France (2008). https://doi.org/10.1145/1416222.1416264
14. Ploplys, N.J., Kawka, P., Alleyne, A.G.: Closed-loop control over wireless network. IEEE Control Syst. **24**(3), 58–71 (2004)

15. Björkbom, M., Nethi, S., Eriksson, L.M., Jäntti, R.: Wireless control system design and co-simulation. Control Eng. Pract. **19**(9), 1075–1086 (2011)
16. Chen, M.: OPNET Network Simulation. Tsinghua University Press, Beijing (2004)
17. Royer, E.M., Toh, C.K.: A review of current routing protocols for ad hoc mobile wireless networks. IEEE Pers. Commun. **6**(2), 46–55 (1999). https://doi.org/10.1109/98.760423
18. Cervin, A., Ohlin, M., Henriksson, D.: TrueTime 1.5 Reference Manual. K. Lund University, Sweden (2007)
19. Hasan, M.S., Harding, C., Yu, H., Griffiths, A.: Modeling delay and packet drop in networked control systems using network simulator NS2. Int. J. Autom. Comput. **2**, 17–194 (2005)
20. Song, Y., Hou, W.Y.: Research of heterogeneous network simulation platform based on MANET and Ethernet. Control Eng. China **22**(3), 465–469 (2015)

Energy Storage Power Control Strategy Based on Adaptive Parameter Virtual Synchronous Generator

Wengen Gao[1,2,3](✉), Minghui Wu[1,3], Lisheng Wei[1,2], Huacai Lu[1,2], and Ye Tang[1,3]

[1] College of Electrical Engineering, Anhui Polytechnic University, Wuhu 241000, China
ahpuchina@ahpu.edu.cn

[2] Key Laboratory of Advanced Perception and Intelligent Control of High-End Equipment, Chinese Ministry of Education, Wuhu 241000, China

[3] AnHui Key Laboratory of Detection Technology and Energy Saving Devices, Wuhu 241000, China

Abstract. According to the electrical model and mechanical model of synchronous generator, the control model of virtual synchronous generator is deduced, and the relationship between control parameters and virtual synchronous generator is analyzed. Based on the power electronics technology, the energy storage power control strategy is given, including the virtual synchronous generator power control loop and the voltage and current double loop control strategy. Combined with the actual operating conditions of energy storage power conversion system, an adaptive adjustment method of virtual synchronous generator parameters is proposed, and the simulation model is built, and the effectiveness of the adaptive parameter control strategy of virtual synchronous generator is verified by simulation analysis.

Keywords: Energy storage system · Adaptive parameter virtual synchronous generator · Power conversion system · Microgrid transient stability

1 Introduction

With the gradual depletion of traditional energy and increasingly serious environmental problems in the world, the traditional power energy structure based on thermal power has been difficult to meet the needs of energy sustainable development [1]. In recent years, the development and utilization of renewable energy has been paid more and more attention, such as wind energy and solar energy [2, 3]. With the continuous increase of renewable energy grid-connected power, its proportion in the power grid is gradually increasing. Because of the randomness and gap of renewable energy generation, the renewable generation energy has fluctuation, and the energy demand of users has space-time randomness, which will lead to the uncertainty of energy sources and the disorder of energy flow in the power grid [4]. Therefore, there are problems with large-scale concentrated renewable energy generation models on large grid impacts, high transmission costs, and energy is not easy to consume, and distributed renewable energy generation has become

K. Li et al. (Eds.): LSMS 2021/ICSEE 2021, CCIS 1468, pp. 418–427, 2021.
https://doi.org/10.1007/978-981-16-7210-1_40

a development trend [5]. Since the power of renewable energy is random and intermittent, it will be a threat to grid stability with increasing distributed renewable energy grid-connected capacity, so it is urgent to propose scientific and reasonable response methods and techniques [6]. Integrating renewable energy generation, energy storage and load to form a microgrid, controlling the charging and discharging of energy storage to suppress the fluctuation of microgrid power from grid connections, and improving the "friendliness" of the microgrid power grid-connected is a feasible solution [7].

The energy flow between microgrid and grid is bidirectional. When microgrid outputs power to grid, it is similar to generator function. With large-scale distributed microgrid grid-connected, the grid stability will be determined by the superposition of many microgrids. Although the capacity of single microgrid is small, its transient characteristics should be considered [8]. In other words, the responsibilities of frequency and voltage regulation should be shared separately, and microgrid should have the ability of frequency and voltage regulation. In the microgrid, energy storage system plays a key role in regulating the transient characteristics, charging and discharging of the energy storage is performed by power conversion system (PCS) [9]. As a typical power electronic interface device, PCS has no inertia and damping characteristics similar to synchronous generator, and can not provide the same frequency and voltage regulation performance as synchronous generator [10]. In view of the above problems, the virtual synchronous generator (VSG) technology has been proposed, which enables the grid-connected power electronic interface to simulate the external characteristics of the synchronous generator, so that it can alleviate the stability problem of power grid with high proportion of renewable energy [11]. Due to the fluctuation of source and load power in microgrid, the output power of energy storage system needs to be adjusted frequently. If energy storage power control strategy of VSG with fixed parameters is adopted, it is difficult to meet the needs of wide range adjustment of energy storage output power. It is easy to cause instability of frequency and voltage in microgrid, reduce power quality, and even cause voltage or frequency out of limit [12].

In this paper, a parameter adaptive VSG control strategy is proposed, which can adjust the VSG control parameters according to the state of energy storage power output, improve the adjustment ability of energy storage power output, and maintain the stability of frequency and voltage of microgrid.

2 Analysis of Relationship Between Dynamic Parameters and Performance of Virtual Synchronous Generator

The model of virtual synchronous generator is a typical second-order system model, which structure of small signal model is shown in Fig. 1 [13].

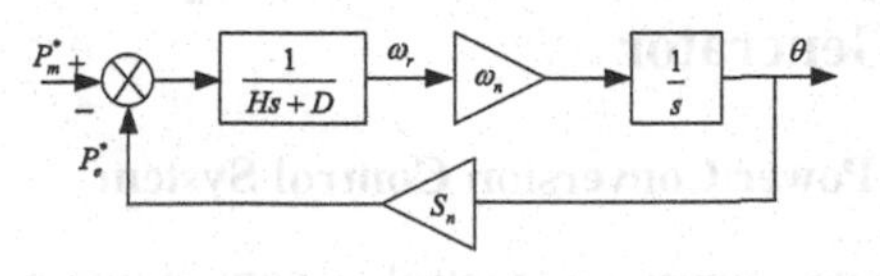

Fig. 1. Virtual synchronous generator small signal model structure diagram.

where, ω_n and ω_r are rated angular frequency and relative shift of angular frequency respectively, S_n is the rated capacity, H is inertia time constant, D is damping coefficient.

The transfer function of the second order model of virtual synchronous generator can be deduced as

$$G(s) = \frac{\omega_0^2}{s^2 + 2\xi\omega_0 s + \omega_0^2}. \tag{1}$$

where, the oscillation angular frequency ω_0 and damping coefficient ξ of the second order system are denoted as

$$\begin{cases} \omega_0 = \sqrt{\frac{\omega_n S_n}{H}} \\ \xi = \frac{D}{2}\sqrt{\frac{1}{\omega_n S_n H}} \end{cases}. \tag{2}$$

The characteristic root of Eq. (1) can be deduced as

$$p_{1,2} = \frac{D}{2H}\left(\pm\sqrt{1 - \frac{4\omega_n S_n H}{D^2}} - 1\right). \tag{3}$$

When $1 - \omega_n S_n H / D^2 > 0$, $p_{1,2}$ is a pair of unequal negative real roots. Let $p_1 > p_2$, and $H \ll D^2 / \omega_n S_n$, p_2 can be ignored, then the model of virtual synchronous generator is simplified to the first order model as

$$G_1(s) \approx \frac{1}{s/p_1 + 1}. \tag{4}$$

According to the step response characteristics of the first-order system, the transition process is generally 3–5 times of the time constant, so the step dynamic transition process time t_{dg} can be expressed as

$$t_{dg} = \frac{2aH}{D(\sqrt{1 - \frac{4\omega_n S_n H}{D^2}} + 1)}, \; a \in [3, 5]. \tag{5}$$

where $H \ll D^2 / \omega_n S_n$, and t_{dg} can be simplified as

$$t_{dg} \approx \frac{aH}{D}, \; a \in [3, 5]. \tag{6}$$

3 Energy Storage Power Control Strategy Based on Virtual Synchronous Generator

3.1 Energy Storage Power Conversion Control System

The energy storage power conversion control system in two-phase rotating coordinate system is shown in Fig. 2. Where, P_{ref} is the reference value of active power, Q_{ref} is the reference value of reactive power, P_0 is the actual output value of active power, Q_0 is actual output value of reactive power, and I_{abc}, U_{abc} is the actual output value of three-phase voltage and three-phase current respectively. The energy storage power conversion control system can be divided into three control loops, they are power control loop, voltage control loop and current control loop. The outer control loop is the power loop, which uses the virtual synchronous generator control strategy to generate the reference value of the voltage loop control. The voltage loop generates the reference value of the current inner loop, the current inner loop generates the reference voltage of PWM control signal, and PWM algorithm generates the three-phase full bridge switch drive signal to implement the inverter control of the energy storage power conversion system.

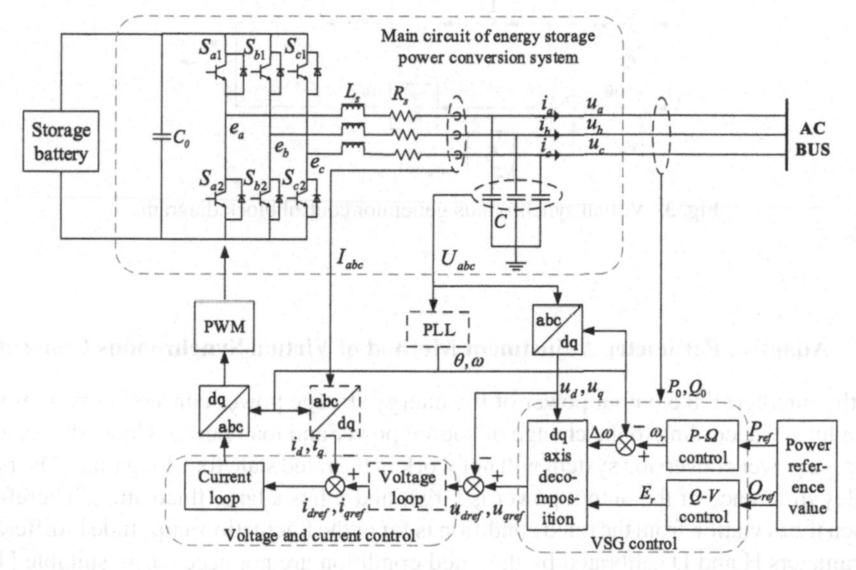

Fig. 2. Energy storage power conversion system control block diagram.

Virtual synchronous generator is similar to PQ and VF control methods, which is a control method of power control loop. The control model of virtual synchronous generator is divided into two parts: *P-f* regulation and *Q-V* regulation, according to the model of virtual synchronous generator in Sect. 2, its control block diagram is shown in Fig. 3.

K_p is the *P-f* droop coefficient, which simulates the frequency regulation function of synchronous generator governor, the regulation principle is expressed as

$$P_m = P_{ref} + K_p(\omega_n - \omega). \tag{7}$$

where, P_m can be understood as the virtual mechanical power of the virtual synchronous generator.

K_q is the *Q-V* droop coefficient, which simulates the excitation voltage regulation function of synchronous generator governor, the regulation principle is expressed as

$$u_{ref} = \frac{1}{K}\int_0^t [K_q(u_n - u) + Q_{ref} - Q_0]dt. \tag{8}$$

where, u_n is the voltage rated value, u is the voltage actual value, and K is the integral coefficient.

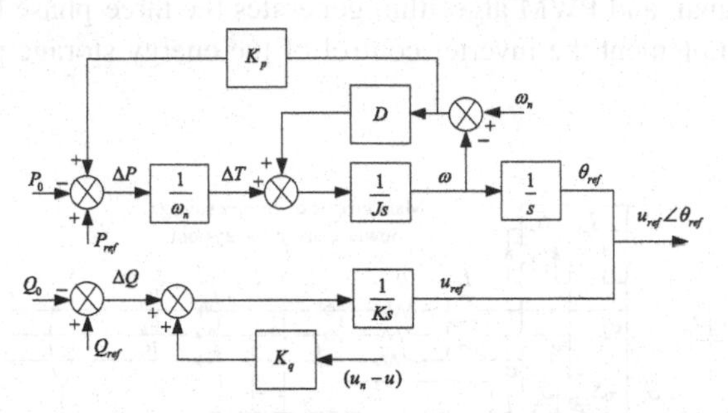

Fig. 3. Virtual synchronous generator control block diagram.

3.2 Adaptive Parameter Adjustment Method of Virtual Synchronous Generator

In the microgrid, the output power of the energy storage power conversion system will be adjusted according to the change of source power and load power. Generally, energy storage power conversion system will not work in the rated state for a long time. The peak valley difference of the output power is large, and it has a large fluctuation. Therefore, when the deviation from the rated condition is far or the fluctuation amplitude is different, parameters H and D calibrated by the rated condition are not necessarily suitable [14]. It will improve the adaptability and performance of the virtual synchronous generator by automatically adjusting the parameters according to the operating conditions.

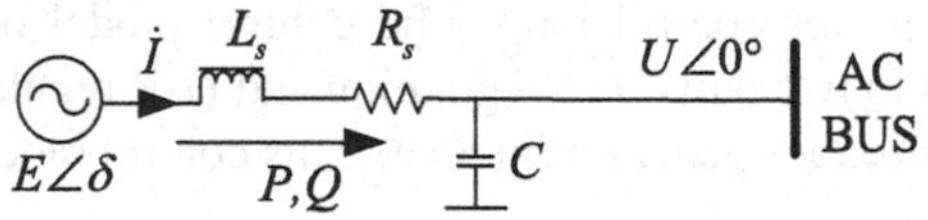

Fig. 4. The single-phase circuit of virtual synchronous generator.

According to Fig. 2, the single-phase circuit of virtual synchronous generator is simplified as Fig. 4. Using circuit theory to analyze the circuit, the rated capacity of the virtual synchronous generator can be obtained as

$$S_n = \left.\frac{\partial P}{\partial \delta}\right|_{\substack{\delta=\delta_0\\E=E_0}} = \left.\frac{EU}{|S|Z}sin(\varphi - \delta)\right|_{\substack{\delta=\delta_0\\E=E_0}}. \tag{9}$$

where, δ_0 and E_0 are the steady-state operation value according to the given power, they are expressed as

$$\begin{cases} \delta_0 = \varphi - \tan^{-1}\left(\frac{Q_{ref}+U^2 sin\varphi/Z}{P_{ref}+U^2 cos\varphi/Z}\right) \\ E_0 = \frac{Q_{ref}Z+U^2 sin\varphi}{Usin(\varphi-\delta_0)} \\ Z = \sqrt{R_s^2+(\omega L_s)^2} \\ \varphi = \tan^{-1}(\omega L_s/R_s) \end{cases}. \tag{10}$$

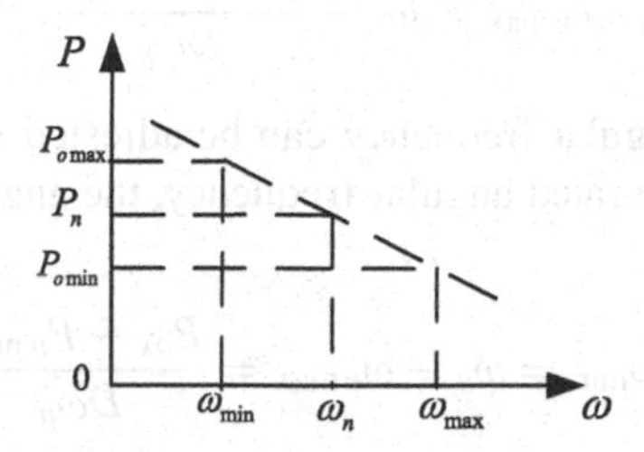

Fig. 5. The frequency regulation diagram of virtual synchronous generator.

In the physical sense, Dampping coefficient D of the virtual synchronous generator is consistent with the *P-f* droop coefficient K_p, and select $K_p = D$. The frequency regulation diagram of virtual synchronous generator is shown in Fig. 5, ω_n is rated angular frequency of the output voltage, ω_{min} is minimum angular frequency, ω_{max} is maximum angular frequency of the output voltage, P_n is rated power, and P_{omin}, P_{omax} are minimum value and maximum value of the output power respectively. Frequency regulation algorithm of virtual synchronous generator is expressed as

$$\begin{cases} P = P_n + D\omega_n(\omega_n - \omega) \\ \Delta P = D\omega_n \Delta\omega \\ \omega_{\min} \le \omega_e \le \omega_{\max} \end{cases}. \tag{11}$$

As shown in Fig. 6, when the output power increases from P_o to P_{ox}, $\omega_x < \omega_{min}$ occurs, which exceeds allowable limit of frequency. The *P-f* characteristic curve is translated to the blue line to make $\omega_e > \omega_{min}$ corresponding to the output power P_{ox}.

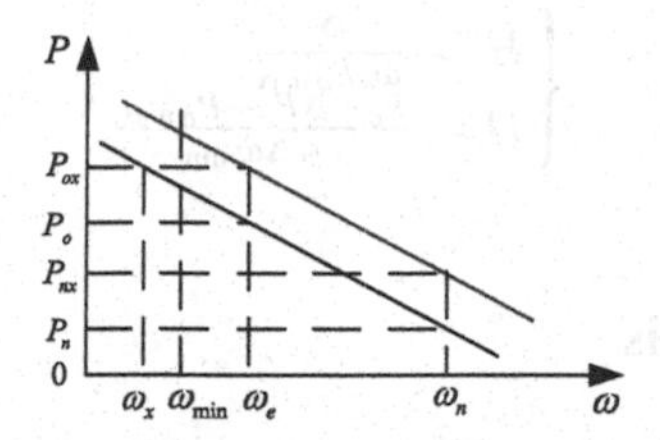

Fig. 6. P-f characteristic shift method.

According to Eq. (11), we obtain

$$\Delta P_n = P_{nx} - P_n = D\omega_n(\omega_e - \omega_x). \tag{12}$$

$$\omega_e = \omega_x + \frac{P_{nx} - P_n}{D\omega_n} = \omega_n - \frac{P_{ox} - P_{nx}}{D\omega_n}. \tag{13}$$

Although the out power of virtual synchronous generator can be adjusted, it is impossible to support unlimited power adjustment due to the limitation of hardware parameters, and the power adjustment has boundary restrictions. The maximum output power of virtual synchronous generator is defined as P_{nmax}, and the limit angular frequency of *P*-*f* characteristic translation regulation is defined as ω_{emax}, which can be expressed as

$$\omega_{e\max} = \omega_n - \frac{P_{ox} - P_{n\max}}{D\omega_n}. \tag{14}$$

When $\omega_{emax} \geq \omega_n$, angular frequency can be adjusted without error. When $\omega_{emax} < \omega_n$, and ω_{emax} closest to rated angular frequency, the angular frequency deviation is minimized

$$\Delta\omega_{\min} = \omega_n - \omega_{e\max} = \frac{P_{ox} - P_{n\max}}{D\omega_n}. \tag{15}$$

When virtual synchronous generator starts up, in order to reduce the regulation overshoot, according to the optimal second-order system, damping ratio is set to $\xi \approx 0.707$. According to the given output power P_o and response time t_r, starting initial values of parameters *H* and *D* are determined as

$$\begin{cases} \xi = \frac{D_0}{2}\sqrt{\frac{1}{\omega_n S_n H_0}} \\ S_n = \frac{EU}{|S|Z} sin(\varphi - \delta)\Big|_{\substack{\delta=\delta_0 \\ E=E_0}} \\ t_r \approx \frac{aH_0}{D_0}, \quad a \in [3, 5] \\ D_0 = \frac{P_{o\max} - P_n}{\omega_n(\omega_n - \omega_{\min})} \end{cases}. \tag{16}$$

After start-up, the output power will be adjusted with the fluctuation of microgrid power and deviate from the rated power point. It is unreasonable to use a fixed S_n to select parameters under this condition. Let $S_n \approx P_{omax}$, according to the output power deviation ΔP_o and angular frequency adjustment deviation $\Delta\omega_{min}$ to automatically adjust parameters *H* and *D*, it is expressed as

$$\begin{cases} H > \frac{D^2}{\omega_n P_{o\max}} \\ D = \frac{P_o + \Delta P_o - P_{n\max}}{\omega_n \Delta\omega_{\min}} \end{cases}. \tag{17}$$

4 Simulation Analysis

The simulation model of virtual synchronous generator is built in Matlab/Simulink, and simulation parameters are shown in Table 1. The simulation steps are as follows:

(1) Given the output active power $P = 40$ kW, the virtual synchronous generator starts to run;

(2) When $t = 1$ s, the output power is suddenly loaded to 80 kW;
(3) When $t = 3$ s, the output power will be reduced to 40 kW;
(4) When $t = 5$ s, the simulation ends.

According to the simulation conditions, the fixed parameter virtual synchronous generator control strategy and adaptive parameter virtual synchronous generator control strategy are compared by simulation, and the simulation results are shown in Fig. 7.

Table 1. Virtual synchronous generator simulation model parameters.

Simulation parameters	Unit	Value
DC side voltage	V	500
RMS of output line voltage	V	310
Rated frequency	Hz	50
Initial value of H	kg.m^2	0.63
Initial value of D	/	14.58
Equivalent inductance	mH	3
Equivalent resistance	Ω	3
Filter capacitor	μF	15

As shown in Fig. 7, when $t = 1$ s, the power jumps from 40 kW to 80 kW, the frequency of the fixed parameter virtual synchronous generator control strategy drops to about 49.5 Hz, which has exceeded the allowable frequency fluctuation range of the power grid, and the frequency out of limit state appears. This out of limit state is maintained until $t = 3$ s, the power changes from 80 kW to 40 kW at this time, the frequency can be adjusted to the normal allowable range. Correspondingly, the voltage amplitude has a jump up at $t = 1$ s and a jump down at $t = 3$ s respectively, the maximum value of line voltage amplitude reaches 350 V, which has exceeded the regulation limit range.

Compared with the fixed parameter virtual synchronous generator control strategy, the frequency of the adaptive parameter virtual synchronous generator control strategy decreases less, the frequency down regulation does not exceed 49.8 Hz, the output voltage amplitude is stable, there is no overshoot state, the frequency and voltage regulation maintain good performance, and there is no out of limit state. In addition, Fig. 7(a) shows that at $t = 1$ s and $t = 3$ s, the frequency regulation process of the fixed parameter virtual synchronous generator takes about 250 ms, and the adaptive parameter virtual synchronous generator control strategy only takes about 50 ms, which greatly reduces the regulation time. Through the comparison of simulation results, it can be proved that the adaptive parameter virtual synchronous generator control strategy has better control performance, makes output more stable and faster response, avoids the risk of frequency and voltage regulation out of limit, and verifies the effectiveness of the adaptive parameter virtual synchronous generator control strategy.

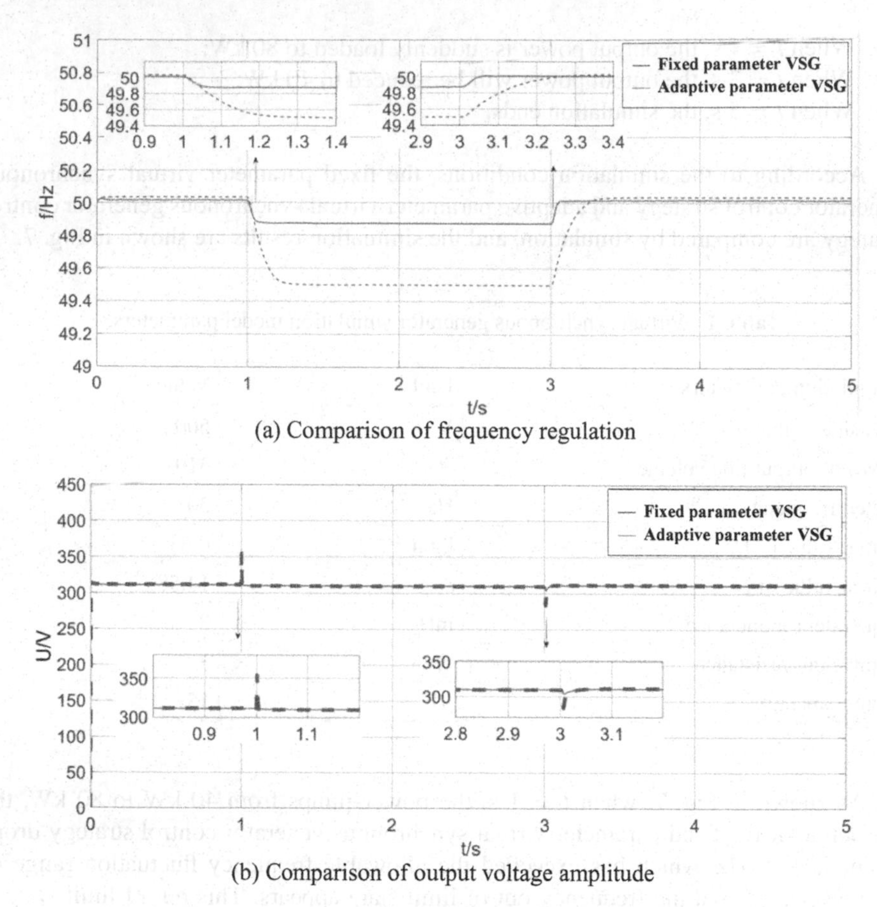

(a) Comparison of frequency regulation

(b) Comparison of output voltage amplitude

Fig. 7. Performance comparison of tow control strategy.

5 Conclusion

For the transient stability of microgrid, an adaptive parameter control strategy of energy storage power is proposed based on virtual synchronous generator. In the paper, the mechanism of the influence of control parameters on the virtual synchronous generator is analyzed. Combined with the actual demand of power regulation of energy storage system in microgrid, by deriving the equation and boundary conditions of *P-f* characteristic shift frequency regulation, the adaptive adjustment method of virtual synchronous generator parameters is designed, It solves the problem that the fixed parameter virtual synchronous generator strategy may cause the energy storage output frequency and voltage to exceed the limit, and improves the performance of energy storage to support the transient stability of microgrid.

Acknowledgments. This work is supported by Natural Science Foundation of Anhui Province (1908085MF215), and Key Research and Development Project of Anhui Province (201904a05020007).

References

1. Chu, S., Majumdar, A.: Opportunities and challenges for a sustainable energy future. Nature **488**, 294–317 (2012)
2. Nikmehr, N., Ravadanegh, S.N.: Optimal power dispatch of multi-microgrids at future smart distribution grids. IEEE Trans. Smart Grid **6**(4), 1648–1657 (2015)
3. Jacob, A.S., Banerjee, R., Ghosh, P.C.: Sizing of hybrid energy storage system for a PV based microgrid through design space approach. Appl. Energy **212**(15), 640–653 (2018)
4. Yang, X., Song, Y., Wang, G., et al.: A comprehensive review on the development of sustainable energy strategy and implementation in China. IEEE Trans. Sustain. Energy **1**(2), 57–65 (2010)
5. Olivares, D.E., Mehrizisani, A., Etemadi, A.H., et al.: Trends in microgrid control. IEEE Trans. Smart Grid **5**(4), 1905–1919 (2014)
6. Ciocia, A., Boice, A.V.A., Chicco, G., et al.: Voltage control in low-voltage grids using distributed photovoltaic converters and centralized devices. IEEE Trans. Ind. Appl. **55**, 225–237 (2018)
7. Carli, R., Dotoli, M., Jantzen, J., et al.: Energy scheduling of a smart microgrid with shared photovoltaic panels and storage: the case of the Ballen marina in Sams. Energy **198**(1), 1–16 (2020)
8. Wu, J., Xing, X., Liu, X., et al.: Energy management strategy for grid-tied microgrids considering the energy storage efficiency. IEEE Trans. Ind. Electron. **65**, 9539–9549 (2018)
9. Yang, L., Yang, Z., Li, G., et al.: Optimal scheduling of an isolated microgrid with battery storage considering load and renewable generation uncertainties. IEEE Trans. Ind. Electron. **66**, 1567–1575 (2019)
10. Wang, K., Yu, J., Yu, Y., et al.: A survey on energy internet: architecture, approach, and emerging technologies. IEEE Syst. J. **99**, 1–14 (2017)
11. Zhong, Q.C., Weiss, G.: Synchronverters: inverters that mimic synchronous generators. IEEE Trans. Ind. Electron. **58**(4), 1259–1267 (2011)
12. Shintai, T., Miura, Y.: ISET: oscillation damping of a distributed generator using a virtual synchronous generator. IEEE Trans. Power Deliv. **29**(2), 668–676 (2014)
13. Zhong, Q.C., Nguyen, P.L., Ma, Z., et al.: Self-synchronized synchronverters: inverters without a dedicated synchronization unit. IEEE Trans. Power Electron. **29**(2), 617–630 (2014)
14. Leng, D., Polmai, S.: Virtual synchronous generator based on hybrid energy storage system for PV power fluctuation mitigation. Appl. Sci. **23**(9), 1–18 (2019)

Intelligent Techniques for Sustainable Energy and Green Built Environment, Water Treatment and Waste Management

An Open Model of All-Vanadium Redox Flow Battery Based on Material Parameters of Key Components

Xin Li, Huimin Zhu(✉), Ya Qiu, Junkang Shao, and Jihan Zhang

College of Electrical and Automation Engineering,
Hefei University of Technology, Hefei 230000, China

Abstract. Based on the component composition and working principle of the all-vanadium redox flow battery (VRB), this paper looks for the specific influence mechanism of the parameters on the final performance of the battery. An open VRB model is built in the MATLAB/Simulink environment, which reflects the influence of the material parameters of electrode, ion exchange membrane, electrolyte and bipolar plate on the battery performance. This model has a guiding significance for the actual assembly of vanadium redox flow battery. In this paper, the final performance of the battery is known in advance through the intuitive model, which effectively avoids a lot of manpower and material resources in the process of assembling the battery.

Keywords: Vanadium redox flow battery · Energy storage · Key materials

1 Introduction

With the development of society, mankind's demand for electricity is increasing year by year. Therefore, it is necessary to constantly find a reasonable way to store and plan electrical energy. All vanadium liquid flow battery is a kind of energy storage medium which can store a lot of energy. It has become the mainstream liquid current battery with the advantages of long cycle life, high security and reusable resources, and is widely used in the power field.

The vanadium redox flow battery is a "liquid-solid-liquid" battery. The positive and negative electrolytes are separated by solid ion exchange membranes to avoid mixing of different liquids on both sides. Establishing an accurate and detailed model can greatly promote the application and promotion of vanadium batteries. At present, in the research of the all-vanadium flow battery model, based on the construction principles from different perspectives, the VRB model is divided into three categories: electrochemical model, circuit model and hybrid model.

Electrochemical model is a model that can simulate the chemical reaction changes in VRB system, which reflects the effect of internal reaction of battery on the charge discharge performance. From the perspective of mass conservation and energy conservation, combined with Nernst equation, the nonlinear differential equation is constructed

K. Li et al. (Eds.): LSMS 2021/ICSEE 2021, CCIS 1468, pp. 431–440, 2021.
https://doi.org/10.1007/978-981-16-7210-1_41

to characterize the effect of electrochemical reaction on VRB electrolyte concentration and stack voltage. Among them, Huang K L et al. [1] considered the influence of the self-discharge reaction caused by the diffusion of vanadium ions across the membrane on VRB, and combined the Nernst equation to establish the self-discharge model and the open circuit voltage mathematical model. From the perspective of mass conservation, Tang A et al. [2, 3] established the variation equations of the concentration of vanadium ions of various valences, and considered the influence of ion diffusion on vanadium ions, perfected the mathematical model of the self-discharge reaction, and studied the electrolyte The influence of flow, transmembrane diffusion and self-reaction on the concentration of vanadium ions of various valences, and the change of voltage is characterized by ion concentration. Sun C et al. [4] studied the diffusion coefficients of vanadium ions of various valences on ion exchange membranes and related influencing factors. Li M H et al. [5] studied the influence of self-discharge reaction on the concentration of vanadium ions by analyzing the electrochemical reaction principle of VRB, and established a charge/discharge process model, which reflects the changing law of vanadium ion concentration of various valences during the operation of VRB, and derives the correlation between the output stack voltage and the charge and discharge voltage. However, the expression is not intuitive enough, the calculation is relatively complicated, and the overall operating state of the battery cannot be displayed. At the same time, the influence of the circulating pump and pipes on the battery is ignored, so it is not suitable for the simulation study of the electrochemical model as the overall battery system.

The circuit model is based on the electrical relationship of VRB components, using related electrical components to simulate the circuit relationship during battery operation, and equivalent components such as voltage sources, current sources, capacitors, and resistors to components in the battery, to establish a reflection of the VRB current, voltage, internal resistance, state of charge (SOC) and other parameters of the relationship equations of the circuit model. Chahwan J et al. [6] compared the internal resistance model, AC impedance method model, RC model, n-RCs model four equivalent circuit models, and based on the 2-RCs model, through improvement, proposed a new RC, etc. Effective circuit model. Skyllas-Kazacos M et al. [7] considered the ohmic impedance of VRB, but ignored the polarization impedance which had little influence on the internal resistance of the battery. The internal resistance model composed of open circuit voltage and internal resistance of the battery can only simulate the steady-state performance of VRB, but it can not reflect the dynamic performance.

The hybrid model is to connect multiple modeling methods with each other, and multiple models are coupled with each other, and a VRB model that meets specific requirements is established according to certain rules and methods. The hybrid model can more comprehensively reflect the influence of electrochemical reactions, fluid mechanics, circuit control, and electrolyte temperature changes on the input and output characteristics of the VRB. It can not only obtain the dynamic parameters of the VRB in real time, but also express its external characteristics. Yi l et al. [8] considered ion diffusion in the equivalent loss model, and established an equivalent circuit model considering ion diffusion. This model couples the equivalent circuit model and the electrochemical model to each other. Ontiveros L J et al. [9] established a transient characteristic model based on the study of the transient characteristics of the electrochemical reaction of VRB. The model calculated the SOC of the electrolyte, the balance voltage of the battery and the superimposed voltage. Zhang Y et al. [10] proposed a VRB comprehensive equivalent circuit model for system-level analysis. This model considers the effects of ion diffusion, electrolyte flow and circuit shunting, and uses the least square method to determine the steady state and transient state of VRB Performance is identified.

Based on the modeling mechanism described above, this paper proposes a new modeling perspective. Starting from the key physical component materials of the all-vanadium flow battery, the parameter characteristics of different component materials are explored, and the specific parameters of the final performance of the battery are found. Influence mechanism, based on MATLAB/Simulink to build an open VRB model, mainly around the four key components of the all-vanadium flow battery. This model has guiding significance for the assembly of the all-vanadium flow battery.

2 The Composition and Chemical Principle of the Battery

The vanadium redox flow battery is mainly composed of four parts: storage tank, pump, electrolyte and stack. The stack is composed of multiple single cells connected in series. The single cells are separated by bipolar plates. Both ends of the stack have a current collecting plate and an end plate, plus auxiliary components such as insulating plates, which are tightened by screws. The electrolyte is stored in a storage tank and a stack, and circulates through the battery under the drive of a circulating pump.

A single battery is composed of auxiliary devices such as ion exchange membranes, electrodes, electrolyte, bipolar plates, and battery casings and sealing materials. The electrode material is generally carbon felt or graphite felt, which is embedded in a flow frame (plate frame) made of polyvinyl chloride plastic. The ion exchange membrane is generally a perfluoro sulfonic acid ion exchange membrane. The electrolyte uses sulfuric acid as the solvent. The vanadium ions in this valence state are the energy storage medium for the active material, and the bipolar plates generally use graphite bipolar plates or composite bipolar plates. The schematic diagram of the battery structure is shown below. The research on the key components of the all-vanadium redox flow battery mainly focuses on four aspects: electrodes, ion exchange membranes, electrolytes and bipolar plates (Fig. 1).

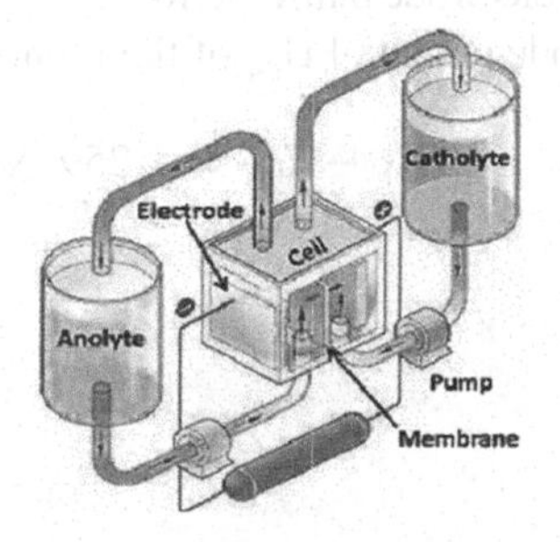

Fig. 1. Structure diagram of all vanadium redox flow battery.

When the external power supply starts to charge the VRB, the vanadium ions undergo charge and discharge oxidation-reduction reactions at the electrodes in the stack. The positive electrode VO^{2+} continuously loses electrons and gradually oxidizes to $VO_2{}^+$. The electrons reach the negative electrode through the circuit, and the negative electrode V^{3+} gets the electrons gradually reduced to V^{2+}. And under the action of the circulating pump pressure, the vanadium electrolyte in the liquid storage tank flows into the stack and evenly flows through the monomers, and continues to react until the charging is completed. When the load VRB is connected to discharge, the $VO_2{}^+$ of the positive electrode gets electrons and is reduced to VO^{2+}, and at the same time, the V^{2+} of the negative electrode continuously loses electrons and oxidizes to V^{3+}. Hydrogen ions permeate the ion exchange membrane to maintain the charge balance inside the battery. During charging, the ion concentration of V^{3+} and VO^{2+} gradually decreases, while V^{2+} and $VO_2{}^+$ increase, and the opposite is true during discharge. Among them, the electrode reaction is:

Positive charge and discharge:

$$VO_2{}^+ + 2H^+ + e^- \underset{charge}{\overset{discharge}{\rightleftarrows}} VO^{2+} + H_2O. \quad (1)$$

Negative charge and discharge:

$$V^{2+} \underset{charge}{\overset{discharge}{\rightleftarrows}} V^{3+} + e^-. \quad (2)$$

Total response:

$$V^{2+} + VO_2{}^+ + 2H^+ \underset{charge}{\overset{discharge}{\rightleftarrows}} VO^{2+} + V^{3+} + H_2O. \quad (3)$$

In the electrochemical reaction, the corresponding standard electric potential will be generated. When the temperature is 298K, the relative electric potential of each valence vanadium ion is as follows:

$$VO_2^+ \overset{1.004\,V}{\longrightarrow} VO^{2+} \overset{0.337\,V}{\longrightarrow} V^{3+} \overset{-0.255\,V}{\longrightarrow} V^{2+}. \quad (4)$$

Then the standard potential of the positive electrode pair/reaction = 1.004V, the standard potential of the negative electrode pair/reaction = −0.255V, that is, at a temperature of 298K, the standard electrode potential U_{eq} of the monomer is:

$$U_{eq} = E_{0+} - E_{0-} = 1.259\,(V). \quad (5)$$

3 The Influence of the Parameters of Each Component on the Battery Performance

3.1 Electrode

The electrode of the all-vanadium flow battery is the place for the charge and discharge reaction of the chemical energy storage system, and the electrode itself does not participate in the electrochemical reaction. The flow battery completes the electrochemical reaction through the active material in the electrolyte solution on the electrode surface to accept or give out electrons to complete the electrochemical reaction, convert electrical energy into chemical energy when charging, and convert chemical energy into electrical energy when discharging, the higher the efficiency of the transformation between electric energy and chemical energy, the better.

All vanadium redox flow battery electrodes have an important impact on the performance of the battery, which is mainly reflected in the effective area for redox reaction, good wettability and electrochemical activity [11]. Its electrical conductivity and catalytic activity directly affect the degree of electrode polarization, current density and thus the energy efficiency conversion of the battery, and the physical and chemical stability of its materials directly affect the battery life and battery operation stability [12].

In order to increase the contact area of the all-vanadium flow battery with the electrolyte, the porous carbon electrode is usually used. The porosity of the porous carbon electrode affects its effective conductivity, and the resistance value of the electrode resistance is related to its conductivity.

$$\sigma_s^{\text{eff}} = (1 - \varepsilon)^{1.5} \sigma_s. \tag{6}$$

In the formula, ε is the porosity of the porous electrode; σ_s is the conductivity of the electrode with zero porosity, and the value is 500 S/m.

$$R_s = \frac{d}{\sigma \cdot S}. \tag{7}$$

In the formula, d is the thickness of the electrode; σ is the effective conductivity of the porous electrode, that is, S is the area of the electrode.

3.2 Ion Exchange Membrane

Ion exchange membrane refers to a polymer membrane with charged groups that can achieve selective permeation of ion species. The ion exchange membrane is one of the key components of the flow battery, and its performance directly affects the performance and life of the flow battery. So far, researchers have researched and developed a variety of ion exchange membranes for flow batteries, mainly including ion exchange membranes and porous ion conductive membranes. According to the type of charged charge of the exchange ions, ion exchange membranes can be divided into cation exchange membranes and anion exchange membranes. The ion exchange membrane not only separates the positive and negative electrolytes of the same single cell to avoid short circuits, but also conducts cations and/or anions to achieve a current loop, which plays a decisive role in the coulombic efficiency and energy efficiency of the vanadium redox flow battery.

Water absorption is a parameter that characterizes the water absorption capacity of the membrane. The membrane water absorption test is generally to dry the membrane to a constant weight (W dry state) in a vacuum oven, then soak the membrane in deionized water at room temperature for 24 h and then take it out. Use filter paper to quickly absorb the moisture on the membrane surface. Heavy (W wet state).

$$\text{Water absorption} = \frac{W_{wet} - W_{dry}}{W_{dry}} \times 100\%. \tag{8}$$

3.3 Electrolyte

The electrolyte of the all-vanadium redox flow battery is the charge and discharge reactant of the all-vanadium redox flow battery. The concentration of vanadium ions in the electrolyte and the volume of the electrolyte affect the power and capacity of the battery. There are four valence states of vanadium ions in the electrolyte. As shown in the figure, vanadium ions in adjacent valence states will not undergo redox reactions. Therefore, the vanadium ions in the positive electrode of the all-vanadium redox flow battery are $VO_2{}^+$, VO^{2+}, and the vanadium ions in the negative electrode are V^{3+}, V^{2+}.

The function of the ion exchange membrane is to prevent the positive and negative active materials from mixing and conducting ions to form the internal circuit of the battery. Because the actual ion exchange membrane cannot completely isolate the positive and negative active materials for the time being, there are corresponding ions for vanadium ions of various valences. The diffusion coefficient of vanadium ions of various valences is mainly affected by the temperature of the electrolyte [13]. The temperature of the electrolyte in the stack will constantly change during the charging and discharging process, so the diffusion coefficient of vanadium ions will also change accordingly.

The relationship between ion diffusion coefficient and temperature change is as follows:

$$k = A_0 e^{-\frac{Ea}{RT_{\text{stack}}}}. \tag{9}$$

In the formula, A_0 is the pre-factor; R is the universal gas constant; Ea is the activation energy of the diffusion process.

3.4 Bipolar Plate

The bipolar plate of the all-vanadium redox flow battery mainly plays the role of collecting current, supporting the electrode and blocking the electrolyte. Good electrical conductivity can ensure the bipolar plate to better collect current; the supporting electrode needs to have a certain mechanical stability; because it needs to block the electrolyte, there are also certain requirements for the sealing performance of the bipolar plate. The bipolar plate works in a strong oxidizing environment, so the bipolar plate material is also required to have a certain degree of oxidation resistance and corrosion resistance. Since there must be a certain contact resistance between the electrode and the bipolar plate in contact, in order to better improve the operating efficiency of the battery, the current bipolar plate generally has an integrated electrode-bipolar plate design.

The influence relationship diagram of bipolar plate performance parameters is as follows (Fig. 2):

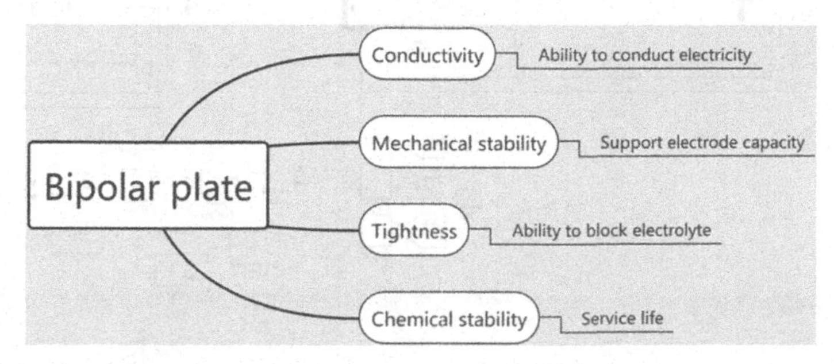

Fig. 2. The influence of bipolar plate material performance parameters.

4 Simulation Implementation and Application

The open model of the all-vanadium redox flow battery can not only realize high-precision simulation calculations, but also meet the requirements of the model's stack-tank separation, while considering the impact mechanism of VRB-related components and materials. In this paper, an open model of an all-vanadium redox flow battery is built in the MATLAB/Simulink environment. The structure is shown in the figure. The key components of VRB, such as electrode, ion exchange membrane, bipolar plate and electrolyte, are used as inputs in the model to simulate the establishment of all vanadium flow battery energy storage system with different requirements (Fig. 3).

All vanadium redox flow batteries are mainly composed of stacks, storage tanks, pipes, circulating pumps, etc. The stacks are mainly composed of electrodes, ion exchange membranes, bipolar plates and other components. The open VRB energy storage system generation module can generate the required type of VRB energy storage system based on the performance parameters of key components such as electrodes, ion exchange membranes, bipolar plates and electrolyte, instead of solely based on battery power and capacity Generate and set the generation interface as shown in the figure (Fig. 4).

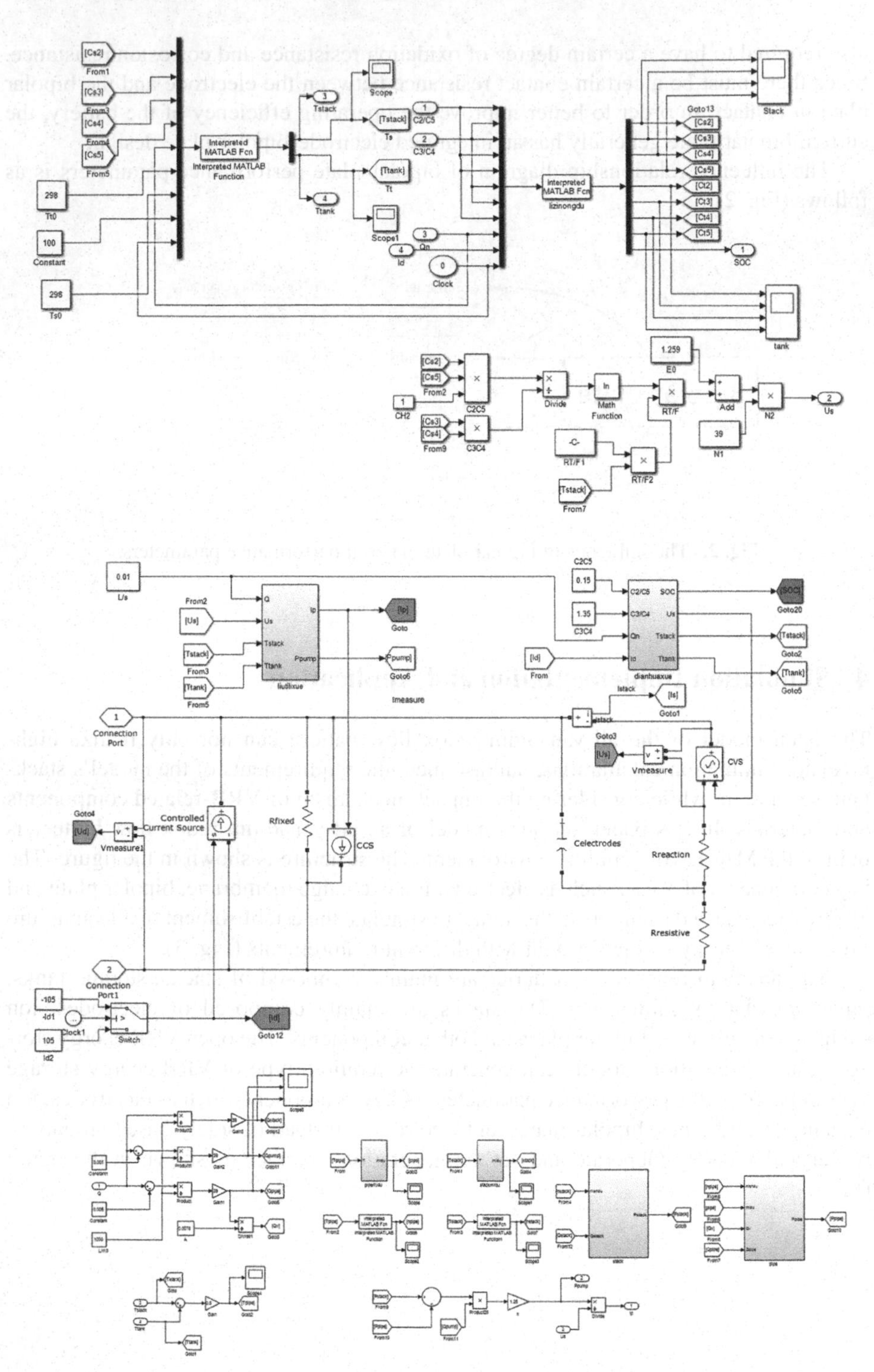

Fig. 3. Simulation diagram of the all-vanadium flow battery model.

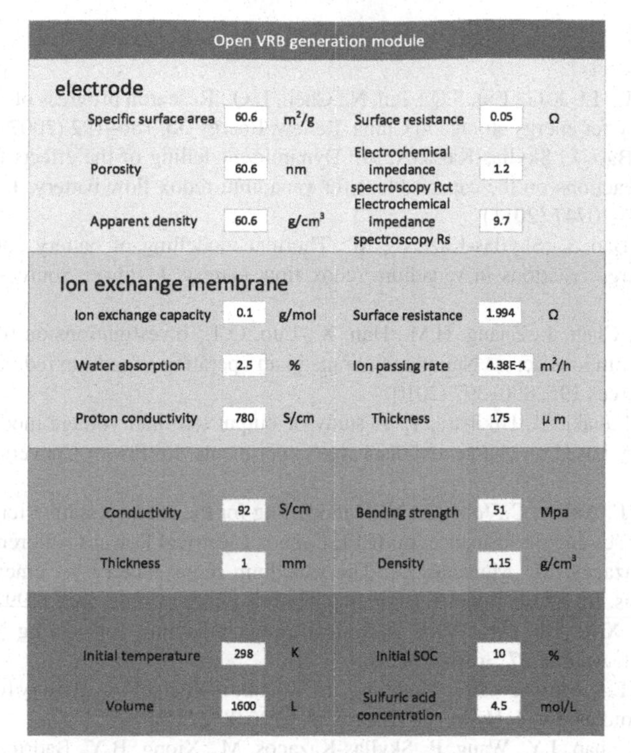

Fig. 4. Model diagram of an open all-vanadium redox flow battery

5 Conclusion

This paper proposes a new open VRB model based on the key component materials of the all-vanadium redox flow battery, which reflects the influence of the parameters of each component on the battery efficiency (performance). Based on the equivalent circuit model with pump loss, an open all-vanadium redox flow battery model is established to reflect the influence of the parameter indicators of the key components of the vanadium redox battery on the battery performance. As the assembly and matching of the various components of the all-vanadium redox flow battery remain at the stage of engineering experience, this paper studies the influence of the key component parameters of the battery on the battery performance, and establishes a parameter based on the key component materials of the all-vanadium redox flow battery. The input open all-vanadium flow battery model has guiding significance for the assembly of the all-vanadium flow battery. The final performance of the battery is known in advance through the intuitive model, which avoids a lot of manpower and material resources in the process of blindly assembling the battery.

Acknowledgement. This work was supported by the National key research and development program No. 2017YFB0903504 and No. JZ2018ZDYF0007.

References

1. Huang, K.L., Li, X.G., Liu, S.Q., Tan, N., Chen, L.Q.: Research progress of vanadium redox flow battery for energy storage in China. Renew. Energy **33**, 186–192 (2007)
2. Tang, A., Bao, J., Skyllas-Kazacos, M.: Dynamic modelling of the effects of ion diffusion and side reactions on the capacity loss for vanadium redox flow battery. J. Power Sources **196**, 10737–10747 (2011)
3. Tang, A., Bao, J., Skyllas-Kazacos, M.: Thermal modelling of battery configuration and self-discharge reactions in vanadium redox flow battery. J. Power Sources **216**, 489–501 (2012)
4. Sun, C.X., Chen, J., Zhang, H.M., Han, X., Luo, Q.T.: Investigations on transfer of water and vanadium ions across Nafion membrane in an operating vanadium redox flow battery. J. Power Sources **195**, 890–897 (2010)
5. Li, H.M., Funaki, T., Hikihara, T.: A study of output terminal voltage modeling for redox flow battery based on charge and discharge experiments. In: Power Conversion Conference (2007)
6. Chawhan, J., Abbey, C., Joos, G.: VRB modelling for the study of output terminal voltages, internal losses and performance. In: IEEE Canada Electrical Power Conference (2007)
7. Skyllas-Kazacos, M., Menictas, C.: The vanadium redox battery for emergency back-up applications. In: International Telecommunications Energy Conference (1997)
8. Yi, L., Li, X.R., Hu, L.H.: VRB equivalent circuit modeling considering ion diffusion. J. Power Syst. Autom. **27**, 36–41 (2015)
9. Ontiveros, L.J., Mercado, P.E.: Modeling of a Vanadium Redox Flow Battery for power system dynamic studies. Int. J. Hydrogen Energy **39**, 8720–8727 (2014)
10. Zhang, Y., Zhao, J.Y., Wang, P., Skyllas-Kazacos, M., Xiong, B.Y., Badrinarayanan, R.: A comprehensive equivalent circuit model of all-vanadium redox flow battery for power system analysis. J. Power Sources **290**, 14–24 (2015)
11. Rao, P., Jayanti, S.: Influence of electrode design parameters on the performance of vanadium redox flow battery cells at low temperatures. J. Power Sources **482**, 14–24 (2015)
12. Li, Y.F., Bao, J., Skyllas-Kazacos, M., Akter, M.P., Zhang, X.N., Fletcher, J.: Studies on dynamic responses and impedance of the vanadium redox flow. Appl. Energy. **237**, 91–102 (2019)
13. Chen, Z., Ding, M., Su, J.H.: Modeling and control for large capacity battery energy storage system. In: 4th International Conference on Electric Utility Deregulation and Restructuring and Power Technologies, pp. 1429–1436. IEEE Press, New York (2011)

Simulating Energy-Saving and Consuming Behaviours in the Design and Manufacturing Process with Adjacent Networks

Bin Ren[1], Lin Wang[1], Xiaoyu Wang[1], and Jiayu Chen[2](✉)

[1] Shanghai Key Laboratory of Intelligent Manufacturing and Robotics, School of Mechatronic Engineering and Automation, Shanghai University, Shanghai 200444, China
binren@i.shu.edu.cn

[2] Department of Architecture and Civil Engineering, City University of Hong Kong, Hong Kong 999077, China
jiaychen@cityu.edu.hk

Abstract. At present, the manufacturing industry's energy consumption accounts for a large proportion of industrial energy consumption. Most of the research is on the manufacturing industry's energy consumption from the aspects of systems and equipment. However, people are an important factor affecting the manufacturing industry's energy consumption, and people's behaviour can be transmitted through the relationship network. Aiming at the energy-saving and consumption-reducing behaviour in manufacturing design and manufacturing process, a simulation algorithm of personnel energy-saving decision-making considering adjacency relation is proposed. The algorithm considers the influencing factors of people's behaviour decision. Based on adjacency relation, the energy consumption model of a random adjacency network is established. The adjacency network propagation of energy-saving behaviour is studied and verified by actual data. The suggestions of energy saving and consumption reduction considering the adjacency relationship are obtained, which is beneficial to reduce the manufacturing industry's energy consumption.

Keywords: Adjacency network · Agent-based · Simulation study · Energy-saving behavior

1 Introduction

With the development of society, China's energy consumption will continue to increase greatly. With the development of new energy technologies and the spread of energy-saving ethos, the total industrial energy consumption has fluctuated and increased. Its proportion to the total national energy consumption has decreased from 71.84% to 65.93%. The manufacturing industry's total energy consumption accounts for about 83% of the total industrial energy consumption. In the research on reducing energy consumption in the manufacturing industry, Hou [1] thinks robot technology in intelligent manufacturing is beneficial to reduce related energy consumption. He studies reducing

K. Li et al. (Eds.): LSMS 2021/ICSEE 2021, CCIS 1468, pp. 441–451, 2021.
https://doi.org/10.1007/978-981-16-7210-1_42

energy consumption and improving the energy efficiency of industrial robots. Chu [2] points out that due to the rising labour cost, the importance of cross-training workers distribution for building an efficient and flexible cellular manufacturing system increased. They proposed a cross-training model with learning and forgetting effects.

Like most functional energy consumption, the operator's behaviour affects the manufacturing workshop's energy consumption. The operator at each station can control the on/off, use mode, and mode of the equipment to affect the production line's energy consumption. Gong [3] thinks that workers are still an indispensable factor in the manufacturing workshop and puts forward a multi-objective optimization model considering employee behaviour, which can reduce the energy cost by scheduling job processing, machine idle mode and labour. The research on increasing energy-saving by changing people's behaviour pays attention to the social network's role [4, 5], including the social-psychological relationship among people [6]. People's behaviour is strongly influenced by people around them, especially when they have a strong relationship network [7]. There is a strong positive correlation between other people's energy-saving behaviour and personal energy-saving behaviour [8]. However, at present, there is insufficient research on the application of individual behaviours in manufacturing.

With the deepening of industrial energy consumption research, researchers continue to study the research scope and subdivision of energy consumption reduction in the manufacturing industry. However, the research on interpersonal relationship and energy-saving behaviour communication is still at an early stage, and the research in this direction still needs to be deepened. Considering the influence of network adjacency, this paper simulates the behaviour of energy saving and consumption reduction in the process of design and manufacture. The decision algorithm and agent-based simulation model of each node in the adjacent network are established. By collecting and analyzing the energy consumption data of the design and manufacturing laboratory of Shanghai University, the validity of the simulation model is basically verified.

2 Decision Algorithm of Energy-Saving and Consumption Reduction Behaviour in a Network Adjacency Relationship

In industrial production, people's personal behaviour greatly influences the energy use of the process. In the research field of personal behaviour and energy consumption, some studies have confirmed that individuals can save energy through behaviour changes, and points out the reasons and motives for individuals to take energy-saving actions [9, 10]. After the energy-saving behaviour meets the conditions of easy and convenient implementation, resources and technology permission and friends and neighbours take actions to protect, the possibility of personal behaviour being affected and changed is greater than a single influencing factor [11]. However, some studies have pointed out that even if some behaviours are highly cost-effective, some people have not adopted the currently available energy-saving behaviours [12]. It is possible to resist energy-saving effects. Therefore, this article considers three behavioural states, which are susceptible, energy-saving and resistant.

Agent-based modelling is a dynamic modelling method for complex systems and complex adaptive systems. It also includes a behaviour model (human or others), which is used to observe the collective effect of agent behaviour and interaction [13]. It can simulate the information exchange between individuals in the network and how this exchange causes behaviour changes, which is especially suitable for reflecting the relationship between human behaviour and factors affecting human behaviour in the environment [14, 15]. In the agent-based simulation [16], the environment of the agent individual is its peer-to-peer network. If the environment changes, the change's feedback will also change the node's energy consumption behaviour. This paper establishes a model, assuming that the energy consumption behaviour of an agent can be affected by the energy consumption behaviour of surrounding agents in varying degrees of probability. The environmental changes will spread through the network and thus have an impact on the whole simulation. Meanwhile, the environment influences the agent's energy consumption behaviour, while the individual influences the overall energy consumption pattern.

3 Simulation of Energy-Saving and Consumption Reduction Behaviour in Design and Manufacturing Process

3.1 Energy Consumption in Design and Manufacturing Process

The energy consumption data of the design and manufacturing process used in this paper is collected from the design and manufacturing laboratory of Shanghai University in April 2021. A simulation model is constructed according to the laboratory's actual situation to simulate the departmental energy consumption of the CRRC Group. The collected data of this experiment can verify the effectiveness of the model developed in this paper.

The design and manufacturing laboratory has seven stations corresponding to different design and manufacturing processes: overall design, 3D modelling, simulation experiment verification, sample trial production, physical experiment, design product inspection and batch manufacturing. Its production equipment is mainly computers and 3D printers, and its production tools are designed drawing software, modelling software, simulation experiment software, etc. To meet the production capacity requirements, each working procedure has different minimum energy consumption. The personnel relationship and location of the design and manufacturing laboratory are shown in Fig. 1.

In this paper, a smart plug is used to obtain each station's energy consumption information and establish a simulation model. Compared with other electricity metering equipment such as smart meters, the smart plug has the characteristics of simple installation, convenient use, real-time display, historical electricity review, control switch and so on, and the error rate is low. Some researchers have used smart plug as the data and control equipment of energy management system and assisted in establishing energy management model through smart plug [17].

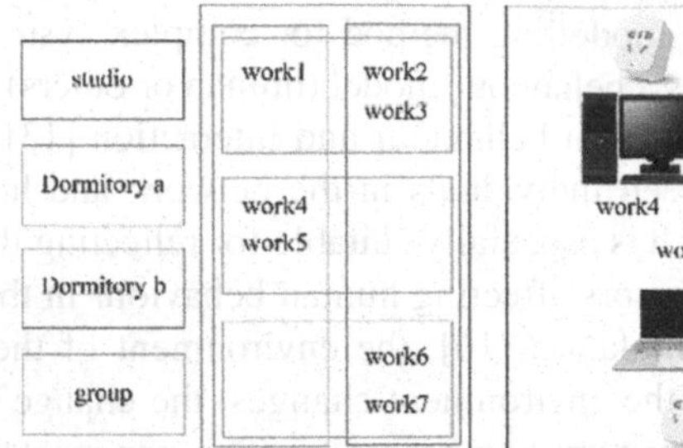

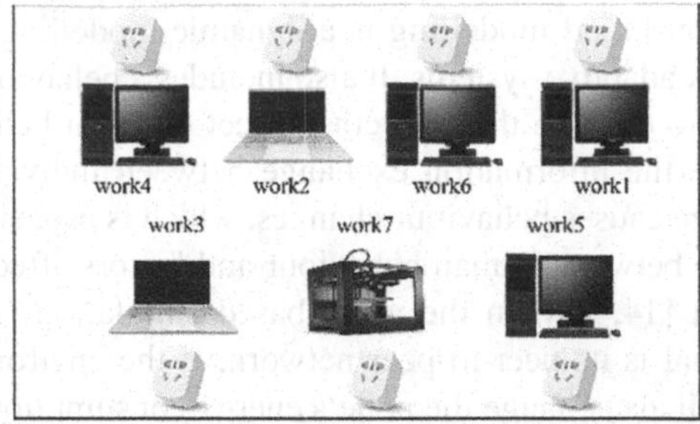

Fig. 1. Induvial interacting network and location diagram of design and manufacturing laboratory.

3.2 Establishment of Network Adjacency Model

Study and establish a simulation model of energy consumption decision based on interpersonal network. The model demonstrates the process of energy-saving behaviour spreading through the interpersonal network, and each node represents a worker. The initial setting of personnel behaviour is daily energy consumption behaviour. The model is set to blue, and those who take energy-saving behaviour to reduce energy consumption are set to green. It can be seen from Fig. 2.

Nodes have the probability (Gain Resistance Rate) to restore the original energy consumption level after a period of energy-saving behaviour and take resistance behaviour when they are no longer affected by energy-saving behaviour. So they will not spread through the network and will not take energy-saving behaviour. The model sets this kind of nodes as grey. The initial setting of the model is to generate a random basic network:

1. Generate n nodes accord to N and giving that random position of each node by using random numbers to obtain a random distribution map of the unconnected node;
2. According to formula 1, the total number of connections is calculated. The nodes with the number of connections less than A are randomly selected and connected to the nearest unconnected nodes until the number of connections between all nodes is num-links. The adjacent node information of the nodes is obtained so that the mean value of each node's adjacency relationship is A.
3. According to I, a corresponding number of nodes are randomly selected, and their status is changed to energy-saving status.
4. Set the check-timer according to E, which is set randomly by E.

$$\text{Num\_links} = A/N. \tag{1}$$

The operation of the network algorithm is mainly divided into the following steps:

1. Initialization is the model initialization design introduced above, for each agent (Onij), obtain its adjacent node (Cnij), and make contact, which is used to establish the model simulation background.

2. Simulation, that is, in Tick < T (ALL Onij(Bsij)! = SAVE), each agent makes behaviour selection, that is, state transition, according to individual decision algorithm, and records the proportion of nodes in each energy consumption state in each Tick.
3. At the end of the simulation, Tick = T (Onij (Bsij)! = SAVE) (no nodes are energy-saving); that is, there is no interaction of energy-saving behaviours, so the simulation is finished. The node proportion of the final energy consumption state is recorded, and the proportion change diagram in the simulation process is drawn.

When the model works, the normal nodes' energy consumption status is normal by default. According to the increase of time steps, every worker who adopts energy-saving behaviour will try to influence the connected workers through the relationship network, and the affected workers will be affected and adopt energy-saving behaviour with the probability given by the Behavioural Spread Chance control. Due to the resistance behaviour, the resistant nodes will not be affected by the energy-saving behaviour.

The feedback design enables all kinds of nodes to have an understanding of their own energy consumption. People in energy-saving behaviour end their energy-saving behaviour with the probability given by Recovery Rate control and take resistance behaviour with the probability given by Gain Resistance Rate.

Setting the Initial Energy-saving Scale is an important step to start the energy-saving model. When some nodes are arranged for energy-saving behaviour, energy-saving behaviour will be generated and spread through the network. The setting of the initial energy-saving scale can also simulate the change of the group's energy-saving behaviour after some people in the group are required to save energy.

This paper studies the design of interpersonal network based on distance and contact number. The network randomly arranges node positions, and sets them based on the proximity between nodes (Euclidean distance), selects a node, connects it to the nearest but not yet connected node, and repeats it until the network has the correct number of links to give the specified average node degree. The Initial Energy-saving Scale slider determines how many nodes' initial behaviour will be energy-saving behaviour.

4 Simulation Results and Verification

4.1 Analog Result

In a single simulation, each energy consumption behaviour state and simulation end state are shown in Fig. 2. In the simulation, each agent will make decisions according to the random network and environment and the decision algorithm.

To detect each parameter's influence on an agent in the network and the fluctuation of energy consumption, the model is used to simulate 36 scheme which settings using three types of N, two types of A, two types of I, and three types of E.

In the simulation process, the system can obtain each node's behaviour between networks and the number of nodes in each state. Suppose it is assumed that a single node is in an energy-saving state, and the energy consumption is reduced by one unit in each TICK. In that case, the data results such as the maximum proportion of energy-saving state, the proportion of final resistant nodes, the proportion of final susceptible nodes,

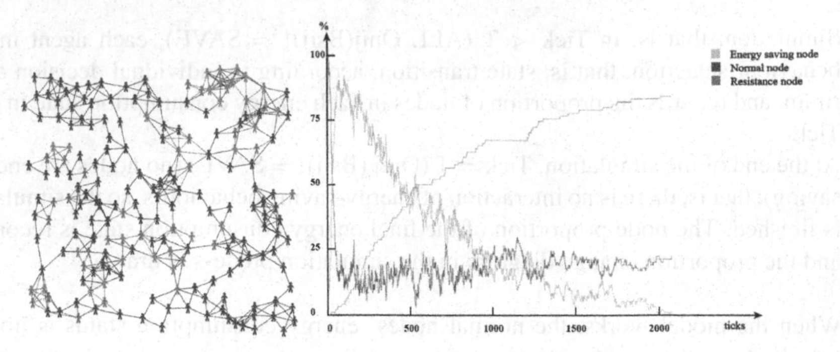

Fig. 2. Simulated network energy consumptions.

the simulation time and the total energy consumption reduction can be obtained for each simulation run under different schemes. They are taking schemes 1, 2, 4, 13 and 25 as examples, as shown in Table 1. The simulation results of each scheme for 50 times are shown in Fig. 3.

Table 1. Simulated operational scenairos.

Plan	N	A	I	B %	E ticks	R %	G %
1	100	5	2	2.5	1	3	5
2	100	5	2	2.5	2	3	5
4	100	5	10	2.5	1	3	5
13	150	5	2	2.5	1	3	5
25	200	5	2	2.5	1	3	5

4.2 Analysis of Simulation Results

Due to the appearance of resistance behaviour, the network connection degree of re-infection after each energy-saving behaviour disappears is reduced. The energy-saving behaviour disappears after a certain time. Due to the different parameter design, different proportions of nodes in the resistance state and normal energy consumption state remain at the simulation. The following five conclusions can be drawn from the specific simulation data:

1. By comparing the changes of four parameters (N, A, I and E) and their influences on the final results, it can be seen from Fig. 3 that the results of scheme 2 are better. That is to say, under the premise that the recovery of resistance behaviour and energy-saving behaviour exists, the higher the feedback frequency is, the better.
2. In Fig. 3, individual points fall to near 0, which means that it is observed that the energy-saving influence with a small probability ends quickly during the simulation

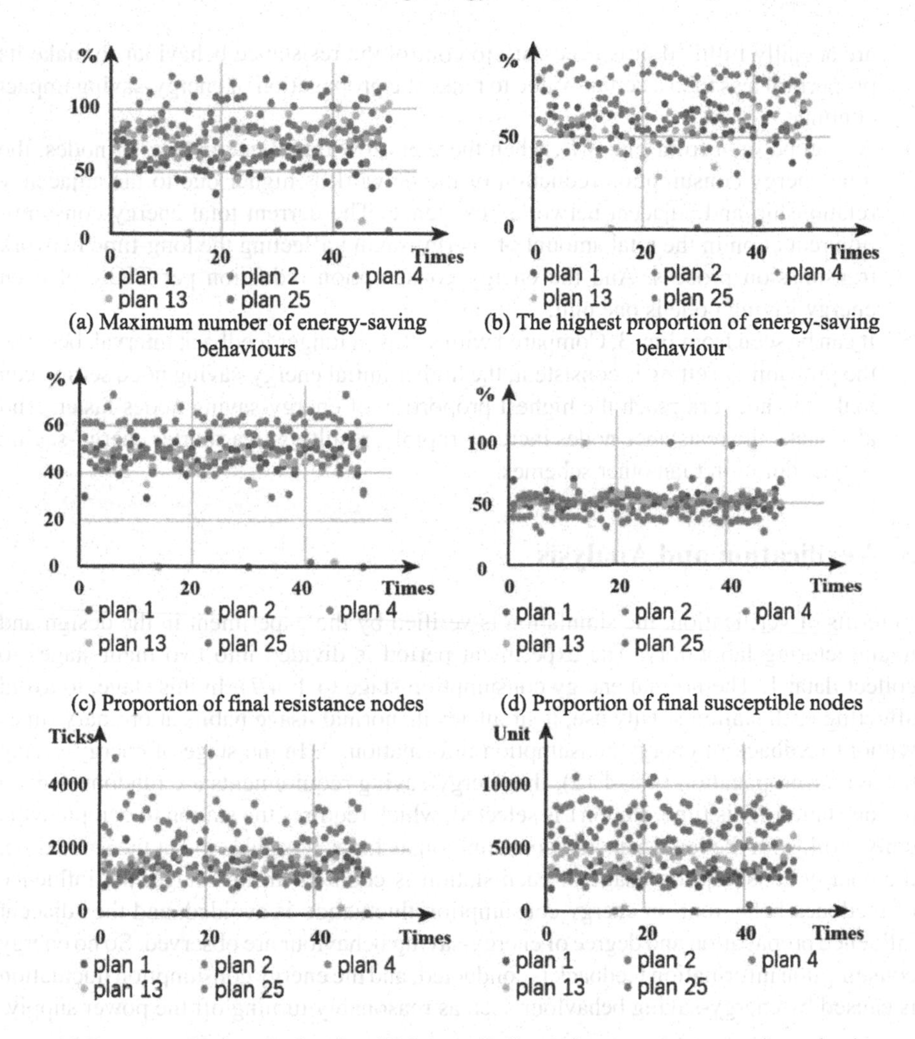

(a) Maximum number of energy-saving behaviours

(b) The highest proportion of energy-saving behaviour

(c) Proportion of final resistance nodes

(d) Proportion of final susceptible nodes

(e) Energy-saving impact duration (Ticks)

(f) Simulation energy consumption reduction

Fig. 3. Comparison of simulation results.

operation. The main reasons leading to its end are as follows: the number of initial energy-saving nodes is small, the number of adjacent nodes is small, the probability of energy-saving behaviour propagation is low, and the probability of resisting behaviour recovering energy consumption behaviour is high. We can further improve the probability of large-scale diffusion of energy-saving behaviour from these five aspects.

3. At the end of the simulation operation, except for scheme 2, as shown in Fig. 3(c), the ratio of resistance nodes in the other four schemes is about 50%. When the node connection rate is 5, and the proportion of resistance nodes is about 50%, the propagation ability of neighbouring networks decreases greatly, and the propagation of energy-saving behaviour is terminated. When the energy-saving requirements

are actually fulfilled, it is necessary to control the resistance behaviour to make its proportion less than a certain value to make the propagation of energy-saving impact continue.

4. As can be seen from Fig. 3(f), when there are few initial energy-saving nodes, the total energy consumption reduction of the network is higher due to the adjacency relationship and adjacent networks' existence. The current total energy consumption reduction in the total amount of energy-saving affecting the long-time network transmission process. And the energy consumption reduction per TICK of each energy-saving node is one unit.
5. It can be seen from Fig. 3, Compared with setting a longer feedback interval, because the probability setting is consistent, the higher initial energy-saving node setting can make the network reach the highest proportion of energy-saving nodes faster. And also make the resistance nodes increase rapidly, resulting in a shorter energy-saving impact duration than other schemes.

5 Verification and Analysis

In terms of verification, the simulation is verified by the experiment in the design and manufacturing laboratory. The experiment period is divided into two main stages to collect data: 1. The normal energy consumption stage (4.1–4.7). In this stage, to avoid affecting each station's daily use, it simulates its normal usage habits at ordinary times without feedback of energy consumption information. 2. In the stage of energy-saving behaviour propagation (4.8–4.12), the energy-saving requirements are randomly given to one station. This time, station1 is selected, which requires the station to complete the daily work with the lowest energy consumption as far as possible, and at the same time, the energy consumption usage of each station is counted. In this stage, the influence of feedback behaviour on energy consumption fluctuation is avoided, and the adjacent influence propagation and degree of energy-saving behaviour are observed. So no energy consumption information feedback is conducted, and the energy consumption fluctuation is caused by energy-saving behaviour such as reasonably turning off the power supply.

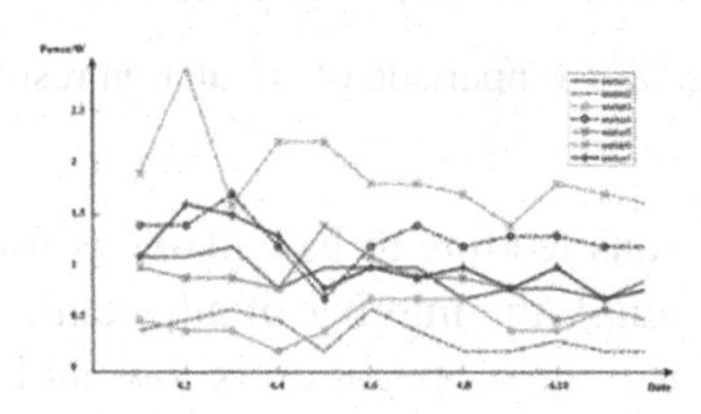

Fig. 4. Trend chart of energy consumption in manufacturing design laboratory.

The results of experimental data are shown in Fig. 4, in which the ordinate is power, the unit of power is Watt, and the abscissa is Date. Due to experimenters' different work contents and production equipment, the daily minimum energy consumption and fluctuation of energy consumption are also different. The work contents and production tools of the laboratory personnel are shown in Table 2.

Table 2. Simulation scenarios.

Worker	Job	Tool(hardware)
Worker1	Overall design	Computer
Worker2	Physical test	Notebook
Worker3	Examination and test of products	Notebook
Worker4	3D modelling	Computer
Worker5	Simulation verification	Computer
Worker6	Simulation operation	Computer
Worker7	Sample making and finished product manufacturing	3D printers

The data in Fig. 4 shows that since April 8th, when the energy-saving requirements were sent to work1, the energy consumption of Workstation 1 decreased and fluctuated at lower energy consumption. On that day, the energy consumption of station 2 and station 6 decreased slightly, in which the adjacency of distance should influence station 6, and the working group influenced station 2. On April 9th, the energy consumption of stations 3 and 7 decreased, and the influencing factors included the adjacent relationship of stations 1, 2 and 6. On April 10th, the energy consumption of station 5 decreased and fluctuated at the lower energy consumption. Due to the task requirements, each station's minimum energy consumption is different due to different tasks. However, after the energy-saving requirements for station1 are issued, the energy-saving behaviour spreads along with the adjacent network, causing the affected stations to have different degrees of energy consumption decline and fluctuate at their low energy consumption. Moreover, the transmission of energy-saving influence is related to the adjacent network, and the data plays a supporting role in the energy-saving behaviour propagation of the adjacent relationship.

In this study, the distance and the average number of node connections are taken as the main factors in the interpersonal network setting. The design mode is mainly related to the process interval of personnel, the energy of making friends and the number of interpersonal contacts. In addition, the current data volume and experimental duration of the experimental data are slightly insufficient, which can roughly verify the energy-saving behaviour transfer mode and needs to obtain more accurate verification through the expansion of data nodes and continuous experimental data statistics. In future research, considering the multi-dimensional connection of adjacent networks and the agent's different behaviour probabilities with the network background.

6 Conclusion

In this paper, a random adjacency network is established, and a simulation algorithm is designed considering the adjacency network. Considering the influencing factors of human decision, the decision simulation algorithm of each node is designed. The algorithm is designed for the decision-making of nodes in adjacent networks and provides algorithm support for the simulation of adjacent networks.

Considering the adjacency of the network, the energy-saving and consumption reduction behaviours in the design and manufacturing process are simulated. The decision algorithm and agent-based simulation model of each node in the adjacent network are established. Because large-scale interpersonal adjacency investigation is difficult, agent-based simulation can better simulate the situation of a large-scale personnel network. It can quantitatively express behaviour patterns, behaviour decisions and specific energy consumption data, and carry out multiple random simulations, thus obtaining relatively practical conclusions. It can provide reference and basis for energy-saving-related decision-making and design and can also verify energy-saving schemes.

By collecting and analyzing the energy consumption data of the design and manufacturing laboratory of Shanghai University, the validity of the simulation model is basically verified. It needs to be further verified by larger actual adjacent network data. In this adjacency model, randomly distributed nodes are composed of adjacency networks based on distance. The complexity of the network can be further improved, which can be applied to the energy consumption requirements of manufacturing adjacent networks under different backgrounds.

Acknowledgements. This research was funded by the National Natural Science Foundation of China (Grant No. 51775325; Funder ID: https://doi.org/10.13039/501100001809); National Key Research and Development Program of China (Grant No. 2018YFB1309200); Young Eastern Scholars Program of Shanghai (Grant No. QD2016033); Hong Kong Scholars Program of China (Grant No. XJ2013015).

References

1. Hou, Q., Yang, D., Guo, S.: Review on energy consumption optimization methods of industrial robots. J. Comput. Eng. Appl. **54**, 1–9 (2018)
2. Chu, X., Gao, D., Cheng, S., et al.: Worker assignment with learning-forgetting effect in cellular manufacturing system using adaptive memetic differential search algorithm. J. Comput. Ind. Eng. **136**, 381–396 (2019)
3. Gong, X., Liu, Y., Lohse, N., et al.: Energy and labor-aware production scheduling for industrial demand response using adaptive multiobjective memetic algorithm. IEEE Trans. Ind. Inf. **15**, 942–953 (2019)
4. Centola, D.: The spread of behavior in an online social network experiment. J. Sci. **329**, 1194–1197 (2010)
5. Stadtfeld, C., Snijders, T.A.B., Steglich, C., et al.: Statistical power in longitudinal network studies. J. Sociol. Methods Res. **49**, 1103–1132 (2020)
6. Meynhardt, T., Brieger, S.A., Hermann, C.: Organizational public value and employee life satisfaction: the mediating roles of work engagement and organizational citizenship behavior. Int. J. Hum. Resour. Manage. **31**, 1560–1593 (2020)
7. Margolis, J.A., Dust, S.B.: It's all relative: a team-based social comparison model for self-evaluations of effectiveness. J. Group Organ. Manage. **44**, 361–395 (2019)
8. Gockeritz, S., Schultz, P.W., Rendon, T., et al.: Descriptive normative beliefs and conservation behavior: the moderating roles of personal involvement and injunctive normative beliefs. J. Eur. J. Soci. Psychol. **40**, 514–523 (2010)
9. Zhang, Y.X., Wang, Z.H., Zhou, G.H.: Antecedents of employee electricity saving behavior in organizations: an empirical study based on norm activation model. J. Energy Policy. **62**, 1120–1127 (2013)

10. Tiefenbeck, V., Staake, T., Roth, K., et al.: For better or for worse? empirical evidence of moral licensing in a behavioral energy conservation campaign. J. Energy Policy **57**, 160–171 (2013)
11. Wang, B., Wang, X.M., Guo, D.X., et al.: Analysis of factors influencing residents' habitual energy-saving behaviour based on NAM and TPB models: egoism or altruism. J. Energy Policy. **116**, 68–77 (2018)
12. Chen, J.Y., Taylor, J.E., Wei, H.H.: Modeling building occupant network energy consumption decision-making: the interplay between network structure and conservation. J. Ener. Buildings **47**, 515–524 (2012)
13. Macal, C.M., North, M.J.: Tutorial on agent-based modelling and simulation. J. Simul. **4**, 151–162 (2010)
14. Mathias, M.: Agent-based systems as a tool of sociological research. J. Sociologia. **48**, 25–47 (2016)
15. Fonoberova, M., Fonoberov, V.A., Mezic, I.: Global sensitivity/uncertainty analysis for agent-based models. J. Reliabil. Eng. Syst. Saf. **118**, 8–17 (2013)
16. Center for Connected Learning and Computer-Based Modeling. http://ccl.northwestern.edu/netlogo/models/VirusonaNetwork
17. Ulloa, F., Garciasantander, L., Carrizo, D., et al.: Towards a home energy management model through a coordinator of smart sockets. Latvian J. Phys. Tech. Sci. **55**, 35–43 (2018)

Optimal Defense Resource Allocation and Geographically Feasible Hexagonal Topology Construction for Power Grid Security

Yifa Liu[1,2] and Long Cheng[1,2](✉)

[1] Institute of Automation, Chinese Academy of Sciences, Beijing 100190, China
long.cheng@ia.ac.cn

[2] School of Artificial Intelligence, University of Chinese Academy of Sciences, Beijing 100049, China

Abstract. Power system faces thousands of physical and cyber attacks which seriously threaten its security. It is noted that most defense methods are only suitable for specific cyber attacks and are not applicable to physical attacks. This paper provides a generic method regardless of different attack types through topological efforts to reduce potential loss of the power grid. In this paper, a proportional loss model is proposed depending on the different attack-defense resource allocations. The optimal allocation strategy can be converted into the solution to a min max problem. In order to further improve the security of the power grid, by taking the geographical feasibility into consideration, a hexagonal construction method is proposed to provide a cost-affordable and geographically-feasible solution for new power grid construction.

1 Introduction

With the fast development of technology and the advancement of infrastructure processes, demands for electricity quality and quantity have increased significantly. Adopting modern sensing, control, measurement technologies, the modern smart power grid has been proposed to satisfy the huge requirement of electrical power [1]. As network complexity increases [2], there are numerous elements in the power system to be protected [3,4]. From the perspective of the attack method, the power system is susceptible to a variety of physical [5] and cyber attacks, such as denial-of-service (DoS) attack [6,7] and false data injection (FDI) attack [8].

There are also plenty of specific defense methods for certain attack in the literature. A data verification method against FDI was introduced in [9]. Countermeasure for the case where the attacker has limited knowledge was also proposed in [10]. A two-layer game theoretical model for FDI attacks against power systems was given in [11]. A list of defense mechanisms against FDI attacks focused on certain devices [12–15]. Analysis of defense against price attacks using game

K. Li et al. (Eds.): LSMS 2021/ICSEE 2021, CCIS 1468, pp. 452–462, 2021.
https://doi.org/10.1007/978-981-16-7210-1_43

theory was given in [16]. An attack-resilient controller and an attack detection mechanism for price attacks were proposed in [17]. Stochastic games were used for fighting against coordinated cyber-physical attacks on power grids [18]. By adding wireless communication, an efficient way of optimizing topology of a wired networked system was proposed in [19]. There are also other methods of reducing damage, for example, a load redistribution way was proposed to reduce the impact on the load after attacks in [20]. It is noted that most papers regarding power grid security focus on cyber attacks and the defense methods can only handle a specific attack. Therefore, a natural question is whether we can find a generic defense way to protect the power grid, or at least reduce the impact of attacks.

In this paper, the power grid is modeled by a graph topology, and each section is treated as a node. The value of each node is equivalent to the total value of users affected by the crash of this node. Then a proportional model is proposed to analyze the probability of a node being crashed depending on different attack-defense resource allocations. For lack of information about the time and location of potential attacks, the defender should adopt a conservative strategy. Then this optimal resource allocation can be converted into a min max problem. It can be found that the total loss function is dependent not only on the resource allocation, but also on the topology of the power grid.

In order to further improve the security of the power grid, the function of topology is taken into account. Adding redundant connections can reduce the harm when some nodes are crashed, which does not conflict with many specific defense methods and can be used with them simultaneously. Moreover, considering the construction feasibility of power grids in reality, a geography-based based hexagonal city planning topology construction method is proposed to provide a novel power grid construction solution.

2 Optimal Resource Allocation Strategies

2.1 Structure of Power Grid

The structure of power grid includes power plants, substations with different capacities, regional dispatching systems at different levels and power transmission lines to users as shown in Fig. 1. Then the power grid can be described by a graph topology, where each vulnerable section is treated as a node. This paper assumes that the direction to the end-users is positive and considers a power grid topology consisting of Q nodes.

2.2 Risk Function

Assumption 1. *The attack resources are quite few compared to the defense resources.*

The attack resources can include personnel or hackers hired to launch the attack, some technological resources such as advanced tools or malwares for

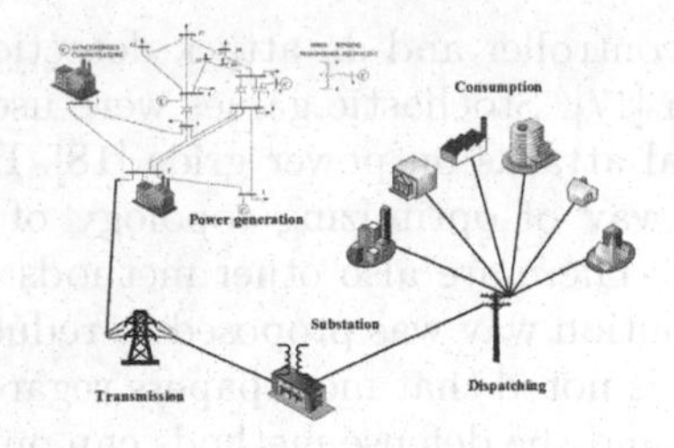

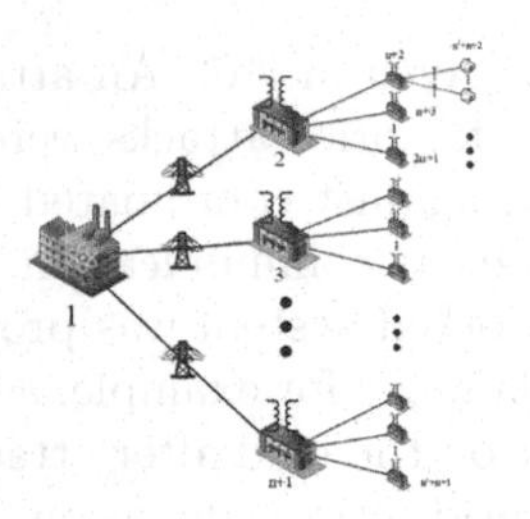

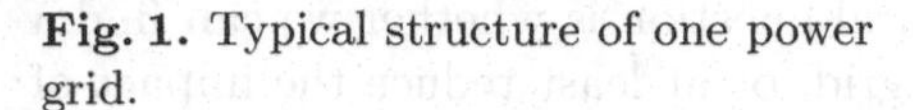

Fig. 1. Typical structure of one power grid.

Fig. 2. Smart grid with a n-ary tree topology

malicious actions, and other economic resources. Similarly, the defense resources can include personnel to reinforce and repair the grid, technological resources such as efficient and effective suites of security tools or softwares to maintain the normal operation of the grid, and other economic resources. If more attack resources are allocated to the node, this node is more likely to be crashed, that is, the node can no longer perform normal work, and vice versa.

Although there have been many serious blackouts to alert humans to the importance of power grid security, and many of them have caused incalculable loss. However, it should still be known that accidents are rare, and most of the power grids are working properly, even if they often face various attacks.

Due to the law of diminishing marginal utility, with the more defense resources already allocated, adding extra defense resources produces fewer effects. By Assumption 1, the attack resources allocated to a single node are small that to add more attack resources does not reduce the value of unit resources, hence, the attack effect can be regarded as a linear function of quantity. Then the probability of node i being crashed is modeled by

$$p_i = \frac{a_i}{d_i}. \tag{1}$$

where p_i denotes the probability of node i being crashed, $0 \leq a_i \leq A$ and $A < d_i < D$ denote the attack and defense resource allocated to node i, $d_i \geq a_i$. $A = \sum_{k=1}^{N} a_k$ and $D = \sum_{k=1}^{N} d_k$, $D \gg A$ denote the total attack and defense resource, respectively.

Remark 1. The attacks discussed in this paper are primarily those that target a single node. This analysis is not applicable to large-scale chain reactions, such as infectious viruses. Unless it can be quickly blocked so that it affects only the area where it occurred.

Assumption 2. *Nodes are independent of each other.*

As long as nodes are crashed, loss is caused. If damaged nodes are independent of each other, then the expectation of total loss of power grid can be expressed by

$$E(L) = \sum_{k=1}^{Q} p_k v_k = \sum_{k=1}^{Q} \frac{a_k}{d_k} v_k. \tag{2}$$

Assumption 3. *The defender would take conservative strategies since it has no information from the attack side.*

Denote $s_a = [a_1\ a_2\ \cdots\ a_N]^T$ and $s_d = [d_1\ d_2\ \cdots\ d_N]^T$ as the strategies of the attacker and defender respectively.

The goal of the attacker is to maximize the total loss by choosing the optimal attack strategy s_a^*, while the task of the defender is to adopt the optimal defense strategy s_d^* and minimize the total loss. Many studies simulated offense-defense scenarios by using game theory [11,13,16], differential game models [12] or multi-agent system [21]. However, neither the attacker nor the defender has global information, which means that the classic 2-player game where both players have perfect information of the payoff matrix and system states is not applicable.

Under Assumption 3, the defender should take the following conservative strategy to avoid the risk of disastrous loss,

$$s_d^* = \underset{s_d}{\operatorname{argmin}} \max_{s_a} E(L(s_a, s_d)) = \underset{s_d}{\operatorname{argmin}} \max_{s_a} \sum_{i=1}^{Q} \frac{a_i}{d_i} v_i, \tag{3}$$

$$s_a^* = \underset{s_a}{\operatorname{argmax}}\, E(L(s_a, s_d^*)). \tag{4}$$

Then the total loss under the optimal attack-defense resource allocation is

$$E(L(s_a^*, s_d^*)) = \min_{s_d} \max_{s_a} \sum_{i=1}^{Q} \frac{a_i}{d_i} v_i = \sum_{i=1}^{Q} \frac{a_i^*}{d_i^*} v_i. \tag{5}$$

2.3 Optimal Resource Allocation Strategy

Considering the aciculate variation $\Delta_{ij} = \delta(\mathbf{e}_i - \mathbf{e}_j)$. Then the difference of any two strategies can be expressed by a linear combination of aciculate variations.

First relax restrictions of $0 \le a_i \le A$ and $A < d_i < D$ to be $a_i \in \mathbb{R}$ and $d_i \in \mathbb{R}$. For any $i, j, i \ne j$, $E(L(s_a^* + \Delta_{ij}, s_d^*)) \le E(L(s_a^*, s_d^*))$, conversely $E(L(s_a^* - \Delta_{ij}, s_d^*)) \le E(L(s_a^*, s_d^*))$. Then it can be obtained that

$$\frac{\delta}{d_i^*} v_i - \frac{\delta}{d_j^*} v_j \le 0, \quad -\frac{\delta}{d_i^*} v_i + \frac{\delta}{d_j^*} v_j \le 0. \tag{6}$$

Therefore, $\delta v_i / d_i^* - \delta v_j / d_j^*$ must be 0.

It can be verified that this conclusion also holds under restrictions $0 \le a_i \le A$ and $A < d_i < D$. Then the optimal defense strategy should satisfy the following condition

$$\frac{d_i^*}{d_j^*} = \frac{v_i}{v_j}, \quad d_i^* = \frac{v_i}{\sum_{k=1}^{Q} v_k} D > 0. \tag{7}$$

It can be found that the amount of defense resources allocated to each node should be proportional to the value of that node. Furthermore, the loss function under the optimal strategies of both sides is shown as follows

$$E(L(s_a^*, s_d^*)) = \frac{A}{D} \sum_{k=1}^{Q} v_k. \tag{8}$$

It is obvious that s_a^* can be any feasible solution, which means s_d^* is sufficiently robust to deal with all attack strategies.

Equation (8) reveals the fact that the loss suffered by the power grid is positively correlated with the attack intensity and the power grid value, and negatively correlated with the amount of defense. However, that value v_i has not yet been assessed.

In reality, Electricity was transmitted from the plant to the users: residents, factories and etc. The power grid distributes electricity to users in a large area. High-voltage transmission (HVT) technology and three-phase transmission technology are adopted to reduce loss in power transmission. Therefore, there are many transformers with different capacities at intermediate levels. This observation makes the tree topology reasonable, where the plant is the root and the users are leaves. Since the completion of the power supply need requires a series of different devices, and the damage of any node can block the process. When a node is crashed, the power supplies of some end-users are affected. From this point of view, v_i is equal to the total value of leaf nodes affected.

Consider a n-ary tree with M, $M > 1$ levels, $Q = (n^M - 1)/(n-1)$. Label each node as shown in Fig. 2. Denote v_{user} as the value of a user, namely the loss caused when a user cannot get power. The loss caused by a crashed leaf node at the M-th level is $v_{\frac{n^{M-1}-1}{n-1}+j} = v_{\text{user}}$, $j = 1, \cdots, n^{M-1}$. There are n^{M-i} leaf nodes in the descendant set of a node on the i-th ($i = 1, 2, \cdots, M-1$) level, and those sets do not intersect. Hence all nodes on the i-th level have the same value $v_{\frac{n^{i-1}-1}{n-1}+j} = n^{M-i} v_{\text{user}}$, $j = 1, \cdots, n^{i-1}$, and the defense resources allocated to them are equal.

Since there are n^{i-1} nodes at the i-th level, the total value of the i-th level is $n^{M-1} v_{\text{user}}$. Then all levels have the same value, which requires that defense resources should be distributed to each level evenly,

$$d^*_{\frac{n^{i-1}-1}{n-1}+j} = \frac{D}{Mn^{i-1}}, \; i = 1, 2, \cdots, M, \; j = 1, 2, \cdots, n^{i-1}. \tag{9}$$

And the expectation of the total loss can be obtained that

$$E(L(s_a^*, s_d^*)) = Mn^{M-1} \frac{A}{D} v_{\text{user}}. \tag{10}$$

3 Improved Loss Function and Allocation Strategies

Assumption 4. *Nodes are relevant to each other.*

Based on Assumption 2, the loss function (2) works only when damaged nodes are irrelevant to each other. However, when any two nodes are at the "high" risk, the relationship between them should be taken into consideration.

When the parent node has been crashed, there is no need to attack any downstream node of this node, and vice versa. Only in the case that all its upstream nodes are intact, the value of the crashed node can be added to the total loss function. Due to the physical isolation, the probabilities of any two nodes being crashed are independent of each other. Therefore, under Assumption 4, the loss function can be improved as

$$E(L) = \sum_{k=1}^{Q}[p_k v_k \prod_{j\in\mathcal{U}_k}(1-p_j)] = \sum_{k=1}^{Q}[\frac{a_k}{d_k} v_k \prod_{j\in\mathcal{U}_k}(1-\frac{a_j}{d_j})], \tag{11}$$

where $\mathcal{U}_i$ is the set of upstream nodes of node i.

This improved loss function reflects one fact that the more attack resources allocated to the upstream node, the less necessity to take downstream nodes into account.

It is noted that loss function (11) is always no greater than (2) under the same strategies of both sides. If any attacked node is not an upstream (downstream) node of other attack targets, loss function (11) degenerates into (2). Therefore, the attacker should try to avoid the path between targets.

In the n-ary tree topology, the optimal attack can be achieved as long as the existence of path between the attacked nodes is avoided, while the defender should distribute defense resources evenly to all levels.

Assume the defender adopts the defense strategy defined by (9). Denote L as the maximal loss under any attack resource allocation. Let $L^* = Mn^{M-1}Av_{\text{user}}/D$, which is equal to the loss in (10), then $L \leq L^*$. Some different attack target selections are shown in Fig. 3. From this figure, it is obvious that the attacker can always avoid the existence of path to get the maximal loss L^* in a tree topology. And nodes at all levels are threaten as the one in Subsect. 2.3.

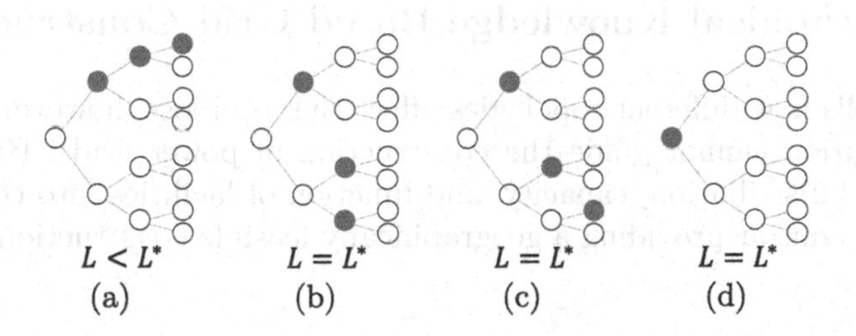

Fig. 3. Loss under different attack target selections in a tree topology.

If the defender does not take the defense strategy defined by (9), there exists a node such that the defense resource allocated to this node is less than that in (10). If the attacker centralizes all attack resources to this node, the loss becomes greater than that of (10). Hence the defense strategy defined by (9) is optimal.

To summarize, under the improved loss function defined by (11), the optimal defense strategy is the same as (9). And the optimal attack strategy can be any feasible strategy as long as there is no path between all attacked nodes. Then the expectation of the total loss can be calculated as

$$E(L(s_a^*, s_d^*)) = \frac{A}{D} M n^{M-1} v_{\text{user}}. \tag{12}$$

In the tree topology, every non-root node has only one parent, which means that there is only one path from the root to the leaf. If any node on the path is crashed, the leaf fails and the loss occurs. If there are backup paths, a node can get the service from an alternative superior in case of an emergency as shown in Fig. 4. Then whatever intermediate node is attacked, the power can still be transmitted from the plant to end-users.

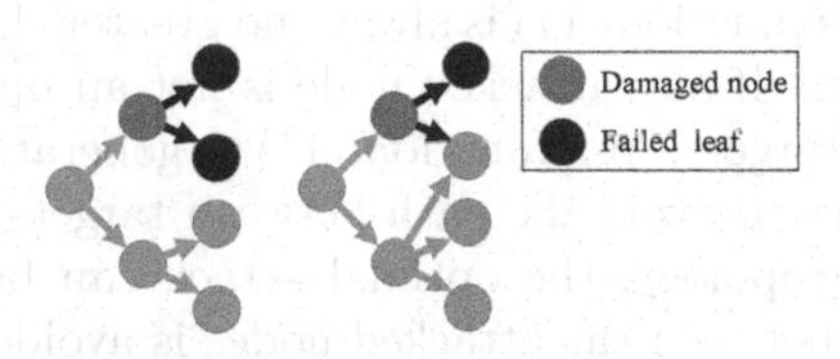

Fig. 4. Node failure situations with and without redundant connections.

It can be revealed that the total loss function is dependent on not only the resource allocation of both sides but also the topology of the power grid. If the connection structure is modified properly, the robustness of the power grid with respect to the attack can be improved.

4 Geographical Knowledge Based Grid Construction

Section 3 tells how different topologies affect the total loss, however, only topology information cannot guide the construction of power grids. By taking the geographical distribution, capacity, and function of facilities into consideration, this section aims at providing a geographically feasible construction solution for power grids.

4.1 Hexagonal City Group

Based on geographical distribution, a feasible topology construction method is developed for the power grid, which is shown in Fig. 5(a). This topology is similar to a 6-ary tree except that six child nodes of one node are connected together to form a ring structure (one child node is linked to its two adjacent siblings).

Within this structure, the capacity of each facility (including the power line) is set to be twice its basic demand. Then even if up to two nodes are damaged, sibling nodes can take the responsibility of damaged nodes and maintain the healthy operation of the entire grid, which is demonstrated in Fig. 6.

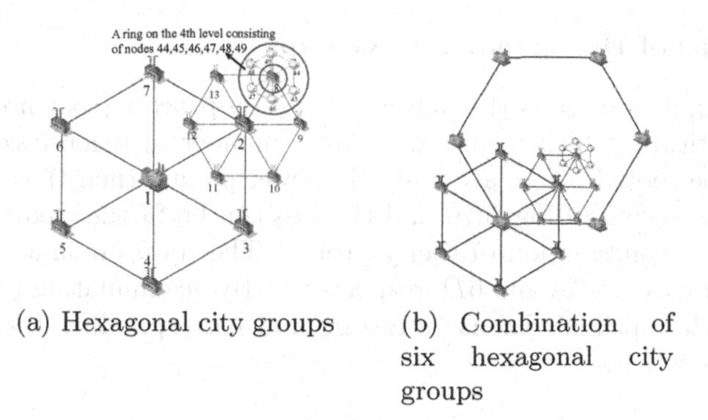

(a) Hexagonal city groups (b) Combination of six hexagonal city groups

Fig. 5. A feasible power grid construction method with the hexagonal city group structure.

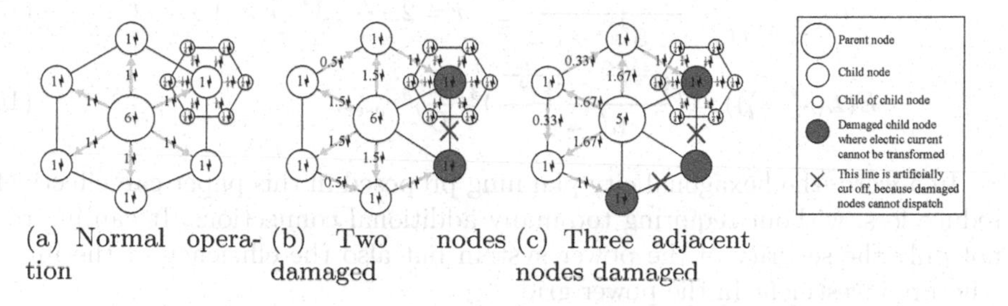

(a) Normal operation (b) Two nodes damaged (c) Three adjacent nodes damaged

Fig. 6. Operation conditions of power transmission under different scenarios: (a) the normal case; (b) although two nodes are damaged, the responsibility of these two nodes can be taken by their sibling nodes and the child nodes of damaged nodes are not affected; and (c) when three adjacent nodes are damaged, the middle damaged node and its descendants are affected, and the loss occurs.

Assume the power grid with the hexagonal structure has M levels, $Q = (6^M - 1)/5$. The total value of all leaf nodes is $6^{M-1}v_{\text{user}}$. According to Fig. 6, to make one node stop working, the attacker must crash this node and its two adjacent nodes to ensure the power cannot pass through this node. Therefore, the probability of any node other than the root stopping working is approximately equal to the probability that the point and its two neighbors are destroyed.

Due to space constraints, the detailed proof is not provided here. By using the same method in Sect. 2, the final optimal defense resource allocation strategy and the corresponding total loss are

$$d_1^* \approx D,\ d^*_{\frac{(6^{i-1}-1)}{5}+l} \approx 6^{-\frac{i+1}{3}} A^{\frac{2}{3}} D^{\frac{1}{3}},\ i = 2, \cdots, M,\ l = 1, \cdots, 6^{i-1}; \tag{13}$$

$$E((s_a^*, s_d^*)) \approx \frac{A}{D} 6^{M-1} v_{\text{user}}. \tag{14}$$

The total loss is reduced to (1/M) compared to that of the original tree structure.

4.2 Groups of Hexagonal City Groups

The most fragile section is the power generation process (root node) because it has no substitute, which results in a large number of defense resources being placed on the root. If there are multiple power plants, then they can be strung into a ring as shown in Fig. 5(b) and the loss can be further reduced.

Consider a combination of 6 city groups. The total amounts of the attack-defense resources are $6A$ and $6D$, respectively. By accumulating the loss on each level, then the optimal defense strategy and corresponding loss can then be calculated as follows

$$d_1^* = d_2^* = \cdots = d_6^* = \frac{6^{\frac{2}{3}} - 1}{6^{\frac{2M}{3}} - 1} D,$$

$$d^*_{\frac{(6^{i-1}-6)}{5}+r} = \frac{6^{-\frac{i-3}{3}} - 6^{-\frac{i-1}{3}}}{6^{\frac{2M}{3}} - 1} D, \; i = 2, \cdots, M, \; r = 1, \cdots, 6^i. \tag{15}$$

$$E(L(s_a^*, s_d^*)) = \left(\frac{6^{\frac{3M-7}{3}} - 6^{\frac{M-7}{3}}}{6^{\frac{2}{3}} - 1}\right)^3 \left(\frac{A}{D}\right)^3 v_{\text{user}}. \tag{16}$$

In a word, the hexagonal city planning proposed in this paper can effectively reduce loss, without requiring too many additional connections. It can improve not only the security of the power system but also the efficiency of the infrastructure investment in the power grid.

5 Conclusion

The operation of the power system requires the coordination of a series of different facilities. As long as any section or any facility fails, the power supply can be cut off. Therefore, concentrating attack resources can produce a better destructive effect. Conversely, the defender should guarantee the reliability of every facility. Redundant connections can make sure that there are other facilities to maintain the operation of the grid after some facilities damaged.

When adding redundant connections, cost, feasibility, and the facility's capacity need to be taken into account. To achieve this goal, based on the city element's real physical location, this paper recommends using the hexagonal city planning method where any facility has the "backup" facilities for replacement in case of damage. Under this method, the expected loss of the entire grid has a cubic-level decay, while the grid construction of the proposed method has high geographic feasibility.

Acknowledgment. This work was supported in part by the National Natural Science Foundation of China (Grants 61633016 61873268 62025307 U1913209) and the Beijing Natural Science Foundation (Grant JQ19020)

References

1. Lu, G., De, D., Song, W.: SmartGridLab a laboratory-based smart grid testbed. In: Proceedings of the First IEEE International Conference on Smart Grid Communications. Gaithersburg, USA, pp. 143–148 (2010)
2. Bella, A.L., Cominesi, S.R., Sandroni, C., Scattolini, R.: Hierarchical predictive control of microgrids in islanded operation. IEEE Trans. Automation Sci. Eng. **14**(2), 536–546 (2017)
3. Bompard, E., Wu, D., Xue, F.: Structural vulnerability of power systems: a topological approach. Electric Power Syst. Res. **81**(7), 1334–1340 (2011)
4. Sridhar, S., Hahn, A., Govindarasu, M.: Cyber-physical system security for the electric power grid. Proc. IEEE **100**(1), 210–224 (2012)
5. Jeler, G.E., Roman, D.: The graphite bomb: an overvies of its basic military applications. Rev. Air Force Acad. **1**(31), 13–18 (2016)
6. Zhang, H., Cheng, P., Shi, L., Chen, J.: Optimal denial-of-service attack scheduling with energy constraint. IEEE Trans. Automatic Control. **60**(11), 3023–3028 (2015)
7. Zhang, H., Cheng, P., Shi, L., Chen, J.: Optimal DoS attack scheduling in wireless networked control system. IEEE Trans. Control Syst. Technol. **24**(3), 843–852 (2016)
8. Kosut, O., Jia, L., Thomas, R.J., Tong, L.: Malicious data attacks on the smart grid. IEEE Trans. Smart Grid **2**(4), 645–658 (2011)
9. Yang, X., Lin, J., Yu, W., Moulema, P., Fu, X., Zhao, W.: A novel en-route filtering scheme against false data injection attacks in cyber-physical networked systems. IEEE Trans. Comput. **61**(1), 4–18 (2015)
10. Deng, R., Liang, H.: False data injection attacks with limited susceptance information and new countermeasures in smart grid. IEEE Trans. Ind. Inf. **15**(3), 1619–1628 (2019)
11. Wang, Q., Tai, W., Tang, Y., Ni, M., You, S.: A two-layer game theoretical attack-defense model for a false data injection attack against power systems. Int. J. Electrical Power Energy Syst. **104**, 169–177 (2019)
12. Srikantha, P., Kundur, D.: A DER attack-mitigation differential game for smart grid security analysis. IEEE Trans. Smart Grid **7**(3), 1476–1485 (2016)
13. Farraj, A., Hammad, E., Daoud, A.A., Kundur, D.: A game-theoretic analysis of cyber switching attacks and mitigation in smart grid systems. IEEE Trans. Smart Grid **7**(4), 1846–1855 (2016)
14. Sridhar, S., Manimaran, G.: Data integrity attack and its impacts on voltage control loop in power grid. In: Proceedings of IEEE Power & Energy Society General Meeting. San Diego, USA, pp. 1–6 (2011)
15. Yang, Q., Yang, J., Yu, W., An, D., Zhang, N., Zhao, W.: On false data-injection attacks against power system state estimation: modeling and countermeasures. IEEE Trans. Parallel Distributed Syst. **25**(3), 717–729 (2014)
16. Esmalifalak, M., Shi, G., Han, Z., Song, L.: Bad data injection attack and defense in electricity market using game theory study. IEEE Trans. Smart Grid **4**(1), 160–169 (2013)
17. Giraldo, J., Crdenas, A., Quijano, N.: Integrity attacks on real-time pricing in smart grids: impact and countermeasures. IEEE Trans. Smart Grid **8**(5), 2249–2257 (2017)
18. Wei, L., Sarwat, A.I., Saad, W., Biswas, S.: Stochastic games for power grid protection against coordinated cyber-physical attacks. IEEE Trans. Smart Grid **9**(2), 684–694 (2018)

19. Wang, H., Zhao, Q., Jia, Q., Guan, X.: Efficient topology optimization for a wired networked system by adding wireless communication. In: Proceedings of 2012 American Control Conference. Montreal, Canada, pp. 448–453 (2012)
20. Wang, J.W., Rong, L.L.: Cascade-based attack vulnerability on the US power grid. Saf. Sci. **47**(10), 1332–1336 (2009)
21. Liu, Y., Cheng, L.: Sampled-data based mean square bipartite consensus of double-integrator multi-agent systems with measurement noises. In: Proceedings of 2018 Chinese Intelligent Systems Conference, pp. 339–349 (2019)

Optimization on the Coefficients of Multi-node Bilateral Control Model

Haisheng Lin, Daxing Zhong(✉), Min Zheng, and Yajian Zhang

School of Mechatronic Engineering and Automation,
Shanghai University, Shanghai 200444, China

Abstract. Bilateral control, which can be realized only by sensors, can suppress the disturbance and ensure the stability of the vehicles. Considering the use of wireless communication between vehicles, bilateral control is extended to multi-node bilateral control, and it's become a vital issue that how to determine the weight coefficient of different vehicle node information in the control strategy. By adjusting the weight of vehicle node in the control strategy, the ability of the system to damp out perturbations can be further improved, and the traffic stream can accomplish steady state more efficiently. In this paper, we compared the properties of different types objective functions and gave the weight coefficients with the least square method to approximate the objective function. Meanwhile, the stability of the system is guaranteed. It is found that bell curve makes the generated coefficients more in line with the optimization direction by adjusting the parameters, i.e., 1) make approximation function smaller; 2) make smaller approximation function cover wider frequency range. The formula of the generating coefficient was obtained, and the parameter range of its stability was determined by mathematical proof. Finally, the performance comparison of traffic stream under different coefficient programs in numerical simulation validated the previous analysis.

Keywords: Multi-node bilateral control · Least square method · Stability analysis · Summation of series

1 Introduction

The development of information acquisition, transmission and exchange technology makes automation and artificial intelligence progress continuously. For vehicles, on-board sensors can help the vehicle to obtain the information of other vehicles on the road, and provide decision-making information for the operation of the vehicles. Due to the progress of communication technology, the information of more vehicles can be collected together, which can better help to complete the control of traffic on the road. The PATH program [1–3] put this into practice and promoted the research of vehicle platoon control. Vehicle platoon control [4] is that adjacent vehicles in a single lane are formed and the longitudinal motion state of the vehicle was adjusted automatically according to the information of adjacent vehicles, and finally the vehicles in the platoon reach the same speed and expected formation. In the face of increasingly serious traffic

K. Li et al. (Eds.): LSMS 2021/ICSEE 2021, CCIS 1468, pp. 463–473, 2021.
https://doi.org/10.1007/978-981-16-7210-1_44

congestion and frequent traffic accidents, vehicle platoon control can increase the road traffic stream, which helps to alleviate traffic jam and reduce traffic accidents [5, 6]. Moreover, the fuel consumption of the vehicles in the platoon can be optimized by setting the controller and adjusting the distance strategy, so as to reduce the energy consumption and exhaust emission [7].

The variety of information stream topologies have become an important part of vehicle platoon control is due to the development of wireless communication technology among vehicles. Zheng [8] discussed the improvement method of stability margin from the perspective of information stream topology. He gave the stability condition of the queue and deduced that the stability margin would decay when platoon size increase under the undirected information stream, but the situation would be improved if the following vehicles behind the lead vehicle were added. Obviously, the platoon control is considered globally, and its control goal is mainly to ensure that the following vehicle tracks the trajectory of the lead vehicle.

For another, the local control strategy will determine the control object as the vehicles in the traffic stream rather than the whole platoon, abandoning the concept of lead vehicle. Intuitively, the local following error can be adjusted through the information stream between vehicles. Among them, the bilateral control has been studied in depth. The platoon with node dynamic model is double integral model under bilateral control is analogous to the mass-spring-damper system in physics, i.e., the disturbance will be attenuated in the process of propagation along the traffic stream, which makes the system easier to reach equilibrium state [9, 10].

Considering the development of wireless communication technology, the simple bilateral control is extended to multi-node bilateral control [11]. At this point, a problem appears, that is, the determination of the weight coefficient of the different nodes in the control decision. For a simple bilateral control, the weight coefficient in the control decision can be set to be equal because only the relative state of the front and rear vehicles with the current vehicle are considered in the design of the controller. However, it is not so simple to identify these for the multi-node bilateral control model. And what principles to follow and how to accurately calculate the value of the coefficient are worthy of further study.

Wang [11] discussed the problem of determining the weight coefficient of node in the multi-node bilateral control model and put forward Taylor series method and least squares method on the premise of ensuring the stability of the system. Compared with the Taylor series method, the least square method is more flexible, and the coefficients are obtained by approximating the objective function. When the frequency is relatively small, the function with higher attenuation rate can be selected as objective function, such as absolute function. Correspondingly, this solution can suppress the low-frequency disturbance waves faster.

Inspired by it, the properties of different types of power functions and exponential functions are compared and they are used as the objective function to generate the weight coefficient by the least squares in this paper. Because the shape of the curve $\beta e^{-\alpha|w|} - \beta$ can be changed from two directions by adjusting the parameters α and β, it is more in line with the optimization direction of the coefficient. The formula of the generating coefficient is obtained, and the parameter range of its stability is determined by mathematical proof. The simulation results show the advantages of the system under the coefficient program, which is to dampen the low frequency disturbance and reach the equilibrium state quickly.

2 Problem Formulation

2.1 System Description

As shown in Fig. 1, the state information of the vehicles on both sides are only considered for the current vehicle with simple bilateral control model, and its control model is

$$a_n = k_d\big[(x_{n-1} - x_n) + (x_{n+1} - x_n)\big] + k_v\big[(v_{n-1} - v_n) + (v_{n+1} - v_n)\big]. \quad (1)$$

where k_d, k_v denotes the controller gains for relative position and relative velocity respectively, and the current node's position, velocity and acceleration are represented by x_n, v_n and a_n. is the acceleration of the current vehicle. The model is always stable for $k_d > 0, k_v > 0$ [12] (Fig. 2).

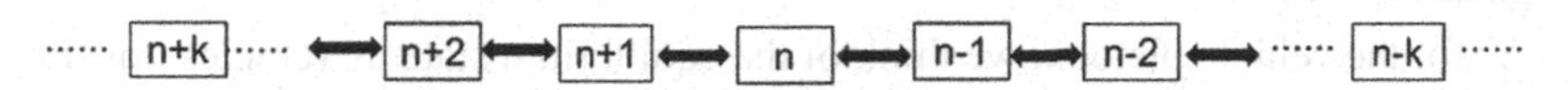

Fig. 1. Information stream under simple bilateral control.

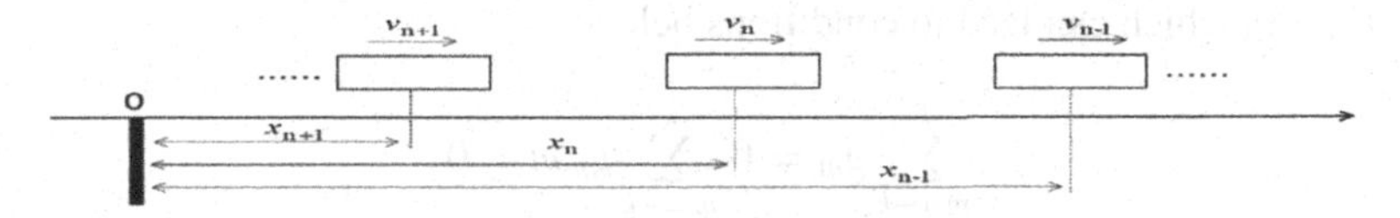

Fig. 2. Bilateral control model.

Figure 3 shows the multi-node bilateral model, and we can see that the control model of the current vehicle is related to the 2k adjacent vehicles. Referring to Eq. (1), the current vehicle's control model is as following:

$$a_n = k_d\Big(\sum_{m=-k}^{k} g_m x_{n-m}\Big) + k_v\Big(\sum_{m=-k}^{k} h_m v_{n-m}\Big). \quad (2)$$

where g_m, h_m are respectively the weights of the position information and speed information of the mth vehicle away from the current vehicle in the control strategy. Obviously, the coefficient set $\{g_m\}$ and $\{h_m\}$ are both $\{1, -2, 1\}$ in the simple bilateral control.

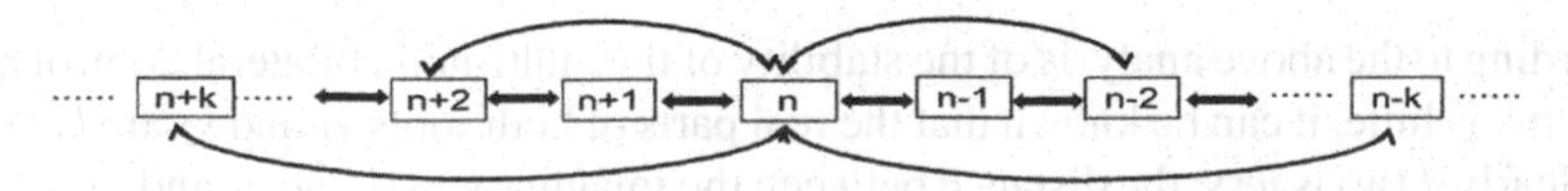

Fig. 3. Information stream under multi-node bilateral control.

2.2 Stability Analysis

It is clear that the model described by the Eq. (2) is linear and shift invariant [13], that means the response of a single frequency wave through the system remains the same frequency. Mathematically, this is $\boldsymbol{x_n(t) = e^{-inw}p_w(t)}$ when $\boldsymbol{x_n(0) = e^{-inw}}$, where $\boldsymbol{p_w(t)}$ denotes the magnitude and phase of the wave. Put $\boldsymbol{x_n(t) = e^{-inw}p_w(t)}$ into Eq. (2), the result is:

$$\ddot{p}_w(t) - k_v f(w)\dot{p}_w(t) - k_d f(w)p_w(t) = 0. \tag{3}$$

To simplify the problem, let $g_m = h_m$, then $f(w) = \sum_{m=-k}^{k} g_m e^{imw}$.
The solution of the ordinary differential Eq. (3) is [16]:

$$p_w(t) = b_1 e^{s_1(w)t} + b_2 e^{s_2(w)t}. \tag{4}$$

where b_1 and b_2 are permanent and the values are resolved by the original states. Meanwhile s_1, s_2 satisfy:

$$s_1 + s_2 = k_v f(w)\ s_1 s_2 = -k_d f(w). \tag{5}$$

As for the vehicle platoon, we think of the equilibrium state as vehicles traveling at the same speed at the same distance. Specific to each car, $x_n(t) = X + ns + Vt$, where X denotes the position of the first car, s indicates the distance between adjacent cars, and V is the speed of each car. And distinctly the equilibrium stated above is one of solutions to the Eq. (2), which can lead to conditions below:

$$\sum_{m=-k}^{k} g_m = 0\ \sum_{m=-k}^{k} g_m m = 0. \tag{6}$$

The stability of the model is discussed, and we think that the fleet model is stable if the traffic stream can accomplish the equilibrium from any status. Combined with the previous discussion of the model, the system is stable when $p_w(t) \to 0$ with $t \to \infty$. Considering periodicity, $0 < |w| < \pi$ was adopted [11]. Due to $p_w(t) = b_1 e^{s_1(w)t} + b_2 e^{s_2(w)t}$, that s_1 and s_2 have negative real parts will result in $p_w(t) \to 0$ when $t \to \infty$ and a natural construct is to make $f(w)$ as a negative real function according to Eq. (5). Wang [11] gave a sufficient condition to ensure the stability of the multi- node bilateral control model, i.e., the system is stable if the symmetrical coefficient $\{g_m\}$ (i.e., $g_m = g_{-m}$) makes the $f(w) < 0$ for all $0 < |w| < \pi$ and $f(0) = 0$.

3 Determination of Weight Coefficient of Node information

3.1 Generating the Coefficients

According to the above analysis of the stability of the multi-node bilateral control model, while $|w|$ is little, it can be known that the real parts of both roots s_1 and s_2 are $k_v f(w)/2$. It is clearly, $f(w)$ is less, the distance between the imaginary axis and s_1 and s_2 is farther, and the $p_w(t)$ converges to zero more quickly. Thus, the coefficients set $\{g_m\}$ has the probability of optimization, and the feasible program are: 1) based on the condition of contenting the stability of the system, take $f(w)$ as small as possible; 2) take smaller $f(w)$ to overlay wider frequency range.

Then, identification of the coefficients set $\{g_m\}$ is through the least square method. Here $z(w)$ is presumed as target function,

$$\arg\min_{\{g_m\}}\{\int_{-\pi}^{\pi}(\sum_{m=-k}^{k} g_m e^{imw} - z(w))^2 dw\}. \tag{7}$$

It's important to note that the previous analysis required $f(0) = 0$, i.e., $\sum_{m=-k}^{k} g_m = 0$. Therefore, we utilize Lagrange multipliers to transform the constrained optimization problem above into an unconstrained optimization problem.

$$\arg\min_{\{g_m\},\lambda}\{\int_{-\pi}^{\pi}(\sum_{m=-k}^{k} g_m e^{imw} - z(w))^2 dw + \lambda \sum_{m=-k}^{k} g_m\}. \tag{8}$$

Next, Eq. (8) was treated with partial differentiation according to $\{g_m\}$ and λ respectively. Simultaneously, the results were set as zero, so

$$g_m = c_m - \frac{\lambda}{4\pi}(m = -k, ..., k). \tag{9}$$

where $c_m = \frac{1}{2\pi}\int_{-\pi}^{\pi} z(w)e^{imw}dw$. Furthermore, considering $z(w)$ as an even real function, we have $c_m = \frac{1}{\pi}\int_{0}^{\pi} z(w)\cos(mw)dw$ and with $c_m = c_{-m}, g_m = g_{-m}$. Combined with previous stability analysis, the stability of the multi-node bilateral control model is secured with the generated coefficients $\{g_m\}$ since $z(w)$ was assumed as a negative even real function.

Substituting Eq. (9) into Eq. (6), $\lambda = \frac{4\pi}{2k+1}\sum_{m=-k}^{k} c_m$ and further.

$$g_m = g_{-m} = c_m - \frac{1}{2k+1}\sum_{n=-k}^{k} c_n. \tag{10}$$

When $z(w)$ is determined as a negative even real function, the stability of the multi-node bilateral control model can be guaranteed. Therefore, next step is to optimize the system performance by designing an objective function $z(w)$. Three different functions are discussed below on the basis of [11].

1) $z(w) = \min\{-w^2, -\sqrt{|w|}\}$
Since the result of the calculation involves the Fresnel integral, the solution is numerical and the $\{g_m\}$ are: 0.0736, −0.0122, 0.1247, −0.0954, 0.2195, −0.5431, 1.9230, −3.3799, 1.9230, −0.5431, 0.2195, −0.0954, 0.1247, −0.0122, 0.0736 (denoted as LSM-7) when $k = 7$ (Fig. 4).

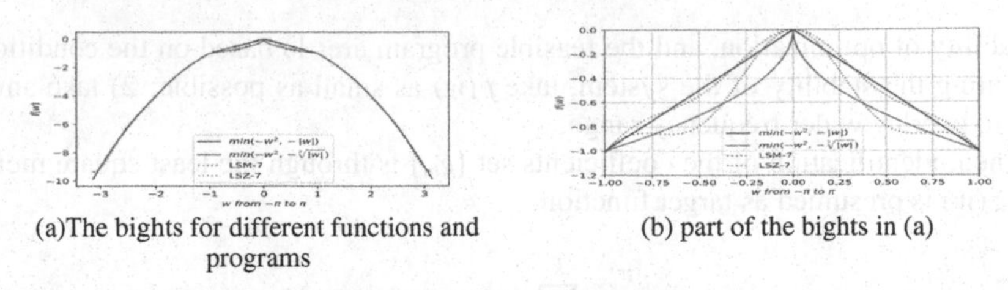

(a)The bights for different functions and programs (b) part of the bights in (a)

Fig. 4. The LSZ-7 program and LSM-7 program.

2) $z(w) = -|w| - w^2 - \sqrt{|w|}$
The coefficient $\{g_m\}$ are: 0.0886, −0.0222, 0.1491, −0.0834, 0.3634, −0.4293, 2.9436, −6.0199, 2.9436, −0.4293, 0.3634, −0.0834, 0.1491, −0.0222, 0.0886 (defined as LSD-7) when $k = 7$ (Fig. 5).

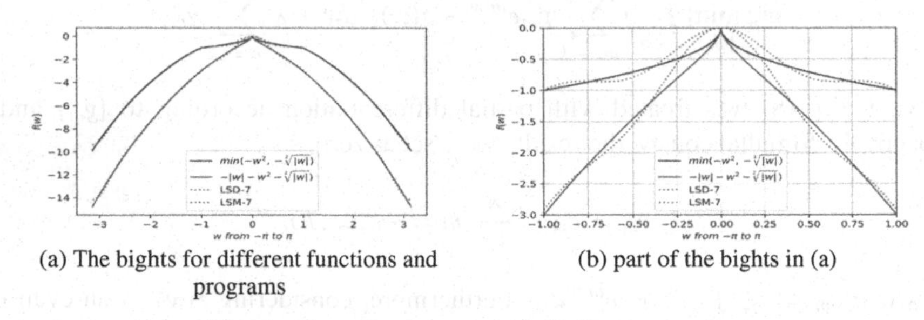

(a) The bights for different functions and programs (b) part of the bights in (a)

Fig. 5. The LSM-7 program and LSD-7 program.

3) $z(w) = \beta e^{-\alpha|w|} - \beta(\alpha\beta > 0)$
For the curve $\beta e^{-\alpha|w|} - \beta(\alpha, \beta > 0)$, the shape can be adjusted by parameter α, β. Substituting $z(w) = \beta e^{-\alpha|w|} - \beta$. into $c_m = \frac{1}{\pi}\int_0^\pi z(w)\cos(mw)dw$, we have

$$c_m = \begin{cases} \frac{\beta(1-\alpha\pi-e^{-\alpha\pi})}{\pi\alpha}, m = 0 \\ \frac{\alpha\beta-\alpha\beta e^{-\alpha\pi}\cos(\pi m)}{\pi(\alpha^2+m^2)}, m \neq 0 \end{cases}. \quad (11)$$

Set $\alpha = 15, \beta = 15$, take $k = 7$, the corresponding coefficients $\{g_m\}$ obtained by formula $g_m = g_{-m} = c_m - \frac{1}{2k+1}\sum_{n=-k}^{k} c_n$ are: 0.9661, 0.9791, 0.9912, 1.0019, 1.0108, 1.0175, 1.0216, −13.9769, 1.0216, 1.0175, 1.0108, 1.0019, 0.9912, 0.9791, 0.9661 (represented by LSE-7($\alpha = \beta = 15$)) (Fig. 6).

In summary, the differences between different types of power function and exponential function as the objective function were systematically compared in the process of generating coefficient by utilizing the least square to approximate objective function int this paper. It is natural that the first objective function is $\min(-w^2, -\sqrt{|w|})$, which is obviously smaller than $\min(-w^2, -|w|)$ in [11]; The second objective function combines the different types of functions mentioned above, taking $z(w) = -|w| - w^2 - \sqrt{|w|}$, which is obviously smaller than the first objective function proposed in this paper in the interval $[-\pi, \pi]$; The third objective function adopts the form of $\beta e^{-\alpha|w|} - \beta(\alpha, \beta > 0)$,

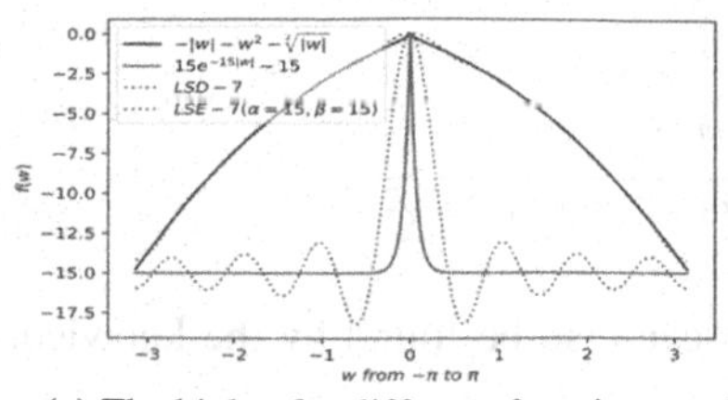

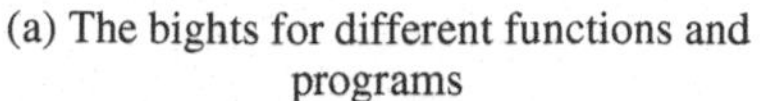
(a) The bights for different functions and programs

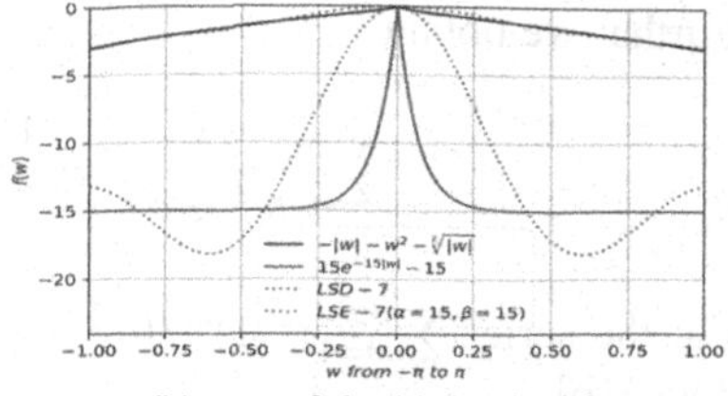

(b) part of the bights in (a)

Fig. 6. The LSD-7 program and LSE-7($\alpha = 15, \beta = 15$) program.

and makes it globally smaller in the interval $[-\pi, \pi]$ by adjusting the parameters α, β. It is worth mentioning that, unlike the numerical results given by the previous two objective functions, an exact analytical solution can be calculated with the third objective function $\beta e^{-\alpha|w|} - \beta(\alpha, \beta > 0)$.

3.2 Testing Stability of the Coefficients

Lemma 1: If $g_m \geq 0$ for all $m = 1, 2, \ldots, k$ and $g_0 < 0$ then the (symmetric) multi-node bilateral control model is stable [11].

Proposition 1: The symmetric multi-node bilateral control model whose coefficient are selected by least square approximation of $\beta e^{-\alpha|w|} - \beta$ when $\alpha > 1/\pi, \beta > 0$ is stable.

Proof: The calculation of cm (11) has been given before, and the key to the proof here is the calculation of the series $\sum_{m=1}^{\infty} \alpha\beta\left(1 - e^{-\alpha\pi}\cos(\pi m)\right)/\pi\left(\alpha^2 + m^2\right)$.

It is known from Theorem 1 in [14],

$$\sum_{n=-\infty}^{\infty} f(n) = -\pi \sum_{k=1}^{m} \operatorname{Res}(f(z)\cot \pi z; z_k). \tag{12}$$

where $\{z_k\}_{k=1}^{m}$ is an isolated singularity of $f(z)$, $Res(f(z)cot\pi z; z_k)$ is the residue of the function at the pole. For the series $\sum_{m=1}^{\infty} 1/\left(\alpha^2 + m^2\right)$, let $f(z) = 1/\left(\alpha^2 + m^2\right)$ and $z = \pm\alpha i$ is the pole of first order of $f(z)$, so

$$\operatorname{Res}(f, \alpha i) = \left.\frac{1}{z + \alpha i}\right|_{z=\alpha i} = \frac{1}{2\alpha i}, \operatorname{Res}(f, -\alpha i) = \left.\frac{1}{z - \alpha i}\right|_{z=-\alpha i} = -\frac{1}{2\alpha i}. \tag{13}$$

As is known from [15], $Res(fg; z_0) = g(z_0)Res(f; z_0)$, so

$$\cot i\alpha\pi = \frac{e^{\alpha\pi} + e^{-\alpha\pi}}{i(e^{\alpha\pi} - e^{-\alpha\pi})}, \cot(-i\alpha\pi) = -\cot i\alpha\pi = -\frac{e^{\alpha\pi} + e^{-\alpha\pi}}{i(e^{\alpha\pi} - e^{-\alpha\pi})}. \tag{14}$$

Obviously, $f(z) = 1/\left(\alpha^2 + m^2\right)$ satisfies the condition of theorem 1 in [14], so

$$\begin{aligned} \sum_{m=-\infty}^{\infty} \frac{1}{\mathrm{m}^2+\alpha^2} &= -\pi\left(-\frac{e^{\alpha\pi}+e^{-\alpha\pi}}{i(e^{\alpha\pi}-e^{-\alpha\pi})} \times \left(-\frac{1}{2\alpha i}\right) + \frac{e^{\alpha\pi}+e^{-\alpha\pi}}{i(e^{\alpha\pi}-e^{-\alpha\pi})} \times \left(\frac{1}{2\alpha i}\right)\right) \\ &= \frac{\pi(e^{\alpha\pi}+e^{-\alpha\pi})}{\alpha(e^{\alpha\pi}-e^{-\alpha\pi})}. \end{aligned} \tag{15}$$

Further we obtain

$$\sum_{m=1}^{\infty} \frac{1}{m^2 + \alpha^2} = \frac{1}{2}\left(\frac{\pi(e^{\alpha\pi} + e^{-\alpha\pi})}{\alpha(e^{\alpha\pi} - e^{-\alpha\pi})} - \frac{1}{\alpha^2}\right). \tag{16}$$

For the series $\sum_{m=1}^{\infty}(-1)^m/(\alpha^2 + m^2)$, It can also be found by the knowledge of residues [16],

$$\sum_{m=1}^{\infty} \frac{(-1)^m}{m^2 + \alpha^2} = \frac{\pi}{\alpha(e^{\alpha\pi} - e^{-\alpha\pi})} - \frac{1}{2\alpha^2}. \tag{17}$$

Further, we find that

$$\begin{aligned}\sum_{-\infty}^{\infty} c_m = c_0 + 2\sum_{m=1}^{\infty} c_m = \frac{\beta(1-\alpha\pi-e^{-\alpha\pi})}{\pi\alpha} + \frac{\alpha\beta}{\pi}\Big(2\sum_{m=1}^{\infty}\frac{1}{\alpha^2+m^2}\Big) - \frac{\alpha\beta e^{-\alpha\pi}}{\pi}\Big(2\sum_{m=1}^{\infty}\frac{(-1)^m}{\alpha^2+m^2}\Big) \\ = \frac{\beta(1-\alpha\pi-e^{-\alpha\pi})}{\pi\alpha} + \frac{\beta(e^{\alpha\pi}+e^{-\alpha\pi})}{e^{\alpha\pi}-e^{-\alpha\pi}} - \frac{\beta}{\pi\alpha} - \frac{2\beta e^{-\alpha\pi}}{e^{\alpha\pi}-e^{-\alpha\pi}} + \frac{\beta e^{-\alpha\pi}}{\pi\alpha} = 0\end{aligned} \tag{18}$$

when $\sum_{-\infty}^{\infty} c_{\mathrm{m}} = 0$, due to $c_{\mathrm{m}} > 0 (m \neq 0)$, $\sum_{n=-k}^{k} c_n < 0$ and $g_m = c_m - \frac{1}{2k+1}\sum_{n=-k}^{k} c_n > 0(m \neq 0)$. It's clear that $c_0 = \beta(1 - \alpha\pi - e^{-\alpha\pi})/\pi\alpha < 0$ and $g_0 < 0$ when $\alpha > \frac{1}{\pi}$. With above analysis, $\alpha > \frac{1}{\pi}, \beta > 0$ meets conditions in Lemma 1. Therefore, the symmetric multi-node bilateral control model whose coefficient are generated through least square approximation of $\beta e^{-\alpha|w|} - \beta$ when $\alpha > \frac{1}{\pi}, \beta > 0$ is stable.

4 Numerical Simulation

In the following, the traffic stream performance of different programs will be compared by simulation. Considering that the programs to be compared include SBC, TS-7, LSS-7, LSA-7 and LSZ-7 of [11], the relevant simulation parameters will not be changed and remain consistent with [11].

Firstly, these eight programs are compared, and the simulation time is designed to be 200 s. Car-following control is used in the first 40 s, and multi-node bilateral control with different coefficient programs is used in the last 160 s. Finally, the fluctuation value of each program is evaluated by average absolute disturbance (AAD) calculated as $E_A(t) = \frac{1}{N}\sum_{n=1}^{N}|d_n(t) - s|$. Figure 7 shows the simulation results.

Ignoring differences between the programs, the overall trend is that the average disturbance value decreases continuously in the last 160 s since and the system under bilateral control during the last 160 s is string stable. For the five programs proposed by Wang [11], the simulation results here are similar, and the average absolute disturbance value of LSZ-7 program whose objective function is $\min\{-|w|, -w^2\}$ decreases fastest. However, comparing all the programs, it is clear that the LSE-7($\alpha = \beta = 15$) program whose objective function is $15e^{-15|w|} - 15$ decreases the fastest, and the final average absolute disturbance value become smallest, basically reaching zero, which confirms the previous analysis.

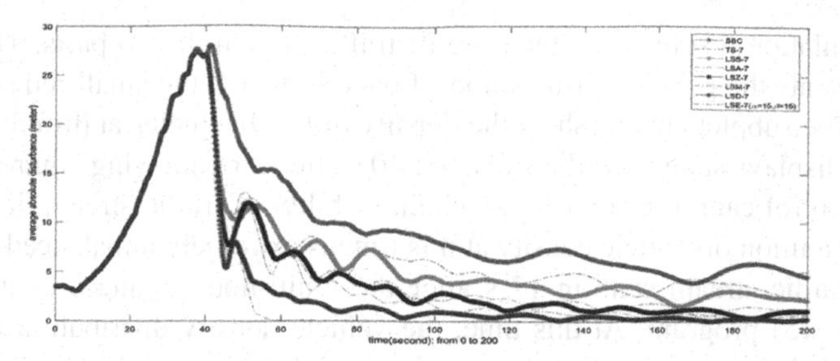

Fig. 7. Average absolute disturbance of different programs.

From the first simulation, we can see that the LSZ-7 program has the best performance among the five programs proposed by Wang [11]. Next, the LSZ program is compared in detail with the three programs LSM-7, LSD-7 and LSE-7 ($\alpha = \beta = 15$) respectively proposed in this paper. The local vehicle density distribution of traffic stream with different coefficient programs was observed when the average absolute disturbance value basically did not change. The local density distribution is expressed as the number of vehicles per mile and is given by $\rho_n(t) = 1609.334/(x_{n-1} - x_n)$. Figure 8 shows the simulation results.

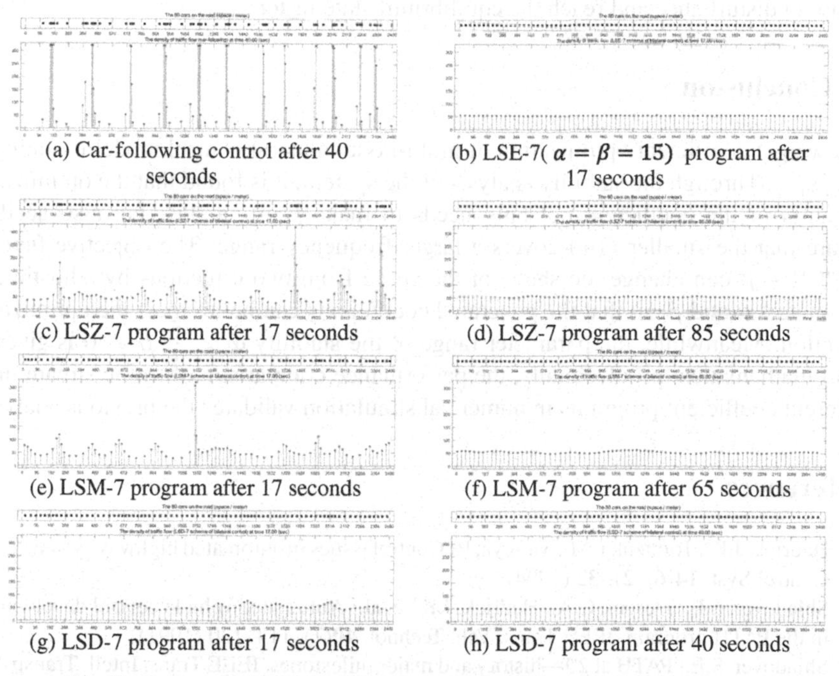

(a) Car-following control after 40 seconds

(b) LSE-7($\alpha = \beta = 15$) program after 17 seconds

(c) LSZ-7 program after 17 seconds

(d) LSZ-7 program after 85 seconds

(e) LSM-7 program after 17 seconds

(f) LSM-7 program after 65 seconds

(g) LSD-7 program after 17 seconds

(h) LSD-7 program after 40 seconds

Fig. 8. Demonstration of traffic stream under different control programs.

The simulation results show the state of traffic stream in two parts. The subgraph above represents the position distribution of cars, in which the small red square represents cars. The subplots below show the density of local vehicles at different locations. Figure 8(a) displays state of traffic stream at 40 s after car-following control. Since car-following control cannot guarantee the chain stability of traffic stream, it can be seen that the distribution of vehicle density at this time is extremely unbalanced. Figure 8(b) shows the traffic stream state in 17 s after the multi-node bilateral control of LSE-7($\alpha = \beta = 15$) program. At this time, the vehicle density distribution is extremely balanced and the system is very close to the equilibrium state. Figure 8(c), Fig. 8(e) and Fig. 8(f) respectively show traffic stream state of 17s after multi-node bilateral control of LSZ-7, LSM-7 and LSD-7. Obviously, the program with the best traffic stream performance is LSE-7($\alpha = \beta = 15$), while the program with the worst traffic stream performance is LSZ-7. Figure 8(d), Fig. 8(f) and Fig. 8(h) respectively display traffic stream state of multi-node bilateral control model of LSZ-7, LSM-7 and LSD-7 when the average absolute disturbance is basically unchanged. The time required with LSE-7($\alpha = \beta = 15$) program is the shortest while that of LSZ-7 is the longest. Moreover, the density of vehicles in the traffic stream with LSE-7($\alpha = \beta = 15$) is the most balanced. This is also consistent with the previous analysis.

In summary, the simulation experiments on the traffic stream performance under different coefficient programs verify the LSE program is more conducive to the system damping disturbance and reach the equilibrium state faster.

5 Conclusion

The weight coefficient optimization of multi-node bilateral control model is studied in this paper. Through the stability analysis of the system, it is found that the optimization direction of the coefficient not only needs to make $f(w)$ smaller, but also needs to ensure that the smaller $f(w)$ covers a larger frequency range. The objective function $\beta e^{-\alpha|w|} - \beta$ can change the shape of the curve from two directions by adjusting the parameters α and β, so that the generated coefficients can better meet the optimization direction. Meanwhile, the parameter range of the stability $\alpha > \frac{1}{\pi}, \beta > 0$ is given by strict mathematical proof. Finally, the performance comparison of traffic stream under different coefficient programs in numerical simulation validates the previous analysis.

References

1. Hedrick, J.K., Tomizuka, M., Varaiya, P.: Control issues in automated highway systems. IEEE Control Syst. **14**(6), 21–32 (1994)
2. Shladover, S.E., Desoer, C.A., Hedrick, J.K., et al.: Automated vehicle control developments in the PATH program. IEEE Trans. Veh. Technol. **40**(1), 114–130 (1991)
3. Shladover, S.E.: PATH at 20—history and major milestones. IEEE Trans. Intell. Transp. Syst. **8**(4), 584–592 (2007)
4. Swaroop, D., Hedrick, J.K., Chien, C.C., et al.: A comparision of spacing and headway control laws for automatically controlled vehicles. Veh. Syst. Dyn. **23**(1), 597–625 (1994)
5. Guanetti, J., Kim, Y., Borrelli, F.: Control of connected and automated vehicles: State of the art and future challenges. Ann. Rev. Control **45**, 18–40 (2018)

6. Li, L., Chen, X.M.: Vehicle headway modeling and its inferences in macroscopic/microscopic traffic flow theory: a survey. Transp. Res. Part C Emerg. Technol. **76**, 170–188 (2017)
7. Li, S.E., Peng, H.: Strategies to minimize the fuel consumption of passenger cars during car-following scenarios. Proc. Inst. Mech. Eng. Part D J. Autom. Eng. **226**(3), 419–429 (2012)
8. Zheng, Y., Li, S.E., Wang, J., Cao, D., Li, K.: Stability and scalability of homogeneous vehicular platoon: study on the influence of information flow topologies. IEEE Trans. Intell. Transp. Syst. **17**(1), 14–26 (2016)
9. Horn, B.K.P., Wang, L.: Wave equation of suppressed traffic flow instabilities. IEEE Trans. Intell. Transp. Syst. **19**(9), 2955–2964 (2018)
10. Horn, B.K.P.: Suppressing traffic flow instabilities. In: 16th International IEEE Conference on Intelligent Transportation Systems (ITSC 2013), pp. 13–20 (2013)
11. Wang, L., Horn, B.: Multinode bilateral control model. IEEE Trans. Autom. Control **64**(10), 4066–4078 (2019)
12. Wang, L., Horn, B.K.P., Strang, G.: Eigenvalue and eigenvector analysis of stability for a line of traffic. Stud. Appl. Math. **138**, 103–132 (2017)
13. Strang, G.: Computational Science and Engineering. Wellesley Cambridge Press, Wellesley (2007)
14. Xu, N.: Method of the complex analysis to some series. J. Nanjing Normal Univ. (Nat. Sci. Ed.) **37** (2014)
15. Conway, J,B.: Functions of One Complex Variable, 2nd ed., pp. 58–127. Springer, New York (1978). https://doi.org/10.1007/978-1-4612-6313-5
16. Zhu, S.: Summation to calculate a class infinite series by using residue theorem. J. Lishui Univ. **27** (2005)

Research on Regional Air Conditioning Temperature Energy Consumption Model Based on Regression Analysis

Jiangyang Wang(✉)

College of Mechanical and Electrical Engineering, China Jiliang University, Hangzhou, China

Abstract. With the increasing improvement of people's living standards, air conditioning as a major energy user has become an important factor affecting household electricity consumption plans. This study first identifies a strong correlation between temperature and air conditioning electricity consumption. Then, a regional air conditioning energy consumption model is developed based on regression analysis to meet the needs of household electricity management systems. First, to obtain a more intuitive relationship between temperature and electricity consumption, an improved k-means clustering algorithm is used to classify regional air conditioning energy consumption data. Second, based on four types of curves obtained by clustering, regression methods are used to propose a multi-class regional air conditioning energy consumption model which can support accurate quantitative analysis. Finally, the validity and accuracy of the proposed model are demonstrated by regional prediction experiments. The results can provide references for scientific electricity consumption planning of households.

Keywords: Air conditioning load · Clustering · Regression analysis · Forecast

1 Introduction

Residential air-conditioning is a commonly used household temperature control device, and it is also a flexible demand response resource. With the improvement of the living standards of the whole society, the amount of air-conditioning owned by residents is constantly increasing. The global abnormal climate phenomenon has intensified, and the extreme weather conditions in summer and winter have also promoted the proportion of air-conditioning energy consumption to increase year by year. As a typical temperature-dependent load, the concentrated operation of air conditioners has become an important factor leading to the increase of peak-to-valley difference. Carry out research on the impact of temperature on air-conditioning energy consumption, establish a regional air-conditioning temperature energy consumption model, and grasp the law of regional electricity consumption.

In recent years, the analysis of air conditioning curves and the establishment of air conditioning models have become a hot spot for researchers. Literature [1] proposed an unsupervised load decomposition method that considers load level differences and

K. Li et al. (Eds.): LSMS 2021/ICSEE 2021, CCIS 1468, pp. 474–484, 2021.
https://doi.org/10.1007/978-981-16-7210-1_45

seasonal changes, and decomposes the whole house energy consumption into residential air conditioning loads (ACLs) and base load components. On this basis, a piecewise analysis method based on the thermal dynamics model of ACLs is proposed, including the constrained regression method of static parameter estimation and the hybrid method of dynamic parameter estimation. Literature [2] first analyzes the correlation between residential cooling power and daily maximum temperature, daily minimum temperature and daily average temperature, and finally establishes a polynomial air-conditioning cold power prediction model with the highest quartic term based on daily average temperature. Literature [3] introduced the residential behavior of residents and the control of indoor environmental temperature through a questionnaire survey of 400 different households in southern China and a field measurement of 100 air-conditioned households. Literature [4] uses Brown's quadratic polynomial exponential smoothing forecasting method to establish a dynamic forecasting model for real-time forecasting of terminal equipment load. The prediction model has dynamic prediction accuracy and rationality. Literature [5] uses the least square method to perform regression analysis on the test data, and obtains the input power-ambient temperature air-conditioning model and the unit performance-refrigerant quantity-ambient temperature air-conditioning model. Literature [6] established an air-conditioning load aggregation model based on a single air-conditioning load model, using the Monte Carlo simulation method and the distribution characteristics of various parameters, and analyzed the influence of the accuracy of the temperature sensor parameters on the total power characteristics of the air-conditioning load. Literature [7] proposed a peak-shift air-conditioning control method suitable for common air-conditioning equipment in existing buildings. Use building and air-conditioning electricity consumption data, air-conditioning operation data and weather data to generate and update simulation models to predict the power demand of buildings and equipment, as well as changes in room temperature.

In the previous experiments, most of the total load data of the household or the direct sampling of the used load were sampled, and a separate instrument was needed to monitor the specific load. The load intelligent sensing data is based on the electrical load intelligent sensing technology to obtain the electricity consumption data of residential users at the electrical level. The load curves directly separated from the total power can be more accurately specified for a certain household at a lower monitoring cost. Discuss the electricity consumption behavior of electrical appliances, and obtain the residents' refined electricity consumption habits and personal electricity consumption characteristics.

Based on the intelligent load perception data in the US Dataport database, this paper uses a clustering algorithm to classify regional air-conditioning energy consumption data, and explores the impact of temperature on residents' use of air-conditioning.

For the electricity consumption data for six months and the whole year, under such a long time scale, the daily temperature change has little effect on the long-term electricity consumption of residents. The literature [8] points out that the average temperature is relative to the maximum temperature and the minimum temperature. The impact on the load is the greatest, so the daily average temperature is selected here for clustering and regression analysis.

The structure of the article is organized as follows: Sect. 1 introduces the current status of research on air conditioning modeling and forecasting methods introduced; Sect. 2 uses an improved clustering algorithm to classify all household air conditioning data in the database, in order to avoid manual misclassification through the algorithm to achieve convergence, and the four types of clustering curves obtained are the focus of the next analysis; Sect. 3 uses the previously obtained clustered households to obtain a quantitative relationship between temperature and air conditioning electricity consumption through regression analysis modeling; Sect. 4 is a prediction experiment, in the same region throughout the year air conditioning data, temperature as training data, model input for the predicted month daily temperature, predicted daily electricity consumption for the month.

2 Improved Clustering Algorithm

Compared with the time-consuming and labor-intensive manual identification of various types of users' electricity consumption patterns, the K-means clustering algorithm [8] can more quickly and labor-savingly classify regional air conditioning energy consumption data and obtain the air conditioning usage patterns of users in the region for further analysis through the calculation of household electricity consumption curves and daily electricity consumption clustering centers. To improve the performance of the clustering algorithm, an improved k-value selection strategy is proposed.

In the classic K-means algorithm, the k value is artificially selected incorrectly in the first step, which may lead to a worse classification effect or more iterations and calculation time. Performing standard K-means clustering in the same database, it can be found that as k increases, the error variance within the cluster decreases because the classification is closer to the original data point, but too large k may require higher computational space and there is the possibility of over-classification. When considering the number of clusters and variance on the image, it can be seen that when k is set to 4, the effect is better. It is found here that the image elbow generally exists where the slope changes greatly. Therefore, it is set when |Slope change rate|$>\Omega$, the current endpoint is considered to be the elbow, and the horizontal coordinate of the point is the k value determined by the improved algorithm (Fig. 1).

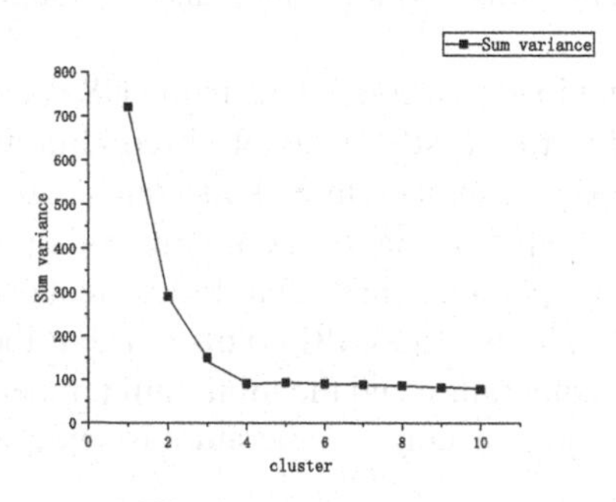

Fig. 1. Elbow diagram.

The algorithm randomly selects a number of initialization points at the beginning. If the selection is inappropriate, it will also cause the aforementioned drawbacks. So after the elbow method determines k, it still randomly selects a cluster center, but the subsequent initial cluster centers should be as far away as possible from each other. Until all initial points are selected.

Finally, an improved k-means clustering algorithm is obtained, which is different from the standard algorithm. It only needs to input a certain slope change rate threshold, and different data can obtain different optimal k values, and the classification effect is better. The flow chart of the improved method is shown in Fig. 2.

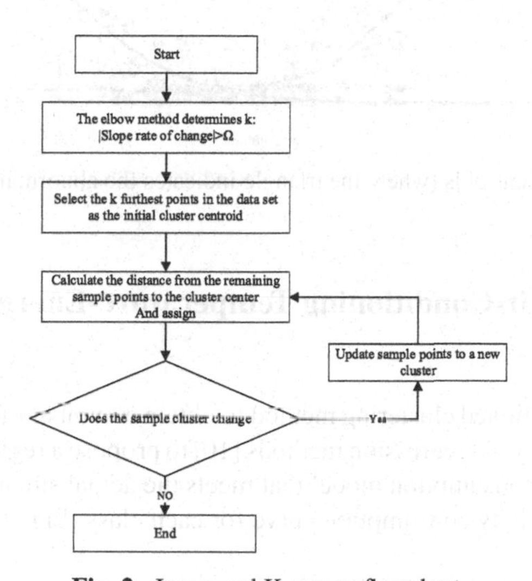

Fig. 2. Improved K-means flowchart.

Table 1. Improve clustering effect comparison.

Method	Iterations	SSE
Standard K-means	4	911.6473
Improved K-means	2	862.4861

The clustering analysis was compared under the same database to obtain the number of iterations and and variance of the two. It can be seen that the optimized clustering algorithm converges faster and the points within the clusters are less offset from the center, which means that the four classes of load clustering curves mentioned by the algorithm, each class of users is closer to the respective cluster center load curve and the electricity consumption habits are more similar, which is ready for the next part of the regression method for accurate analysis (Table 1).

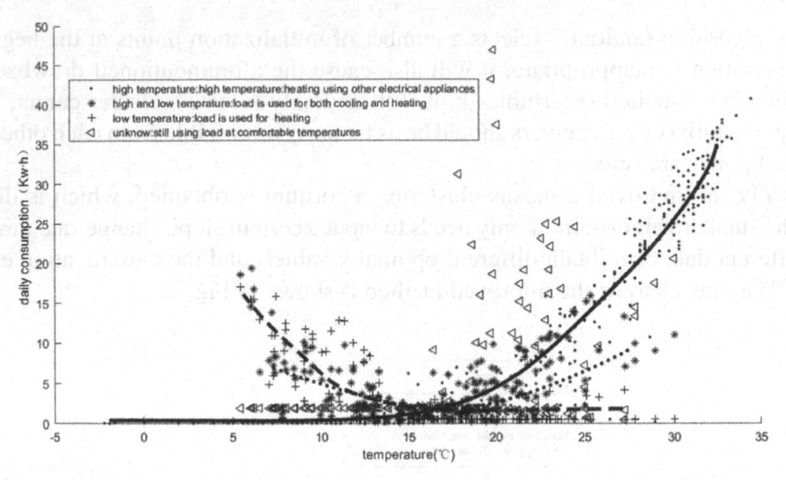

Fig. 3. Four types of centroids (where the triangle indicates the abnormal tendencies not to give the cluster centroids).

3 Household Air-Conditioning Temperature-Energy Consumption Model

After the above-mentioned clustering method to obtain a number of classes of users, it is also necessary to carry out regression methods [10] to propose a regional air conditioning temperature energy consumption model that meets the actual situation, and to carry out analysis of the electricity consumption curve for each class (Fig. 3).

3.1 Selection and Classification of Typical Houses

The database is the US dataport database, which contains data on 31 households. Filter in the database, select the households that use air-conditioning energy, the improved clustering method obtains an adaptive k value of 4. After clustering, four types of users are obtained: high temperature tendency, unknown tendency, high and low temperature tendency, and low temperature tendency. The second regression model of the standard model in this area can describe the real situation more accurately. In other regional models, the cluster ratio varies according to the local regional climate, the proportion of household components, and the level of household consumption. The standard model The order of may also be different (Tables 2 and 3).

Table 2. Datapot dataset timezone.

	Time start	Time end
Timezone	2018-01-01	2018-12-31

Table 3. Resident type ratio.

Types	Regional proportion	Total consumption per household (kW·h)
High temperature	78.79%	3999.66
High and low temperature	15.15%	2838.76
Abnormal tendency	3.03%	723.18
Low temperatlow	3.03%	472.13

3.2 Typical Energy Consumption Model

The regression analysis of the predicted results for high temperature households is a segmented regression, where the linear part is basically a straight line with a value of basically 0 (off state), while the right part is a typical parabolic curve. The analysis of the actual situation is that: such households with electricity exist electric furnace, floor heating or fireplace other heating equipment, low temperature conditions, users prefer to use the above three types of equipment rather than air conditioning for heating, when the temperature continues to rise, cooling equipment is limited, the choice of air conditioning for indoor cooling, so the daily energy consumption increased, a parabola on the right side of the axis of symmetry. This type of user is sensitive to temperature and tends to use a lot of heating and cooling loads, especially in extreme high temperature situations with strong dependence on air conditioners (Figs. 4 and 5).

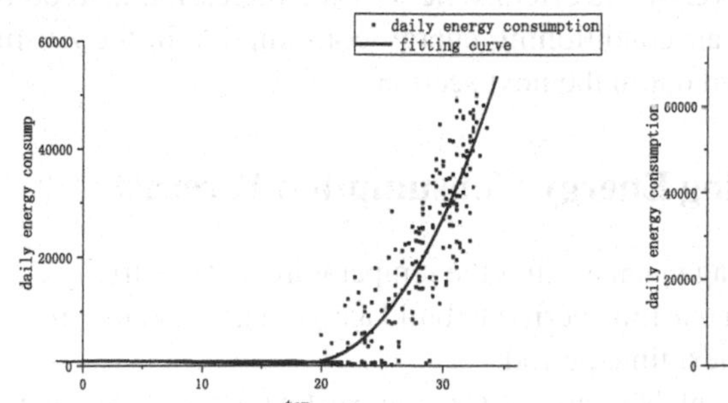

Fig. 4. High temperature tendency.

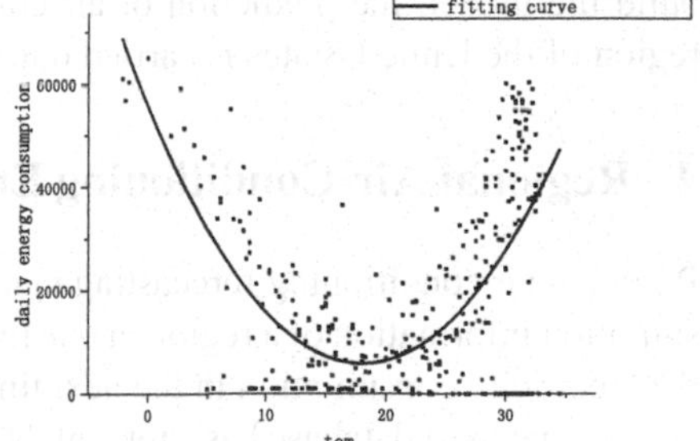

Fig. 5. High and low temperature tendency.

In the range of 0~15 °C, the air-conditioning energy consumption is high, and the three indicators are the lowest among all the trends. The overall temperature rise curve shows a parabolic downward trend. The lowest air-conditioning energy consumption at 17 °C is the lowest point of the parabola, and the total energy consumption is very small. The main purpose of use is heating.

The energy consumption of household air conditioners with abnormal tendency is relatively random, which can be regarded as the random distribution of particles in the plane. This type of user has certain consumption in the human body comfort temperature range of 15~25 °C, and most of the horizontal axis has a very low power consumption (standby state). From this analysis: at the data level, this type of user is not sensitive to temperature throughout the year, and the use of air conditioners is less correlated with temperature (Figs. 6 and 7).

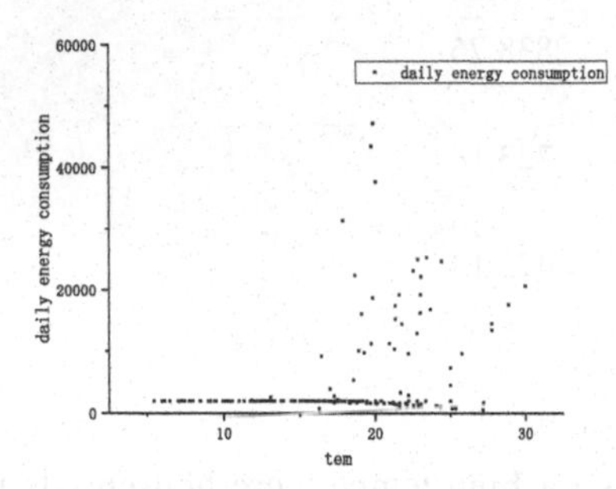

Fig. 6. Abnormal tendency.

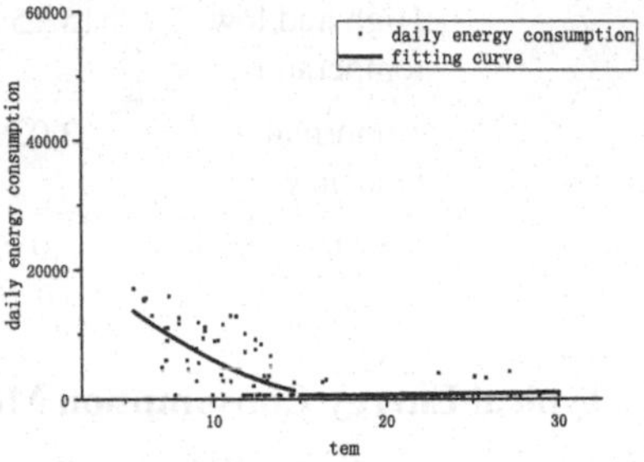

Fig. 7. Low temperature tendency.

This type of user is highly interpretable in the image, and the curve shows that there is a strong relationship between temperature and energy consumption, and the daily energy consumption of air conditioning increases in a parabolic trend as the temperature changes to two extreme values. This type of user is sensitive to temperature and will use air conditioning as an option to regulate temperature both at high and low temperatures, so the response in the image is a parabolic regression model.

Within the database, these four types of customers with varying electricity consumption patterns exist, but not all of these types are necessarily included within the same region. For the prediction of regions, it is also necessary to model the regional air conditioning energy consumption. To verify the effectiveness of the regression method to build the model, the prediction of air conditioning energy consumption in the Austin region of the United States is carried out in the next section.

4 Regional Air-Conditioning Energy Consumption Forecast

Regional air conditioning forecasting means: using the temperature and electricity consumption information of a region in one time period to build a regional model to forecast the electricity consumption in the next time period.

The dataport database has a total of 24 households in the Austin area, which contains two categories of users with high and high and low temperature propensities as described above. To verify the validity of the model, both summer August and winter December were selected for air conditioning electricity consumption prediction (Table 4).

4.1 Regional Energy Consumption Modeling

The regional model was obtained by modeling with all household data of Ausin. Among them, 96% are high temperature inclined households and 4% are high and low temperature inclined. Since in July and August both belong to summer, high temperature days

Table 4. Database residents.

Region	High temperature	High and low temperature
Houses	23	1
Rition	95.8%	4.17%

account for most of the regression model, which is basically the same on the right side of the regression model, and the prediction test results are similar (Figs. 8 and 9).

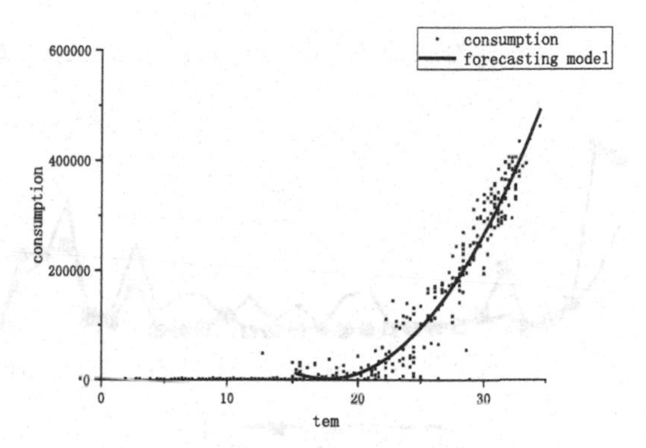

Fig. 8. Regional temperature model.

4.2 Regional Electricity Consumption Forecast and Analysis

For the forecast, the input is the August daily average temperature, and the August daily regional energy use is obtained. The predicted results are as follows (Table 5).

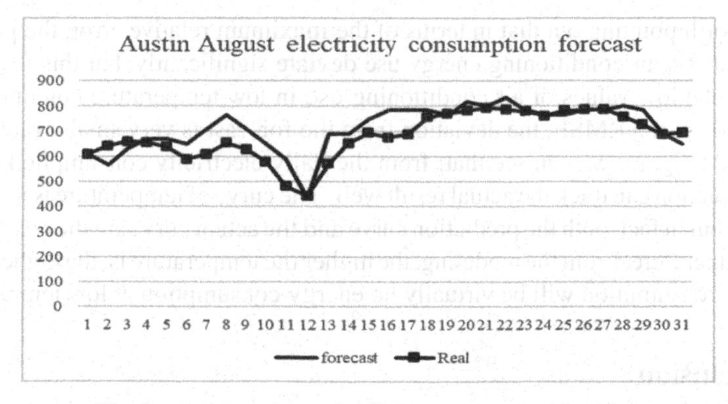

Fig. 9. Austin August electricity consumption forecast.

Table 5. Predicted results for August.

Error Mean	RMSE	Maximum relative error
33.87	59.41	14.35%

Similarly, to forecast the electricity consumption of air conditioners in the region in December, the input is the daily flat temperature of the month, and the output is the daily electricity consumption of the whole month (Fig. 10).

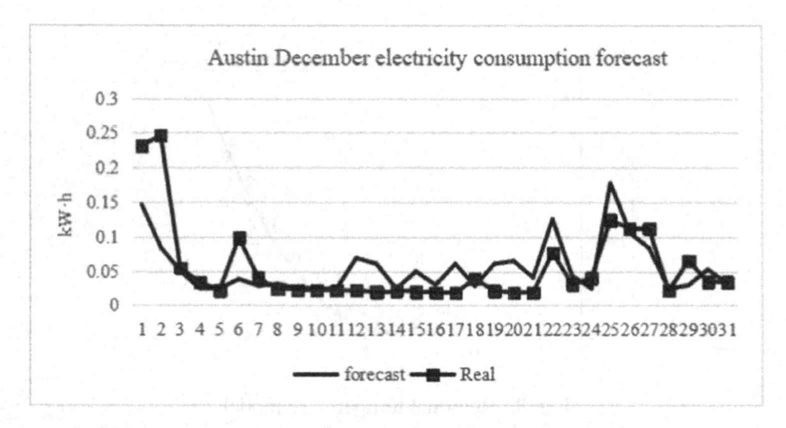

Fig. 10. Austin December electricity consumption forecast.

Table 6 Predicted results for December.

Error mean	RMSE	Maximum relative error
0.71	4.57	18.62%

It is worth pointing out that in terms of the maximum relative error, the predictions for December's air conditioning energy use deviate significantly, but this is a problem caused by the low values of air conditioning use in low temperature conditions. If we look directly at the RMSE, the deviation from the forecast is very small (Table 6).

From the figure, we can see that: from the daily electricity consumption trend, the model prediction can track the actual result well. The curve of temperature is not drawn in the figure, but in fact, both the prediction curve and the actual curve are the result of tracking the temperature. From the modeling, the higher the temperature is, the higher the daily electricity consumption will be virtually no energy consumption at low temperatures.

5 Conclusion

Air conditioning load is an important factor affecting household electricity consumption. In this paper, we mainly apply the load sensing data to study the regional residential air conditioning load, use the improved clustering method to process and then perform regression analysis, propose a regional air conditioning temperature energy consumption model, and use the algorithm comparison experiment and prediction experiment to verify the model.

The experimental results show that:

(1) Under the same database conditions, the improved clustering algorithm can accelerate the convergence speed and the result error is relatively smaller.
(2) The established regional air conditioning temperature energy consumption model can better describe the relationship between air conditioning energy consumption and average daily temperature.
(3) The regional air conditioning temperature energy consumption model can effectively predict the electricity consumption in a small area.

For households, the model can be used to analyze household electricity consumption based on temperature, and improve electricity consumption plans to reduce electricity expenses according to their own and regional levels; for power supply departments, the prediction model can help model inter-regional air conditioning, help analyze the portrait of electricity users, predict peak periods of electricity consumption and electricity consumption based on temperature, expand the data base of power management departments for grid regulation and control, optimize regional power supply strategies, and provide a theoretical analysis basis for the development of more flexible electricity pricing strategies.

References

1. Qi, N., Cheng, L., Xu, H.: Smart meter data-driven evaluation of operational demand response potential of residential air conditioning loads. J. Appl. Energ. **279**, 115708 (2020)
2. Xiang, K.L., Chen, B.J, Yu, X.: Air conditioning cooling power prediction of provincial residents based on regional comprehensive temperature. In: 2019 IEEE Innovative Smart Grid Technologies-Asia (ISGT Asia), pp. 2472—2475. IEEE (2019)
3. Wang, Y., Wu, J., Xie, F.: Survey of residential air-conditioning-unit usage behavior under south china climatic conditions. In: 2011 International Conference on Electric Information and Control Engineering, pp. 2711—2714. IEEE (2011)
4. Xu, X., Huang, G., Liu, H.: The study of the dynamic load forecasting model about air-conditioning system based on the terminal user load. J. Energ. Buildings **94**, 263–268 (2015)
5. Li, H., Zhang, M., Qi, X.: Performance tests research of residential central air-conditioning. In: 2011 International Conference on Electric Information and Control Engineering, pp. 1532—1534. IEEE (2011)
6. Liu, M., Chu, H., Liu, J.: Aggregation model of air conditioning load considering temperature sensor accuracy. In: 2017 IEEE Conference on Energy Internet and Energy System Integration (EI2), pp. 1–5. IEEE (2017)
7. Sato, F., Kawano, H., Kobayashi, N.: An air conditioning control method for peak power reduction using heat capacity based room temperature constraints. In: 2015 IEEE International Telecommunications Energy Conference (INTELEC), pp. 1–4. IEEE (2015)
8. Shi, J., Zhou, Q., Tan, J., Yang, J.Y., Li, H., Zhu, L.: Summer air conditioning load characteristic mining and temperature sensitivity identification of Jiangsu power grid. J. Electr. Power Eng. Technol. **37**(03), 28–32 (2018)

9. MacQueen, J.: Some methods for classification and analysis of multivariate observations. In: Proceedings of the fifth Berkeley symposium on mathematical statistics and probability, vol. 1, no. 14, pp. 281–297 (1967)
10. Draper, N.R., Smith, H.: Applied Regression Analysis. Wiley, Manhattan (1998)

Parameter Optimization Method of EWM Rectangular Reducer Based on RSM-MOGA

Feiyan Qi, Liping Liang(✉), Lingbin Chai, and Kun Wei

School of Electrical Engineering and Automation,
Hefei University of Technology, Hefei 230009, China

Abstract. Electromagnetic water meter (EWM) with a rectangular reducer is widely used because of its low power consumption and high measurement accuracy. In order to improve the measurement performance of EWM under different working conditions, it is necessary to optimize its rectangular reducer structure parameters to satisfy high weight function uniformity and low-pressure loss, which are two contradictory objectives. Thus, this paper proposes a method to optimize EWM's rectangular reducer structure parameters based on the response surface method and multi-objective genetic algorithm (RSM-MOGA). Taking weight function uniformity and pressure loss as optimization objectives, the method establishes response surface function models between optimization objectives and rectangular reducer structure parameters. Then, the multi-objective genetic algorithm is used to optimize response surface function models of the two objectives, and the Pareto optimal solution set is obtained. Finally, according to the pressure loss constraint, the optimized structure parameters are selected from the Pareto optimal solution set. Results show that, under the constraint of 40 kPa pressure loss, the weight function uniformity of EWM rectangular reducer structure optimized by this method is increased by 4.20%, and the pressure loss is reduced by 10.84%. Therefore, the proposed method can improve the measurement performance in practice.

Keywords: Electromagnetic water meter · Rectangular reducer · Response surface method · Multi-objective genetic algorithm

1 Introduction

Electromagnetic water meter (EWM) is widely used in flow measurement because of its low power consumption, high measurement accuracy and large measurement range. In order to meet the requirements of low power consumption and high measurement accuracy at the same time, EWM often uses a reducer (such as a rectangular reducer) to improve the flow rate in the measurement area, so as to enhance the induced electromotive force signal under small flow. However, the rectangular reducer structure will increase pressure loss of EWM, even exceed the limit of national standard, making it unable to use. Moreover, there is often no ideal straight pipe section in the upstream and downstream of the water meter, due to the narrow installation space in reality. The flow

K. Li et al. (Eds.): LSMS 2021/ICSEE 2021, CCIS 1468, pp. 485–496, 2021.
https://doi.org/10.1007/978-981-16-7210-1_46

field in measurement area of the pipeline is usually non ideal, which makes the measurement value of EWM deviate from the real value, seriously affecting the measurement accuracy. In order to improve the measurement performance of EWM under non ideal flow field, it is necessary to study appropriate rectangular reducer structure parameters to improve weight function distribution uniformity in the water meter. Therefore, how to carry out research on optimization of rectangular reducer structure parameters under the constraints of actual pressure loss to improve weight function distribution uniformity of the measurement area, so as to improve the measurement accuracy of EWM with rectangular reducer in non-ideal flow field, is an urgent problem to be solved.

At present, scholars at home and abroad have carried out relevant research on the optimization of EWM rectangular reducer structure. Li et al. [6] theoretically analyzed the rectangular tube electromagnetic flowmeter (EMF) and proved that the induced electromotive force of rectangular tube EMF is larger than that of circular tube EMF. Heijnsdijk et al. [7] changed EMF's original circular tube into a reduced diameter reducer of different shapes such as a rectangle to make the excitation coil closer, so as to improve EMF's measurement sensitivity. Liu et al. [8] simulated the three-dimensional flow field of electromagnetic flow sensor with rectangular reducer, and found that the velocity distribution in the electrode cross section of rectangular measuring area is more uniform, which is beneficial to measure flow more accurately. Chen et al. [9] used Fluent software to simulate the flow field of EMF reducing pipe, and studied the influence on the pressure loss and the central section's average velocity, when the length, width and height of rectangular reducing pipe section are changed individually. However, this study didn't consider the influence on pressure loss when the length, width and height were changed simultaneously. Ge et al. [10] carried out a three-dimensional simulation of electromagnetic flow sensor with rectangular reducer, and studied influence of rectangular section's length, width and height on the sensor's weight function distribution. However, the function model between weight function uniformity and rectangular section's length, width and height has not been established. Liang et al. [11] constructed the nonlinear function model between pressure loss and EWM rectangular reducer's length, width and height by the method combining orthogonal experiment and approximate linear modeling, and carried out experimental verification. However, the approximate linearization modeling method based on step-by-step fitting is not suitable for parameters of large range. Moreover, the function model between weight function uniformity and rectangular reducer's length, width and height is not established through the same experimental design samples. To sum up, the weight function and pressure loss are calculated by finite element simulation in the existing research thesis, which is more convenient than the traditional prototype making method. However, due to the large number of geometric model variables, large number of finite element simulation samples takes a long time, which is inefficient and is not conducive to the subsequent optimization process. If the constraint object such as pressure loss restriction changes, all simulation and calculation processes need to be restarted, which is time-consuming and labor-intensive. Moreover, the high weight function uniformity and low pressure loss constraint in the optimization process are a pair of contradictory performance objectives. Rectangular reducer structure with more uniform weight function distribution tends to make the pressure loss larger [10, 11]. At present, there is no literature to study and solve the problem of simultaneously optimizing the two objectives of high-weight function uniformity and low-pressure loss constraints of EWM with rectangular reducer. Duc to the complexity of reducer structure, it is difficult for experienced technicians to obtain the suitable optimal rectangular reducer structure parameters according to their experience.

In order to solve this problem, this thesis proposes an optimal design method for rectangular reducer structure parameters based on the response surface method and the multi-objective genetic algorithm. First, taking the measuring tube of DN100 EWM with rectangular reducer as the optimization object, and finite element simulation model of the measuring tube is established. Then, mathematical models of weight function uniformity and pressure loss are established by response surface method based on central composite design [12, 13, 14], revealing the essential characteristics between the two objects (weight function uniformity, pressure loss) and the rectangular reducer structure parameters. Finally, the multi-objective genetic algorithm (NSGA-II) [15] is used to obtain the Pareto optimal solution set that satisfies both weight function uniformity and pressure loss requirements [2]. In this way, even if there are different pressure loss constraint requirements, the Pareto optimal solution set can be used to quickly obtain the optimal rectangular reducer structure parameters which meet the pressure loss constraints and high weight function uniformity requirements.

2 Description of Structure Optimization Problem

2.1 Three-Dimensional Modeling and Verification of Rectangular Reducer

In order to study the influence of EWM rectangular reducer structure parameters on weight function uniformity and pressure loss, a three-dimensional finite element model of rectangular reducer was established, as shown in Fig. 1. The Pipe diameter D of water meter is 100 mm, and the total pipe length L is 250 mm. The measuring pipe is mainly composed of a reducing section and two transition sections. The reducing section is a rectangular structure, the length of rectangular structure is l, the width is w, and the height is h. The finite element calculation is performed on the EWM measuring tube according to actual boundary conditions, and response values of weight function uniformity U [10] and pressure loss ΔP [11] of EWM are obtained. The smaller U is, the more uniform weight function is. The smaller ΔP is, the smaller EWM measuring tube's pressure loss is.

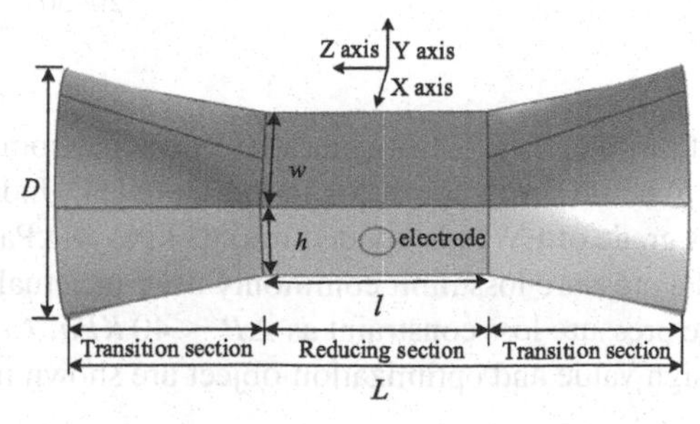

Fig. 1. Three-dimensional structure of EWM measuring tube.

The accuracy of EWM's weight function simulation model has been verified [10]. The curve comparison between simulation results of EWM pressure loss model and experimental results is shown in Fig. 2. It can be seen that under different flow points, simulation results of pressure loss model are highly consistent with experimental results. Therefore the pressure loss simulation model has high accuracy and can be used for the subsequent optimization design of the rectangular reducer structure.

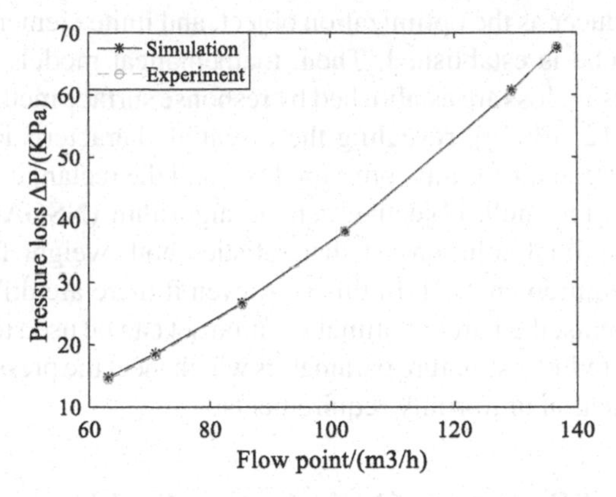

Fig. 2. Comparison of simulated pressure loss and experimental pressure loss of EWM.

2.2 Setting of Design Variables, Optimization Goals and Constraints

The length *l*, width *w*, and height *h* of the rectangular reducer's rectangular section are used as design variables. The initial value and value range are shown in Table 1.

Table 1. Initial value and value range of rectangular section's length, width and height.

Design variable	Initial value/mm	Value range/mm
Length *l*	80	60–120
Width *w*	80	60–90
Height *h*	25	20–50

The economical flow rate of EWM is generally 3 m/s, and common flow rate of the sensor is determined to be 100 m^3/h according to literature [1]. Under this common flow rate, the pressure loss grade of EWM is divided into 63 kPa, 40 kPa, 25 kPa, 16 kPa and 10 kPa. Taking 40 kPa pressure loss limit commonly used in actual working conditions as an example, set the pressure loss constraint as $\Delta P \leq 40\,KPa$. At this time, the output response's initial design value and optimization object are shown in Table 2.

Table 2. Output response.initial design and optimization object.

Category	Output response	Initial design	Optimization object
Object responses	Weight function uniformity U	0.9368	Minimize
	Pressure loss ΔP	44.672 kPa	Minimize
Constraint response	Pressure loss ΔP	44.672 kPa	$\leq$ 40 kPa

2.3 Optimization Design Idea

Length l, width w, and height h of the rectangular reducer's rectangular section simultaneously affect the weight function uniformity U and pressure loss ΔP of EWM. The two performance requirements of high weight function uniformity (U minimized) and low pressure loss (ΔP minimized) are mutually contradictory. The rectangular reducer structure that makes weight function distribution more uniform, makes pressure loss larger. In order to obtain the rectangular reducer structure that satisfies requirements of U and ΔP at the same time, in this thesis, EWM magnetic circuit structure parameters' optimal design process with weight function uniformity and pressure loss as objectives or constraints is shown in Fig. 3. Firstly, establish a three-dimensional finite element

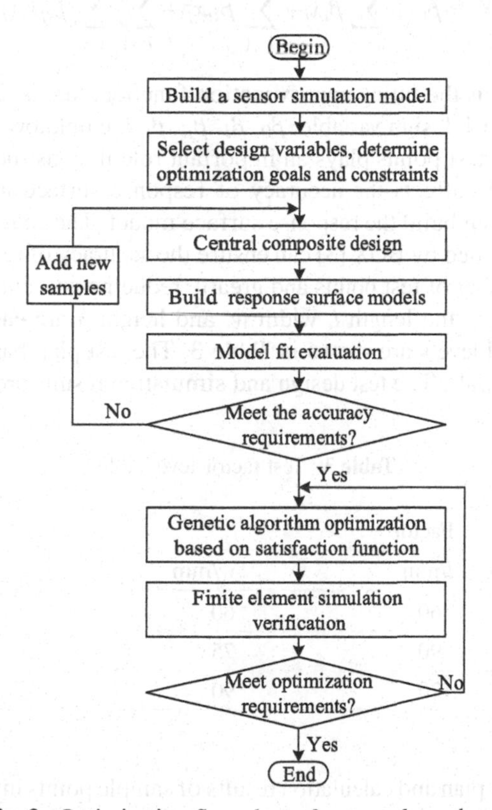

Fig. 3. Optimization flow chart of rectangular reducer.

simulation model of EWM rectangular reducer, and perform finite element calculation on weight function uniformity U and pressure loss ΔP of rectangular reducer. Secondly, response surface method is used to establish nonlinear response surface function models of U and ΔP. Finally, through multi-objective genetic algorithm optimization, the Pareto optimal solution set that make U and ΔP meet the requirements is obtained. From this solution set, the corresponding optimal rectangular reducer structure parameters can be screened conveniently and quickly according to different pressure loss constraints.

3 Establishment and Verification of Response Surface Model

3.1 Model Building

Response Surface Methodology (RSM) was first proposed by Box and Wilson [16]. It is a mathematical statistical method to solve experimental design with multivariate problems, which uses multiple regression equations to fit the functional relationship between factors (design variables) and response values (design goals).In order to get a accurate curved response surface, response surface function represented by a quadratic polynomial with cross terms as the main body is usually used.

$$Y = \beta_0 + \sum_{i=1}^{n} \beta_i x_i + \sum_{i=1}^{n} \beta_{ii} x_i^2 + \sum_{i=1}^{n-1} \sum_{j=1}^{n} \beta_{ij} x_i x_j. \tag{1}$$

In the Eq. (1), Y is the fitting approximation function, that is, the response surface; x_i is the n-dimensional design variable; β_0, β_i, β_{ii}, β_{ij} are unknown coefficients.

The selection of test points plays an important role in construction of the response surface, they directly affects the accuracy of response surface model. Unreasonable test points cannot even build the response surface model. The central composite design (CCD) method proposed by BOX [3] can ensure the accuracy of response surface while minimizing the number of test points and greatly reducing amount of calculation.

In the trial design, the length l, width w, and height h are each taken at 3 levels. The trial factors and levels are shown in Table 3. The test plan based on CCD method includes 20 sets of trials. The test design and simulation results are shown in Table 4.

Table 3. Test factor level table.

Level	Factor		
	l/mm	w/mm	h/mm
Level 1	60	60	20
Level 2	90	75	35
Level 3	120	90	50

Based on the test plan and calculation results of sample points in Table 4, the response surface function models of U and ΔP are established, as shown in Eq. (2–3).

Table 4. Experimental design and simulation results.

Test number	Factor level bar			Simulation result bar	
	l/mm	w/mm	h/mm	Weight function uniformity U	Pressure loss ΔP/ KPa
1	90	60	35	1.0265	37.938
2	60	90	20	0.8365	51.937
3	90	75	35	1.1521	20.366
4	90	75	35	1.1521	20.366
5	90	75	35	1.1521	20.366
6	120	60	50	1.3339	13.094
7	120	90	20	0.8134	65.539
8	90	75	50	1.4768	4.640
9	60	75	35	1.1694	16.696
10	120	75	35	1.1569	23.145
11	90	75	35	1.1521	20.366
12	60	60	20	0.6598	140.004
13	120	60	20	0.6481	186.378
14	60	90	50	1.5905	0.987
15	90	75	35	1.1521	20.366
16	60	60	50	1.3334	9.600
17	90	90	35	1.2521	10.324
18	90	75	35	1.1521	20.366
19	90	75	20	0.7342	101.916
20	120	90	50	1.5679	2.786

$$
\begin{aligned}
U = &-0.60263 - 1.30532 \times 10^{-3} l + 0.014960 w + 0.033016 h - 9.60900 \times 10^{-6} lw \\
&+ 8.28102 \times 10^{-5} wh + 9.97136 \times 10^{-6} l^2 - 6.62339 \times 10^{-5} w^2 - 2.16517 \times 10^{-4} h^2 .
\end{aligned} \quad (2)
$$

$$
\begin{aligned}
\Delta P = &\exp(8.95010 - 0.014782 l - 0.029858 w - 0.049975 h + 1.87087 \times 10^{-4} lw \\
&+ 2.30317 \times 10^{-4} lh - 9.91902 \times 10^{-4} wh) .
\end{aligned} \quad (3)
$$

3.2 Model Fit

Use R^2 (coefficient of determination) and R^2_{adj} (adjusted coefficient of determination) to test the fit of response surface function models of U and ΔP, as shown in Eq. (4) and (5). R^2 and R^2_{adj} are between 0.9 and 1.0, which means that the response surface function model is close to the actual model [4, 5].

$$R^2 = 1 - \frac{\sum_{i=1}^{n} (y_i - \hat{y}_i)^2}{\sum_{i=1}^{n} (y_i - \overline{y})^2}. \tag{4}$$

$$R^2_{adj} = 1 - \left(\frac{n-1}{n-p}\right)\frac{\sum_{i=1}^{n} (y_i - \hat{y}_i)^2}{\sum_{i=1}^{n} (y_i - \overline{y})^2}. \tag{5}$$

In Eq. (4–5), y_i-sampling point's actual value, $\hat{y}_i$-sampling point's predicted value, $\overline{y}$-predicted samples' average value, n – samples' number, and p – factors' number.

The fitting metric values of U and ΔP response surface function models are shown in Table 5, it can be seen that accuracy of the established models is better. In order to assess accuracy of U and ΔP response surface function models, a set of rectangular section's length l, width w and height h are randomly selected for finite element simulation, and simulation results are compared with calculation results of the response surface function models as shown in the Table 6. The errors between calculation results of U and ΔP response surface function models and the finite element simulation results are within ± 3.36% and ± 5.87%, respectively, indicating that the response surface function models of U and ΔP are accurate and reliable.

Table 5. Fitting metrics.

Fitting metrics	U	ΔP
Coefficient of determination R^2	0.9997	0.9965
Adjusted coefficient of determination ${R_{adj}}^2$	0.9995	0.9948

Table 6. Accuracy of response surface function model verified by finite element simulation.

Structural parameters /mm			Weight function uniformity U			Pressure loss ΔP/KPa		
l	w	h	Finite element model	Response surface model	Error /%	Finite element model	Response surface model	Error /%
60	75	35	1.1694	1.1688	−0.05	16.560	16.388	−1.04
70	90	20	0.8246	0.8270	0.29	52.524	51.641	−1.68
80	70	40	1.2649	1.2317	−2.63	13.857	14.671	5.87
90	70	45	1.3347	1.3309	−0.28	9.369	9.657	3.08
100	80	25	0.9063	0.9183	1.32	48.552	50.530	4.07
110	65	50	1.3876	1.3769	−0.77	9.562	9.634	0.75
120	85	40	1.2917	1.3351	3.36	10.164	9.783	−3.75

4 Structure Optimization Method Based on Pressure Loss Constraint

EWM rectangular reducer structure optimization is a multi-objective optimization problem. In the multi-objective optimization problem, although the two goals of weight function uniformity U and pressure loss ΔP can be achieved optimal in theory, respectively. But the two performance requirements are mutually contradictory, that is, rectangular reducer structure parameters with high weight function uniformity U will cause the pressure loss ΔP to increase, and generally there is no optimal solution to achieve all the goals at the same time. Therefore, in order to find the optimal parameters that satisfy both high-weight function uniformity (U minimization) and low-pressure loss constraints, this thesis uses a multi-objective genetic algorithm.

4.1 Multi-objective Optimization Based on Genetic Algorithm

Set the goal of minimizing U (that is, the most uniform weight function) and minimizing pressure loss ΔP, as shown in Eq. (6). The design range of pressure loss and parameters l, w, and h are respectively taken as constraints, as shown in Eq. (7).

The mathematical model representing the multi-objective optimization problem of EWM rectangular reducer structure is described as.

$$\text{Objects}:\begin{cases} Find & l, w, h \\ Minimize & U, \Delta P \end{cases}. \tag{6}$$

$$\text{Constraints}:\begin{cases} \Delta P \le 40\,KPa \\ 60\,\text{mm} \le l \le 120\,\text{mm} \\ 60\,\text{mm} \le w \le 90\,\text{mm} \\ 20\,\text{mm} \le h \le 50\,\text{mm} \end{cases}. \tag{7}$$

In Eq. (6) and (7), l, w, and h are design variables; U and ΔP are response surface functions of weight function uniformity and pressure loss, respectively.

The NSGA-II multi-objective genetic algorithm [17] is used to obtain the Pareto optimal solution set of the multi-objective optimization problem. The MATLAB simulation parameters are set as follows: initial population is 1000, maximum number of iterations is 1000, crossover probability is 0.9, and mutation probability is 1/3.

The Pareto optimal solution set obtained after iterative calculation is shown in Fig. 4. The horizontal and vertical coordinates respectively correspond to the response surface function values of weight function uniformity U and pressure loss ΔP.

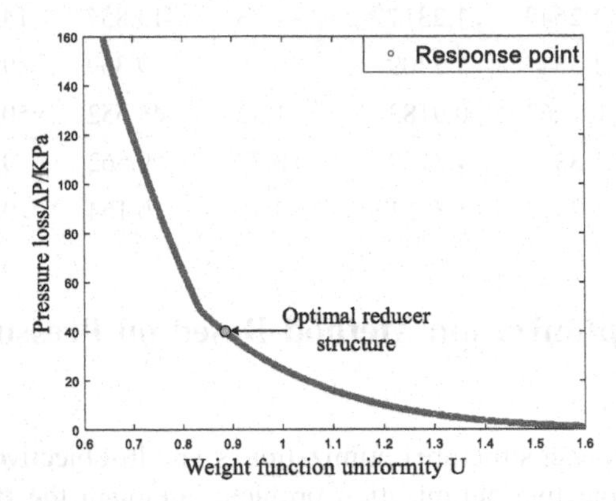

Fig. 4. Pareto optimal solution set of weight function uniformity U and pressure loss ΔP.

It can be seen from Fig. 4 that the two goals of high weight function uniformity (U minimization) and low pressure loss (ΔP minimization) are contradictory. The more uniform weight function distribution (the smaller U), the larger pressure loss ΔP. However, the non-dominated solution in this Pareto optimal solution set has the least conflicts compared with other solutions, which provides a better choice space for decision-makers. Therefore, reasonable design parameters can be selected from the Pareto optimal solution set according to design requirements. According to the performance optimization requirements of Table 2 (when pressure loss $\Delta P \leq 40\,KPa$, make U the smallest (that is, weight function is most uniform)), a set of optimal rectangular reducer structure parameter combination is selected from the Pareto optimal solution set in Fig. 4: $l = 60.08$ mm, $w = 90.00$ mm, $h = 21.55$ mm, the parameter combination is rounded to get: $l = 60$ mm, $w = 90$ mm, $h = 22$ mm.

4.2 Simulation Verification of Optimization Results

The three-dimensional modeling of the optimal rectangular reducer structure is carried out, and values of U and ΔP are simulated, and compared with the rectangular reducer's simulation results before optimization, as shown in Table 7. The comparison results show that, after optimization, U is reduced by 4.20% (that is, weight function uniformity is increased by 4.20%), and the pressure loss ΔP is reduced by 10.84%, which meets the pressure loss constraint ($\Delta P \leq 40\,KPa$).

Table 7. Comparison of results before and after optimization of structure parameters.

Parameter	l/mm	w/mm	h/mm	U	ΔP/KPa
Structural parameters before optimization	80	80	25	0.9368	44.672
Structural parameters after optimization	60	90	22	0.8975	39.831

5 Conclusion

In order to solve the problem of EWM rectangular reducer structure parameters optimization with high weight function uniformity under the constraint of actual pressure loss, this thesis proposes a new rectangular reducer structure optimization design method. This method combines the response surface method and the multi-objective genetic algorithm to solve the EWM rectangular reducer structure optimization problems with multi-variable constraints, multi-objective constraints, and nonlinear difficulties. The research conclusions are as follows:

(1) Based on central composite design method, non-linear response surface function models of weight function uniformity U and pressure loss ΔP are established. The theoretical relationship between performance objectives and structural parameters is revealed, which provides a new research perspective.
(2) In the process of multi-objective optimization, although the sub-objective can't achieve optimal at the same time, the global Pareto optimal solution set is obtained by using multi-objective genetic algorithm. Through the solution set, the ideal structural parameters can be obtained quickly according to different pressure loss requirements.
(3) Results show that, after parameters optimization, the pressure loss of rectangular reducer decreases and meets the constraint requirements, and weight function uniformity is improved, which is helpful to improve measurement accuracy of water meter. This method can also be used to optimize the structure of other flow meters.

References

1. OIML, Water meters for cold potable water and hot water Part1: Metrological and technical requirements. In: OIML R 49–1 Edition (2013) (2013)
2. Jiao, L.C.: Multi Objective Optimization Immune Algorithm, Theory and Application. Science Press, Beijing (2010)
3. George, E.P., Box, Norman R.: Empirical model-building and response surfaces. M. Wiley, New Jersey (1987)
4. Zhao, G.F., Jin, W.L., Gong, J.X.: Structural Reliability Theory. China Construction Industry Press, Beijing (2000)
5. Zhao, Y.G., Ono, T.: Moment methods for structural reliability. J. Struct. Saf. **23**, 47–75 (2001)
6. Li, B., Yao, J., Xia, L.: The analysis and application of the rectangular electromagnetic flowmeter. In: Instrumentation & Measurement Technology Conference, IEEE (2003)

7. Heijnsdijk, A.M., Van, Willigen A.L., Lodge, G.R., et al.: Magnetoinductive flowmeter and method for producing a magnetoinductive flowmeter. US (2007)
8. Liu, T.J., Gong, T.S., Chen, Y.J.: Research on low power electromagnetic flowmeter with reducing measuring tube. J. Sens. Technol. **26**, 348–352 (2013)
9. Chen, Y.J., Liu, T.J., Xie, D.L.: Simulation study on flow field of electromagnetic flowmeter in reducing pipe. J. China Univ. Metrol. **23**, 227–231 (2012)
10. Ge, Y.S., Liang, L.P., Xu, K.J., et al.: Study on weight function distribution of electromagnetic flow sensors with reducer. Sens. Microsyst. 25–28+32 (2019)
11. Liang, L.P., Ge, Y.S., Xu, K.J., et al.: Design method for flow tube structure of electromagnetic water meter with shrunk measurement tube based on pressure loss-flow restriction. Flow Meas. Instrum. **74**, 101778 (2020)
12. Esfe, M.H., Mahian, O., Hajmohammad, M.H., et al.: Design of a heat exchanger working with organic nanofluids using multi-objective particle swarm optimization algorithm and response surface method. Int. J. Heat Mass Transfer. **119**, 922–930 (2018)
13. Subasi, A., Sahin, B., Kaymaz, I., Multi-objective optimization of a honeycomb heat sink using response surface method. Int. J. Heat Mass Transf. **101**, 295–302 (2016)
14. Jiang, H., Guan, Y.S., Qiu, Z.C., et al.: Dynamic and static multi-objective optimization of a vertical machining center based on response surface method. Chinese J. Mech. Eng. **47**, 125–133 (2011)
15. Deb, K., Pratab, A., Agarwal, S., et al.: A fast and elitist multi-objective genetic algorithm: NSGA-II. IEEE Trans. Evol. Comput. **6**, 182–197 (2002)
16. Box, G.E.P., Wilson, K.B.J.R.: On the experimental attainment of optimum conditions. J. Roy. Stat. Soc. **13**, 1–45 (1951)
17. Xuan, G.N.: Genetic Algorithm and Engineering Optimization. Tsinghua University Press, Beijing (2004)

A Study on Forest Flame Recognition of UAV Based on YOLO-V3 Improved Algorithm

Zhen Wang(✉), Huidan Zhang, Muxin Hou, Xiaoting Shu, Jianguo Wu, and Xiaoqian Zhang

School of Electrical and Energy, Engineering Nantong Institute of Technology, Nantong, Jangsu, China

Abstract. In recent years, all regions of China have constantly paid attention to forest fire prevention, which however is still restricted to onsite observation carried out by forest ranger and basic satellite resource survey. The use of UAV system for forest fire monitoring is still in its infancy. To bridge the gap, this study trains the YOLO-V3 algorithm for forest fire detection based on UAV collected data. Traditional flame detection models are commonly based on RGB colors. They can suffer low accuracy and detection speed, and it is still difficult for the YOLO-V3-based model to detect small flames. In this paper, the YOLO-V3 model is improved to support multi-feature detection. Specifically, 208208 smaller resolution feature scales are added to allow the model learning shallow features of flame images. In this way, the learning ability of the proposed model for shallow image information is improved in the feature extraction stage, which can facilitate the dentification of small flame regions. In addition, the prior box is optimized to further improve detection precision. In the experiment, the mAP value can reach 67.6% with detection speed of 190FPS.

Keywords: Forest fire · UAV · YOLO-V3 · Flame identification

1 Introduction

Forest resources have an extremely important impact on human beings, not only to cultivate a biodiversity environment, but the forest industry also provides a lot of employment opportunities. However, China's forest resources are very scarce, and forest fires occur frequently. And the forest fire is also the main cause of destroying the forest ecological environment [1]. Natural disasters such as bare burned forests, floods, landslides and debris flows are also counted, which will also cause serious difficulties to the production and life of people living in the forest areas. Therefore, forest fire monitoring is imperative.

Over the past few decades, forest fire detection relies mainly on ground fire monitoring systems, but there are different technical and practical problems in these systems, low spatial-temporal resolution, poor flexibility, expensive prices, environmental interference, and artificial omissions. Advanced scientific and technological means need to be improved. Visual-based detection techniques can capture and provide intuitive and high

K. Li et al. (Eds.): LSMS 2021/ICSEE 2021, CCIS 1468, pp. 497–503, 2021.
https://doi.org/10.1007/978-981-16-7210-1_47

real-time information, and can easily cover a wide range of observations while reducing development costs. The rapid development of electronics, computer science, and digital camera technology has made it possible to apply computer vision-based systems to fire monitoring. Vision-based systems have become an important part of the UAV's forest fire detection system [2, 3].

At present, the forest fire monitoring method usually includes ground patrol, observation tower garrison, air patrol, remote video monitoring and satellite remote sensing. The common scope of artificial and observation tower technology is limited, the safety of personnel is not guaranteed, manual inspection affects the line of sight, the workload is large and low, it is difficult to achieve real-time information transmission.

In the studies already conducted, most researchers tend to combine the color characteristics of the flames with the dynamic properties to provide more reliable identification results [4]. They used a flame detection scheme based on support vector machines (SVM), by using brightness maps, moving unless the flame pixel region. Effective extraction of fire pixels is solved using the advantages of Lab color model that can obviously display fire color features [5]. To further improve the accuracy of forest fire detection, many researchers have taken the image texture analysis method to propose some theoretical analysis of forest fire wavelets, and simply introduce an image bandpass filter based on previous work. Often, video-based forest flame detection techniques use fixed cameras to facilitate separating the moving flame from a stationary background. However, due to the motion properties of the UAV itself, the method of extracting the forest fire area in the image changes the force unchanged from the heart [6]. The fire itself has the unknown, complexity, frequency, rapid and other characteristics, once it occurs, easy to cause serious economic losses and casualties. Accurate flame warning can greatly reduce the loss. In video-based flame detection methods, flame detection using color is the earliest method. However, the method requires a large range of flame area. In order to improve the accuracy of flame recognition, the researchers added the flame movement characteristics based on the color characteristics, improving the accuracy of flame recognition, but the error detection rate is still high.

This paper will focus on deep learning-based forest fire detection methods. Deep learning is a very rapid research field in recent years, and has achieved great achievements in image processing. This paper combines UAV with deep learning technology and introduces YOLO-V3 detection method to realize real-time detection of forest fire. Figure 1 shows a forest fire visual detection system based on the UAV as a carrier [7].

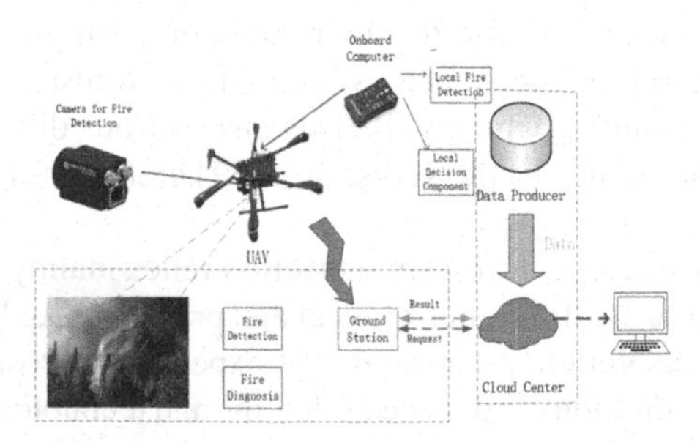

Fig. 1. UAV-based forest fire detection system.

2 Flame Recognition Algorithm

2.1 Flame Recognition Algorithm

YOLO-V3 introduces the residual module and improves the target detection accuracy and rate by reducing the deep network training complexity. Among them, Darknet-53(as shown in Fig. 2) as the backbone network of YOLO-V3, uses convolution 32,16,8 times downsampling of 2, conducts target detection, splicing feature information, integrates the characteristics of three different scales: $13 \times 13, 26 \times 26$ and 52×52, connects deep and shallow features, and achieves the effect of learning multi-scale features at the same time. Despite the strong YOLO-V3 target detection capability, the recognition capabilit of small targets is still limited, and during flame detection, it is required to prevent flame spreading. For the problems of YOLO-V3 in video image flame detection, this paper proposes the improved real-time flame detection algorithm of YOLO-V3, and entally verifies the feasibility of this method in flame detection and the accurate identification of in small flame area [3].

When Darknet-53 network extracts feature information, shallow feature network division is small, mainly provides location information; deep feature network division is large, mainly provides semantic information. In order to utilize shallow position information and improve target detection accuracy, YOLO-V3 uses 3 scale fusion methods ($13 \times 13, 26 \times 26, 52 \times 52$) for multi-scale detection. The characteristic scale with the smallest resolution is 52×52, and the grid division relative to the 416×416 pixel input image is still not fine enough. After layers of convolution calculation, some information will be lost, resulting in a waste of shallow characteristic information, resulting in its detection accuracy of the small flame area is not high [8, 9].

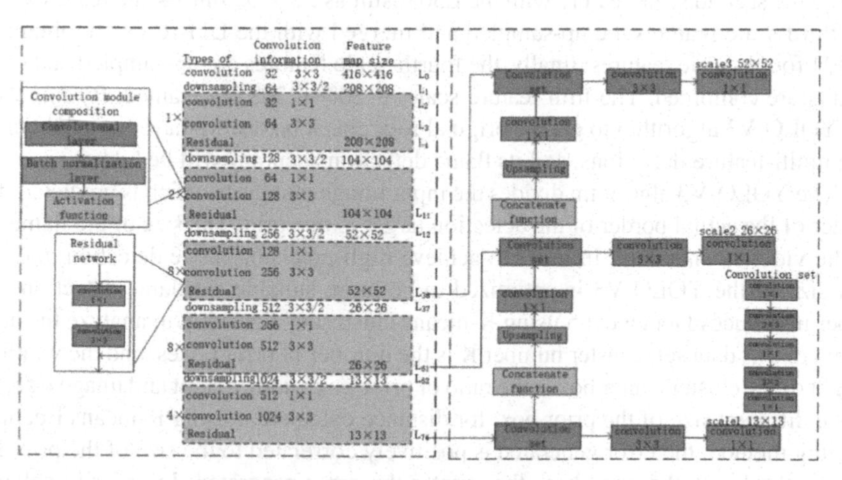

Fig. 2. YOLO-V3 network structure.

In order to improve the network fusion multi-scale feature capabilities and accurately identify small flame regions, this paper improves the YOLO-V3 multi-scale detection network and adds feature scales with smaller resolution to fully learn the shallow features. Because the new feature scale will increase the network complexity, pull down the detection rate, comprehensively consider the flame detection accuracy and rate requirements, add 208 × 208 scale, and improve to 5 scale detection. The input image size is 416 × 416 pixels, and the improved network model is shown in Fig. 3.

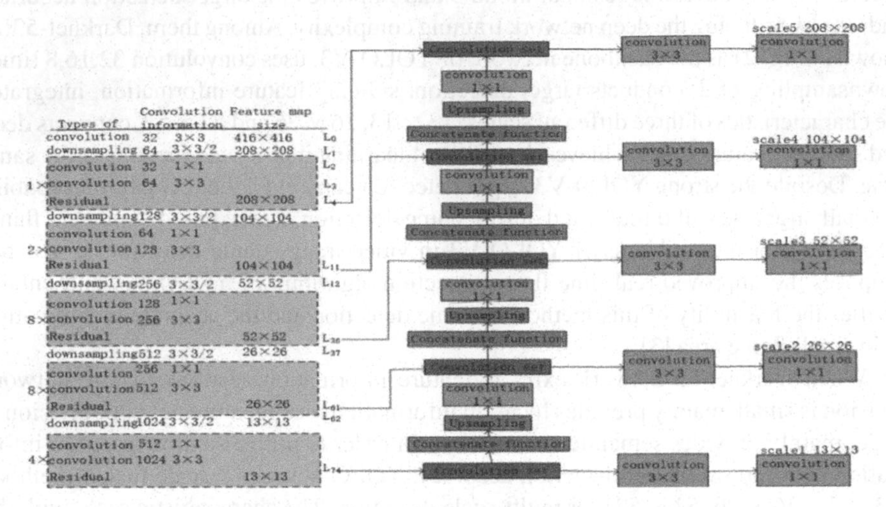

Fig. 3. Improved Multi-scale Detection Network.

First, take L74 (74 layers) as the first scale feature 13 × 13, perform convolution and upsampling operations, and merge with L61 to obtain a 26 × 26 s scale feature; secondly, merge the second scale feature with the L36 result as 52 × 52 third-scale features; then, the third-scale features are up-sampled and merged with the L11 results to obtain 104 × 104 fourth-scale features; finally, the fourth-scale features are up-sampled and the L4 results are combined, The fifth feature scale of 208 × 208 is obtained. By improving the YOLO-V3 algorithm to get the original four check boxes, we have turned them into five multi-feature detections, and its flame detection accuracy will be higher.

The YOLO-V3 algorithm divides the input image into grids which is predicted if the center of the actual border of the detection target is in a grid. The size of the flame area in the video is uncertain. In order to achieve high-precision flame detection, the prior box size of the YOLO-V3 is optimized to be more suitable for flame detection. This paper introduces Document 5 using K-means clustering algorithm to analyze the actual frame of this data set. Cluster number K is the number of prior boxes, and the width and height of the cluster center box is the ratio of prior box width, height and image size. Due to the different size of the prior box, for distance calculation using K-means Euclidean ranging method, the error generated is positively correlated to the size of the prior box, that is, the larger the prior box, the greater the error generated. In order to solve the above problems, and the intersection between the prior box and the actual border of the detection target is larger, this paper calculates the improved distance formula (1) and uses the I value to calculate the distance between the prior boxes. The larger the value, the smaller the distance between the prior boxes. The distance between the prior boxes is:

$$d(b, c) = 1 - I(b, c). \tag{1}$$

In the formula, b is any prior box; c is the center of the corresponding prior box; I is the ratio of the overlap area and the total area between the prior box. The optimized K-means algorithm obtains the relationship between the number of prior boxes K and I. With the increase of K value, the I value increases. When K = 12, the I value flatten, considering the rate and accuracy of network flame detection, the number of prior boxes is set at 12, and the 12 initial cluster centers are randomly generated. The calculation is iterated by formula (1) until the cluster center distance change is less than the threshold (0.003 in this article), and the final cluster center coordinates are obtained. The improved target detection network has five dimensional feature maps, presented using the YOLO-V3 algorithm, with each one matched with three prior boxes.

3 Analysis of the Experimental Results

3.1 Experimental Environment

The experimental dataset used in this paper is 416 × 416 pixels. The architecture, training, and testing of the improved YOLO-V3 model are performed on the TensorFlow deep learning framework Keras, implemented on GPU with the Win10 system NVIDIA GeForce GTX1650 Ti.

Table 1. Data set description/sheet.

Dataset name	Flame image	Non-fire image	Total number
Training set	2000	500	2500
Validation set	500	200	700
Test set	15667	6418	22085
Total number	18167	7118	24285

The 2000 images obtained from the disclosed flame dataset Dunnings, Andy, and 500 images downloaded via the Internet, constitute the training set of this experiment, as shown in Table 1. The training set data is divided into flame images and non-fire images.

A verification set consisting of 500 flame data and 200 non-fire data downloaded from the Internet. The 30,000 images, extracted from the Cristiano Rafael Steffens, constitute the test set of this experiment. Test set Fig. 4 video is flame video.

3.2 Comparison of the Experimental Results

Using the PASCAL VOC 2012 dataset as the test and training dataset, the improved YOLO-V3 algorithm is compared with the original YOLO-V3, Faster R-CNN, SSD, and

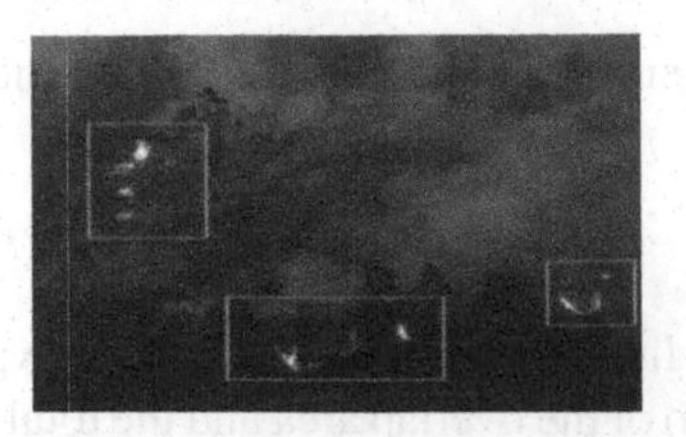

Fig. 4. Video

the experimental results take the mAP and FPS values as the evaluation indicators.FPS represents the number of pictures per second the model can detect, and the greater the FPS value, the faster the model detects. The mAP index is the mean of various AP s, and the formula is shown in (2):

$$mAP = \frac{\sum_{i}^{C} AP}{C}. \tag{2}$$

In formula (3), C is the number of sample types; AP is the average accuracy of the sample.

The mAP and FPS for 4 algorithm models are shown in Table 2. As can be seen from Table 2, the mAP value of the improved YOLO-V3 algorithm is higher than the original YOLO-V3 algorithm, by about 2.4% but lower than the Faster R-CNN and SSD300 algorithms; where the value is lower compared with the original YOLO-V3 algorithm, but far beyond the Faster R-CNN and SSD300 algorithms. This is because although the backbone network selects the Darknet-49 network with fewer layers than the original network layer, it also increases the number of predicted and output layers of the network, having a certain impact on the detection speed, resulting in the FPS value reduction. Overall, the improved YOLO-V3 algorithm mAP, whose detection speed is close to the original algorithm [10].

Table 2. mAP and FPS of 4 algorithm models.

Method	mAP /%	FPS
Faster R-CNN	73.3	7.0
SSD300	74.4	48.0
YOLO-V3	65.1	200.0
Improved after YOLO-V3	67.6	190.0

The AP value of the YOLO-V 3 algorithm before and after the improvement shows that the m AP value of the improved YOLO-V3 is increased by about 2.5%, and the AP value of the corresponding categories is improved to a certain extent. The large scale picture itself has achieved high accuracy in the corresponding AP value before and after the improvement.

4 Conclusion

This article was funded by the Nantong Science and Technology Bureau Project JC2019096.With the intelligent development of visual UAV system, real-time flame detection is beneficial to control the spread of fire. This paper optimizes the multi-scale detection network in the original YOLO-V3 algorithm, applies the optimized algorithm to the video image flame recognition field, strengthens the network's ability to identify the small flame area, improves the video image flame recognition accuracy and reduces the error detection rate, but also realizes real-time detection. The effectiveness of the present algorithm is demonstrated through multiple group experiments, which can be more efficiently applied to video flame detection as compared to existing techniques.

References

1. Li, Q., Hou, J., He, N., Xu, L., Zhang, Z.: Changes in leaf stomatal traits of different aged temperate forest stands. J. For. Res. **32**(3), 927–936 (2020)
2. Nuri, N.S., Endah, S.: A seed rain community in a reforested post-agricultural fieldand adjacent secondary forest of Mount Papandayan nature reserve, west java, Indonesia. J. For. Res. **34**(03), 1013–1023 (2021)
3. Jin, Y.H., Wang, W.K., Lee, W.H., Jaegeol, Y.: An indoor location-based positioning system using stereo vision with the drone camera. Mob. Inf. Syst. **3**, 5160543–5516559 (2018)
4. Michael, F., Jeremy, B.: US Navy "brings down" Iranian UAV threatening LHD in Strait of Hormuz. J. Jane's Defence Weekly **56**(30), 4 (2019)
5. Frizzi, S., Kaabi, R., Bouchouicha, M.: Convolutional Neural Network for Video Fireand Smoke Detection. In: Conference of the IEEE Industrial Electronics Society, pp. 529--542. IEEE Press, Italy (2016)
6. El Bouhissi, M., Bouidjra, S.E.B., Benabdeli, K.: GIS, forest fire prevention and risk matrix in the national forest of Khoudida, Sidi Bel Abbes, Algeria. Open J. Ecol. **10**(06), 356–369 (2020)
7. Shi, R., Li, T., Ya, M.: An attribution-based pruning method for real-time mango detection with YOLO network. Comput. Electron. Agric. **169**, 105214–105219 (2020)
8. Shaikh, S.A., Abdul, Kadir, K.S.: Object detection and tracking using YOLO v3 framework for increased resolution video. Int. J. Innov. Technol. Exploring Eng. **9**(6), 141–148 (2020)
9. Zhao, Y.Y., Zhu, J., Xie, Y.K., Li, W.L., Guo, Y.K.: Improved YOLO-V3 video image flame real-time detection algorithm. Inf. Sci. Ed. **6**, 78–82 (2021)
10. Huang, Z.C., Wang, J.L., Fu, X.S., Yu, T.G., Wang, Y.Q.: DC-SPP-YOLO, dense connection and spatial pyramid pooling based YOLO for object detection. Inf. Sci. **3**, 551–556 (2020)

Water Quality Prediction Based on Data Mining and LSTM Neural Network

Liqin Zhao[1], Senjun Huang[2], and Jiange Jiao[1](✉)

[1] College of Mechanical and Electrical Engineering, China Jiliang University, Hangzhou 310018, China
careerjiao@cjlu.edu.cn

[2] Power China Huadong Engineering Corporation Limited, Hangzhou 311122, China

Abstract. Water pollution exacerbates water shortages affecting human health and quality of life. Water quality prediction is of great significance in the future water quality management. In this thesis, the internal relations among dissolved oxygen, temperature, pH and turbidity were revealed by using grey correlation analysis method. Furthermore, the LSTM neural network was used to predict dissolved oxygen in water. The results showed that dissolved oxygen is closely related to temperature and pH. Temperature and dissolved oxygen are negatively correlated, and pH is positively correlated with dissolved oxygen. Dissolved oxygen, which affects the key indicators of water quality, has a good prediction effect, with an accuracy of more than 90%. The research results provided valuable references in water pollution control and water resources management.

Keywords: Water quality prediction · Grey relational analysis · LSTM

1 Introduction

The era of big data has produced a large amount of data which is closely related to our lives. Can solve conventional problems, provide new ideas and new methods by mining and analyzing these data. Data mining has been applied to various fields, such as analysis of diseases, accidents analysis, is a highly precise method of analysis. Zhang et al. [1] used K-mean algorithm to classify historical data. Based on the improved BP neural network, wavelet analysis was introduced to construct the wavelet neural network. At the same time, genetic algorithm was used to conduct global optimization of the initial parameters of the network to obtain the optimal parameters; Babbar et al. [2] proposed to use data mining technology to quickly predict river water quality grade, and established water quality classification and prediction model based on pollution index by using K-nearest neighbor, decision tree, Naive Bayes and artificial neural network and other commonly used classification technologies such as rules and support vector machine; Kusiak [3] used multi-layer perceptron, classified regression tree, multiple adaptive regression and random forest to predict the change of dissolved oxygen in the next 5 days for CBOD; Deng et al. [4] proposed a general analytical framework based on time series data mining technology to discover potential patterns or information of historical water quality data;

K. Li et al. (Eds.): LSMS 2021/ICSEE 2021, CCIS 1468, pp. 504–512, 2021.
https://doi.org/10.1007/978-981-16-7210-1_48

Wang et al. [5] proposed a water pollution density prediction model based on data mining technology, the model consists of three stages: first, the rough set theory and genetic algorithm are used to select the corresponding water pollution density predictor variables; The second is to use the data mining technology to implement the artificial neural network training model similar to the prediction; Thirdly, the artificial neural network is used to predict the density of water pollution.

For the analysis and prediction of water quality, domestic and foreign scholars adopt different methods, such as: Wang et al. [6] combined the improved dynamic clustering algorithm with machine learning thought and forward dynamic regression neural network to design the CA-NARX algorithm for water quality prediction; Zhou et al. [7] used CNN-LSTM model to predict water quality. Linear interpolation method was used for missing values to ensure the integrity of data. Continuous feature map was constructed for the dissolved oxygen data according to the time sliding window as input; Zhou et al. [8] put forward the prediction research of reservoir water quality based on grey model; Kuo et al. [9] proposed an applied study of water quality in Jingjiang Port based on the moving average weighted Markov model; Sun et al. [10] proposed the long and short term memory network (LSTM) time series prediction model (W-LSTM) based on wavelet decomposition to predict water quality; Peng [11] corrected and quantified the uncertainties in daily water quality prediction of large lakes by adopting Bayesian joint probability model, which helped to better realize the water quality prediction ability derived from the dynamic model; Rajib et al. [12] used remote sensing big data for watershed modeling and MODIS LAI to improve hydrology and water quality prediction. This study filled in the existing knowledge gap on vegetation dynamics in the watershed model and proved that assimilation of MODIS LAI data in the watershed model could effectively improve hydrology and water quality prediction. Analysis of quality prediction method for the domestic and foreign, this thesis mainly adopts the gray correlation analysis in data mining and the LSTM neural network prediction model in neural network to predict water quality.

The main research objectives of this article: First of all, the grey correlation method is used to reveal that temperature and pH are correlated with dissolved oxygen, while turbidity has no special relationship with dissolved oxygen; Second, LSTM neural network prediction of dissolved oxygen in water brings a new method for prevention and control of future water pollution.

2 River Forecasting Model

2.1 Influencing Factors of River Water Quality

There are many factors that affect the water quality of the river. Different seasons will lead to changes in water quality, because as the season changes, the temperature will also change; residential areas near the river will also change the water quality of the river. At present, the available water resources are not enough for all mankind in the world, and there are droughts and water emergencies in many areas. Therefore, the prediction of water quality will enable relevant departments to focus on considerable indicators and take corresponding measures to prevent water pollution incidents in the coming days.

2.2 Influencing Factors of River Water Quality

Firstly, the data set is processed, the data set processing steps need not be too cumbersome, since the data set is continuous; then, the dissolved oxygen and temperature, PH, turbidity are analyzed by the method of grey correlation analysis; Finally, the dissolved oxygen and its related characteristics are predicted by using the LSTM neural network. The general flow of river water quality prediction model is as follows (Fig. 1):

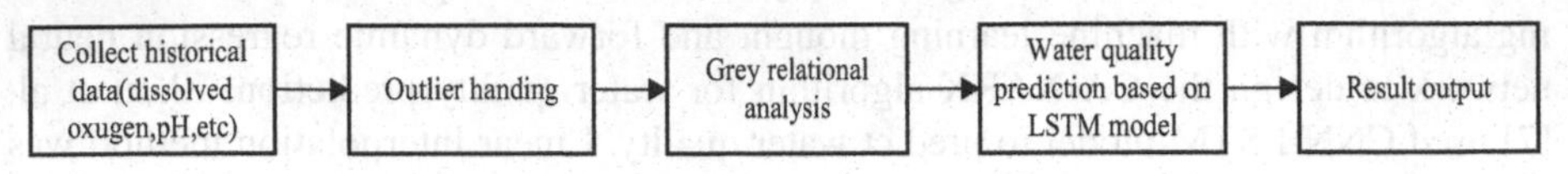

Fig. 1. Work flow chart.

3 Empirical Method

3.1 Grey Correlation Method

Grey relational analysis is an important branch of grey system theory, which judges the degree of correlation by comparing the geometric shape of reference data column with that of data column.

The detailed calculation steps of grey relational analysis are as follows:

Step 1: Determine the analysis sequence.

Determine the reference sequence that reflects the behavior characteristics of the system and the comparison sequence that affects the behavior of the system.

(1) List of references:

$$Y = Y(k)|k = 1, 2 \dots n\,. \tag{1}$$

(2) Comparative breakdown:

$$X_i(k)|k = 1, 2 \dots n; i = 1, 2 \dots m\,. \tag{2}$$

Step 2: Dimensionless of variables.

The data in the columns of the factors in the system may have different dimensions, making it difficult to obtain more accurate conclusions when comparing. Therefore, when performing gray correlation analysis, it is generally necessary to perform non-dimensional processing on the data first.

Initial value processing:

$$\frac{x_i(k)}{x_1(k)} = x_i(k), k = 1, 2 \dots n; i = 1, 2 \dots m\,. \tag{3}$$

Step 3: Find the gray correlation coefficient between the reference series and the comparison series:

$$\gamma x_o(k), x_i(k) = \frac{\min\limits_i \min\limits_k |x_o(k) - x_i(k)| + \zeta \max\limits_i \max\limits_k |x_o(k) - x_i(k)|}{|x_o(k) - x_i(k)| + \zeta \max\limits_i \max\limits_k |x_o(k) - x_i(k)|}.$$

$\rho \in (0,\infty)$, called the resolution coefficient; the smaller ρ is, the greater the resolution, and vice versa; ρ average value range is 0.5.

Step 4: Calculate the degree of relevance:

$$\gamma(X_o, X_i) = \frac{1}{n}\sum\nolimits_{k=1}^{n} \gamma(x_o(k), x_i(k)). \tag{5}$$

Based on the grey correlation analysis method, to eliminate the water quality index that is not related with dissolved oxygen for improve the prediction precision. The relationship between temperature, pH, turbidity and dissolved oxygen is analyzed through Matlab. The experimental results show that the (Fig. 2), as the change of dissolved oxygen, temperature and PH curve has correlation with dissolved oxygen, and the turbidity is no correlation.

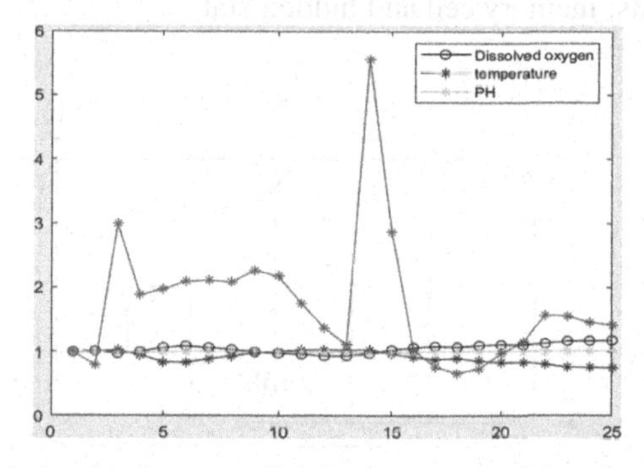

Fig. 2. The relationship between dissolved oxygen and the other three indexes.

3.2 Linear Regression Model (LRM)

Linear regression is a method of using regression analysis in mathematical statistics to determine the correlation between two or more variables.

The correctness of the above grey correlation analysis method is verified by linear regression, and it is confirmed that there is a certain relationship between temperature and pH and dissolved oxygen. In addition, it can be found from the linear regression that there is a negative correlation between temperature and dissolved oxygen (Fig. 3), and a positive correlation between pH and dissolved oxygen (Fig. 4).

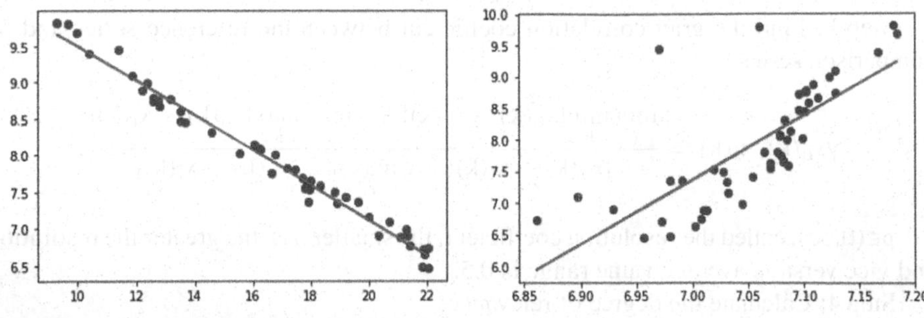

Fig. 3. Temperature versus dissolved oxygen. **Fig. 4.** PH versus dissolved oxygen.

3.3 LSTM Neural Network

The Long Short Term Memory network retains thc advantages of the traditional cyclic neural network and overcomes the difficulties in the traditional cyclic neural network. The traditional cyclic neural network has the phenomenon of gradient vanishing and gradient descent, and the learning ability is also inferior. LSTM neural network model is an adjusted model for these problems. The structure of LSTM neural network is shown in Fig. 5, which is mainly composed of forgetting gate, input gate and output gate, as well as two basic units: memory cell and hidden state.

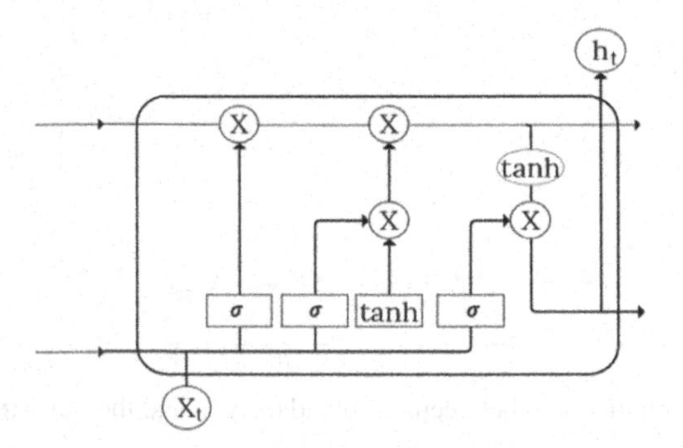

Fig. 5. The structure of LSTM neural network.

(1) The calculation formula of the output control vector of the forgetting gate is as follows:

$$f_{(t)} = \sigma\left(G_f\left[\frac{s_{(t-1)}}{x_{(t)}}\right] + b_f\right). \tag{6}$$

$f_{(t)}$ is the output control quantity of the forgetting gate at time t, $\sigma()$ is the logical function, G_f is the weight matrix of the forgetting gate, $\left[\frac{s_{(t-1)}}{x_{(t)}}\right]$ is the row concatenation matrix of $s_{(t-1)}$ and $x_{(t)}$, b_f is the bias item of the forgetting gate.

(3) The formula for calculating the output control vector of the input gate is as follows:

$$i_{(t)} = \sigma\left(G_i\left[\frac{s_{(t-1)}}{x_{(t)}}\right] + b_i\right). \tag{7}$$

$i_{(t)}$ is the output control vector of the input gate at time t, G_o is the weight matrix of the input gate, b_i is the offset item of the output.

(2) The formula for calculating the output control vector of the output gate is as follows:

$$o_{(t)} = \sigma\left(G_o\left[\frac{s_{(t-1)}}{x_{(t)}}\right] + b_o\right). \tag{8}$$

4 Simulation Example and Experimental Demonstration

The data in this paper came from the Shared Service Platform of National Geoscience Data Center. The data from 2015 to 2018 were used as the training set, and the data after 2018 were used as the test set (Fig. 6). The LSTM neural network is used to predict the dissolved oxygen, as shown in the figure (Fig. 7). The curve of the test data is highly consistent with the curve of the predicted data, indicating that the prediction effect is fine and it can be applied in water quality prediction. In addition, the prediction effect of pH (Fig. 8) and temperature (Fig. 9) is not alright, and these two characteristics are not suitable for water quality prediction. Table 1 shows that the prediction effect of dissolved oxygen is the acceptable, so dissolved oxygen is used for water quality prediction.

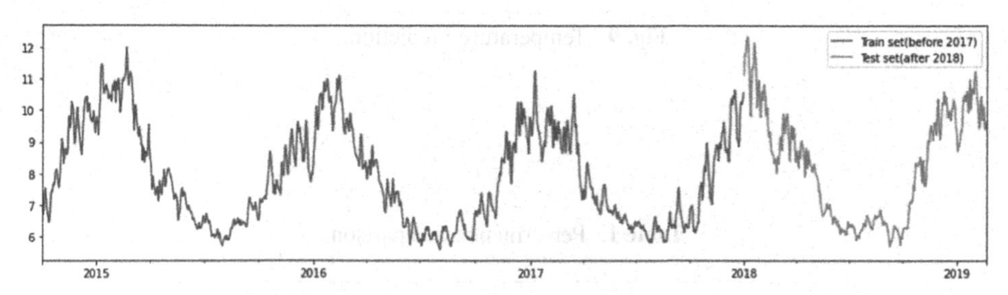

Fig. 6. Divide the training set and the test set.

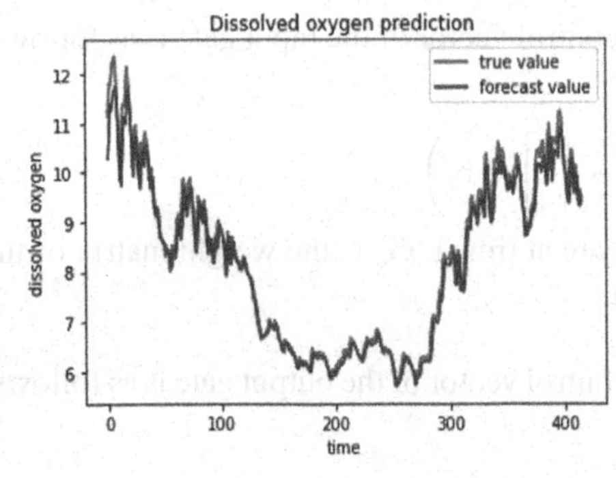

Fig. 7. Dissolved oxygen prediction.

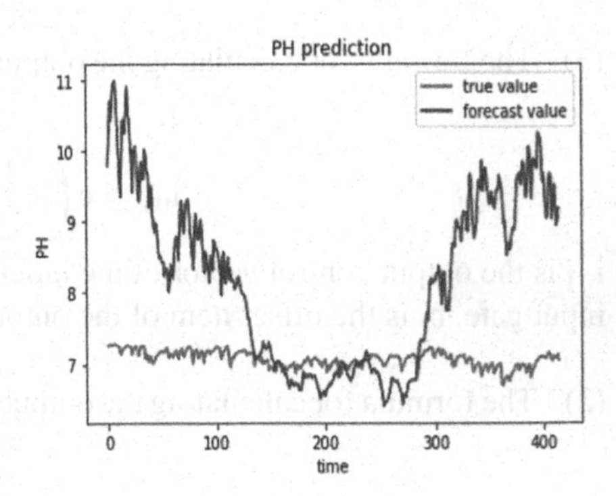

Fig. 8. PH prediction.

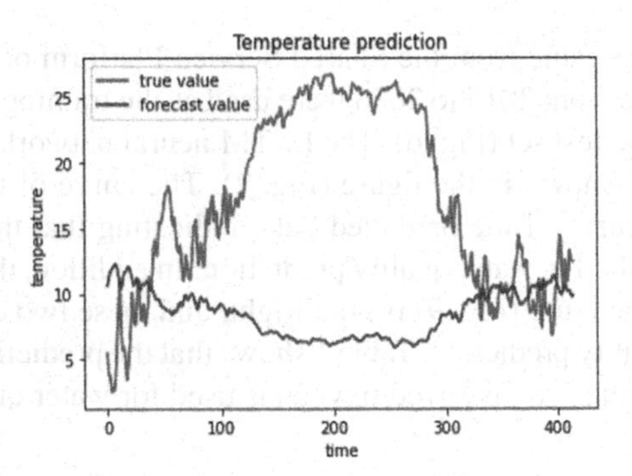

Fig. 9. Temperature prediction.

Table 1. Performance comparison.

Performance evaluation	Dissolved oxygen	Temperature	PH
MSE	0.081090	147.980.996	2.660966
RMSE	0.284763	12.164744	1.631247
MAPE	0.207150	9.813717	1.270892

5 Conclusion

Based on the reality, this thesis uses the prediction of water quality indicators to protect the water environment, and makes use of the advantages of grey correlation analysis and LSTM model to predict the water quality indicators. The results are acceptable, which can provide a method for subsequent water quality management.

However, there are still some deficiencies in this thesis:

(1) There are not enough data and few water quality parameters in the data set;
(2) The LSTM model depends on a large amount of data, and the convergence rate will be slow as the number of iterations increases.

Therefore, improvements will be made to solve these two problems in the future by adding data sets and optimizing the LSTM model.

The experimental results show that:

(1) Grey correlation analysis method can quickly identify the relationship between dissolved oxygen and temperature, pH, turbidity.
(2) LSTM neural network model can effectively predict dissolved oxygen in water, and the prediction accuracy is high.

The method adopted in this thesis can bring a new idea for future water pollution control and water environment management. The prediction model is used to analyze the dissolved oxygen in water and predict the content of dissolved oxygen in the future. When the predicted value is greater than the normal index set by the Environmental Protection Bureau, corresponding measures are taken to optimize the strategy of water pollution governance and provide theoretical analysis basis for the development of more flexible water quality management.

References

1. Zhang, C., Bai, J.B., Kang, L., et al.: Prediction of photovoltaic power generation based on data mining and genetic wavelet neural network. Acta Energiae Solari Sinica. **42**(03), 375–382 (2014)
2. Babbar, R., Babbar, S.: Predicting river water quality index using data. Mining techniques. Environ. Earth Sci. **76**(14), 504 (2017)
3. Kusiak, A., Verma, A., Wei, X.P.: A data-mining approach to predict influent quality. Environ. Monit. Assess. **185**(3), 2197–2210 (2013)
4. Deng, W.H., Wang, G.Y.: A novel water quality data analysis framework based on time-series data mining. J. Environ. Manage. **196**, 365–375 (2017)
5. Wang, C.Y., Liu, B.S., Qiu, E.W.: The prediction of river water pollution density based on data mining technology. Adv. Mater. Res. **956**, 1285–1288 (2010)
6. Wang, J., et al.: Water quality prediction of water sources based on meteorological factors using the CA-NARX approach. Environ. Model. Assess. **26**(4), 529–541 (2021).
7. Zhou, C.M., Liu, M.P., Wang, J.W., et al.: Water quality prediction model based on CNN-LSTM. Hydropower Energy Sci. **39**(03), 20–23 (2021)

8. Zhou, Y., Hu, J.R., Xie, M.F.: Research on reservoir water quality prediction based on grey model. In: IOP Conference Series: Earth and Environmental Science, vol. 621, no. 1, pp. 012120–12120 (2021)
9. Jiao, K., et al.: Prediction studies of river water quality based on moving average weighted Markov model: a case study of Shiwei port, Jingjiang City. In: IOP Conference Series: Earth and Environmental Science, vol. 631, no. 1 (2021)
10. Sun, M., Wei, S.K., Wang, Y.J., et al.: Water quality prediction model of LSTM based on wavelet decomposition. Comput. Syst. Appl. **29**(12), 55–63 (2020)
11. Peng, Z.L., Hu, Y., Liu, G.: Calibration and quantifying uncertainty of daily water quality forecasts for large lakes with a Bayesian joint probability modelling approach. Water Res. **185**, 116162 (2020)
12. Rajib, A., Kim, I.L., Golden, H.E., et al.: Watershed modeling with remotely sensed big data: Modis leaf area index improves hydrology and water quality predictions. Remote Sens. **12**(13), 2148 (2020)

Real-Time Adjustment and Control of Three-Phase Load Balance Based on Change-Phase Switch

Xuewen Liu[1], Tingzhang Liu[1], and Chen Xing[2](✉)

[1] School of Mechatronic Engineering and Automation, Shanghai University, Shanghai 200444, China
liutzh@staff.shu.edu.cn

[2] School of Electronic and Electrical Engineering, University of Leeds, Leeds LS29JT, England
elcx@leeds.ac.uk

Abstract. In order to solve the problem of unbalanced three-phase load in low-voltage distribution network, the method of using three-phase load automatic adjustment device which is the type of the change-phase switch and the intelligent change-phase terminal to govern the three-phase load imbalance in the distribution station area is studied. First of all, make a load balancing real-time control strategy, choose suitable objective function and optimal algorithm. In order to verify the effectiveness of the algorithm, establishing a distribution station area model and use genetic algorithm to calculate the optimal change-phase instruction which can reduce the three-phase load imbalance based on the simulated distribution network load data. Afterwards make use of the change-phase control switch and the intelligent change-phase terminal to form a three-phase real-time load balance control system, make a real-time continuous governance of three-phase load imbalance, the results of experiments verify the effectiveness of the control system.

Keywords: Three-phase load imbalance · Change-phase switch · Genetic algorithm · Intelligent change-phase terminal

1 Introduction

With the continuous improvement of the national economy, electricity consumption in all walks of life is also increasing, the power industry is facing more and more problems of power quality [1], among them, three-phase load imbalance is an important factor affecting power quality. The long-term unbalanced operation of the three-phase load of the distribution network will cause a series of problems such as:1) The unbalance of the three-phase load leads to a decrease in the utilization rate of the generator capacity, thereby reducing the power conversion efficiency. 2) The three-phase load imbalance of the transformer not only shortens the life of the one-phase winding with a larger load due to overheating. 3) Lead to unbalanced three-phase output voltage, causing damage to electrical equipment. 4) Causes an increase in the line loss of the power grid [2].

K. Li et al. (Eds.): LSMS 2021/ICSEE 2021, CCIS 1468, pp. 513–524, 2021.
https://doi.org/10.1007/978-981-16-7210-1_49

At present, there are three main ways to solve the problem of three-phase load imbalance: manual adjustment of change-phase, automatic adjustment of capacitive/power electronic three-phase load and use three-phase load automatic adjustment which is the type of change-phase control switch [3].

Literature [4] studied the composition and working principle of three-phase load automatic adjustment device which is the type of change-phase control switch, and designed a change-phase control strategy based on fuzzy C-means clustering algorithm. Literature [5] proposed a solution method based on vector gene genetic optimization algorithm to reduce the three-phase load imbalance in the distribution station area, and use simulation to verify the effectiveness of the method. Literature [6] proposed a pre-calculation algorithm to find the optimal load phase sequence adjustment plan, and reasonably and evenly distribute the three-phase load. The research work described is mainly to adjust the three-phase load imbalance at an independent point in time, which has certain limitations.

This paper mainly studies the real-time three-phase load imbalance under the long-term continuous operation of the low-voltage distribution network in the distribution station area, and proposes a method of real-time balance control for the three-phase load in the distribution station area.

2 System Composition and Working Principle

2.1 Main Components of the System

The real-time governance system is mainly composed of two parts: the change-phase control switch and the intelligent change-phase terminal. The change-phase control switch is installed at the inlet end of each user, used to collect the data of current, phase and other information of each user, and upload the data information to the intelligent change-phase terminal by using the RS485 communication line. The intelligent change-phase terminal is installed on the low-voltage outlet side of the distribution transformer, its function is receiving the data of current, phase and other data that uploaded by each change-phase control switch, calculate the optimal change-phase command and feed it back to the change-phase control switch, the change-phase control switch can quickly switch the load phase line without impact online, no power off during the switching process. Finally realized the phase sequence adjustment of the electric load under the load condition, and reduce the unbalance of the three-phase load. The system composition is shown in Fig. 1.

2.2 Change-Phase Control Switch

The main function of the change-phase control switch is to quickly switch the phase of each bus load in real time and without impact under load conditions, that is, the phase sequence of the load is switched from one phase to the other smoothly and quickly, the switching process does not affect the power supply. In addition, the change-phase control switch can collect the user's current and phase data. The block diagram of the change-phase control switch is shown in Fig. 2, including two parts: the control circuit and the main circuit.

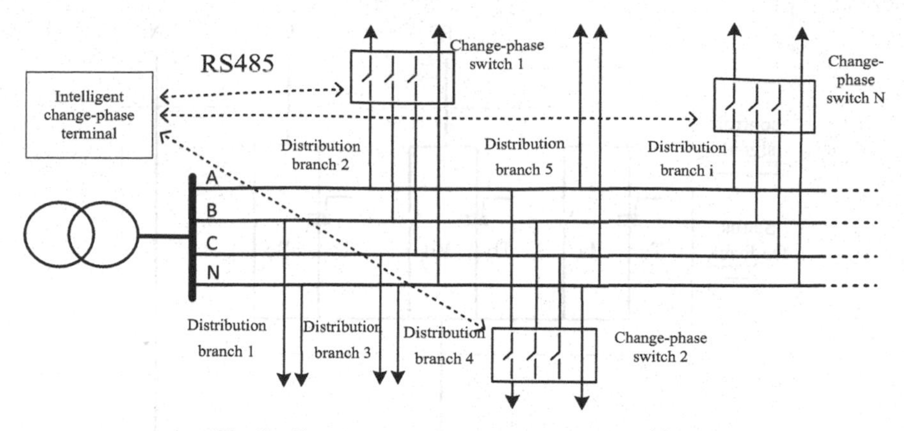

Fig. 1. Governance system structure composition.

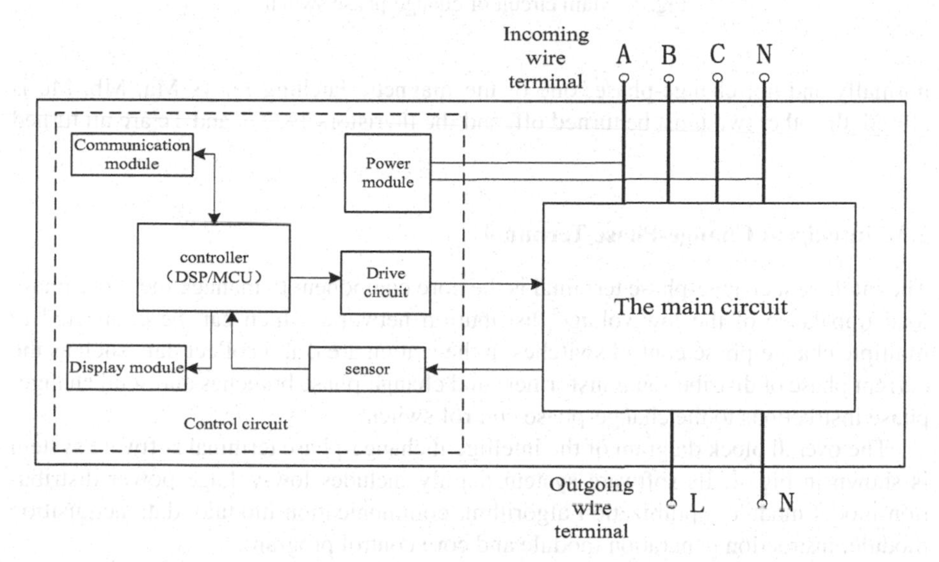

Fig. 2. Change-phase control switch.

The main functions of the control circuit include supplying power for the change-phase control switching system, external communication, status display, main circuit drive, main circuit information acquisition and control of the main circuit work. The control circuit mainly includes power supply module, change-phase module, drive circuit, display module, sensor and other parts.

The block diagram of the main circuit design is shown in Fig. 3. Mainly composed of thyristor (SCR) and magnetic latching relay, A, B, C three-phase each group contains a group of thyristors (SCR) and a parallel circuit of a magnetic latching relay, this design combines the advantages of fast switching speed of thyristor, large current carrying capacity of magnetic latching relay and self-holding switching state. When working

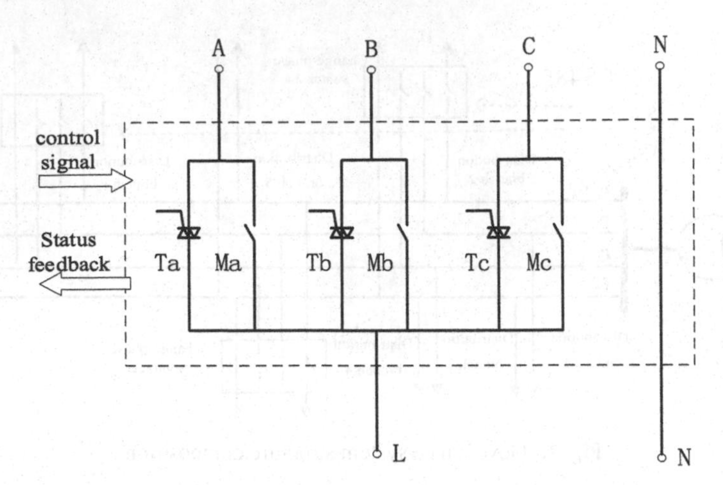

Fig. 3. Main circuit of change-phase switch.

normally and not change-phase, one of the magnetic latching relays Ma, Mb, Mc is closed, the other two must be turned off, and the thyristors Ta, Tb, and Tc are all turned off.

2.3 Intelligent Change-Phase Terminal

The intelligent change-phase terminal is the core component to manage the three-phase load imbalance of the low-voltage distribution network, which can be connected to multiple change-phase control switches in the station area, and collect data such as the current phase of distribution transformers and change-phase branches and send change-phase instructions to the change-phase control switch.

The overall block diagram of the intelligent change-phase terminal software system is shown in Fig. 4. Its software system mainly includes low-voltage power distribution model module, optimization algorithm, communication module, data acquisition module, instruction generation module and core control program.

The distribution network model module is related to the low-voltage distribution topology and is closely related to the optimization algorithm module, according to the real-time data collected on the low-voltage side current of the distribution transformer, the current of the change-phase control switch branch, and the phase sequence, the optimal change-phase instruction is calculated by the optimization algorithm. The communication module mainly includes communication within switchboard area. The data acquisition module is connected with the communication module to collect the current and phase data of the change-phase switch in real time. The optimal change-phase command calculated by the intelligent change-phase terminal of the command generation module which controls each change-phase control switch to complete the change-phase operation through the communication module.

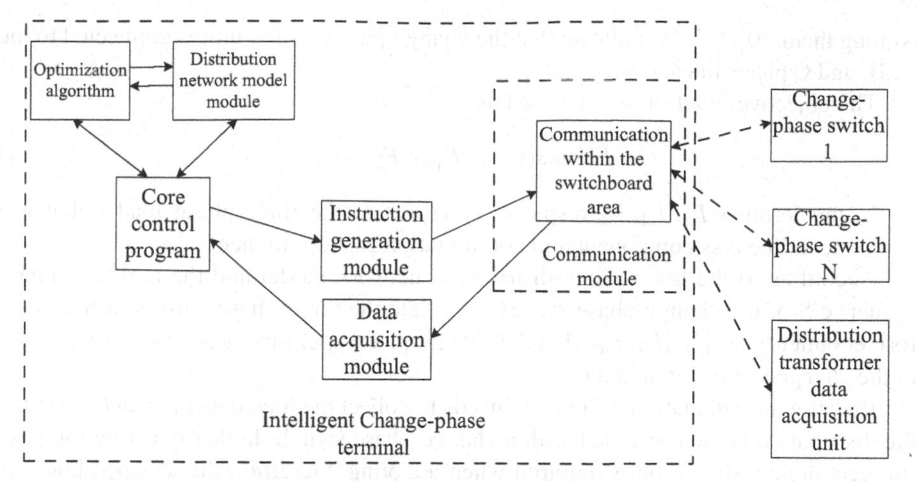

Fig. 4. Overall block diagram of intelligent change-phase terminal software.

3 Load Balancing Real-Time Control Strategy

Automatic change-phase control algorithm is the core part of real-time control strategy for load balancing. The control strategy in this article is mainly to obtain the real-time phase sequence of the change-phase switch according to the change-phase action's feedback of the change-phase control switch by the intelligent change-phase terminal, then combine the built-in low-voltage distribution network model and real-time load data to calculate the optimal change-phase instruction, and control each change-phase control switch for change-phase operation. Automatic change-phase control algorithm mainly includes two parts: objective function module and optimization algorithm module.

3.1 Establishment of Objective Function

The goal of this article to manage the three-phase load imbalance of the low-voltage distribution network is to minimize the three-phase load imbalance at the outlet of the distribution transformer and the line loss of the low-voltage distribution network. Therefore, the design idea of the objective function is mainly to quantify the additional cost of the three-phase load imbalance and change-phase operation to the operation and maintenance of the distribution network, and then add an appropriate penalty value according to the requirements of safe operation.

Suppose there are N change-phase switches, the change-phase command is $C = [C_1, C_2, \ldots, C_N]^T$, The value space of C_i is {0,1,2}, Where '0' means no change-phase operation is performed, '1' means that the phase moves 1 bit in the positive order of A, B, and C(For example, the change-phase switch unit is switched from the A phase line to the B phase line, and the C phase line is switched to the A phase line), '2' means that the phase moves 2 bits in the positive order of A, B, and C(For example, the change-phase switch unit is switched from the A phase line to the C phase line, and the C phase line is switched to the B phase line). The real-time phase sequence of the change-phase switch unit at a certain moment is S, $S = [S_1, S_2, \ldots, S_N]^T$, the value space of S_i is {0,1,2}.

Among them, '0', '1', '2' indicate that the change-phase switch unit is connected to the A, B, and C phase lines, respectively.

The objective function is expressed as:

$$Fitness(C) = F_1 + F_2 + F_3. \quad (1)$$

In the formula F_1, F_2, F_3 respectively represent the three-phase load imbalance component, line loss component and commutation cost component.

According to the low-voltage distribution network model and the real-time phase sequence S of the change-phase switch, calculate the three-phase current at the transformer outlet $I_M = [I_{MA} I_{MB} I_{MC}]^T$ if the change-phase operation is performed according to the change-phase command C.

Because the calculation of line loss needs to collect the load data of each busbar and the distribution branch installed with a change-phase switch, higher construction cost, full economic justification is required when adopting, therefore, the dropped line loss component F_2 can be removed in this article. Only need to analyze the three-phase load imbalance component F_1 and the change-phase action component F_3.

(1) three-phase load imbalance component

For the three-phase transformer of Yyn0 connection mode, when asymmetry in size of the three-phase load causes the three-phase load unbalanced, the additional copper loss Δp on the secondary side of the distribution transformer when the three-phase load is unbalanced is about:

$$\Delta p \approx \frac{(I_{MA} - I_{MB})^2 + (I_{MB} - I_{MC})^2 + (I_{MC} - I_{MA})^2}{3} R_2. \quad (2)$$

R_2 is the DC resistance of the secondary winding of the transformer.

Two characteristics can be obtained from the relationship between the copper loss on the secondary side of the transformer and the total three-phase load: 1) Under the same three-phase load imbalance, the greater the load, the greater the copper loss. 2) Under the same load, the greater the three-phase load imbalance, the greater the copper loss value. According to the above two characteristics, the three-phase load unbalance component is designed as:

$$F_1 = k_1 * R \frac{(I_{MA} - I_{MB})^2 + (I_{MB} - I_{MC})^2 + (I_{MC} - I_{MA})^2}{3}. \quad (3)$$

R is the DC resistance of the secondary winding of the transformer, and k_1 is the coefficient.

(2) change-phase component

The rated working current of the change-phase control switch is generally not less than 100A, one change-phase operation needs to switch 2 times of magnetic latching relay and 4 times of thyristor. Although frequent change-phase operation can improve the treatment effect, it will also accelerate the damage of the change-phase control switch. Therefore, the number of commutations is mainly considered when designing the change-phase action component.

Calculate the total number of 1 and 2 in the change-phase command C to get the commutation frequency of the change-phase command C, so the change-phase action component is shown in Eq. (5):

$$M(x) = \begin{cases} 0 & x = 0 \\ 1 & x = 1\ or\ 2 \end{cases} \tag{4}$$

$$F_3 = k_3 * \sum_{i=1}^{N} M(C_i). \tag{5}$$

k_3 is the coefficient.

3.2 Optimization Algorithm to Solve the Model

The objective function in this paper is a multi-objective nonlinear optimization function, and it is planned to use genetic algorithm which is a global optimization algorithm to solve it. The main advantages of genetic algorithm are simple, universal, robust, suitable for parallel distributed processing and a wide range of applications [7, 8].

The main steps of the genetic algorithm in this paper are as follows:

(1) Gene coding: the process of converting the parameters of the problem space into a chromosome or individual composed of genes in a certain structure in the genetic space. This article encodes the change-phase action into three values of 0, 1, 2.
(2) Selection operation: for a population M with a size of N, the selection probability $P(x_i)$ of each chromosome $x_i \in M$ is used to determine the selection opportunity, randomly select several chromosomes from M in N times and copy them. The relationship between $P(x_i)$ and fitness $f(x_i)$ is:

$$P(x_i) = \frac{f(x_i)}{\sum_{j=1}^{N} f(x_j)} \tag{6}$$

(3) Crossover operation: the operation of replacing part of the structure of two parent individuals to generate a new individual.
(4) Mutation operation: mutation operator is to make changes to the gene values on some genomes of individuals in the population.

4 Algorithm Simulation and System Experiment Analysis

4.1 Algorithm Simulation Analysis

The college building of a school is shown in Fig. 5, and an experimental model of low-voltage distribution network is established on this basis. The model has 11 distribution buses and 44 single-phase distribution branches.

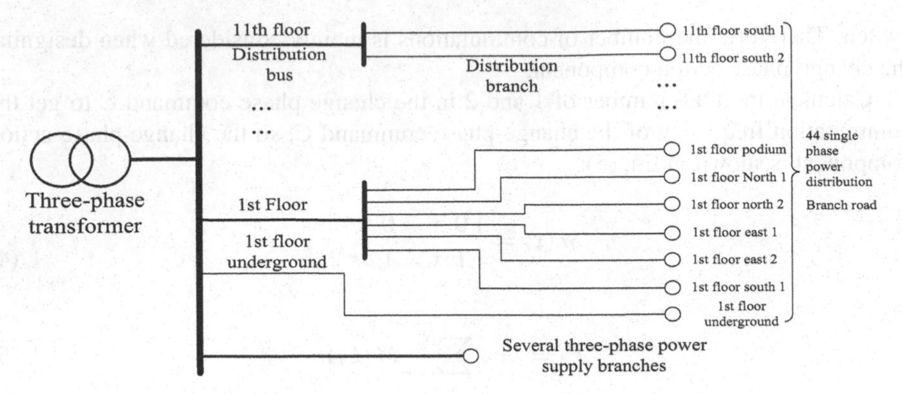

Fig. 5. Distribution area model.

Establish the above distribution station model in the intelligent change-phase terminal, and based on the actual load and phase data of the building during a certain period of April 2019, the change-phase control algorithm simulation experiment is carried out.

Suppose that 9 change-phase control switches are selected to install on the appropriate distribution branch in the distribution station area, and simulation experiments are carried out with and without the change-phase control algorithm to verify the effectiveness of the change-phase algorithm. Using the load and phase data of the distribution station area model from April to June 2019 for simulation. Taking the average of the three-phase load imbalance every hour on a certain day, and plotting the three-phase load imbalance with time before and after treatment, as shown in Fig. 6. Taking the average of the three-phase load imbalance every day, and plotting the three-phase load imbalance with time before and after treatment, as shown in Fig. 7.

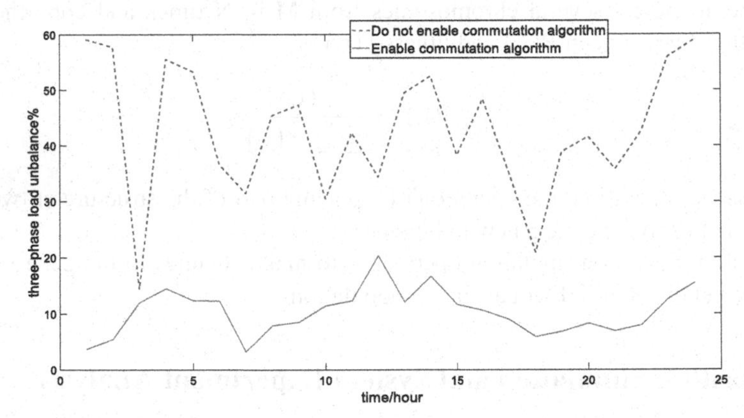

Fig. 6. Three-phase load real-time balance management simulation comparison chart (hour by hour).

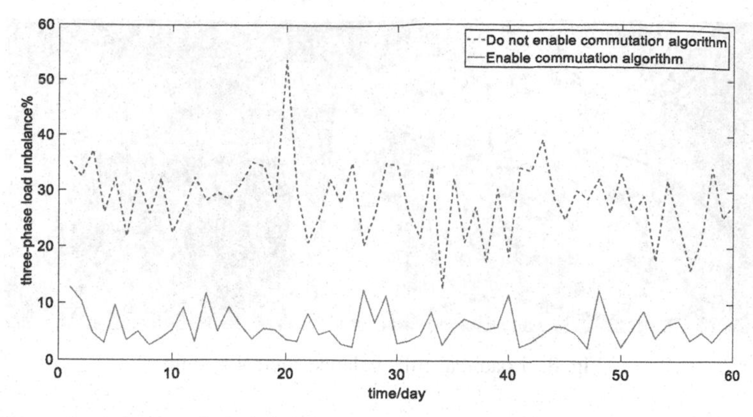

Fig. 7. Three-phase load real-time balance management simulation comparison chart (day by day).

It can be seen from Fig. 6 that under the same load data, the average three-phase unbalance degree on a certain day is 42.7% and 10.3% when the change-phase control algorithm is used and when the change-phase control algorithm is not used.

It can be seen from Fig. 7 that under the same load data, the average three-phase load imbalance between April and May when the change-phase control algorithm is used and when the change-phase control algorithm is not used is 28.7% and 5.9%, respectively. The governance effect is relatively significant, which verifies the effectiveness of the algorithm.

4.2 Load Balance Control System Experiment

Under the existing conditions, a single change-phase control switch and an intelligent change-phase terminal are used to carry out the experiment of the three-phase load balance control system. Connect the change-phase control switch to a certain distribution bus, assuming that another 8 change-phase controls are connected to other suitable power distribution buses, the change-phase control terminal and the change-phase control switch are connected by RS485 communication lines, to realize the interaction and data transmission between the two and adjust the three-phase load imbalance in real time.

The load real-time balance control system is shown in Fig. 8.

Setting the initial phase of the change-phase control switch to phase A, then run the intelligent change-phase terminal, read the load, phase and other data of the change-phase control switch in real time, and the intelligent change-phase terminal will add it to the low-voltage distribution network model after obtaining the data. Calculating the unbalance degree under the current situation, if it exceeds the threshold, calculate the optimal change-phase instruction, and the change-phase control switch will obtain the change-phase instruction to perform the change-phase operation.

The load real-time balance control experiment is shown in Fig. 9, mainly shows the change-phase process of the change-phase control switch in the regulation process.

As shown in Fig. 9(a), the initial phase of the change-phase control switch is phase A, after real-time load balance adjustment, the change-phase control switch receives the

Fig. 8. Load real-time balance control system.

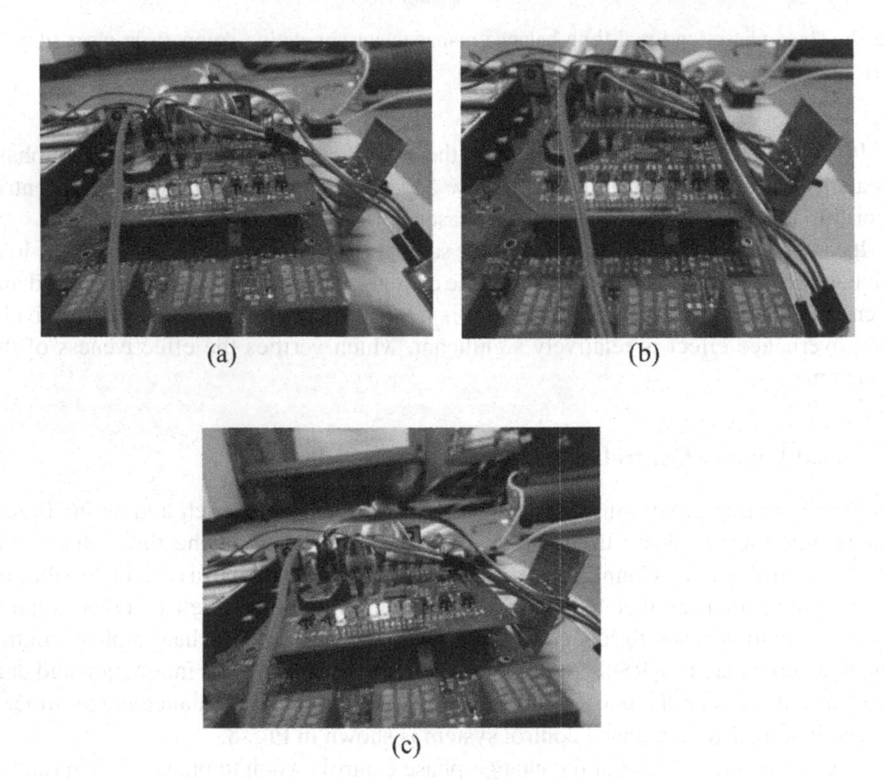

Fig. 9. Control system software flow chart.

change-phase command ‘1’, then it changes the phase to phase B, as shown in Fig. 9(b). The pressure loss time during the change-phase process does not exceed 20ms. In the next period of load balance control, the change-phase switch receives the change-phase command ‘2’, it will change the phase to phase C, as shown in Fig. 9(c).

Before and after change-phase control operation, the current at the outlet of the simulated distribution transformer and the unbalanced degree of the three-phase load are shown in Table 1.

Table 1. Specific environment configuration.

	I_n/A	I_a/A	I_b/A	I_c/A	Unbalance Factor
Before change phase	36.4	63.0	69.1	102.3	38.4%
After change phase	7.5	76.1	75.0	83.2	9.8%

It can be seen from Table 1 that the load balance control system plays a significant role in controlling the three-phase load imbalance in the distribution station area, improving the power supply quality of the distribution station area and reducing losses.

5 Conclusion

This paper focuses on the change-phase control switch and genetic algorithm to design the load real-time balance control strategy, establish a low-voltage distribution network model and use it to carry out simulation experiments of change-phase control algorithms. The simulation results verify the effectiveness of the change-phase control algorithm.

Combining the actual change-phase control switch and the intelligent change-phase terminal to establish a load balance control system. The experimental operation results show that the system can effectively reduce its three-phase load imbalance in a long-term continuous operation of the distribution station area.

It demonstrates the significant role of the system in the treatment of three-phase load imbalance, which is of great significance for improving the power supply quality and economic level of the low-voltage distribution network.

References

1. Yang, Y.L., Wang, F.Q.: Additional loss and voltage deviation caused by unbalanced operation of distribution transformer and countermeasures. Power Syst. Technol. **28**(8), 73–76 (2004)
2. . Lin, H.X.: Three-phase Unbalance of Power System. China Electric Power Press, Beijing (1998). (in Chinese)
3. Hu, Y.H., Wang, J.Z., Ren, J.J., et al.: Balance component decomposition and compensation method for unbalanced load. Proc. CSEE. **32**(34), 98–104 (2012). (in Chinese)
4. Lu, H.B., Xu, Y., Wu, Y.X., et al.: Research on three-phase unbalance treatment device based on change-phase technology. Power Capacitor React. Power Compensation **37**(6), 64--69 (2016). (in Chinese)
5. Fang, H.F., Sheng, W.X., Wang, J.L., et al.: Research on the method for real-time online control of three-phase unbalanced load in distribution area. Proc. CSEE. **35**(9), 2185–2193 (2015)
6. Li, Y.X., Gong, Y.L., Zhao, Y.Y., et al.: Three-phase unbalance pre-calculation control strategy. Proc. CSU-EPSA. **32**(03), 20–24 (2020)
7. Wu, G.H., Wen, Y., Le, M.F.: Genetic algorithm and its application. Chin. J. Appl. Mech. **23**(6), 9–10 (2000). (in Chinese)

8. Srinivas, M., Patnaik, L.M.: Adaptive probabilities of crossover and mutation in genetic algorithms. IEEE Trans. Syst. Man Cybern. **24**(4), 656–667 (1994)
9. Zhi, J., Liu, T.Z.: Research on optimal dispatching and governance technology for unbalanced three-phase load in distribution network. Ind. Control Comput. **32**(02), 149–150+153 (2019). (in Chinese)
10. Ghaeb, J.A., Aloquili, O.M.: High performance reactive control for unbalanced three-phase load. Eur. Trans. Electr. Power. **20**(6), 710–722 (2010)

Multiphysics Simulation Research on Microwave Drying of Municipal Sludge

Gongqin Wang[1(✉)], Shengjun Xu[2], Hongru Pang[1], Jiadi Lian[1], Long Deng[1], and Binqi Rao[1(✉)]

[1] College of Mechanical and Electrical Engineering, China Jiliang University, Hangzhou 310018, China

[2] Xingyuan Environmental Technology Co., Ltd., Hangzhou 311199, China

Abstract. The COMSOL Multiphysics software was used to perform numerical simulation analysis on the microwave drying process of pressurized sludge, and the influence of sludge thickness, moisture content, microwave power and sludge position on the electric field and sludge temperature field in the microwave cavity was studied. The results show that the microwave power has a great influence on the electric field and temperature field, while the sludge thickness, moisture content and position of the sludge does little. With the increase of the microwave power, the intensity of the electric field in the microwave cavity and the heating rate of the sludge increases. The optimal process conditions for the microwave drying of the pressurized sludge after numerical simulation are the microwave power of 1000 W, and the sludge is in the center of the cavity.

Keywords: Municipal sludge · COMSOL multiphysics · Microwave radiation · Temperature distribution · Electric field distribution

1 Introduction

Municipal sludge is a by-product of sewage treatment and water purification processes, and its main components flocs which composed of various microorganisms, organic or non-polar particles. Depending on the source of the sewage, the sludge also contains heavy metal substances, germs [1]. In 1995, the World Water Environmental Federation changed the name of sludge to biosolids, which is a kind of urban pollutants containing organic matter that can be effectively recycled. Thus, sludge drying technology can reduce the moisture content and helps to reutilize effectively.

Sludge drying is a process of surface water vaporization and internal water diffusion. Generally, it can decrease the moisture content of sludge to 10% or even 5% [2, 3]. With the increasing requirements for environmental protection and energy conservation, sludge drying technology has also been continuously developed, improved, and innovated in terms of the diversity of heat source forms [4], operational stability, safety, energy conservation [5] and environmental protection [6]. Microwave drying technology has attracted much attention around the world.

K. Li et al. (Eds.): LSMS 2021/ICSEE 2021, CCIS 1468, pp. 525–534, 2021.
https://doi.org/10.1007/978-981-16-7210-1_50

Microwave radiation refers to any electromagnetic radiation with the frequency between 300 MHz and 300 GHz [3, 7]. The dipole will be rotated to produce heat rapidly by the alternating electromagnetic field of microwave radiation. Thus, temperature of sludge under microwave radiation will be dramatically increased, and the moisture will be evaporated efficiently. However, due to the high energy consumption, microwave drying technology must be further optimized [3].

The purpose of this article is to study the influence of microwave power, sludge location, thickness (T_c) and moisture content (M_c) on electric field and temperature distribution through numerical simulation. The simulation work is completed by COMSOL Multiphysics software.

2 Numerical Building and Experimental Process

2.1 Model Constraints

Although there are certain differences between the actual situation and the numerical model, some parameters that cannot be obtained in the real case can be obtained in the numerical simulation. Such as temperature distribution and electric field distributions in microwave oven. In order to reduce the calculation capacity and interference factors, the following assumptions are made to simplify the model reasonably [8–11]:

Assumption 1. Sludge sample is homogeneous and isotropic.
Assumption 2. Thermo-physical is constant.
Assumption 3. The sludge sample temperature is similar to that of the air and the overall temperature of the sludge is evenly distributed.
Assumption 4. There is little mass transfer which is ignored.
Assumption 5. The dielectric loss of glass plate and air is 0.
Assumption 6. The material of waveguide and cavity wall is copper.
Assumption 7. The heat transfer between the air and the glass plate is not considered.
Assumption 8. Organic content of the sludge samples is 50%.
Assumption 9. Imaginary part of dielectric constant of sludge is always -20j.

2.2 Geometry

The geometry of the model is shown in Fig. 1. The cavity is 260 × 270 × 188 mm. The microwave source is 50 × 78 × 18 mm. Glass plate radius is 113.5 mm. Glass plate height is 6 mm.

The relative density of dry sludge is obtained by the formula:

$$\rho_d = \frac{250}{100 + 1.5P_v}. \tag{1}$$

Where P_v represents the proportion of organic matter in the sludge, %, and ρ_d represents the relative density of dry sludge, kg/m^3. After obtaining the relative density of dry

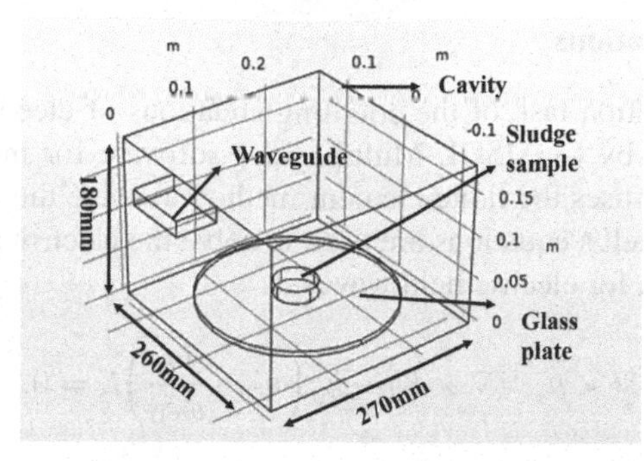

Fig. 1. Numerical simulation geometric model of microwave heating.

sludge, ρ is the relative density of wet sludge under different moisture contents is obtained by the following formula:

$$\rho = \frac{100\rho_d}{W \cdot \rho_d + (100 - W)}. \tag{2}$$

Where W is the moisture content of the wet sludge, %. ρ is the relative density of wet sludge, kg/m^3.

The dielectric constant (ε_r) can be calculated by the following formula [12]:

$$\varepsilon_r = \frac{\left[2\varepsilon_w(\varepsilon_w - \varepsilon_d)\frac{\rho_d}{\rho_w} - 3\varepsilon_w\varepsilon_d\right]W + 3\varepsilon_w\varepsilon_d}{3\varepsilon_w + \left(\frac{\rho_d}{\rho_w}(\varepsilon_w - \varepsilon_d) - 3\varepsilon_w\right)W}. \tag{3}$$

ρ_w represents the relative density of water, and ε_w is the dielectric constant of water, and ε_d is the dielectric constant of dry sludge. Generally, ε_w is 78.5 and ε_d is 3 [13].

The parameters including moisture content (M_c), thickness of sludge cake (T_c), specific heat capacity (H_c), sludge sample density (D_s) and real part of dielectric constant (ε_r) are shown in Table 1. In addition, the frequency of microwave radiation used in this article is 2.45 GHz and the action time is 2 min.

Table 1. Experimental data of sludge samples.

Experimental number (N)	M (%)	T_C (mm)	H_C (J/kg·K)	D_s (kg/m^3)	ε_r
N1	35.1	1.6	2.01	1240	33.60
N2	47.7	4.5	2.44	1190	48.40
N3	53.3	20.7	2.63	1160	55.90
N4	65.5	17.1	3.04	1160	74.67
N5	73.6	37.8	3.31	1090	89.40

2.3 Governing Equations

The numerical simulation task of the coupling equations of electromagnetic and heat transfer is calculated by COMSOL Multiphysics software for microwave heating of sludge. This software uses the finite element method to solve the numerical model. In the simulation, Maxwell's equations are used to solve the electromagnetic propagation problem. The formula for electric field waves is:

$$\nabla \times \mu_r^{-1}(\nabla \times E) - k_0^2\left(\varepsilon_r - \frac{j\sigma}{\omega\varepsilon_0}\right)E = 0. \tag{4}$$

where ω is the frequency of angular, ε_0 is the vacuum permittivity (8.85×10^{-12} F/m), μ_r is the relative permeability, ε_r is the relative permittivity, k_0 is the wave number in free space and σ is the electrical conductivity. k_0 is given:

$$k_0 = \omega\sqrt{\varepsilon_0\mu_0} = \frac{\omega}{c_0}. \tag{5}$$

The heat transfer equation is:

$$\rho C_p\frac{\partial T}{\partial t} + \rho C_p u \cdot \nabla T = Q + \nabla \cdot (k\nabla T). \tag{6}$$

where ρ is the density, kg/m^3. C_p is the specific heat capacity, J/kg·K. k is the material thermal conductivity, W/m^2·K^{-1}. T is the temperature, K. and Q is the heat source.

2.4 Boundary Conditions

The symmetry boundary was defined by the boundary of a perfect magnetic conductor, which is the equation shown below:

$$n \times H = 0. \tag{7}$$

Since the field only penetrates a small scope outside the boundary, the equation for the waveguide and the cavity wall is [3]:

$$\sqrt{\frac{\mu_0\mu_r}{\varepsilon_0\varepsilon_r - j\frac{\sigma}{\omega}}}n \times H + E - (n \cdot E)n = (n \cdot E_s)n - E_s. \tag{8}$$

where μ_0 is the vacuum permeability.

The setting of the port boundary condition depends on the position where the microwave from the microwave emission source enters the cavity. In this simulation, the microwave power is 1000 W and the rectangular port is in TE_{10} mode which excites a frequency of 2450 MHz.

The following expression shows the calculation method of the propagation constant:

$$\beta = \frac{2\pi}{c_0}\sqrt{v^2 - v_c^2}. \tag{9}$$

where β is the propagation constant. v is the frequency of microwave. v_c is the cutoff frequency.

The heat flux without cross-domain boundary is the adiabatic boundary condition, which is depicted by the definition:

$$-n \cdot (-k\nabla T) = 0. \tag{10}$$

It can be clearly seen from the above formula that the temperature gradient is zero, that is, the temperature on both sides of the boundary remains the same.

2.5 Mesh Quality

Select the physical control mesh to mesh the entire geometry. The element size is set to be finer. In order to obtain accurate results, the maximum component size is reduced to one-tenth of the microwave wavelength. The largest unit is 0.02447 m, and the smallest unit is 7.342×10^{-4} m. The physical control mesh is shown in Fig. 2.

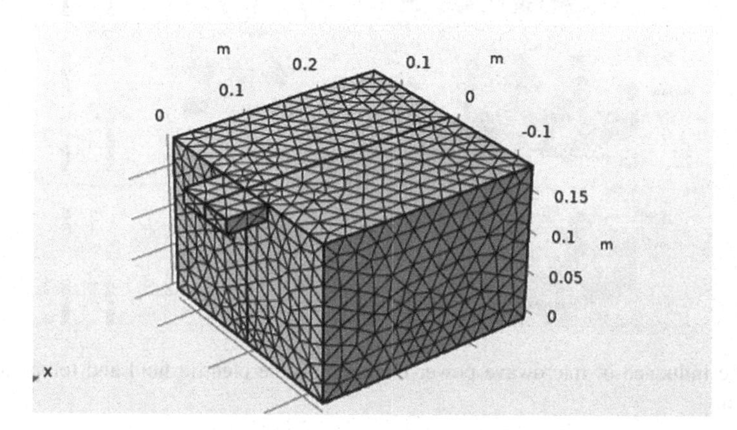

Fig. 2. Geometric grid diagram.

3 Results and Discussion

3.1 Influences of Microwave Power

Figure 3 shows the simulation of electric field and temperature distribution of the sludge microwave chamber under the action of five kinds of microwave power, which is 75 w, 150 w, 350 w, 535 w, 750 w and 1000 w. Although the microwave power increases and the electric field intensity and temperature both rise. There are little changes in electric field

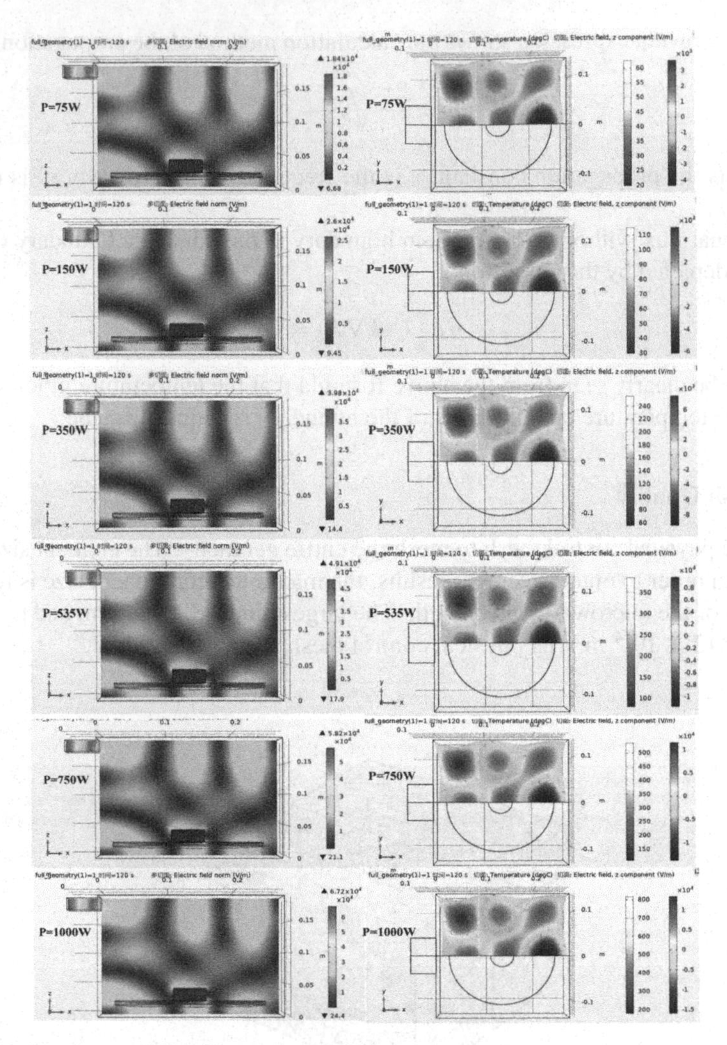

Fig. 3. The influence of microwave power changes on the electric field and temperature field distribution.

distribution and temperature distribution. When the microwave drying power of sludge is increased and the drying time is shortened, the corresponding energy consumption also increases. In the simulation, the humidity and heat transfer of the air are ignored. In the actual microwave drying process, it is still necessary to explore ways to improve the microwave energy utilization rate.

3.2 Influences of Sludge Content and Sludge Thickness

From Fig. 4, when the thickness of the sludge ranges from 1.6 mm to 20.07 mm, the electric field distribution varies little. However, when the thickness reaches 37.8 mm, the

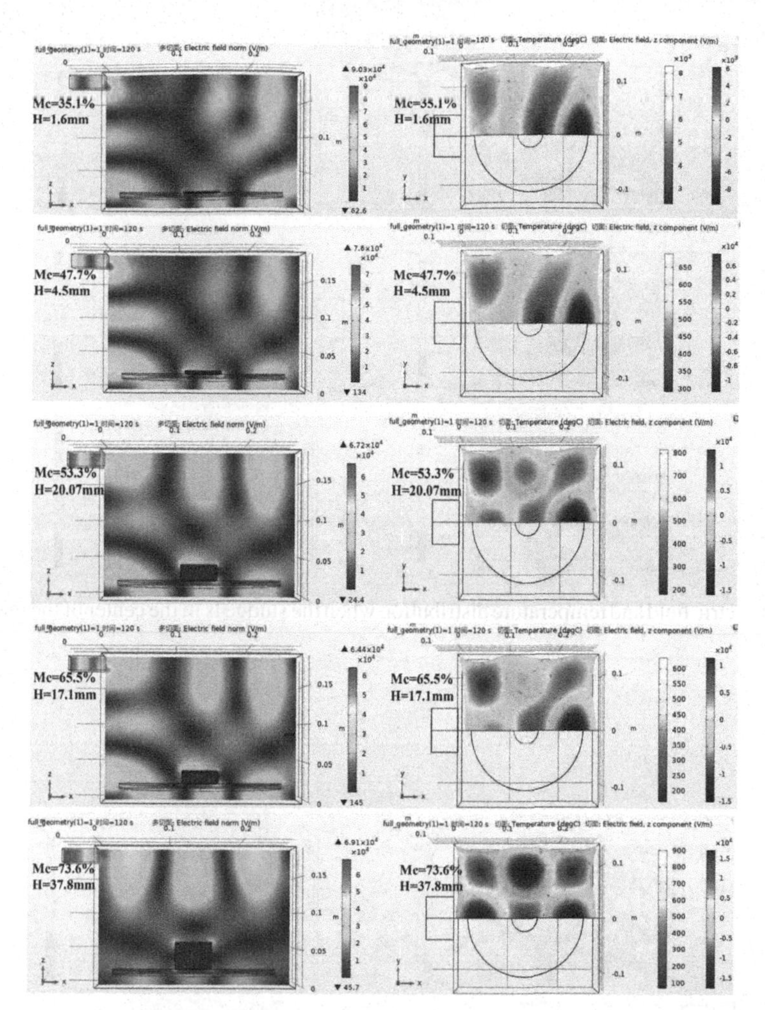

Fig. 4. The influence of sludge moisture content and sludge thickness on the distribution of electric field and temperature field.

electric field distribution changes significantly. While the sludge thickness is 1.6 mm, the temperature rises faster and the temperature is higher than others. The reason for this situation may be that the sludge is so thin that electromagnetic waves can completely penetrate the sludge. Therefore, the temperature of the sludge rises faster. When the sludge thickness is greater than 1.6 mm, the higher the water content in sludge, the faster the temperature of the sludge rises. Overall, there may be an optimal sludge drying thickness when microwave drying sludge. On the basis of the optimal sludge drying thickness, exploring the influence of sludge moisture content on microwave drying plays an important role in sludge drying.

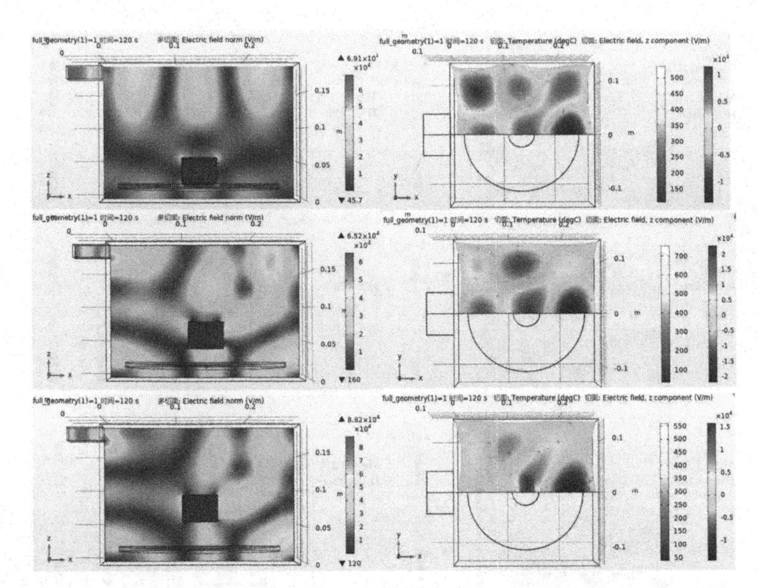

Fig. 5. Electric field and temperature distribution when the sludge is in the center of the microwave chamber.

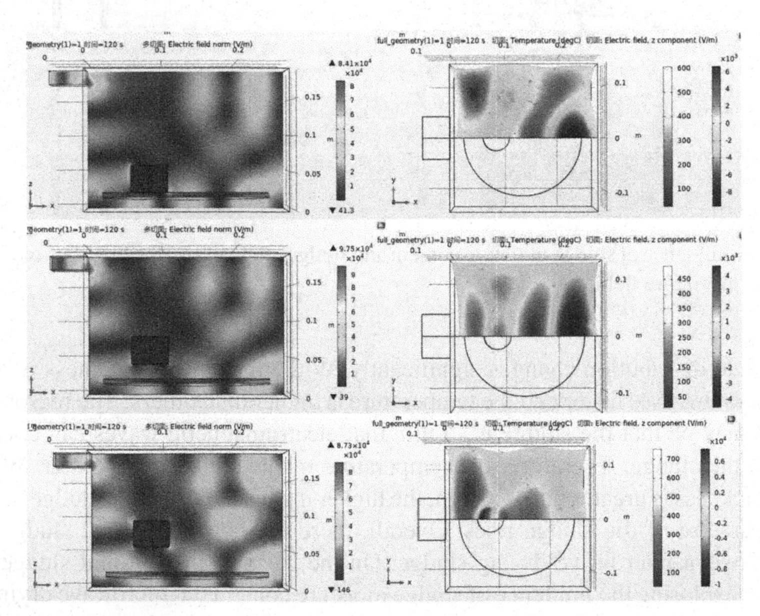

Fig. 6. Electric field and temperature distribution when the sludge is close to the microwave source.

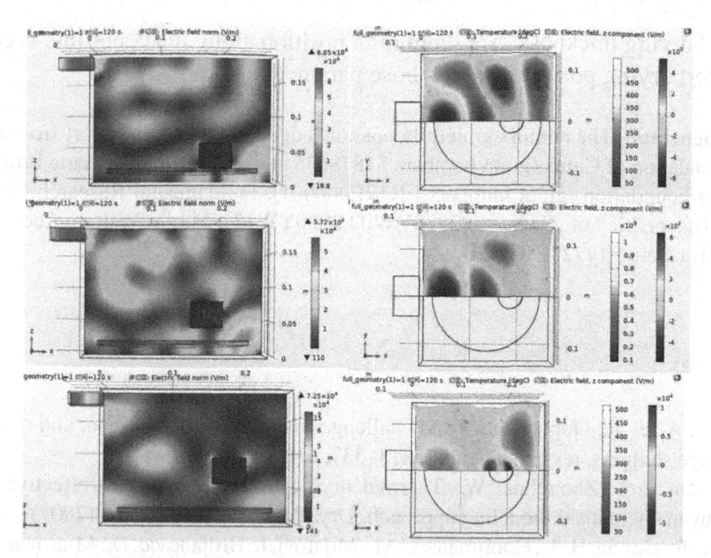

Fig. 7. Electric field and temperature distribution when the sludge is far away from the microwave source.

3.3 The Influence of Sludge Position in the Cavity

In order to obtain the electric field distribution and temperature distribution of the sludge with the corresponding position change. The first experiment we place the sludge in the center of the chamber, and move the sludge up 14 mm and 24 mm from the center respectively. As shown in Fig. 5, the temperature of the sludge reaches the peak as the sludge moves up 14 mm along the centerline. When the sludge approaches the microwave source, the sludge moves 14 mm and 24 mm in the vertical direction, which was depicted in Fig. 6. Temperature of the sludge gradually decreases with increasing height. When the sludge is far away from the microwave source, we set the sludge at 14 mm and 24 mm high respectively to obtain Fig. 7. Under this condition, the electric field and temperature distribution of the sludge varies greatly. The position change of the sludge has a great influence on the electric field distribution and temperature distribution of microwave drying. It can be concluded that the best position to dry a sludge cake is 14 mm high above the center of chamber's bottom.

4 Summary

The microwave drying simulation coupled electromagnetic and heat transfer equations were successfully solved by COMSOL Multiphysics software. Microwave power affected the temperature and heating performance of sludge sample. The greater the microwave power, the higher the efficiency of sludge drying, the greater the energy consumption. Therefore, further improvement is still needed in terms of microwave power energy saving. And it is found that there is an optimal drying thickness and it can still be optimized, and the optimal drying moisture content can be discussed on the basis of the optimal drying thickness. A 14 mm high position along the centerline is considered to be the best drying position among those parameters.

Acknowledgments. The authors gratefully acknowledge the financial support from the Natural Science Foundation of China (grant numbers 51878635 and 51876196), Zhejiang Provincial Natural Science Foundation (grant numbers LR19E060001), Fundamental Research Funds for the Provincial Universities of Zhejiang (2020YW02,2020YW13), Public Welfare Technology Key Projects of Huzhou City (2018GZ26).

References

1. Raheem, A., et al.: Opportunities and challenges in sustainable treatment and resource reuse of sewage sludge: a review. Chem. Eng. J. **337**, 616–641 (2018)
2. Mujumdar, A.S., Zhonghua, W.: Thermal drying technologies—cost-effective innovation aided by mathematical modeling approach. Dry. Technol. **26**, 145–153 (2007)
3. Kocbek, E., Garcia, H.A., Hooijmans, C.M., Mijatovi, I., Brdjanovic, D.: Microwave treatment of municipal sewage sludge: Evaluation of the drying performance and energy demand of a pilot-scale microwave drying system. Sci. Total Environ. **742**, 140541 (2020)
4. Bennamoun, L.: Solar drying of wastewater sludge: a review. Renew. Sustain. Energy Rev. **16**, 1061–1073 (2012)
5. Hza, B., Lin, S.A., Tai, L.B., Kd, A.: Coupling heat pump and vacuum drying technology for urban sludge processing. Energy Procedia. **158**, 1804–1810 (2019)
6. Eom, H., Jang, Y.H., Lee, D., Kim, S.S., Cho, E.M.: Optimization of a hybrid sludge drying system with flush drying and microwave drying technology. Chem. Eng. Res. Des. **148**, 68–74 (2019)
7. Chen, Z., Afzal, M.T., Salema, A.A.: Microwave drying of wastewater sewage sludge. J. Clean Energy Technol. **2**, 282–286 (2014)
8. Kun, L., Jun, Z., Lei, W., Jingjing, Z.: Numerical simulationon microwave pyrolysis of low-rank coal. Coal Conv. **43**, 20–28 (2020)
9. Hong, Y.D., Lin, B.Q., Li, H., Dai, H.M., Hao, Y.: Three-dimensional simulation of microwave heating coal sample with varying parameters. Appl. Thermal Eng. **93**, 1145-–1154 (2016)
10. Bennamoun, L., Chen, Z., Afzal, M.T.: Microwave drying of wastewater sludge: experimental and modeling study. Dry. Technol. **34**, 235–243 (2016)
11. Thiagarajan, R.C.: Multiphysics Simulations of Granular Sludge on the Optimization of Effluent Treatment Plant. Uk.comsol.com (2011)
12. Graham, B., Ryan, D., Philip, A.S., Lyn, A., Mohan, V.J.: Dielectric propertis of sewage biosolids: measurement and modeling. J. Microwave Power Electromagn. Energy **48**, 147–157 (2016)
13. Serdyuk, V.M.: Dielectric study of bound water in grain at radio and microwave frequencies. Progr. Electromagn. Res. **84**, 379–406 (2008)

Quantitative Data Cleaning and Analysis for Large Databases of Passenger Car Fuel Consumptions

Hongliang Wang[1], Ping Du[2], Xinchao Liu[3], Ruili Song[4], Hui Fu[4], Dongxiao Yu[5], Qing Lan[6(✉)], and Sicong Zhu[7(✉)]

[1] Shijiazhuang Highway and Bridge Construction Group Co., Ltd., Shijiazhuang, Hebei 050070, China

[2] Department of Transportation of Hebei Province, Highway Administration Bureau, Shijiazhuang, Hebei, China

[3] Hebei Express Group Limited Preparation and Construction Office, Baoding, Hebei, China

[4] Shijiazhuang Highway and Bridge Construction Group Co., Ltd., Shijiazhuang 050070, Hebei, China

[5] Beijing Zhongjiaoruida Technology CO., LTD., Haidian District, Beijing, China
DongxiaoYu@highwaytech.com.cn

[6] Hebei University of Water Resources and Electric Engineering, Cangzhou, China
QingLan@highwaytec.com.cn

[7] Beijing Jiaotong University, Beijing, China
sczhu@bjtu.edu.cn

Abstract. This paper presents the work a development of data cleaning and analysis for on-board diagnostic system (OBD II) data. The OBD device can collect speed and mass air flow via ECU, from which the instantaneous fuel consumptions are also calculated, along with the vehicle trajectories collected by an integrated GNSS module. Therefore, a large vehicle trips survey collects trajectories and instantaneous fuel consumption of over 8000 vehicles. The large trajectory database consists a number of trips whose positioning failures are significantly higher than the average. The datasets are proceeded with quantitative analysis to identify the faulty trip data by fitting with the power law distribution. After the cleanup, the analysis results demonstrate that the datasets can deliver a reliable estimation of the average daily passenger car fuel consumption across Beijing.

Keywords: Fuel consumption · OBD · Power law distribution

1 Introduction

With the increasing interests in environmental problems, the energy conservation and emission reduction have become a pressing topic. In China, more than 1/3 of the petroleum products are consumed by transportation, and it can produce a large amount of emission pollution [1]. Both the volatile fossil fuel price and exacerbating environmental problems make it necessary to monitor the vehicles emission and fuel consumption and analyze these data to find the solution of the problems.

K. Li et al. (Eds.): LSMS 2021/ICSEE 2021, CCIS 1468, pp. 535–551, 2021.
https://doi.org/10.1007/978-981-16-7210-1_51

There are growing evidences overseas and in China of that the gap between laboratory test findings and real-world carbon dioxide (CO_2) emissions and fuel consumption. Tietge et al. [2] analyzed on-road fuel consumption data for more than 1.5 million vehicles from four vehicle markets: EU, U.S., China, and Japan, and the result of his study indicates the divergences between official and real-world CO_2 emissions and fuel consumption are growing. Based on the conclusions, different sources of real-world CO_2 emissions and fuel consumption data for petrol light-duty vehicles (LDVs) conduct in-depth analysis of real-world patterns that are crucial to support regulatory program development. In Beijing metropolitan areas, the statistics were primarily supported by self-reported car owner experience data. However, more rigorous data are needed to start with an accurate summary of fuel consumption data that are representative of fleet characteristic. The new analyses are conducted using detailed vehicle fuel consumption testing data collected from real-world testing with OBD devices in the field. The objective of this research is to explore quantitative approaches that can improve real-world fuel consumption data quality, especially for the fuel consumption survey on regional level.

Recently, the concept of Big Data and associated technologies have been widely applied in the field of transportation research. In this fuel consumption case study, the research object is not fully overlapped with the concept of the Big Data. As one popular interpretation of big data refers to extremely large datasets. The concept of Big Data has been defined as a large magnitude of datasets and the amounts of data range from terabytes to petabytes [3]. The daily fuel consumption survey datasets collected in Beijing are significantly smaller than the abovementioned industry standard. The sensors of the OBD device collected the fuel consumption and vehicle trajectory information in a predetermined format, and their update rates are roughly 1 Hz. Hence, the current fuel consumption databases show critical differences with the concept of big data in the terms of volume, velocity, and/or variability. Unlike the Big data, the database does not rely on a specialized scalable IT architecture for storage, manipulation, analysis, and presentation.

Despite the importance of data collection, data quality remains a pervasive and critical problem for this fuel consumption survey. The presence of incorrect or inconsistent data can significantly distort the results of analyses, often negating the potential benefits of information-driven approaches. As a result, there has been a variety of research over the last decades on various aspects of data cleaning. And this paper will deliver a reliable cleanup methodology for large datasets and a useful insight into a large lump of data.

Initially, this paper presents the existing OBD based trip surveys, especially the applications in the field of fuel consumption survey. The paper focuses on the data cleaning methods that errors in quantitative attributes of large databases. The object is targeted at professional practices that manage large databases of quantitative information and develop auditing tools for practitioner. Because of the focus on quantitative data, the data quality of dataset is evaluated by statistical analysis with an emphasis on faulty vehicle trip detection. In addition, the implementation of the proposed algorithms should be cost-effective in large datasets, which are easy to understand and visualize graphically. A procedure is proposed to identify errors automatically or semi-automatically in large datasets. After the cleanup, an analysis is followed to present the fuel consumption of passenger cars in Beijing.

2 Literature Review

In last decade, PEMS was implemented extensively to assess instantaneous tailpipe emissions. However, because it has a series of shortcomings, namely, affordability, continuous maintenance and calibrations, cumbersome volume and weight, and its fragile components. It cannot use in large scale travel survey [4]. Thus, several on-board devices which overcome the above shortcomings of PEMS are developed, and they are successful to collect engine operation parameters via OBD. These affordable on-board devices have increasingly penetration rate [5–9]. The OBD is an on-board diagnostic system, it has a CAN bus for intra-subsystems communication. Via a 16-pin OBD II port, information from the ECU (Engine Control Unit) can be acquired directly from standard PIDs for the users. Moreover, the OBD data will be available on connected and automatic vehicles [10, 11]. However, the instantaneous fuel consumption does not have a standard PIDs, it has to be calculated from measurable airflow and/or air press with ECU outputs.

$$\mathrm{Fuel_{con}} = \frac{\mathrm{FuelFlow}}{\mathrm{Speed}}. \tag{1}$$

Where

$\mathrm{Fuel_{con}}$ is the fuel consumption;
FuelFlow is a measurement of vehicle fuel consumption, measured in liters per hour (L/h);
Speed is the velocity of the vehicle.
Although the fuel flow has a defined PID, some car cannot access it. Another method to calculate fuel consumption is following:

$$\mathrm{Fuel_{con}} = \frac{\mathrm{MAF}}{\mathrm{AFR} \times \mathrm{D} \times \mathrm{V}}. \tag{2}$$

Where

$\mathrm{Fuel_{con}}$ is the fuel consumption;
MAF is the grams of air per second (g/s);
AFR is the ratio of air and fuel;
D is the density of fuel measured in grams per cubic decimeter (g/dm3);
V is the speed of the vehicle measured in (Km/h).

The data are collected by the devices are manufactured and operated by a company in Beijing named ZhiJia Box [12]. The data collected by the devices have been used for travel survey and fuel consumption extensively. It consists of three components, namely, communication, OBD, and GNSS. The data used for the analysis presented in this paper were extracted from an in-service fleet for fuel consumption study. The fleet consisted of approximate 8000 private passenger vehicles, traveling across Beijing. The data collected the whole fleet in one week beginning from 04/09/2018 to 04/15/2018, and this week is free from public holiday and major events. The fuel consumption rates of the fleet were recorded second by second in addition to the instantaneous speed. Although the OBD component is dead reckoning, the GNSS component is subject to environmental effects. When using GNSS receivers in street canyons with tall buildings within Beijing metropolitan area, the shadowing and multipath effects may contribute to poor GNSS signal reception, so vehicles trajectories maybe incomplete due to the reasons mentioned above. Prior to the analysis, the integrity of the fuel consumption estimation was reconfirmed and was implemented in the test [13].

To analyses the fuel consumption data collected by OBD devices, it is primary to identify and process the faulty data with the incomplete trajectories. For the missing trajectories data, there are three ways to deal with this problem, ignoring, imputing or removing. Ignoring the missing data cannot may dilute the quality of the data. Imputing the missing data is a common method. There were numerous researches focusing on the imputation data by matching trajectories data to digital maps is a method for missing data completion. Greenfeld [14] proposed a map matching algorithm based on assessing the similarity between the characteristics of the street network and the positioning pattern of the user. Noland et al. [15] developed a Map Matching (MM) algorithm to accurately display vehicle location on a GIS-based digital map, and probabilistic approaches were applied for both identification of the actual road segment on which the vehicle was traveling. Yin et al. [16] introduced the weight-based snapping algorithm to solve the problem of snapping a trajectory to the road network after the trip is over. Lou et al. [17] proposed a new map-matching algorithm called ST-Matching to match low-sampling-rate GPS data onto a digital map.

There were also other methods to impute the missing trajectories. Shen et al. [18] proposed a STZ algorithm to impute the missing trajectories based on the area between any two possible routes, and it had a well performance. May et al. [19] used survival analysis to rate arbitrary poster campaigns in outdoor advertising using GPS mobility data that is affected by GPS dropout behavior. Li et al. [20] proposed a borrowed sample tracking technique to concatenate dense sub-trajectories into a complete trajectory at fine level, and this approach does not rely on existing map, yet still produces quality trajectory completion from sparse GPS samples. Zheng et al. [21] proposed a history based route inference system to solve the problem of reducing uncertainty of a given low-sampling-rate trajectory. Zhang et al. [22] developed a introduced a trajectory completion approach based on frequent pattern, and a method for evaluating trajectory completion algorithms. Huang et al. [23] proposed an integrated FCM imputation method based on the GA to estimate the missing taxi GPS values.

Thus, there are various efficient methodologies to impute the missing trajectories data. But there are limited researches on the method or rule for removing failure data. Although the method of removing may lose some useful information, the large quantity of data can make up this shortcoming. This paper researches a method of data cleanup to exclude invalid trip survey data from the large datasets.

3 Data Alignment and Failure Mode

3.1 A Subsection Sample

Literally, the OBD device collects the vehicle operation information of a whole trip from engine starting to vehicle parking. It collects speed, fuel consumption, vehicle kilometers travelled, and trajectory. The relational databases are managed by MySQL and split into two data tables, namely, track and trip. The track table stores the disaggregated trajectory information of active cars, and it includes time, position, and instantaneous gas consumption data. The trip table collects the aggregate statistics of each trip. The fields information for these two tables are shown from Table 1 to Table 2.

Table 1. Field information for Track Table

Table	Field name	Variable type	Remarks
Track	**trip_id_track**	**Double**	**ID for each trip**
	time_stamp	int(11)	Time stamp
	distance_stamp	int(11)	Distance stamp
	gas_stamp	int(11)	Gas consumption stamp
	speed	int(11)	Instantaneous velocity
	longitude	Float	Current longitude
	latitude	float	Current latitude
	direction	int(11)	Direction of vehicle
	time	text	Current time
	pos_type	int(11)	Positioning method of the car

Table 2. Field information for Trip Table

Table	Field name	Variable type	Remarks
Trip	**trip_id**	**double**	**ID for each trip**
	car_id	text	ID for each active car
	start_t	text	Strat time stamp
	end_t	text	End time stamp
	start_lon	float	Longitude of starting point
	start_lat	float	Latitude of starting point
	end_lon	float	Longitude of end point
	end_lat	float	Latitude of end point

(*continued*)

Table 2. (*continued*)

Table	Field name	Variable type	Remarks
	trip_t	int(11)	Trip time
	trip_d	int(11)	Trip distance
	gas	float	Gas consumption stamp
	gas_per_hm	int(11)	Gas consumption per hundred km
	trip_v	Float	Average velocity

The primary keys of two tables are highlighted in bold. By the primary keys of two tables, the trajectory information of a trip can be searched by the track table. Therefore, the data quality of trip can be evaluated on disaggregated level in a quantitative way. To be pointed out, the gas consumption rate in terms of liter per km is based on the mileage collected by GNSS component. In the database, the trajectory information that have longitude or latitude equal to 0 indicate the invalid data. A variable named failure ratio is an indicator of GNSS data quality, it is calculated as following.

$$\text{Failure ratio} = \frac{t_{invalid}}{t_{trip}}. \tag{3}$$

Where

$t_{invalid}$ is the accumulated time of invalid trajectory data in a trip;
t_{trip} is the duration of a trip.
By using failure ratio, this paper distinguishes the causes for abnormal trajectory data, and the method of excluding invalid data from large datasets is proposed.

3.2 Failure Mode

The data obtaining from the OBD device have some abnormal positioning data. It leads to the errors associated with the fuel consumption rate estimation. For further analysis of fuel consumption data, it is necessary to classify varied abnormal situations and find the corresponding solutions.

There are various causes of invalid positioning data such as satellite related errors, signal propagation related errors and GNSS receiver related errors. Satellite related errors can be classified by the failure origins as: satellite ephemeris error, satellite clock error, relativistic effect, signal propagation related errors, and GNSS receiver related errors [24]. This paper categories these errors by failure modes: accidental failure, Phase-Locked Loop (PLL) loss, and equipment failure [24]. An accidental failure means that some invalid position data occur accidentally. A GNSS PLL loss suggests the receiver loses the track of satellite signals. An equipment failure means the positioning data of a trip are almost lost because of positioning component failure or signal blockage. The Fig. 1 demonstrates three cases by failure modes. In Fig. 1, the trajectory data equal to 1 indicating valid data, and the data equal to −1 indicating invalid data. The GNSS PLL loss is illustrated in the Fig. 1 (b) and Fig. 1(c). The accidental failure and the GNSS PLL loss occur randomly, and the duration of the latter is significant longer than the former. This paper uses the power-law distribution to analyze this random phenomenon to find the failure ratio boundary between the GNSS PLL loss mode and the accidental failure mode.

4 Fitting Power-Law Model to Data

This section will analyze the failure ratio of trip survey data and determine whether it is drawn from the power-law distribution. This section will take the failure ratio from a certain day as a case study to show the method of analyzing the data.

Analysis of power law data generally consists of four steps as follow:

(1) Preliminarily check the data validity of following power-law distribution. The method is described in Sect. 3.1.
(2) Estimate the parameters x_{min} and scaling parameter using the method described in Sect. 3.2.
(3) Compute the goodness-of-fitness between the data and the power-law model using the method described in Sect. 3.3.
(4) Compare power-law model with other alternative distributions that are similar with the power-law model using the method described in Sect. 3.4.

4.1 Preliminary Check

The probability distribution of power-law is:

$$p(x) = Cx^{-\alpha}. \tag{4}$$

Where

C is a constant;
α is a coefficient of the distribution known as the exponent or scaling parameter.
Taking logarithm of two sides of Eq. (4), the equation can be changed as following:

$$\ln p(x) = \alpha \ln x + C. \tag{5}$$

The Eq. (5) suggests a power-law PDF should be a straight line on a log-log plot. Therefore, the power-law distribution is necessary to check preliminarily by plotting histogram in double logarithm coordinate if it follows the power-law distribution. Figure 2(a) is the histogram of positioning failure ratio of trajectory on April 9, 2018 as a case study. Most of the trip failure ratios concentrate in the left part. And the large frequency of this part of data make it difficult to fit with any distribution. But the rate of failure ratio in this part is very low, it indicates the accidental failure happens rarely. Its influence on the data quality is ignorable. As the failure ratio increases, the frequency shows a decreasing trend, and the histogram skews and fluctuates at the tail of the data. The frequency increases as the failure ratio is greater than 0.8 in Fig. 2(b), this trend

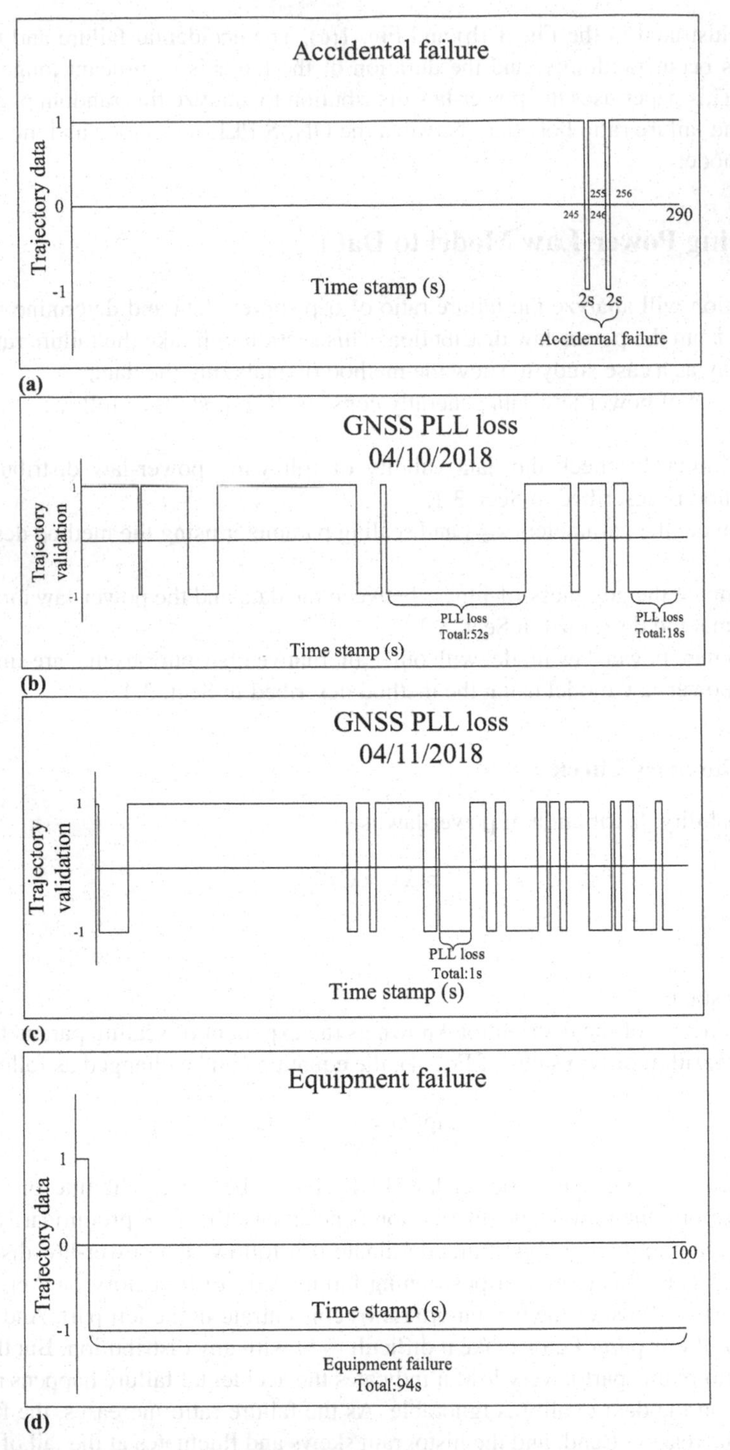

Fig. 1. State of positioning data in different failure mode, accidental failure (a), GNSS PLL loss on 04/10/2018 (b), GNSS PLL loss on 04/11/2018 (c) and equipment failure (d)

does not follow the power-law distribution, so this part of trip data are removed from the fuel consumption survey data base. For convenient observing, this paper displays the corresponding frequency of each interval of the histogram as scattered points in Fig. 2(c). After excluding the data in the tail of the distribution, the left part and middle part of the data distribution approximately follows a straight line on a log-log plot to some extent. It indicates that this part of data is highly probable to follow the power-law distribution. The varied patterns of failure ratio suggest the origins of positioning failure are the different. The large quantities of low failure ratio data belong to the accidental failures which can be negligible for fuel rate estimation. On the other end, the data of high failure ratio (>0.8) indicates the utter failures of positioning component that is meaningless for the purpose of the research. Particularly, neither accidental failure nor equipment failure can explain the phenomena of long tail. If the accidental failures occur only, the failure ratio should largely follow the law of large numbers that the results tend to be close to the expected value with very few extreme large values. In contrast, the GNSS PLL loss results in the long tail due to the long duration of GNSS malfunction, and this phenomenon also indicates few trips take up a large share out of the total faulty positioning data. The underestimated mileages of the survey trips can deliver biased estimations of the fuel consumption rate. The analysis of this unequal distribution phenomena dates back to Pareto's work on the distribution of wealth at the close of the 20th century [25].

4.2 Parameter Estimation

The second step is the estimation of scaling parameter α and lower bound x_{min}. In the Sect. 3.1, data on log-log plot follow a straight line approximately. The simplest method of estimating scaling parameter is taking the absolute slope of probability distribution on log-log plot as α. But this simple linear method often causes estimation errors [26]. Maximum Likelihood Estimators (MLEs) is a useful tool for parameter estimation, it is more precise than the method of moments [27]. This paper uses this method to estimate the parameters of power-law distribution.

This paper applies a package named powerlaw using the R language. This package uses the method proposed by Clauset et al. [28], and it can use MLE to estimate the scaling parameter and x_{min}.

The quantitative indicator failure ratio is continuous data. The parameters estimation for continuous case is below, and the result of Eq. (6) is the estimation of the scaling parameter but not the true value.

$$\hat{\alpha} = 1 + n\left[\sum_{i=1}^{n} \ln \frac{x_i}{x_{min}}\right]^{-1}. \quad (6)$$

Where

x_i is failure ratio, $i = 1, 2\ldots, n$, and $x_i \geq x_{min}$.

In practice, empirical data follow the power-law distribution for values of x above the lower bound x_{min}. The accuracy of the estimation of x_{min} has a great impact on the accuracy of the estimation of α. If the value of x_{min} is too low, the power-law model may

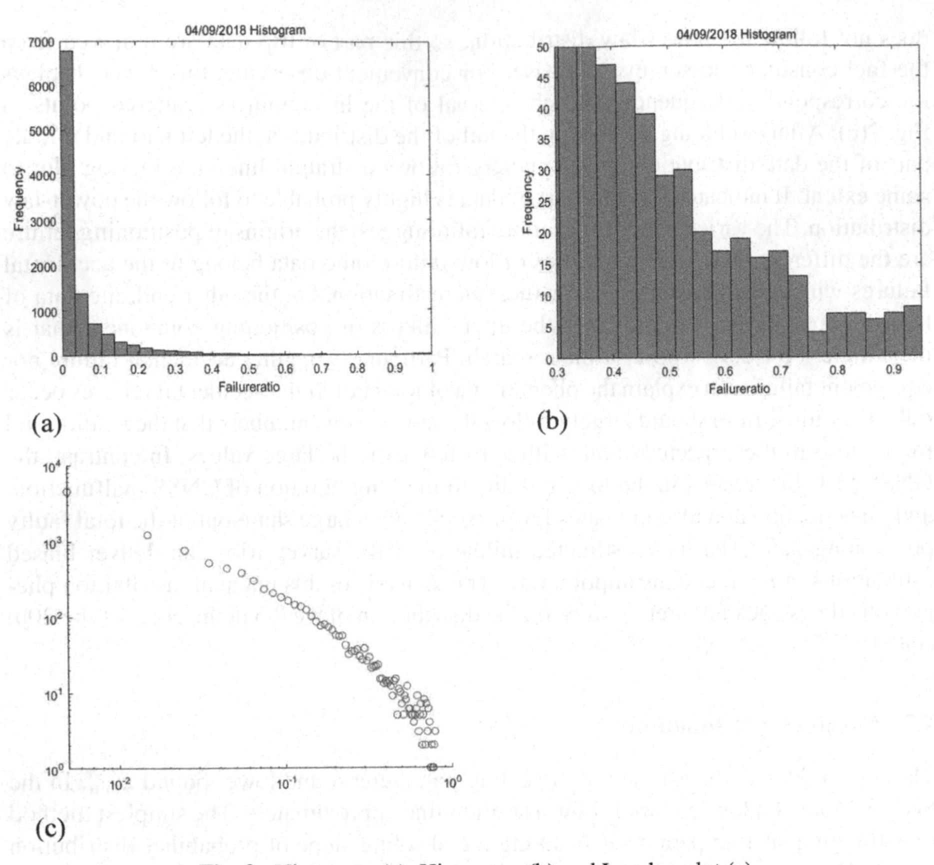

Fig. 2. Histogram (a), Histogram (b) and Log-log plot (c)

be fitted to non-power-law data. If the value of x_{min} is too high, the estimation of scaling parameter may cause errors because some data that follow power-law distribution are neglected.

The method of selecting lower bound proposed by Clauset, Young and Gleditsch [29] is presented as following: For each x_{min} over reasonable range, estimate α that larger than x_{min} and calculate the non-parametric Kolmogorov-Smirnov (K-S) statistic distance between the data and the theoretical power-law model with parameter x_{min} and α. The x_{min} selected is the one where the distribution can yield the best fit of the data. The K-S statistic is the maximum distance between the empirical CDF of data and the CDF of the fitting model, the formula is shown as the Eq. (7):

$$D = \max_{x \geq x_{min}} |S(x) - P(x)|. \tag{7}$$

Where

D is K-S statistic;
S(x) is the CDF of data lager than x_{min};

P(x) is the CDF of theoretical model.
The estimation of $\hat{x}_{min}$ is the value that can minimize the value of D. According to the above method, after removing the blunder data ($x > 0.8$), the results of parameter estimation are following: $x_{min} = 0.453$, $\alpha = 41.4$.

4.3 Goodness-Of-Fit Test

This section tests the goodness-of-fit of the power-law model estimated in Sect. 3.2 by K-S test. The K-S test derives a p-value that quantifies the plausibility of the hypothesis [28]. A low p-value indicates the hypothesis is rejected. This paper sets 0.05 as the significance level of the hypothesis test.

To complete two-sample K-S test, the data of theoretical power-law model are needed. This paper uses a function of the package to generate random samples from the model estimated in Sect. 3.2. According to Muthén and Muthén [30], the random sample size, should be no less than 100,000. Thus, more than 100,000 samples are generated to check against original failure ratio data using K-S test.

According to Fig. 2(c), the data between 0.15 and 0.5 approximately fit the straight line well which means the data in this interval have a large possibility of following power-law distribution. The p-value of the test result in this interval is 7.93e–10 which means the data in this interval can reject the null hypothesis. Considering the fluctuations of the empirical data, the interval should be adjusted appropriately and the test results are shown in Table 3.

Table 3. The results of K-S tesst

Failure ratio	p-value
0.15–0.5	7.93E–10
0.15–0.48	8.59E–10
0.17–0.48	0.764

In Table 3, the data between 0.17 and 0.48 derives the p-value larger than 0.05, and the power-law distribution cannot be rejected.

4.4 Comparisons with Other Distributions

In Sect. 4.3, the K-S test suggests the data follow the power-law distribution. Other alternative distributions also need to be tested. In this section K-S test and A-D test (Anderson–Darling test) is used to compare the goodness-of-fit of the same interval as Sect. 4.3 between the power-law distribution and the alternative distribution. The test result is shown in Table 4.

In Table 4, it indicates that the alternative distributions are ruled out by K-S test while the power-law distribution gives a better fit to the data than the alternative distributions.

This paper also uses another method A-D test. It is used to test the null hypothesis H_0 that the data come from the alternative distributions. The distribution model is rejected

Table 4. The K-S test and A-D results of alternative distribution

Data	Interval	K-S test		A-D test	
		Ln	Exp	Ln	Exp
		p-value	p-value	p-value	p-value
04/09/2018	0.17–0.48	3.22E–06	5.53E–12	6.45E–07	6.4E–07

Note:
Ln is lognormal distribution;
Exp is exponential distribution;

when the value of p-value is smaller than 0.05. If accepted, the alternative distribution is placed into the candidature models pool [31]. The result of the A-D test in Table 4 is to demonstrate the values of p-value. The result indicates the dataset follows the alternative distributions is rejected, the data do not follow exponential or lognormal distribution.

4.5 Fitting Results of One-Week Trip Survey Datasets

Using the analysis method proposed above, the estimation of parameters, the K-S test results of power-law and alternative distributions, and the A-D test results for alternative distributions are shown in Table 5.

Table 5. The fitting result of other data

Data	Parameters estimation		K-S test				A-D test	
			Power-law		Ln	Exp	Ln	Exp
	x_{min}	α	Inteval	p-value	p-value	p-value	p-value	p-value
04/09/2018	0.453	41.4	0.17–0.48	0.764	3.22E–06	5.53E–12	6.45E–07	6.4E–07
04/10/2018	0.377	7.69	0.15–0.54	0.0527	2.21E–05	2.8E–22	5.3E–07	5.29E–07
04/11/2018	0.362	4.32	0.15–0.8	0.0841	8.77E–07	7.09E–45	4.89E–07	4.89E–07
04/12/2018	0.445	54.5	0.15–0.47	0.883	2.17E–05	2.86E–14	5.6E–07	5.6E–07
04/13/2018	0.437	56.5	0.13–0.46	0.749	2.49E–06	2.82E–13	4.39E–07	4.39E–07
04/14/2018	0.406	316	0.1–0.41	0.105	2.68E–09	8.66E–17	3.01E–07	3.01E–07
04/15/2018	0.468	120	0.15–0.48	0.897	2.34E–06	2.87E–15	5.15E–07	5.15E–07

Note:
Ln is lognormal distribution;
Exp is exponential distribution;

From Table 5, the interval above x_{min} derive p-value that is greater than 0.05 in K-S test. Thus, the power-law distribution cannot be ruled out in the K-S test of each dataset. Comparing with the power-law distribution fitting, the K-S tests for lognormal distribution and exponential distribution both generate low p-values, implying the alternative distributions can be rejected. The A-D test results also reconfirm the conclusion that the alternative distributions are rejected. Overall, the power-law distribution gives a good fitting with the data that are greater than x_{min}. So the data cleaning method applies the fitting of power-law distribution for the long tail data to find the boundary between the accidental failure and GNSS PLL loss. Table 6 demonstrates the quality of the trajectories data which contain GNSS PLL loss. After extracting all trajectory data in which GNSS PLL loss occurred, the time proportion of continuous positioning failure (lager than 3 s) is collected in a trip dataset, afterwards the mean value of the proportion in a day is calculated. Taking the data of 04/09/2018, 04/12/2018, 04/14/2018 as an example, the daily average portion is shown in Table 6. The average portions of continuous failure are around 40%, and using the data cleaning criteria proposed in this paper can efficiently identify errors and clean these data.

Table 6. The average proportion of continuous failure

Data	04/09/2018	04/12/2018	04/14/2018
Average proportion	45.31%	35.14%	42.78%

After removing the fuel consumption data with a failure ratio larger than x_{min}, the histograms of the fuel consumption data for the whole week are shown in Fig. 3.

In Fig. 3, it is clear that long tail phenomenon of the fuel consumption data almost disappears. The vast majority of the extreme high values are removed, while some high values are still kept due their integrity of trajectories. Therefore, the proposed method is "surgical" to identify faulty trips out of numerous trips. Therefore, the fuel consumption estimation can be more accurate and reliable.

5 Fuel Consumption Data Analysis

The previous studies [32, 33] have shown that the traffic pattern in weekdays and weekends has the similar double peak pattern. And the comparison of daily average speed across different locations is shown in Fig. 4.

Using the data cleaning method, the power-law distribution is fitted with the data, and the lower bound x_{min} is estimated. The boundary between accidental failure and GNSS PLL loss of failure ration data is determined. Therefore, the invalid data caused by GNSS PLL loss and equipment failure can be removed for mileage estimations. The remaining data is used to calculate the weighted mean of fuel consumptions, and the weight is the mileage of the corresponding trip. The results are shown in Table 7. And the pattern of the fuel consumption rate is consistent with the traffic operation in Beijing.

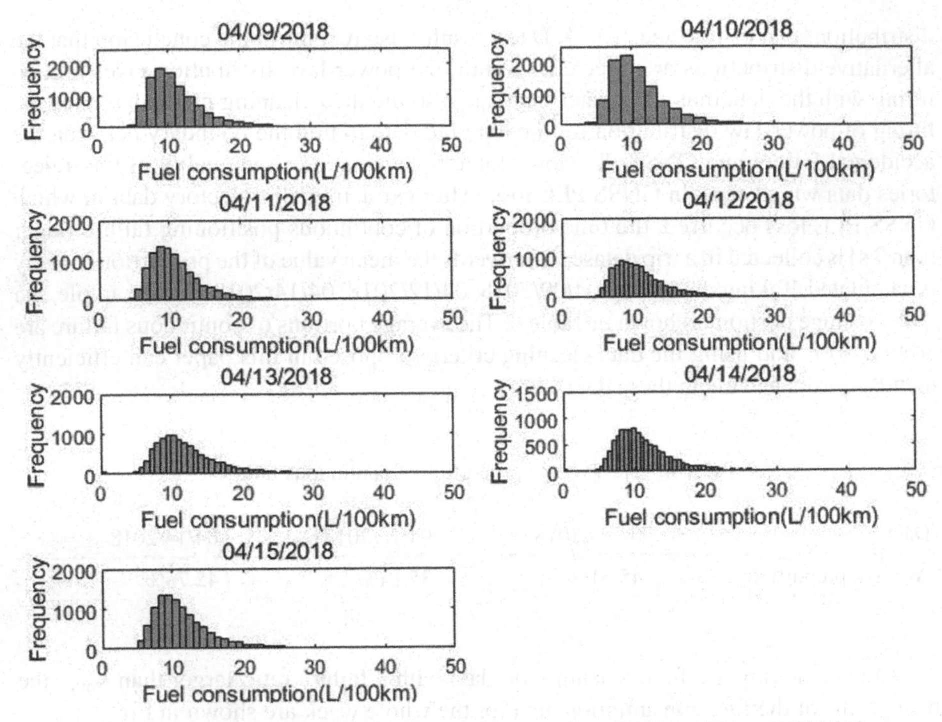

Fig. 3. The histogram of fuel consumption rates from 04/09/2018 to 04/15/2018

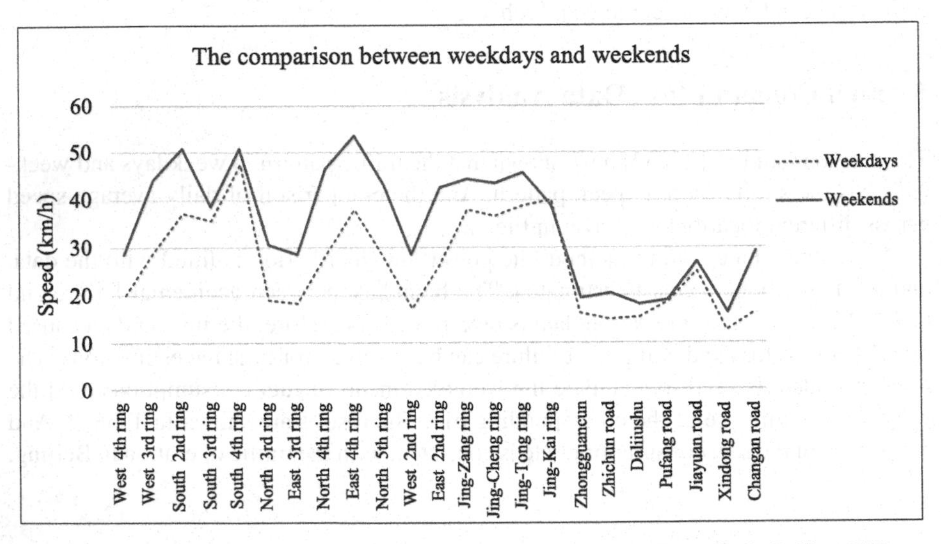

Fig. 4. The comparison between weekdays traffic and weekends traffic in Beijing [33]

Table 7. The weighted mean of fuel consumption

Data	Fuel consumption (L/100 km)
04/09/2018	9.51
04/10/2018	9.41
04/11/2018	9.39
04/12/2018	9.57
04/13/2018	10.25
04/14/2018	9.48
04/15/2018	9.49

6 Conclusion

Fuel consumption monitor is a necessary task of energy conservation and emission reduction, and using the OBD device to acquire vehicle fuel consumption data is a mature method. To monitor the disaggregated passenger car fuel consumptions, the OBD device can be a sound solution in terms of affordability, easy access, and acceptable accuracy. Because the GNSS component of the device is subject to the environment, some trip information with invalid position data need to be excluded. This paper classifies the causes of these abnormal data by failure modes and uses the failure ratio to represent them. After preliminary check of the distribution of trip failure ratio on log-log plot, this paper spots the long tail phenomenon caused by GNSS PLL loss and uses the power-law distribution to fit the data to find the lower failure ratio boundary of the faulty trips for the fuel consumption research. The results of K-S tests and A-D tests show that the empirical failure ratio data can have a good fitting with the power-law distribution, while the alternative distributions can be ruled out. So the faulty data caused by GNSS PLL loss and equipment failure can be identified and cleaned by removing the trips whose failure ratio is greater than the boundary x_{min}. The poor quality of the data which are diluted by PLL loss are also shown. The small proportion of trip data affected by GNSS PLL loss take up a significant share of trajectory errors. Therefore, the cleaning method is more efficient to clean date comparing with the imputation. After the effective cleanup the valuable information can be extracted from the large datasets. The survey data can deliver the fuel consumption estimations of passenger cars for the transportation professionals. The composite fuel consumption of private passenger cars in Beijing is approximate 9.6L/100 km. The average fuel consumption data fluctuates little in the whole week, and it agrees with the traffic operation in Beijing where weekends have the similar traffic pattern and degree of congestion with weekdays.

Acknowledgments. The research is supported by Beijing Municipal Commission of Transport and Zhi-jiaxing Technology Co.,Ltd. This work was supported in part by Hebei Province Talent Project Training Fund (#A2017002019), in part by the Hebei Jitong Highway and Bridge Construction Company Ltd. under Grant 2021130915000005, 2021130915000006).

References

1. Zhou, X., Huang, J., Lv, W., Li, D.: Fuel consumption estimates based on driving pattern recognition. In: 2013 IEEE International Conference on Green Computing and Communications and IEEE Internet of Things and IEEE Cyber, Physical and Social Computing, pp. 496—503. IEEE (2013)
2. Tietge, U., Mock, P., Franco, V., Zacharof, N.: From laboratory to road: Modeling the divergence between official and real-world fuel consumption and CO2 emission values in the German passenger car market for the years 2001–2014. Energy Policy **103**, 212–222 (2017)
3. Gandomi, A., Haider, M.: Beyond the hype. Int J Inf. Manag. **35**(2), 137–144 (2015)
4. Ortenzi, F., Costagliola, M.A.: A new method to calculate instantaneous vehicle emissions using OBD data. In: SAE Technical Paper (2010)
5. Kim, H., Jang, S., Jang, J.: A study on development of engine fault diagnostic system. Math. Probl. Eng. **2015**, 1–6 (2015)
6. Malekian, R., Moloisane, N.R., Nair, L., Maharaj, B.T., Chudeokonkwo, U.A.K.: Design and implementation of a wireless OBD II fleet management system. IEEE Sensors J. **17**(4), 1154–1164 (2017)
7. Meseguer, J. E., Calafate, C.T., Cano, J.C., Manzoni, P.: Assessing the impact of driving behavior on instantaneous fuel consumption. In: 12th Annual IEEE Consumer Communications and Networking Conference (CCNC), pp. 443—448. IEEE (2015)
8. Yang, Y., Chen, B., Su, L., Qin, D.: Research and development of hybrid electric vehicles CAN-bus data monitor and diagnostic system through OBD-II and android-based smartphones. Adv. Mech. Eng. **5**, 741240 (2013)
9. Zaldivar, J., Calafate, C.T., Cano, J., Manzoni, P.: Providing accident detection in vehicular networks through OBD-II devices and Android-based smartphones. In: IEEE 36th Conference local computer networks, pp. 813–819. IEEE Press (2011)
10. Baek, S., Jang, J.: Implementation of integrated OBD-II connector with external network. Inf. Syst. **50**, 69–75 (2015)
11. Fernandes, B., Alam, M., Gomes, V., Ferreira, J., Oliveira, A.S.R.: Automatic accident detection with multi-modal alert system implementation for ITS. Veh Commun. **3**, 1–11 (2016)
12. Beijing Zhijia Travel Technology Co., L. Zhijia Trip. http://www.zhijiaxing.net/
13. Sentoff, K., Aultmanhall, L., Holmen, B.A.: Implications of driving style and road grade for accurate vehicle activity data and emissions estimates. Transp. Res. D-Tr E. **35**, 175–188 (2015)
14. Greenfeld, J.: Matching GPS Observations to Locations on a Digital Map. In Transportation Research Board 81st Annual Meeting, Washington D.C (2002)
15. Noland, R.B., Quddus, M., Ochieng, W.Y.: Map-matching in complex urban road networks. Revista Brasileira De Cartografia **55**(2), 1–14 (2003)
16. Yin, H., Wolfson, O.: A weight-based map matching method in moving objects databases. In: Statistical and Scientific Database Management, pp. 437--438. IEEE Press (2004)
17. Lou, Y., Zhang, C., Zheng, Y., Xie, X., Wang, W., Huang, Y.: Map-matching for low-sampling-rate GPS trajectories. In: ACM SIGSPATIAL Conference on Advances in Geographic Information Systems, pp. 352--361 (2009)
18. Shen, G., Zhang, C., Tang, B., Yuan R.: An area-based method for missing trajectory completion: a STZ algorithm. In: International Symposium on Computational Intelligence & Design (2015)
19. May, M., Korner, C., Hecker, D., Pasquier, M., Hofmann, U., Mende, F.: Handling missing values in GPS surveys using survival analysis: a GPS case study of outdoor advertising. In: International Workshop on Data Mining and Audience Intelligence for Advertising, pp. 78–b84 (2009)

20. Li, Y., Li, Y., Gunopulos, D., Guibas, L.J.: Knowledge-based trajectory completion from sparse GPS samples. In: 24th ACM SIGSPATIAL International Conference on advances in geographic information systems, vol. 33, pp. 1--10 (2016)
21. Zheng, K., Zheng, Y., Xie, X., Zhou, X.: Reducing uncertainty of low-sampling-rate trajectories. In: International Conference on Data Engineering, pp. 1144--1155 (2012)
22. Zhang, Z., Rao, W., Di, X., Zhao, P., Abdesslem, F.B.: Frequent pattern-based trajectory completion. In the 16th ACM on Embedded Networked Sensor Systems, pp. 311--312 (2018)
23. Huang, J.: An integrated fuzzy C-means method for missing data imputation using taxi GPS Data. Sensors-Basel **20**(7), 19 (2020)
24. Kaplan, E., Hegarty, C.: Understanding GPS: Principles and Applications. Artech house, Boston (2005)
25. Arnold, B.C.: A flexible family of multivariate Pareto distributions. J. Stat. Plan. Infer. **24**(2), 249–258 (1990)
26. Goldstein, M.L., Morris, S.A., Yen, G.G.: Problems with fitting to the power-law distribution. Eur. Phys. J. B. **41**(2), 255–258 (2004)
27. Rice, J.A.: Mathematical statistics and data analysis. Technometrics **37**(1), 127 (1995)
28. Clauset, A., Shalizi, C.R., Newman, M.E.J.: Power-law distributions in empirical data. SIAM Rev. **51**(4), 661–703 (2009)
29. Clauset, A., Young, M., Gleditsch, K.S.: On the frequency of severe terrorist events. J. Conflict Resolut. **51**(1), 58–87 (2007)
30. Muthen, L.K., Muthen, B.: How to use a monte carlo study to decide on sample size and determine power. Struct. Equ. Model. **9**(4), 599–620 (2002)
31. Ma, Z., Ferreira, L., Mesbah, M., Zhu, S.: Modeling distributions of travel time variability for bus operations. J. Adv. Transp. **50**(1), 6–24 (2016)
32. Wang, Y., Huang, Y.: Analysis of Beijing traffic congestion evaluation index based on big data. Transp. Syst. Eng. Inf. **16**(4), 231–240, 246 (2016)
33. Zhao, H.: Analysis of Traffic Congestion Characteristics and Its Influencing Factors in Beijing. Jiaotong University, Beijing (2017)

Author Index

Printed in the United States
by Baker & Taylor Publisher Services